High Resolution NMR of Macromolecules

High Resolution NMR of Macromolecules

FRANK A. BOVEY

BELL LABORATORIES
MURRAY HILL, NEW JERSEY

1972

ACADEMIC PRESS New York and London

ACADEMIC PRESS, INC.
111 Fifth Avenue, New York, New York 10003

United Kingdom Edition published by
ACADEMIC PRESS, INC. (LONDON) LTD.
24/28 Oval Road, London NW1 7DD

LIBRARY OF CONGRESS CATALOG CARD NUMBER: 74-185030

PRINTED IN THE UNITED STATES OF AMERICA

CONTENTS

CHAPTER II. Isomerism in Polymer Chains

CHAPTER III. Configurational Sequences and Their Observation: Polymers of α,α′-Disubstituted Vinyl Monomers

CHAPTER IV. Polyacrylates, Polyacrylonitrile, and Polyvinyl Alcohol and Its Derivatives

CHAPTER V. Haloethylene Polymers

CHAPTER XI. Diene Polymers and Copolymers

CHAPTER XII. Ring-Opening Polymers, Polyamides, Polyesters, and Miscellaneous Polymers

CHAPTER XIII. Polypeptides

CHAPTER XIV. Proteins

CHAPTER XV. Nucleic Acids

PREFACE

The study of polymers by nuclear magnetic resonance can be roughly divided into two fields: (1) the measurement of nuclear (usually proton) relaxation by pulse methods or by direct observation of "wide-line" spectra; and (2) the study of high resolution spectra. The first of these commonly employs polymers in the solid state with the object of obtaining information concerning morphology and molecular motion. The second, which is the province of this book, deals (with rare exceptions) with polymers in solution and is directed toward the elucidation of the structure and stereochemistry of polymer chains. It is the younger field of the two. The first studies of solid polymers were reported only about a year after the original observations of NMR in bulk matter in 1946; by 1958 there was a sufficient accumulation of data to justify a lengthy review.[1] In 1957, the first proton spectrum of a natural polymer (lysozyme) was published, and in the following year the first report of a synthetic polymer spectrum (polystyrene) was published. Since that time the field has expanded rapidly, partly as the result of the observation in 1960 that NMR is a powerful means for the investigation of the stereochemistry of vinyl polymers.

Chapter I is a description of the fundamentals of NMR, very similar to that in Chapter I of my more general book.[2] This discussion is intended to make the book more or less self-contained. For more detailed descriptions of spectral

[1] W. P. Slichter, *Fortsch. Hochpolym. Forsch.* **1**, 35 (1958).

[2] F. A. Bovey, "Nuclear Magnetic Resonance Spectroscopy," Academic Press, New York, 1969.

analysis and the dependence of chemical shifts and J couplings on structure, one may wish to consult the treatises listed at the end of Chapter I (this is a very partial list, in both senses of the word). Chapter II is a discussion of isomerism in polymer chains without special reference to NMR. Chapters III–XII describe the interpretation of synthetic polymer spectra in terms of structure, stereochemical configuration, conformation, and chain growth mechanism. The last three chapters (XIII–XV) deal with biopolymers (excluding carbohydrates) and their model compounds. This is the most exciting area of application of NMR to polymer science and is at present growing with explosive rapidity, largely as the result of the more general accessibility of instruments of high magnetic field (50 kG and upwards). Indeed, it may be said that in general the fruitfulness of the NMR study of polymers has paralleled the development of instruments of increasing field strength. For biopolymer studies, 50-kG instruments have proved virtually indispensable.

ACKNOWLEDGMENTS

I wish to thank Mrs. Anne I. Brewster for running a number of spectra obtained specifically to illustrate this book. I am grateful also to Dr. R. G. Shulman and Dr. W. D. Phillips for critically reading Chapters XIII and XIV and Dr. Tetsuo Yamane and Dr. Dinshaw Patel for performing a similar service with respect to Chapter XV. Thanks are also due to many others, particularly Dr. Dennis Torchia, for helpful discussions.

High Resolution NMR
of Macromolecules

Chapter I

FUNDAMENTALS OF NUCLEAR MAGNETIC RESONANCE†

1. Introduction

The phenomenon of nuclear magnetic resonance depends on the fact that some atomic nuclei possess (in addition to charge and mass, which are of course common to all nuclei) spin, or angular momentum. A spinning charge generates a magnetic field, and so there is associated with this angular momentum a magnetic moment. Nuclear spin was first proposed by Pauli[1] as an explanation for certain hyperfine structural features of atomic spectra. The phenomenon of nuclear magnetic resonance has long been known in molecular beams and has yielded much fundamental information on nuclear properties.[2] This information was of little significance to chemists, however, until the observation of nuclear magnetic resonance in bulk matter. This was achieved by two independent groups in 1945—one with Purcell at Harvard and the other with Bloch at Stanford. Purcell *et al.*[3] reported resonance in solid paraffin while Bloch *et al.*[4] reported it in liquid water. With the observation of three magnetically nonequivalent types of protons in ethyl alcohol,[5] nuclear magnetic resonance began to be primarily the province of the chemist, as it is today.

† As noted in the Preface, much of the material in this chapter is similar to that in Chapter I of the author's more general book "Nuclear Magnetic Resonance Spectroscopy," Academic Press, New York, 1969.

2. Nuclear Spin

The angular momentum of a nucleus possessing a spin (or of any particle or system having angular momentum) is a half-integral or integral multiple of $h/2\pi$, where h is Planck's constant. The maximum value of the angular momentum is I, which is called the spin quantum number or more commonly simply "the spin." Each nuclear ground state is characterized by just one value of I. If $I = 0$, the nucleus has no angular momentum and no magnetic moment. If I is not zero, the nucleus will possess a magnetic moment μ, which is taken as parallel to the angular momentum vector. The permitted values of the vector moment along any chosen axis are described by means of a set of magnetic quantum numbers m given by the series:

$$m = I, \quad (I-1), \quad (I-2), \quad \ldots, \quad -I \tag{I-1}$$

Thus, if I is $\frac{1}{2}$, the possible magnetic quantum numbers are $+\frac{1}{2}$ and $-\frac{1}{2}$. If I is 1, m may take on the values 1, 0, and -1 and so on. In general, then, there are $2I + 1$ possible orientations or states of the nucleus. In the absence of a magnetic field, these states all have the same energy. In the presence of a uniform magnetic field H_0, they correspond to states of different potential energy. For nuclei for which I is $\frac{1}{2}$, the two possible values of m, $+\frac{1}{2}$ and $-\frac{1}{2}$, describe states in which the nuclear moment is aligned with and against the field H_0, respectively, the latter state being of higher energy. The detection of transitions of magnetic nuclei (often themselves referred to as "spins") between these states is made possible by the nuclear magnetic resonance phenomenon.

The magnitudes of nuclear magnetic moments are often specified in terms of the ratio of the magnetic moment and angular momentum, or *magnetogyric ratio* γ, defined as

$$\gamma = \frac{2\pi\mu}{Ih} \tag{I-2}$$

A spinning spherical particle with mass M and charge e uniformly spread over its surface can be shown to give rise to a magnetic moment $eh/4\pi Mc$, where c is the velocity of light. For a particle with the charge, mass and spin of the proton, the moment should be 5.0493×10^{-24} erg/G on this model. Actually, this approximation is not a good one even for the proton, which is observed to have a magnetic moment about 2.79 times as great as the over-simplified model predicts. No simple model can predict or explain the actual magnetic moments of nuclei. However, the predicted moment for the proton serves as a useful unit for expressing nuclear moments and is known as the *nuclear magneton*; it is the analog of the Bohr magneton for electron spin.

Observed nuclear moments can be specified in terms of the nuclear magneton by

$$\mu = g \frac{ehI}{4\pi M_p c} \tag{I-3}$$

where M_p is the proton mass and g is an empirical parameter called the nuclear g factor. In units of nuclear magnetons, then,

$$\mu = gI \tag{I-4}$$

In Table I-1, nuclear moments are expressed in these units. It will be noted that some nuclei have negative moments. This is of no great practical significance to NMR spectroscopists; its theoretical meaning will be evident a little

TABLE I-1

NUCLEI OF MAJOR INTEREST FOR POLYMER NMR SPECTROSCOPY

Isotope	Abundance (percent)	NMR frequency (Hz) in 10 kG field	Relative sensitivity[a]	Magnetic moment (μ)[b]	Spin[c] (I)
^{1}H	99.9844	42.577	1.000	2.7927	$\frac{1}{2}$
^{2}H(D)	0.0156	6.536	0.0096	0.8574	1
^{10}B	18.83	4.575	0.0199	−1.1774	3
^{11}B	81.17	13.660	0.165	1.8006	$\frac{3}{2}$
^{13}C	1.108	10.705	0.0159	0.7022	$\frac{1}{2}$
^{14}N	99.635	3.076	0.0010	0.4036	1
^{15}N	0.365	4.315	0.0010	−0.2830	$\frac{1}{2}$
^{19}F	100	40.055	0.834	2.6273	$\frac{1}{2}$
^{31}P	100	17.235	0.0664	1.1305	$\frac{1}{2}$

[a] For equal numbers of nuclei at constant H_0.
[b] In multiples of the nuclear magneton, $eh/4\pi M_p c$.
[c] In multiples of $h/2\pi$.

later. It should also be noted that the neutron, with no net charge, has a substantial magnetic moment. This is a particularly striking illustration of the failure of simple models to predict μ. Clearly, the neutron must contain separated charges (for at least a part of the time) even though its total charge is zero.

Although magnetic moments cannot be predicted exactly, there are useful empirical rules relating the mass number A and atomic number Z to the nuclear spin properties:

1. If both the mass number A and the atomic number Z are even, $I = 0$.
2. If A is odd and Z is odd or even, I will have half-integral values: $\frac{1}{2}$, $\frac{3}{2}$, $\frac{5}{2}$, etc.
3. If A is even and Z is odd, I will have integral values: 1, 2, 3, etc.

Thus, some very common isotopes, such as ^{12}C, ^{16}O, and ^{32}S, have no magnetic moment and cannot be observed by NMR. This is fortunate in a way, however, for if these nuclei did have magnetic moments, the spectra of organic molecules would be much more complex than they are.

3. The Nuclear Magnetic Resonance Phenomenon

Nuclei with spins of $\frac{1}{2}$, such as protons, are often likened to tiny bar magnets. But because of their small size and because they spin, their behavior differs in some ways from the ordinary behavior of macroscopic bar magnets. When placed in a magnetic field, the spinning nuclei do not all obediently flip over and align their magnetic moments in the field direction. Instead, like gyroscopes in a gravitational field, their spin axes undergo *precession* about the field direction, as shown in Fig. I-1. The frequency of this so-called Larmor

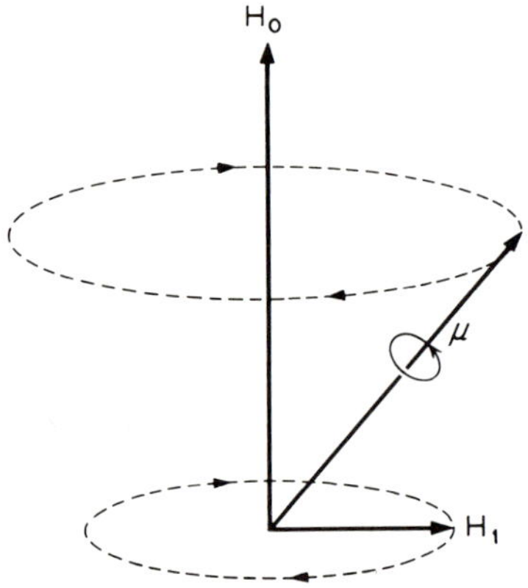

FIG. I-1. Nuclear moment in a magnetic field.

precession is designated as ω_0 in radians per second or ν_0 in cycles per second or hertz, abbreviated Hz ($\omega_0 = 2\pi\nu_0$). If we try to force the nuclear moments to become aligned by increasing H_0, they only precess faster. They *can* be made to flip over, however, by applying a second magnetic field, designated as H_1, at right angles to H_0 and causing this second field to rotate at the precession frequency ν_0. This second field is represented by the horizontal vector in Fig. I-1, although in practice (see Section 4) it is actually very much smaller in relation to H_0 than this figure suggests. It can be seen that if H_1 rotates at a frequency close to but not exactly at the precession frequency, it will cause at most only some wobbling or *nutation* of the magnetic moment μ. If, however, H_1 is made to rotate exactly at the precession frequency it will cause large oscillations in the angle between μ and H_0. If we vary the rate of rotation of H_1 through this value, we will observe a resonance phenomenon as we pass through ν_0.

One might suppose that the Larmor precession of the nuclear moments could itself be detected by some means without the need to invoke a resonance phenomenon. This, however, is not possible because each nucleus precesses with a completely random phase with respect to that of its neighbors and there is therefore no macroscopic property of the system which changes at the Larmor frequency.

By a well-known relationship, the Larmor precession frequency is given by

$$\omega_0 = \gamma H_0 \tag{I-5}$$

or, from Eq. (I-2),

$$h\nu_0 = \mu H_0/I \tag{I-6}$$

The result of this classical treatment can also be obtained by a quantum mechanical description which is in some ways a more convenient way of regarding the resonance phenomenon. It is best, however, not to try to adopt either viewpoint to the exclusion of the other, since each provides valuable insights. In quantum mechanical terms the quantity $h\nu_0$ is the energy separation ΔE between the magnetic energy levels in a magnetic field H_0, as shown in Fig. I-2. For a nucleus of spin $\frac{1}{2}$, ΔE will be $2\mu H_0$, and, as we have seen, only

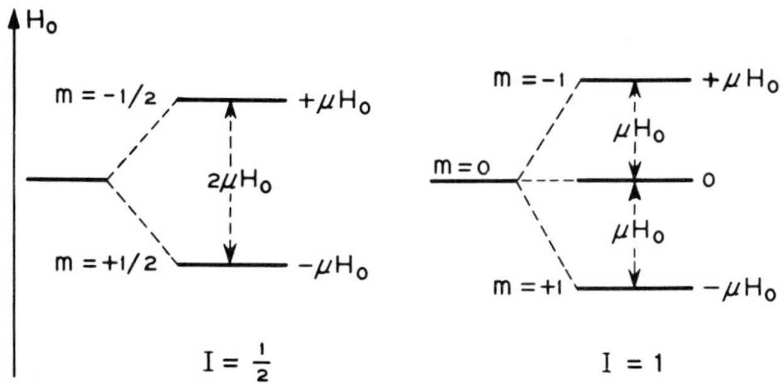

FIG. I-2. Magnetic energy levels for nuclei of spin $\frac{1}{2}$ and 1.

two energy levels are possible. For a nucleus of spin 1, there are three energy levels, as illustrated in Fig. I-2. The quantum mechanical treatment gives us an additional result for such systems with $I \geqslant 1$ which the classical treatment does not: it tells us that only transitions between adjacent energy levels are allowed, i.e., that the magnetic quantum number can only change by ± 1. Thus, transitions between the $m = -1$ and $m = 0$ levels and between the $m = +1$ and $m = 0$ levels are possible, but transitions between the $m = -1$ and the $m = +1$ levels are not possible.

4. The Detection of Resonance

Let us now consider how the rotating magnetic field H_1 is to be provided and how the resonance phenomenon is to be observed. Our discussion will be a simplified one, since the treatment of instrumental detail is outside the scope of this book. (Suggestions for further reading are given at the end of the chapter.)

In the schematic diagram of the NMR spectrometer (Fig. I-3), the sample is contained in tube A, commonly about 5 mm in diameter, and is placed in

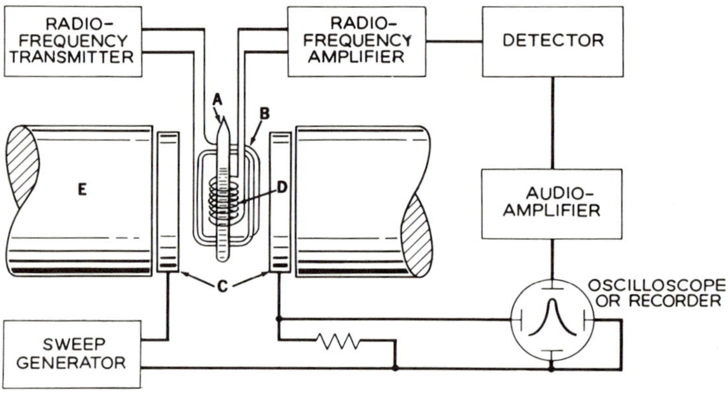

FIG. I-3. Block diagram of nuclear magnetic resonance spectrometer.

the field of the electromagnet E, which in present-day spectrometers may have a field strength of 14.1 to 52 kG. We find from Table I-1 [or from Eq. (I-6)] that the Larmor precession frequency in a field of 14.1 kG is 60 MHz (i.e., 6×10^7 cycles/sec) while in a field of 52 kG it is 220 MHz. The radio-frequency transmitter applies an rf field of this frequency by means of the exciting coil B, wound with its axis perpendicular to the magnetic field direction. The magnetic field is customarily taken as being in the negative z direction, as designated in Fig. I-4. The magnetic vector of the rf field oscillates along the y direction (i.e., in the direction parallel to the axis of the sample tube and perpendicular to H_0; see Fig. I-4); this oscillating magnetic field is appropriate for flipping over the nuclear spins, for it may be thought of as being composed of two equal magnetic vectors rotating in phase with equal angular velocities, but in opposite directions. (An exact analog is the decomposition of plane polarized light into two equal and opposite circularly polarized vectors.) The precessing magnetic moments will pick out the appropriate rotating component of H_1 in accordance with Eq. (I-6) and the sign of μ (see Table I-1). The other rotating component is so far off resonance that it has no observable effect. (Note that if H_1 were circularly polarized, one would be able to determine the sign of μ.)

To display the resonance signal, the magnetic field of the smaller "sweep" coils C is is increased slowly until Eq. (I-6) is satisfied and resonance occurs. (Alternatively, H_0 may be held constant and the frequency of H_1 swept.) At resonance, the nuclear magnetic dipoles in the lower energy state flip over

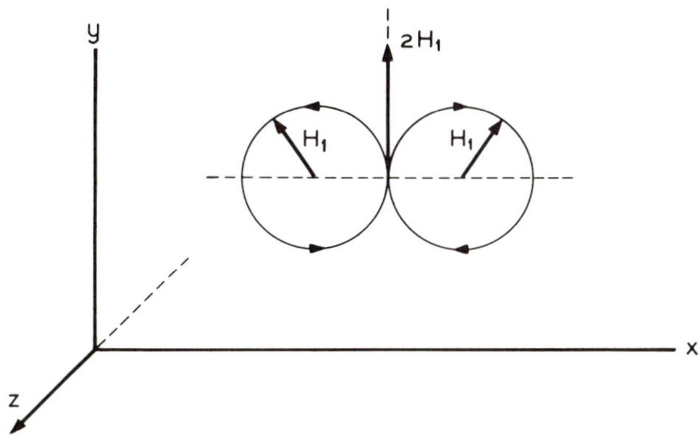

FIG. I-4. Decomposition of the magnetic vector H_1 into two counterrotating vectors.

and in so doing induce a voltage in a second coil (D), which is placed so that its axis is at right angles to both the magnetic field and the exciting coil. This induced voltage is amplified and recorded. One may regard the rotating field H_1 as having given the precessing spins a degree of coherence, so that now there is a detectable *macroscopic* magnetic moment, precessing at a rate ν_0. The resonance phenomenon observed in this way is termed *nuclear induction*, and is the method originally used by the Bloch group. Alternatively, both transmitting and receiving coils can be combined into one and the nuclear flipping detected as an *absorption* of energy from the rf field.

Commercial spectrometers are of both two-coil and one-coil design, and may have a number of features which cannot be discussed in detail here. One of the most important of these is the "field-lock" device, which by means of secondary oscillating circuits ensures that Eq. (I-6) is obeyed within narrow limits (1 in 10^8 to 10^9 ppm), and that spectra are thus highly reproducible. (See Chapter II of Bovey and Chapter III of Becker in suggested further readings at the end of this chapter.) Further experimental details are considered in Section 17.

5. Spin–Lattice Relaxation

Even in the highest magnetic fields now attainable, the separation of magnetic energy levels is very small. For example, for protons in a field of 14.1 kG, it is of the order of only 10^{-2} cal, and $2\mu H_0/kT$ is about 10^{-5}. Even in the absence of the rf field, there is usually a sufficiently rapid transfer of spins from the lower to the upper state and vice versa (for reasons to be explained shortly) so that an equilibrium population distribution is attained within a few seconds after H_0 is applied. If this distribution of spins is given by the Boltzmann factor $\exp(2\mu H_0/kT)$, it can be expressed with sufficient accuracy by

$$\frac{N_+}{N_-} = 1 + \frac{2\mu H_0}{kT} \tag{I-7}$$

where N_+ and N_- represent the spin populations of the lower and upper states, respectively, and T is the Boltzmann *spin temperature*. It can thus be seen that even for protons, the net degree of polarization of the nuclear moments

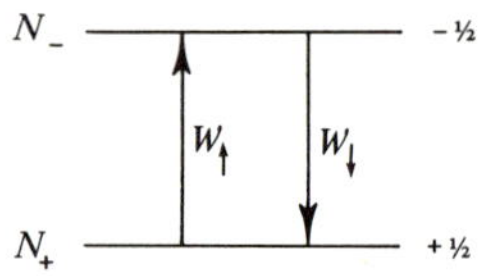

in the magnetic field direction is only of the order of a few parts per million; for all other nuclei (with the exception of tritium), it is even smaller. It is also clear that the nuclear magnetic energies cannot be expected to perturb the thermal energies of the molecules to an observable degree, except possibly at temperatures close to 0°K.

For nuclei with spins greater than $\frac{1}{2}$, the $2I+1$ equally spaced magnetic energy levels will be separated in energy by $\mu H_0/I$; the relative populations of adjacent levels will be given by expressions analogous to Eq. (I-7). To simplify the subsequent discussion, we shall confine our attention to nuclei with spins of $\frac{1}{2}$, recognizing that what is said for such two-level systems will apply also to any pair of adjacent levels in systems of spin greater than $\frac{1}{2}$.

We have seen that in the presence of the field H_1, there will be a net transfer of spins from the lower energy state to the upper energy state. In time, such a process would cause the populations of the levels to become equal, or nearly so, corresponding to a condition of *saturation* and to a very high Boltzmann spin temperature, unless there were some means by which the upper level spins could relax to the lower level. Such a process transfers energy from or *cools* the spin system. Similarly, when a system of spins is first thrust into a magnetic field, the populations of spins in the upper and lower energy states are equal, and the same relaxation must occur in order to establish an equilibrium spin population distribution and permit a resonance signal to be observed. It should be realized that since the energies involved are very small and since the nuclei are, as we shall see, usually rather weakly coupled thermally to their surroundings, i.e., the thermal relaxation is a slow process, the spin temperature may readily be made very high with little or no effect on the actual temperature of the sample as ordinarily observed. We may say that the heat capacity of the nuclear spin system is very small.

The required relaxation can occur because each spin is not entirely isolated from the rest of the assembly of molecules, commonly referred to as the "lattice," a term employed in dealing with both liquids and solids. The spins and the lattice may be considered to be essentially separate coexisting systems with a very inefficient but nevertheless very important link by which thermal energy may be exchanged. This link is provided by molecular motion. Each nucleus "sees" a number of other nearby magnetic nuclei, both in the same molecule and in other molecules. These neighboring nuclei are in motion with respect to the observed nucleus, and this motion gives rise to fluctuating magnetic fields. The observed nuclear magnetic moment will be precessing about the direction of the applied field H_0, and will also be experiencing the fluctuating fields of its neighbors. Since the motions of each molecule and of its neighbors are random or nearly random, there will be a broad range of frequencies describing them. To the degree that the fluctuating local fields have components in the direction of H_1 and at the precession frequency ν_0, say 60 MHz, they will induce transitions between energy levels because they provide fields equivalent to H_1. In solids or very viscous liquids, the molecular motions are relatively slow and so the component at ν_0 will be weak. The frequency spectrum will resemble curve (a) in Fig. I-5. At the other extreme, in liquids of very low viscosity, the motional frequency spectrum may be very flat, i.e., the magnetic noise may be nearly "white," and so no one component, in particular, that at ν_0, can be very intense [curve (c) in Fig. I-5]. We are then led to expect that at some intermediate condition, probably that of a moderately viscous liquid [curve (b)], the component at ν_0 will be at a maximum and thermal relaxation of the spin system can occur with optimum efficiency.

The probability per unit time of a downward transition of a spin from the higher to the lower magnetic energy level, $W_\downarrow$, exceeds that of the reverse transition $W_\uparrow$ by the same factor as the *equilibrium* lower state population exceeds the upper state population [Eq. (I-7)]. We may then write

$$\frac{W_\downarrow}{W_\uparrow} = 1 + \frac{2\mu H_0}{kT} \tag{I-8}$$

This must be the case, or equilibrium could not be maintained. (For isolated spins, i.e., those not in contact with the thermal reservoir provided by the lattice, the transition probabilities are exactly the same in both directions.) Let the spin population difference at any time t be given by n; let the equilibrium population difference (i.e., in the presence of H_0, but in the absence of

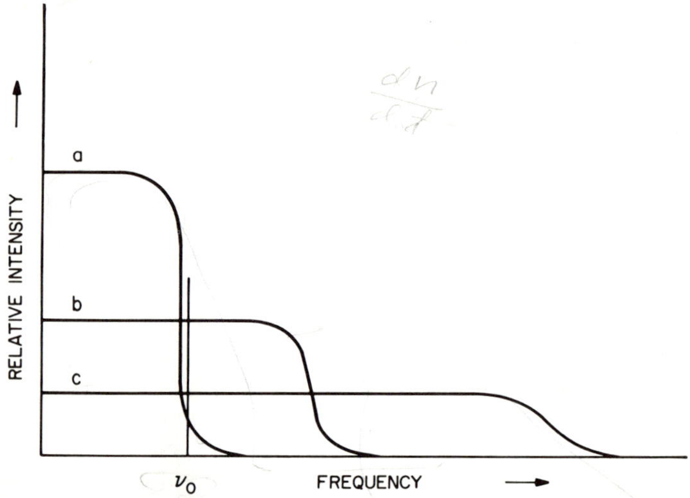

FIG. I-5. Frequency spectrum at (a) high viscosity, (b) moderate viscosity, and (c) very low viscosity.

the rf field) be given by n_{eq}; and let us define a quantity T_1, in units of seconds, the reciprocal of which is the sum of the probabilities (per second) of upward and downward transitions, i.e., the total rate of spin transfer in both directions:

$$1/T_1 = W_\downarrow + W_\uparrow \tag{I-9}$$

The approach to equilibrium will be described by

$$dn/dt = 2N_- W_\downarrow - 2N_+ W_\uparrow \tag{I-10}$$

N_- and N_+ being the populations of the upper and lower levels, respectively, as in Eq. (I-7); this represents the *net* transfer of spins from the upper state. (The factors of 2 arise because each transfer of one spin changes the population difference by 2.) If we designate the total spin population $(N_+ + N_-)$ by N, we can express this rate in terms of N and n as

$$dn/dt = (N - n)W_\downarrow - (N + n)W_\uparrow \tag{I-11a}$$

$$= N(W_\downarrow - W_\uparrow) - n(W_\downarrow + W_\uparrow) \tag{I-11b}$$

At equilibrium, we find from Eq. (I-11b) that

$$\frac{n_{eq}}{N} = \frac{W_\downarrow - W_\uparrow}{W_\downarrow + W_\uparrow} \tag{I-12}$$

and so, using Eq. (I-9), we may rewrite Eq. (I-11b) as

$$\frac{dn}{dt} = \frac{1}{T_1}(n_{eq} - n) \tag{I-13}$$

which gives upon integration,

$$n = n_{eq}(1 - e^{-t/T_1}) \tag{I-14}$$

We see than that, as might perhaps have been intuitively expected, the rate of approach to thermal equilibrium depends upon how far away from equilibrium we are. The first-order rate constant (in the language of chemical kinetics) for this process is $1/T_1$. The quantity T_1 is called the *spin–lattice relaxation time*; it is the time required for the difference between the excess spin population and its equilibrium value to be reduced by a factor *e*.

Except in a static collection of nuclei, i.e., a solid near 0°K, spin–lattice relaxation through molecular motion always occurs, although it may be very slow. There are two other possible contributing causes to spin–lattice relaxation which may occur under some conditions, and which are of importance to the chemist. These are:

1. Spin–lattice relaxation by interaction with unpaired electrons in paramagnetic substances. The mechanism is fundamentally exactly the same as that just discussed, but is much more effective because the magnetic moment of an unpaired electron is close to the magnitude of the Bohr magneton, $eh/4\pi M_e c$, and of the order of 10^3 times greater than the nuclear magneton because of the small mass of the electron, M_e.
2. Spin–lattice relaxation by interaction of the *electric quadrupole moments* of nuclei of spin 1 or greater with electric fields within the tumbling molecule. This can be a very potent cause of thermal relaxation and will be dealt with in Section 10.

Spin–lattice relaxation is often termed *longitudinal* relaxation because it involves changes of energy and therefore involves the component of the nuclear moment along the direction of the applied magnetic field. It also of course shortens the lifetimes of the spins in both the upper and lower energy states. This leads to an uncertainty in the energies of these states, and to a broadening of the resonance lines. This can be estimated from the Heisenberg relation $\delta E \cdot \delta t \simeq h/2\pi$, from which the uncertainty in the frequency of absorption is given by

$$\delta\nu \simeq 1/(2\pi \cdot \delta t) \tag{I-15}$$

Under the present-day operating conditions, line widths are at least 0.3 Hz, which, if arising from uncertainty broadening, would correspond to a spin lifetime of about 0.5 sec. Since T_1 is usually of the order of 1–10 sec in mobile liquids, spin–lattice relaxation ordinarily does not contribute observably to line broadening. As the temperature is lowered and viscosity increases, the component of the local magnetic noise spectrum at the Larmor frequency will increase, pass through a maximum, and decrease again. The value of T_1 will correspondingly decrease, pass through a minimum, and increase again. It is useful to define a correlation time t, which we shall understand to be the average time required for a rigid, approximately spherical molecule to rotate through an angle of one radian, and therefore for a line joining any two nuclei in such a molecule to rotate one radian. For such molecules, t is given to a good degree of approximation by

$$t = t_D/3 = 4\pi\eta a^3/3kT \qquad \text{(I-16)}$$

t_D being the correlation time used in the Debye theory of dielectric dispersion, η the viscosity of the liquid, and a the effective radius of the molecule. For most nonassociated molecules, T_1 is governed mainly by interactions of nuclei within the same molecule, rather than by motions of neighboring molecules. Under these conditions, it is found from the detailed theory of Bloembergen *et al.*[6] (as modified by Kubo and Tomita[7] and by Solomon[8]) that the rate of spin–lattice relaxation for a given nucleus in a molecule composed of j identical spin-$\frac{1}{2}$ nuclei will be given by

$$\frac{1}{T_1} = \frac{96}{5}\left(\frac{\pi}{h}\right)^2 \cdot \mu^4 \cdot \sum_j r_j^{-6}\left[\frac{t}{4\pi^2\nu_0{}^2t^2} + \frac{4t}{1+16\pi^2\nu_0{}^2t^2}\right] \qquad \text{(I-17)}$$

where μ is the magnetic moment, and r_j are the distances from the observed nucleus to each of its magnetic neighbors. We shall assume all the nuclei to be protons in the subsequent discussion. Figure I-6 represents a log–log plot of t versus T_1, as given by Eq. (I-17), which expresses quantitatively the dependence of T_1 on t discussed earlier in qualitative terms. (This plot also shows the behavior of T_2, which we shall discuss in the next section.) Ordinarily, however, we cannot measure t directly. From Eq. (I-16), it can be seen that for experimental purposes η/T can be taken as a measure of t for a given liquid system. Experiment has shown that plots of this sort do indeed have the predicted form.

Equation (I-17) can be simplified if we consider only pairs of protons in a liquid at room temperature or above, where we can be sure that $t \ll 1/2\pi\nu_0$. Under these circumstances, it reduces to

$$\frac{1}{T_1} = 96\left(\frac{\pi}{h}\right)^2 \cdot \mu^4 \cdot \frac{t}{r^6} \qquad \text{(I-18)}$$

from which, on combining with Eq. (I-16), we obtain

$$\frac{1}{T_1} = \frac{128\pi^3}{h^2} \cdot \mu^4 \cdot \frac{a^3}{r^6} \cdot \frac{\eta}{kT} \tag{I-19a}$$

$$= (\text{const}) \frac{\mu^4}{r^6} \cdot V \cdot \frac{\eta}{kT} \tag{I-19b}$$

where V is the molecular volume. This equation tells us at least three things that have practical implications:

(i) For a given liquid, T_1 will vary inversely with η/T over the usual temperature ranges employed in NMR spectroscopy, i.e., those over which the sample remains reasonably mobile. At any given temperature, narrower lines may be expected in solutions of low viscosity. For any given solution, the lines will become narrower as the temperature is increased, to the extent that they are dependent on T_1.

(ii) Other things being equal, T_1 will vary inversely as the molar volume. Thus for reasonably dilute solutions of a series of solutes in a given solvent of viscosity η, spin–lattice relaxation will increase in proportion to the molar

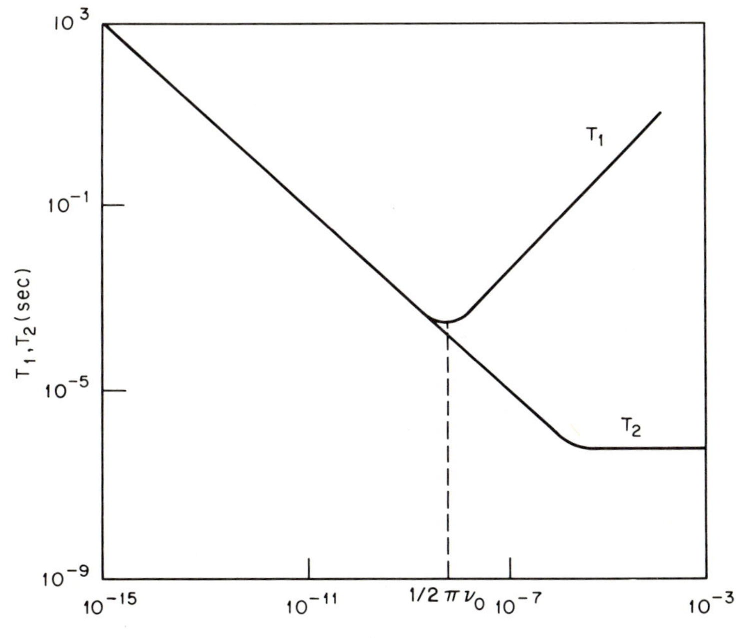

FIG. I-6. Theoretical dependence of T_1 and T_2 on correlation time t, assuming all interactions to be characterized by the same value of t.

volumes of the solutes, and may for sufficiently large molecules become an important contributor to line broadening even though η remains small.

(iii) Substances in which the magnetic nuclei are relatively far apart or have small moments will exhibit longer relaxation times, since T_1 is a sensitive function of both μ and r.

The equations above contain severe approximations and cannot be expected to hold for most polymer molecules either in the solid state or in solution, since polymers are in general far from spherical and are usually composed of different groups and structural units, each with its own characteristic relaxation. Since long polymer chains usually move segmentally, T_1 is much longer than one would estimate from the molecular volume using Eq. (I-19b). For example, for polystyrene of high molecular weight in carbon tetrachloride solution, the main-chain protons are observed to have a T_1 of the order of 0.03 sec; for the *ortho* protons of the phenyl ring, T_1 is about 0.08 sec and for the *meta* and *para* protons, about 0.2 sec.[9] The T_1 of liquid benzene is about 20 sec near room temperature. The ratio T_1 (benzene)/T_1 (polymer *meta* proton) is substantial but much less than the ratio of molar volumes. For a polyisobutylene of 1,129,000 molecular weight in benzene, T_1 at room temperature is about 0.1 sec.[10] The viscosities of solutions of such polymers may be very high, and they will clearly violate Eq. (I-19), with respect to both V and η.

6. Dipolar Broadening and Spin–Spin Relaxation

A cause of broadening of NMR spectral lines usually more important than spin–lattice relaxation is the so-called "dipolar" broadening. To understand this effect, let us first consider not the liquid samples which the chemist normally examines in the NMR spectrometer, but rather a proton-containing solid, such as a solid polymer. If the protons are sufficiently removed from each other that they do not feel the effects of each other's magnetic fields, the resonant magnetic field at the nucleus will be essentially equal to H_0 (actually very slightly smaller because of the shielding effects of local electrons, to be discussed in Section 8); therefore, if by careful design of the magnet, the field H_0 can be made very homogeneous over the volume occupied by the sample, the width of the absorption peak may be less than 10^{-4} G, i.e., of the order of 1 part in 10^8 or a few tenths Hz. In most such substances, the protons are actually near enough to each other so that each is appreciably influenced by the magnetic fields of its neighbors. Let us first imagine that we are dealing with a static system of isolated pairs of protons, i.e., each member of a pair experiences the field of the other member but not those of the other pairs. The field felt by each proton will be made up of the applied field H_0, plus this small additional field, H_{loc}. The sign and magnitude of this increment will depend upon the distance apart of the nuclei r and upon θ, the angle between the line joining the nuclei and the vector representing the direction of H_0. We suppose that all the pairs have the same values of these parameters. The equation expressing the functional dependence of the separation of the protons' magnetic energy levels upon r and θ is given by

$$\Delta E = 2\mu(H_0 \pm H_{loc}) = 2\mu[H_0 \pm \tfrac{3}{2}\mu r^{-3}(3\cos^2\theta - 1)] \qquad \text{(I-20)}$$

The $\pm$ sign corresponds to the fact that the local field may add to or subtract from H_0 depending upon whether the neighboring dipole is aligned with or

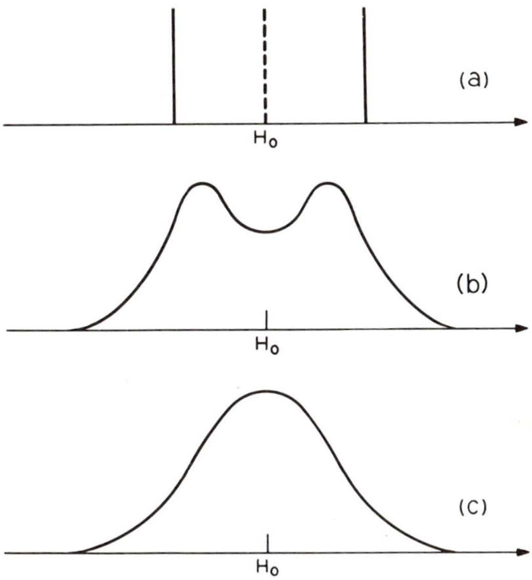

FIG. I-7. Schematic NMR spectra of arrays of protons: (a) two protons; (b) semi-isolated pairs of protons; (c) a random or nearly random array of protons.

against the direction of H_0. We recall that the net polarization of nuclear moments along H_0 is ordinarily only a few parts per million, and so in a large collection of such isolated pairs the probabilities of a neighboring dipole being aligned with or against H_0 are almost equal. Equation (I-20) then expresses the fact that the spectrum of this proton-pair system will consist of two equal lines [Fig. 1-7(a)] whose separation, at a fixed value of θ, will vary inversely as r^3. Only when the orientation of the pairs is such that $\cos^2\theta = \frac{1}{3}$ ($\theta = 54.7°$, often called the "magic angle") will the lines coincide to produce a single line.

In many solids, for example, certain hydrated crystalline salts, we actually have pairs of protons in definite orientations and may expect to find a twofold NMR resonance. This is indeed observed[11]; the lines, however, are not narrow and isolated as in Fig. I-7(a), but are broadened and partially overlapping as in Fig. I-7(b). The separation of the maxima is found to depend on the orientation of the crystal in the magnetic field. In such substances, the magnetic interactions of the members of each pair dominate, but the interactions between pairs are not negligible. These many smaller interactions, varying with both r and θ, give rise in effect to a multiplicity of lines whose envelope is seen as the continuous curve of Fig. I-7(b). [Actually, the simple picture suggested here, while qualitatively useful, does not correctly give the numbers and intensities of the lines composing this envelope in a multispin system. The correct solution of the problem involves quantum mechanical considerations; in fact, the factor $\frac{3}{2}$ in Eq. (I-20) is a result of this more correct treatment.[12, 13]] In most organic solids, including polymers, pairing of protons does not occur to this degree and instead we find a nearly Gaussian distribution of magnetic interactions, such as one would expect if the protons were randomly distributed throughout the substance. For such a complex array of nuclei, the NMR spectrum is a single broad peak, as in Fig. I-7(c). In most rigid-lattice solids, where the protons are of the order of 1 Å apart, the distribution of local field strengths about each proton is such that the half-height width of the peak is of the order of 10 G, or about 0.5 to 1.0 × 10^5 Hz.

We have so far assumed that the nuclei are fixed in position. If *molecular motion* is allowed to take place, as by raising the temperature of the solid, by melting it, or by dissolving it in a mobile solvent, the variables in Eq. (I-20) become functions of time. If we assume that r is constant and only θ varies with time, as would be true for pairs of protons within a rigid molecule, then the time-averaged local field will be given by

$$H_{\text{loc}} = \mu r^{-3} \cdot T_2^{-1} \int_0^{T_2} (3\cos^2\theta - 1)\,dt \tag{I-21}$$

where T_2 is the time that the nucleus resides in a given spin state. If θ may vary rapidly over all values, this time average can be replaced by a space average[14]:

$$H_{\text{loc}} = \mu r^{-3} \int_0^{\pi} (3\cos^2\theta - 1)\sin\theta\,d\theta = 0 \tag{I-22}$$

Thus, if the correlation time t [Eq. (I-16)] is so short that this space averaging is valid, then the net effect of the neighboring magnetic nuclei is effectively erased, and the line will be drastically narrowed (by a factor of 10^4 to 10^5) compared to its rigid lattice value. If we begin with a solid at a temperature so low that it is virtually a rigid lattice, and then permit molecular motion (chiefly rotation) to occur with increasing frequency by raising the temperature, narrowing of the resonance line will begin when

$$1/t \geqslant 2\pi\delta\nu \tag{I-23}$$

where $\delta\nu$ is the static line width (in Hz). In liquids of ordinary viscosity, molecular motion is so rapid that Eq. (I-22) holds; the local variations in magnetic field strength have become so short-lived that *motional averaging* is complete. But for large molecules and for viscous liquids this line narrowing effect may not be complete and broadening may still be observable. For high polymers, and even for molecules of 400–500 molecular weight (particularly if they are rigid), this broadening can remain conspicuous [see Section 11 and Fig. I-12(b)].

There is another aspect of the interaction of neighboring magnetic dipoles that is closely related to the effects we have just considered, and must also be considered in relation to line broadening. We recall that nuclear spins are not merely small static magnetic dipoles, but that even in a rigid solid they are precessing about the field direction. We may resolve a precessing nuclear moment (Fig. I-8) into a *static component* along the direction of H_0, which

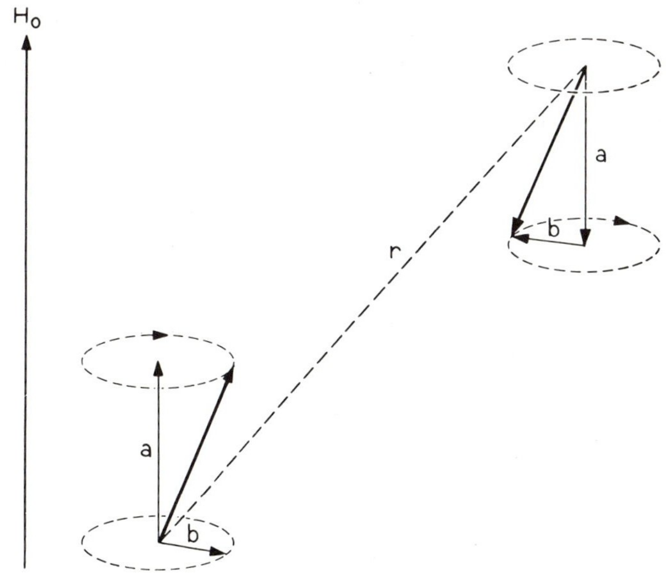

FIG. I-8. Static and rotating components of a precessing nuclear moment.

we have considered so far, and into a *rotating component*, whose effect we must now also consider. This component constitutes the right type of magnetic field, as we have seen (Section 4), to induce a transition in a neighboring nucleus if this is precessing at the same frequency. If this spin exchange or flip-flop occurs, the nuclei will exchange magnetic energy states with no overall change in the energy of the system, but with a shortening of the lifetime of each. The magnitude of local field variations may be taken as μ/r^3 [an approximation of H_{loc} from Eq. (I-20)], and consequently the relative phases of the nuclei will change in a time of the order of hr^3/μ^2, the "phase memory time." From Eq. (I-15), we expect an uncertainty broadening of about μ^2/hr^3, i.e., of the same form and order of magnitude as that produced by the interaction of the static components of the nuclear moments. It has become customary to include both effects in the quantity T_2, which we defined above as the spin lifetime. Thus T_2 is an inverse measure of the broadening of the spectral lines:

$$T_2 = 1/\pi\delta\nu \tag{I-24}$$

(The basis of this relationship will become more evident in Section 7.) It is called the *spin–spin relaxation time* or *transverse relaxation time*. A detailed theory of its dependence on molecular correlation time has been given by Bloembergen *et al.*[6]; Fig. I-6 shows this theoretical dependence. Note that at short correlation times, as in mobile liquids (where "phase memory" is long) it becomes equal to T_1, but after T_1 passes through its minimum, T_2 continues to decrease as molecular motion becomes slower and finally levels out as the system begins to approach a rigid lattice.

Spin–spin exchange and dipolar broadening should not be considered as merely two alternative ways of looking at the same phenomenon, closely interrelated though they are. For example, in a lattice composed of magnetic nuclei *a* containing a dilute distribution of a different magnetic nuclear species *b*, spin exchange between nuclei *a* and *b* cannot occur, since they precess at greatly different frequencies, but nevertheless dipolar broadening will be present. These considerations become very real in the observation of the ^{13}C spectra of polymers, to be discussed in later chapters.

Spin–spin relaxation is associated with a decay of the macroscopic nuclear moment in the *xy* plane; spin–lattice relaxation is associated with a decay of the macroscopic moment along the *z* direction. We may suspect that these relaxation rates, although they can be quite different in magnitude (as Fig. I-6 shows), are nevertheless closely associated, and that both T_1 and T_2 must be considered in describing resonance signal shapes and intensities. This will be discussed in Section 7.

7. The Bloch Equations

In treating the experimental observation of nuclear magnetic resonance, it is convenient to adopt the approach of Bloch,[15] and to consider the assembly of nuclei in macroscopic terms. We define a total moment M, which is the resultant sum, per unit volume, of all the individual nuclear moments in an assembly of identical nuclei, with magnetogyric ratio γ and $I=\frac{1}{2}$. We consider that M is not collinear with any of the axes x, y, and z, as in Fig. I-9. The

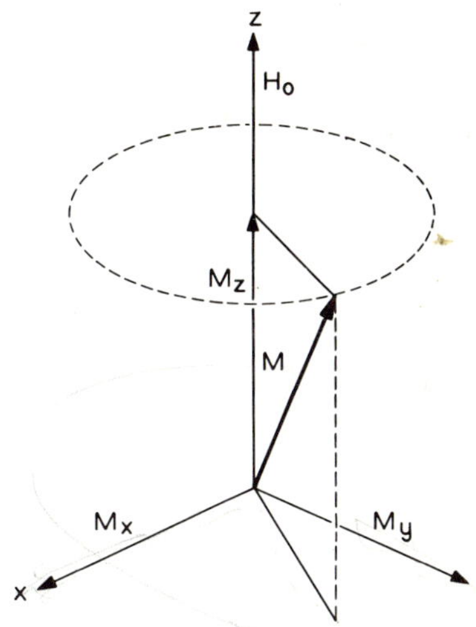

FIG. I-9. The macroscopic nuclear moment M and its components along the x, y, and z axes.

static field H_0 is in the z direction, and M, like the individual moments, will precess about z with an angular frequency ω_0. In the absence of relaxation effects and the rotating field H_1, the projection of M on the z axis M_z would remain constant:

$$\frac{dM_z}{dt}=0 \tag{I-25}$$

The magnitudes of the x and y projections M_x and M_y will, however, vary with time as M precesses, and, as can be seen from Fig. I-9, will be 180° out of phase, since when the projection of M along the x axis is a maximum the projection will be zero along the y axis, and vice versa. This time dependence can be expressed by

$$\frac{dM_x}{dt}=\gamma M_y \cdot H_0=\omega_0 M_y \tag{I-26}$$

$$\frac{dM_y}{dt}=-\gamma M_x \cdot H_0=-\omega_0 M_x \tag{I-27}$$

(Note that the inclusion of γ in macroscopic expressions causes no difficulty if dimensions are consistent.) In addition to the fixed field H_0, we must now consider the rotating field H_1, which we recall from earlier discussion will be provided by one of the two counterrotating magnetic vectors of a linearly polarized rf field. This vector H_1 rotates in the xy plane, i.e., perpendicularly to H_0, with frequency ω (equal to ω_0 only at exact resonance, i.e., at the center of the absorption peak). Consideration of the effect of H_1 upon the magnitudes of M_x, M_y, and M_z, in accordance with the basic laws describing the tipping of magnetic vectors in magnetic fields, leads to the following modifications of Eqs. (I-25)–(I-27):

$$\frac{dM_x}{dt} = \gamma[M_y H_0 - M_z(H_1)_y] \tag{I-28}$$

$$\frac{dM_y}{dt} = -\gamma[M_x H_0 + M_z(H_1)_x] \tag{I-29}$$

$$\frac{dM_z}{dt} = \gamma[M_x(H_1)_y - M_y(H_1)_x] \tag{I-30}$$

where $(H_1)_x$ and $(H_1)_y$ are the components of H_1 along the x and y axes, and are given by

$$(H_1)_x = H_1 \cos \omega t \tag{I-31}$$

$$(H_1)_y = -H_1 \sin \omega t \tag{I-32}$$

We have so far omitted any consideration of the relaxation of the components of M in the x, y, and z directions. By putting Eq. (I-13) in phenomenological terms, we may express the relaxation of the z component toward its equilibrium value M_0 as

$$\frac{dM_z}{dt} = \frac{M_z - M_0}{T_1} \tag{I-33}$$

The transverse relaxations may be expressed similarly, but with T_2 as the time constant.

$$\frac{dM_x}{dt} = -\frac{M_x}{T_2} \tag{I-34}$$

$$\frac{dM_y}{dt} = -\frac{M_y}{T_2} \tag{I-35}$$

These relaxation processes differ in that they go to zero rather to equilibrium values. Upon adding these terms to Eqs. (I-28)–(I-30) and using Eqs. (I-31) and (I-32), we obtain the complete Bloch equations:

$$\frac{dM_x}{dt} = \gamma(M_y H_0 - M_z H_1 \sin \omega t) - \frac{M_x}{T_2} \tag{I-36}$$

$$\frac{dM_y}{dt} = -\gamma(M_x H_0 - M_z H_1 \cos \omega t) - \frac{M_y}{T_2} \tag{I-37}$$

$$\frac{dM_z}{dt} = -\gamma(M_x H_1 \sin \omega t + M_y H_1 \cos \omega t) - \frac{M_z - M_0}{T_1} \tag{I-38}$$

A clearer insight into the significance of these equations in experimental terms is gained if the frame of reference is changed from fixed axes, x, y, and z to a set of axes imagined to rotate with H_1, i.e., with angular velocity $-\omega$ about the z axis. In the rotating frame, both H_0 and H_1 are fixed. We may then resolve the projection of M on the xy plane into components u and v, which are along and perpendicular to H_1, respectively, and will accordingly be in-phase and out-of-phase with H_1. See Fig. I-10. To transform to this

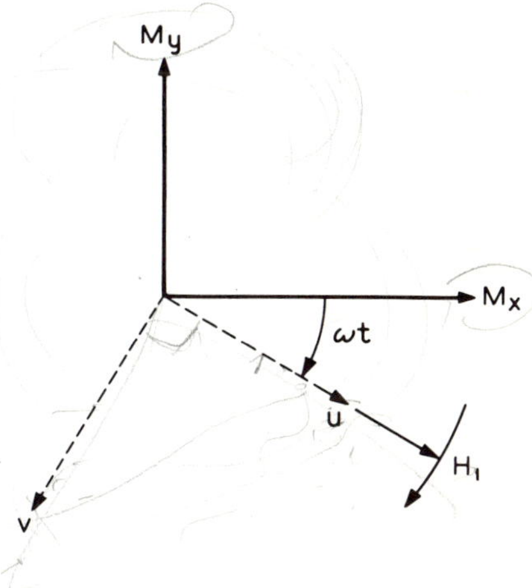

FIG. I-10. The components of the transverse macroscopic moment in a fixed plane (M_x, M_y) and in a plane rotating at the Larmor frequency about the static field H_0, taken as normal to the xy plane (and upward); the rotating frame components u and v are along and perpendicular to H_1, respectively.

new frame, we note that

$$M_x = u \cos \omega t - v \sin \omega t \tag{I-39}$$

$$M_y = u \sin \omega t - v \cos \omega t \tag{I-40}$$

and also that $\gamma H_0 = \omega_0$; we then may replace Eqs. (I-36) and (I-37) by:

$$\frac{du}{dt} + \frac{u}{T_2} + (\omega_0 - \omega)v = 0 \tag{I-41}$$

$$\frac{dv}{dt} + \frac{v}{T_2} - (\omega_0 - \omega)u + \gamma H_1 M_z = 0 \tag{I-42}$$

$$\frac{dM_z}{dt} + \frac{M_z - M_0}{T_1} - \gamma H_1 v = 0 \tag{I-43}$$

Here, the terms $(\omega_0 - \omega)$ are a measure of how far we are from exact resonance. We see from Eq. (I-43) that changes of M_z, i.e., changes in the energy of the spin system, are associated only with v, the out-of-phase component of the macroscopic moment, and not with u. We may then anticipate that absorption signals will be associated with the measurement of v. The component u will be associated with "dispersion-mode" signals.

Under the normal conditions of experimental observation, we are dealing with a steady state in which u, v, and M_z are constant in the rotating frame. We pass through the resonance peak, varying ω (or in practice H_0) so slowly that u, v, and M_z always have time to reach these steady values. Under such "slow passage" conditions we obtain Eqs. (I-44)–(I-46):

$$u = M_0 \frac{\gamma H_1 T_2^2(\omega_0 - \omega)}{1 + T_2^2(\omega_0 - \omega)^2 + \gamma^2 H_1^2 T_1 T_2} \tag{I-44}$$

$$v = -M_0 \frac{\gamma H_1 T_2}{1 + T_2^2(\omega_0 - \omega)^2 + \gamma^2 H_1^2 T_1 T_2} \tag{I-45}$$

$$M_z = M_0 \frac{1 + T_2^2(\omega_0 - \omega)^2}{1 + T_2^2(\omega_0 - \omega)^2 + \gamma^2 H_1^2 T_1 T_2} \tag{I-46}$$

We see from Eq. (I-45) that under conditions when $\gamma_2 H_1^2 T_1 T_2 \ll 1$, i.e., when H_1 is a few mG and T_1 and T_2 are no greater than a few seconds, the absorption or "v-mode" signal should be proportional to

$$\frac{\gamma H_1 T_2}{1 + T_2^2(\omega_0 - \omega)^2}$$

This describes what is known as a Lorentzian line shape, as shown in Fig. I-11(a), and is the type of signal normally used in high resolution spectra.

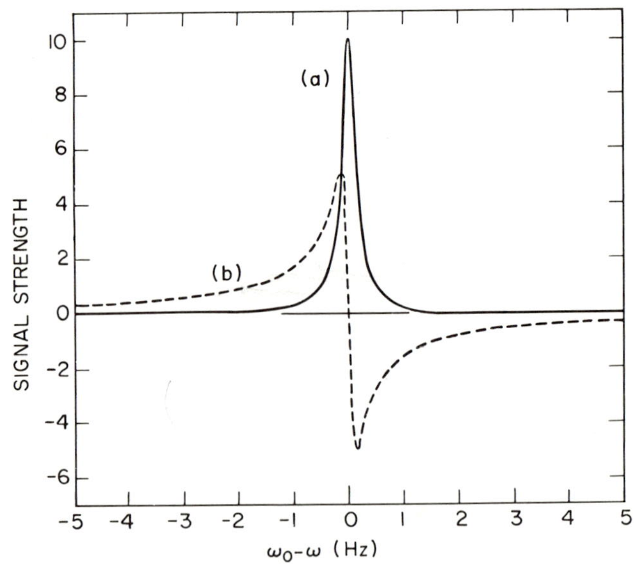

FIG. I-11. (a) Absorption (v-mode), and (b) dispersion (u-mode) resonance signals.

At the center, when the resonance condition is exactly fulfilled, $\omega_0 - \omega$ becomes zero and the signal height is proportional to $\gamma H_1 T_2$. It follows that the width must be inversely proportional to T_2, as we have already seen [Eq. (I-24)]; if we define the peak width $\delta\omega$ or $\delta\nu$ in the customary way as its width at half the maximum height, it is easily shown that

$$\delta\omega = 2/T_2 \tag{I-24a}$$

or

$$\delta\nu = 1/\pi T_2 \tag{I-24b}$$

where Eq. (I-24b) is a transposition of Eq. (I-24).

For some purposes, such as the observation of solid polymers it is preferable to employ the dispersion or "u-mode" signal, which under the same conditions will be proportional to $\gamma H_1 T_2{}^2(\omega_0 - \omega)/[1 + T_2{}^2(\omega_0 - \omega)^2]$. This will give the line shape shown in Fig. I-11(b). The maximum and minimum are separated by $1/\pi T_2$ for Lorentzian signals.

8. Saturation

Equations (I-44)–(I-46) allow us to express the phenomenon of saturation, discussed qualitatively on p. 8, in more exact terms. Partial saturation will begin to become observable when the term $\gamma^2 H_1{}^2 T_1 T_2$ is no longer much less than unity, as assumed in the preceding discussion. The field H_1 is now beginning to cause a net transfer of lower-state spins to the upper state at a rate that cannot be neglected in comparison to that of the spin–lattice relaxation. Equations (I-13) and (I-14) are no longer adequate to describe the relaxation. The new equilibrium population difference n'_{eq} will be given by

$$n'_{eq} = \frac{n_{eq}}{1 + \gamma^2 H_1{}^2 T_1 T_2} \tag{I-47}$$

The ratio n'_{eq}/n_{eq} will be less than unity and has been termed the saturation factor Z_0:

$$Z_0 = 1/(1 + \gamma^2 H_1{}^2 T_1 T_2) \tag{I-48}$$

For ordinary, mobile, proton-containing liquids under good experimental conditions, principally good field homogeneity, $T_1 T_2$ will be of the order of 10; γ^2 being approximately 10^8 for protons, we have

$$Z_0 = 1/(1 + 10^9 H_1{}^2)$$

We thus find that observable saturation effects can be expected when H_1 is no greater than 0.01–0.05 mG.

For polymers in solution, T_1 and T_2 are controlled by the same molecular motions and are usually nearly equal.[10] They are substantially shorter than for liquids—in particular, for the solvent employed—and so higher values of H_1 may be employed without approaching saturation. In solid polymers, the behavior of T_1 is often complex; T_2 becomes short at low temperatures, but because of the breadth of the signal the spin system is exposed to the rf field for extended intervals and saturation can readily occur.

9. Measurement of T_1 and T_2

The actual measurement of nuclear relaxation rates is not of immediate interest in this book, since we are primarily concerned with structure determination. We shall therefore indicate only briefly the types of methods employed.

For mobile liquids, T_1 may be obtained directly by high resolution methods by observing the growth of the signal either immediately after inserting the sample in the magnetic field or after saturating it with a high H_1. This is done by a succession of rapid scans. When T_1 is less than 2–3 sec, it is not feasible to sweep with the required rapidity.

As indicated by Eq. (I-24), T_2 may be obtained from the line width only if it is less than about 1 sec, for otherwise the contribution of the magnetic field homogeneity to the line width is relatively too great, and T_2 will be seriously underestimated. For liquids, $T_2 \simeq T_1$ and this method cannot be employed. For spectra of polymers in solution, line widths are generally 5–10 Hz or greater and should provide a reasonably accurate estimate of T_2, provided one can be sure that there are no other contributions to the line width, such as unresolved multiplicity.

For $T_1 < 2$–3 sec or $T_2 > \sim 1$ sec, it is customary to employ *pulse* techniques, in which a strong rf field is applied for a time τ short enough so that there is no relaxation during the time of its application. The effect of such a pulse is to rotate the macroscopic moment M toward the xy plane with an angular velocity γH_1. If $\gamma H_1 \tau$ is so chosen that M is just rotated into this plane, this is called a 90° pulse. The resulting signal will decay with a time constant T_2 ("free induction decay") which still, however, contains a contribution from the magnetic field inhomogeneity. The true T_2 can be determined by the "spin echo" method, in which the effect of field inhomogeneity is ingeniously nullified. (For a description of this technique, the reader is referred to pp. 46–49 of Pople *et al.* in the list of suggested further reading at the end of this chapter.) Measurements of T_1 may be made over a wide range by turning the moment back along the z axis with a second 90° pulse. If the time between pulses is less than T_1, the z component of magnetization will not yet have had time to increase to the equilibrium value M_0. By measuring M_z at varying pulse intervals, one will be able to obtain T_1 [Eq. (I-33)].

10. Nuclear Electric Quadrupole Relaxation

Nuclei with spins of $\frac{1}{2}$ possess a spherical distribution of the nuclear charge and are therefore not affected by the electric environment within the molecule. Nuclei with spins of 1 or greater, however, are found to have *electric quadrupole moments*, and their charge distribution can be regarded as spheriodal in form; the nucleus is to be regarded as spinning about the principal axis of the spheroid. The quadrupole moment may be *positive*, corresponding to a prolate spheroid, or *negative*, corresponding to an oblate spheroid. The energies of spheroidal charges will depend upon their orientation in the molecular electric field gradient. In certain classes of molecules, where an approximately spherical or tetrahedral charge distribution prevails, as for example in the ammonium ion NH_4^+, little or no electric field gradient will be present and the nuclear quadrupole moment will not be disturbed by the tumbling of the molecule. But in most molecules, substantial electric field gradients are present, and can interact with the nuclear quadrupoles. The spin states of such nuclei may thus be rapidly changed by the tumbling of the molecular framework. This furnishes an additional pathway for energy exchange between the spins and the lattice, i.e., an important contribution to spin–lattice relaxation, and can broaden the resonance peaks very markedly. The resonance of nuclei such as ^2H or ^{14}N (Q positive) or ^{17}O, ^{35}Cl, and ^{36}Cl (Q negative) may be so broad as to be difficult or impossible to observe. Nuclear quadrupole relaxation can also affect the resonance even of nuclei of spin $\frac{1}{2}$ when these are in sufficiently close proximity to nuclei of spin 1 or greater. We shall discuss this further in Chapter XIII.

11. Magnetic Shielding and Chemical Shift

From what we have said so far, it might be supposed that at any particular radio frequency ν_0, all nuclei of a given species, say, all protons, would resonate at the same value of H_0. And indeed this is true within the limits of the significant figures with which nuclear properties are expressed in Table I-1. If this were *strictly* true, NMR would be of little interest to chemists. But about five years after the first demonstration of NMR in condensed phases, it was found that the characteristic resonant frequency of a nucleus depends to a very small but measurable extent upon its chemical environment. It was found that the protons of water do not absorb at quite the same frequency as those of mineral oil,[16,17] the difference being only a few parts per million. For heavier nuclei, much larger effects are noted, up to 2% for certain metals.[18] In 1951, Arnold *et al.*[5] observed for alcohols separate spectral lines for chemically different nuclei within the same molecule, a discovery which may be said to have opened a new era for organic chemistry. The total range of variation of H_0 for protons is about 13 ppm or about 800 Hz in a 14,100 G field. The spectrum of ethyl orthoformate, $HC(OCH_2CH_3)_3$, shown in Fig. I-12, shows

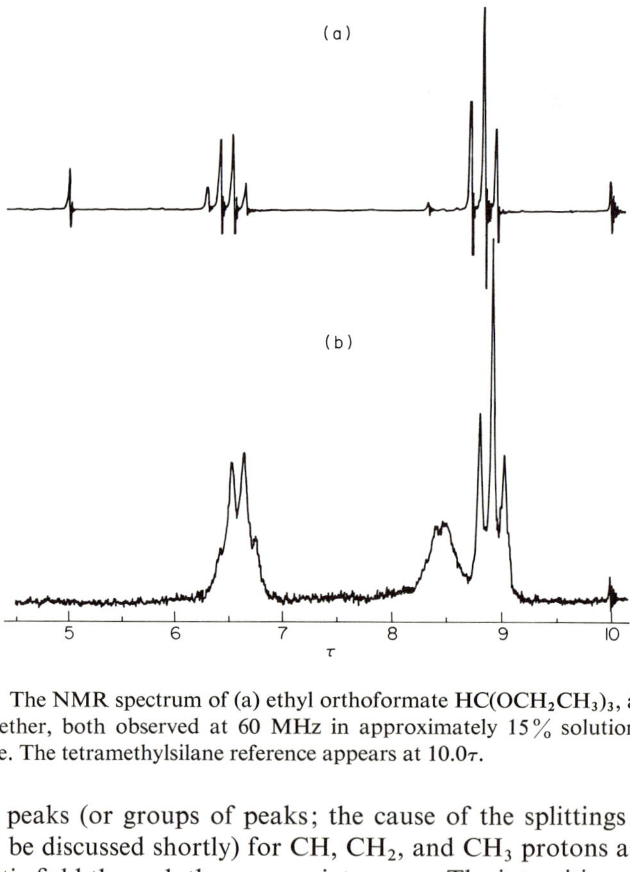

FIG. I-12. The NMR spectrum of (a) ethyl orthoformate $HC(OCH_2CH_3)_3$, and (b) polyvinyl ethyl ether, both observed at 60 MHz in approximately 15% solutions in carbon tetrachloride. The tetramethylsilane reference appears at 10.0τ.

successive peaks (or groups of peaks; the cause of the splittings within the group will be discussed shortly) for CH, CH_2, and CH_3 protons as we sweep the magnetic field through the appropriate range. The intensities of the peaks are proportional to the numbers of protons of each type.

The origin of this variation in resonant field strength is the cloud of electrons about each of the nuclei. When a molecule is placed in a magnetic field H_0, orbital currents are induced in the electron clouds, and these give rise to small local magnetic fields, which, in accordance with Lenz's law, are always proportional to H_0 but opposite in direction.† Such behavior is common to all molecules and gives rise to the universally observed diamagnetic properties of matter. Each nucleus is in effect partially shielded from H_0 by the electrons and requires a slightly higher value of H_0 to achieve resonance. This can be expressed as follows:

$$H_{loc} = H_0(1 - \sigma) \tag{I-49}$$

where H_{loc} is the actual local field experienced by the nucleus and σ is the *screening constant*, expressing the reduction in effective field; σ is independent of H_0 but highly dependent upon chemical structure. Equation (I-6) can then be modified to

$$h\nu_0 = \mu H_0(1 - \sigma)/I \tag{I-50}$$

The effect of nuclear screening is to decrease the spacing of the nuclear magnetic energy levels. It will be seen that at constant rf field frequency ν_0, an increase in σ, i.e., an increase in the magnetic shielding of the nucleus, means that H_0 will have to be *increased* to achieve resonance. Thus, if, as in Fig. I-12, resonance peak positions are expressed on a scale of magnetic field strength increasing from left to right, as is now almost universally done, the peaks for the more shielded nuclei will appear on the right-hand side of the spectrum. If, on the other hand, we hold H_0 constant and vary ν_0, as is now done in some NMR spectrometers, Eq. (I-50) tells us that the frequency ν_0 must be *decreased* to achieve resonance. In this case, the spectra are still presented with the more shielded nuclei on the right-hand side, but frequency will be *decreasing* from left to right.

Variations in nuclear shielding may be thought of as arising from:

a. Variations in local electron density. Protons attached to or near electronegative groups or atoms such as O, OH, halogens, CO_2H, NH_3^+, and NO_2 experience a lower density of shielding electrons and resonate at lower values of H_0. Protons removed from such groups appear at higher values. Such inductive effects are the principal cause of chemical shifts. The spectrum of ethyl orthoformate (Fig. I-12) furnishes an illustration of these effects. The methyl groups, being farthest removed from the oxygen atoms, appear on the right-hand side of the spectrum; the methylene resonance is further "downfield"; the formyl proton, being on a carbon atom bearing three oxygen atoms, is the least shielded.

b. Special shielding effects produced by certain groups and structures that allow circulation of electrons only in certain preferred directions within

† It is assumed that all the electrons in the molecule are paired off, so that they have no net magnetic moment *per se*. If this is not the case, i.e., if the molecule is a free radical, the much stronger magnetic field arising from the unpaired electron will be *independent* of H_0.

the molecule, i.e., which exhibit diamagnetic anisotropy. Benzene rings, and aromatic rings generally, show this behavior very strongly. When such compounds are placed in a magnetic field, the six π-electrons circulate in two parallel doughnut-shaped orbits on each side of the ring (Fig. I-13). The resulting local magnetic field opposes H_0 in a cone-shaped zone of excess shielding extended along the hexad axis, but reinforces H_0 in a zone of deshielding extending from the edge of the ring. In aromatic compounds, the deshielding zone is more commonly occupied than the shielding zone. Thus, protons on aromatic rings appear at much lower field (2 to 3τ) than olefinic protons

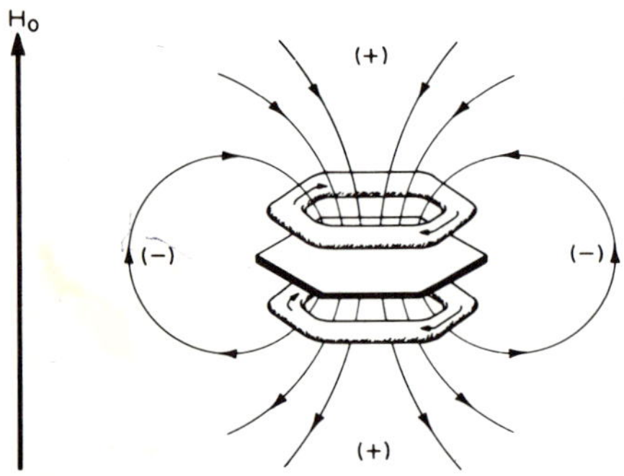

FIG. I-13. Electron density, ring currents, and magnetic lines of force about a benzene ring.

(4–5τ). In acetylenes (Fig. I-14) the current circulates in such a way that the shielding zone extends along the bond direction, and the acetylenic protons appear at unexpectedly high fields (7–8τ). In C═C and C═O double bonds, the deshielding zone seems to extend along the bond direction; even C—C single bonds show marked deshielding effects in this same sense. We shall discuss these matters in more detail when we deal with the spectra of polymers in which they have substantial effects.

Anisotropic shielding effects are in general smaller than inductive effects, but nonetheless can be very marked and can provide valuable structural clues.

Protons, because of their low density of screening electrons, show smaller variations due to (a) than do other nuclei, but for the same reason exhibit relatively greater effects from (b) than do heavier nuclei.

c. Effects of intramolecular reaction fields.[19] An atom having a spherical electron distribution will experience a distortion of the electron cloud when placed in an electric field E. This distortion will be proportional to E^2 and will result in a reduction of the shielding of the nucleus. Even in molecules with no electric dipole moment local distortions can occur. Field effects are of greater significance for heavier nuclei such as ^{19}F and ^{13}C than for the proton.

d. Effects of unpaired electron spin. Large chemical shifts may occur in complexes of paramagnetic metals in which there is a donation of electron spin density from the metal to the ligand. These are called *contact* or hyperfine shifts. In addition the electron magnetic dipole may cause *pseudo-contact* shifts. These effects are considered in detail in Chapters XIII and XIV.

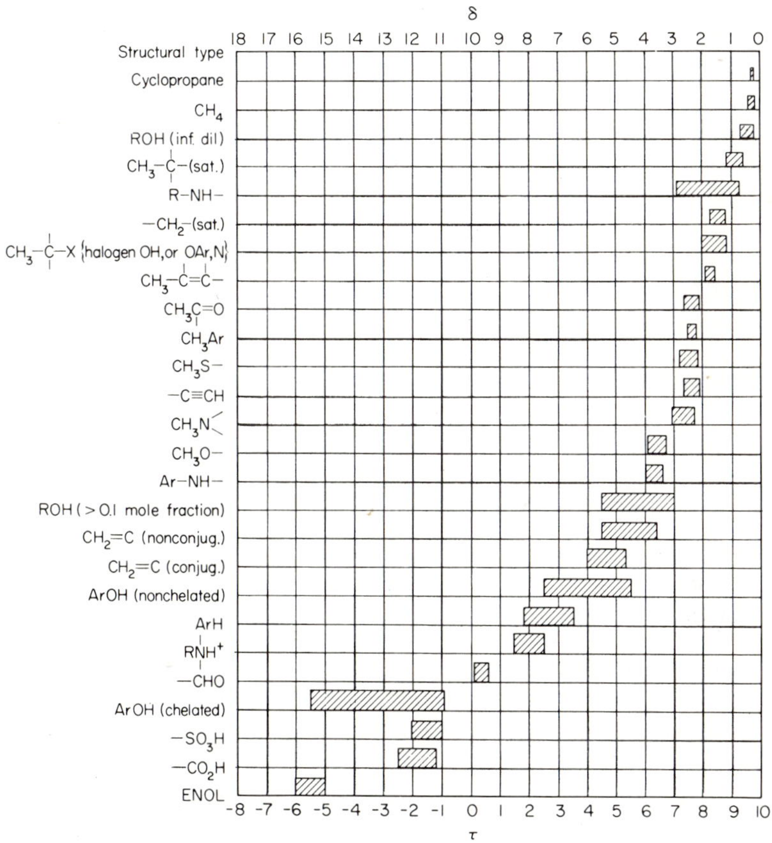

FIG. I-14. Proton chemical shifts.

Because nuclear shielding is proportional to the applied magnetic field, it of course follows that the spacing between peaks (or groups of peaks) corresponding to different types of nuclei is also proportional to the magnetic field. Thus, in NMR spectroscopy, in contrast to optical spectroscopy, there is no natural fundamental scale unit; the energies of transitions between quantum levels are proportional to the laboratory field. There is also no natural zero of reference. For practical purposes, these difficulties are evaded by the following devices:

a. Using parts-per-million relative change in H_0 (or less commonly v_0) as the scale unit.

b. Using an arbitrary reference substance dissolved in the sample and referring all displacements in resonance, called *chemical shifts*, to this "internal" reference.

The use of a dimensionless scale unit has the great advantage that the chemical shift values so expressed are independent of the value of H_0 of any particular spectrometer, and so a statement of chemical shift does not have to be accompanied by a statement of the frequency (or field) employed, as is the case when actual magnetic field strength (in gauss) or frequency (in hertz) is used as the scale unit. A standard reference substance for proton spectra, tetramethylsilane has been proposed[20] and widely accepted. It is the basis of the τ scale. On this scale, tetramethylsilane is assigned a value of 10.000τ rather than zero, since its protons are more shielded than those of nearly all other organic compounds. In Fig. I-12 peak positions are expressed in τ units. The great majority of proton chemical shifts fall between 0 and 10 on this scale. Figure I-14 shows the chemical shifts for a number of types of protons in organic structures and functional groups.

Some authors prefer the so-called δ scale, which is the inverse of the τ scale, and on which the tetramethylsilane resonance is at zero; thus

$$\delta = 10 - \tau$$

12. Electron-Mediated Coupling of Nuclear Spins

We have seen in Section 6 that nuclear spins may be directly coupled to each other through space, giving rise to the phenomenon of dipolar broadening. In addition to this direct coupling, magnetic nuclei may also transmit information to each other indirectly through the intervening chemical bonds. This interaction occurs by slight polarizations of the spins or orbital motions of the valence electrons and, unlike the direct couplings, is not affected by the tumbling of the molecules. It is independent of H_0. If two nuclei of spin $\frac{1}{2}$ are so coupled, each will split the other's resonance to a doublet, for in a collection of many such pairs of nuclei, there is an almost exactly equal probability of each finding the other's spin to be oriented with $(+\frac{1}{2})$ or against $(-\frac{1}{2})$ the applied field. If one nucleus is coupled to a second group of two identical nuclei, the possible combinations of orientations of the latter will be as shown in case (b). It is therefore to be expected that the resonance of the first nucleus

(a)		+		−	
(b)	++		+− −+		−−
(c)	+++	++− +−+ −++		+−− −+− −−+	−−−
(d)		+	0	−	
(e)	++	+0 0+	00 +− −+	0− −0	−−

will appear as a 1:2:1 triplet, while that of the group of two identical nuclei will be a doublet. Three equivalent neighboring spins would split the single nucleus' resonance to a 1:3:3:1 quartet, as in (c). Generalizing, we may say that if a nucleus of spin $\frac{1}{2}$ has n equivalently coupled neighbours of spin $\frac{1}{2}$, its resonance will be split into $n+1$ peaks, corresponding to the $n+1$ spin states of the neighboring group of spins. The peak intensities are proportional to the coefficients of the binomial expansion. A convenient mnemonic device for these is the Pascal triangle (Fig. I-15).

```
                  1
                1   1
              1   2   1
            1   3   3   1
          1   4   6   4   1
        1   5  10  10   5   1
      1   6  15  20  15   6   1
```

FIG. I-15. The Pascal triangle.

The spectrum of the ethyl groups of ethyl orthoformate [Fig. I-12(a)] illustrates these features of NMR spectra. The methyl resonance at 8.84τ is a triplet because of coupling to the CH_2 protons, which appear as a quartet centered at 6.48τ. The formyl proton is too distant to experience observable coupling. The strength of the coupling is denoted by J and is given by the spacing of the multiplets, expressed in Hz. In Fig. I-12(b) is shown the spectrum of polyvinyl ethyl ether (observed at 60 Hz in $CDCl_3$ solution). The ethyl

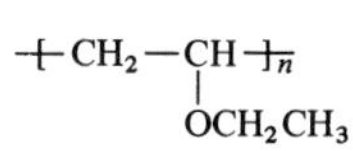

group triplet and quartet, at about 8.8 and 6.5τ, respectively, can be observed, but the peaks are poorly resolved because of dipolar broadening (Section 6), a feature common to all polymer spectra, and also because of other effects to be described in later chapters. The polymer spectrum also exhibits a main-chain CH_2 resonance at 8.4τ and a main-chain CH resonance under the ethyl methylene resonance. As we shall see, main-chain nuclei generally show greater line width than side-chain nuclei because of their slower reorientation.

The coupling of nuclei with spins of 1 or greater is more rarely dealt with, deuterium (^2H) and ^{14}N being the most frequently encountered. For nuclei of spin 1, we have seen that three spin states are possible. One such nucleus will split another nucleus (or group of equivalent nuclei), of whatever spin, into a 1:1:1 triplet [(d) above]; two will give a 1:2:3:2:1 pentuplet (e), and so on. We may generalize further, and say that coupling to n equivalent nuclei of spin I will give rise to $2nI + 1$ lines.

It is particularly important to understand that the occurrence of spin–spin coupling does not depend upon the presence of the magnetic field H_0, and that the magnitude of J, unlike that of the chemical shift, is independent of the magnitude of H_0. Thus, the ratio $J/\Delta\nu$, where $\Delta\nu$ is the chemical shift difference between two given nuclei or two groups of equivalent nuclei, decreases as H_0 increases. The multiplets move further apart on a field-strength or hertz scale (but not on the τ or δ scales); however, the spacings within each multiplet (in hertz) remain the same.

13. Spin Systems and Spectral Analysis

An important aspect of nuclear coupling which concerns the spectra of polymers as well as small molecules is the calculation of J values and chemical shifts for systems of spins in which the couplings are comparable in magnitude to the chemical shift differences, $\Delta\nu$, both expressed in hertz. The spectra of such molecules often appear as complex systems of lines having spacings exhibiting no obvious regularity and intensities which deviate widely from binomial. It is not within the scope of this book to consider such spin systems in detail, particularly since there exist many sources of information concerning them. (See the suggested further reading list at the end of this chapter.) We shall consider briefly here some of the characteristics of such systems, including one type in particular which is often important in the spectra of vinyl polymers.

The analysis of spectra characterized by relatively large values of $J/\Delta\nu$ (called "strong-coupled" even though the absolute value of J may not be particularly large) does not require the assumption of any particular physical model, i.e., we do not have to know what the molecule is, although we must know the number of spins involved. The task is to compute by quantum mechanical methods the energy levels and stationary-state wave functions of a system of coupled spins in a static magnetic field, and then to use perturbation methods and selection rules to give the probabilities of transitions occurring between these levels when a resonant rf field is applied. The line positions will be functions of the energy level separations and their relative intensities will correspond to the transition probabilities. If a suitable choice of parameters is made, the calculated spectrum will usually match the observed spectrum very closely. From the chemical shifts and couplings thus deduced, one may hope to deduce the structure of a new molecule or polymer chain. Or if, as is usually the case in polymer studies, the basic chain structure is known, one may deduce information concerning its configuration and conformation.

We note that J couplings may be either positive or negative. The signs of the couplings affect the energy levels of the coupled spins, and in systems of three or more spins, their relative signs can affect the appearance of the spectrum and will therefore be determinable. Consult the suggested further reading list at the end of this chapter for a discussion of the quantum mechanical basis of these facts.

For describing systems of nuclear spins within molecules, it is convenient to adopt an alphabetical notation.[21] Nonequivalent nuclei of the same species which have chemical shift differences comparable to their mutual couplings are designated A, B, C, etc. Another group of nuclei, separated from this group by large chemical shifts but among themselves by chemical shifts comparable to the couplings, are designated X, Y, Z, etc. In some cases, still a third group having intermediate chemical shifts may be present and would be denoted by K, L, M, etc. Nuclei within a group are commonly designated in alphabetical progression in order of increased shielding, but this practice is not always adhered to.

In these terms, the ethyl groups in ethyl orthoformate or polyvinyl ethyl ether would each be designated as an A_2B_3 system, there being no observable coupling to other protons or between ethyl groups. It is observed that, although $J/\Delta\nu$ is only about 0.05, the peak intensities of the ethyl spectra at 60 MHz deviate noticeably from binomial in such a way as to "lean" toward each other. However, the multiplet positions and spacings give the chemical shifts and couplings correctly within experimental error, and so the spin system can be designated A_2X_3 or "first-order" for all practical purposes. If observed at 100 or 220 MHz, this approximation would become progressively more exact since, as we have seen, $\Delta\nu$ will be proportionately increased with no change in J.

Systems of two coupled spins are often important in polymer spectra, and can be readily treated even when strong-coupled. When $J/\Delta\nu$ is small (<0.01)

the spectrum of this system will consist of two well separated doublets, the members of which will be of equal intensity. Their spacing gives J exactly and their separation gives $\Delta\nu$ within experimental error, i.e., the system can be designated as AX. As $J/\Delta\nu$ increases to about 0.1 [Fig. I-16(a)], the inner

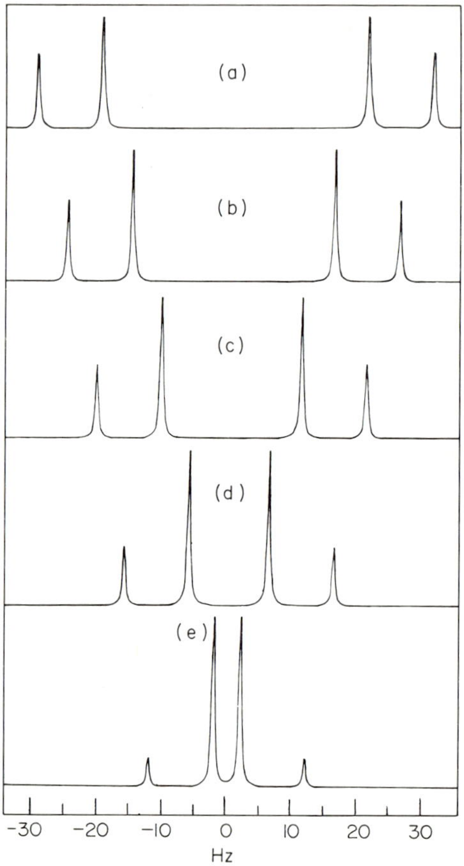

FIG. I-16. Calculated AB spectra for $J = 10$ Hz and $J/\Delta\nu$ equal to (a) 0.20, (b) 0.25, (c) 0.33, (d) 0.50, and (e) 1.00.

peaks become observably greater than the outer, but the spectrum can still be treated as AX and solved by inspection. When $J/\Delta\nu$ exceeds 0.1–0.2, the peak intensities become more perturbed and the doublet centers do not correspond to the chemical shifts; the doublet spacings give J exactly, however, no matter what its magnitude. Such a pair of spins is designated an AB system. The spectrum can be easily solved for $\Delta\nu$. i.e.,

$$\Delta\nu = (\Delta\nu_i \, \Delta\nu_o)^{1/2} \tag{I-51}$$

where $\Delta\nu_i$ is the spacing of the inner peaks and $\Delta\nu_o$ the spacing of the outer peaks. The relative intensities of the inner and outer peaks are inversely proportional to their spacings, and so as $J/\Delta\nu$ is increased without limit the quartet will approach a singlet. The outer peaks are reduced to vanishingly small satellites. This means that as $\Delta\nu$ vanishes we no longer can observe the coupling. It is observed and can be shown from quantum mechanical principles that in systems such as A_2, A_3, A_2B, A_3B, A_nB etc., the coupling J_{AA} has no influence on the spectrum and hence cannot be deduced from it. Spin groups such as A_2, A_3, ..., A_n are termed "equivalent." The CH_3 and CH_2 groups in the ethyl spectrum are each examples of such equivalent sets of spins. Later, we shall see (Chapter III) that nuclei which are geometrically equivalent and therefore have the same chemical shift are not necessarily equivalent in this sense; the couplings between such spins *do* affect the spectrum, and can therefore be obtained from it.

Although certain patterns become recognizable with experience, proper solutions of strong-coupled spectra involving more than two spins require quantum mechanical calculations, as we have seen. These become very laborious as the number of nuclei is increased and are best carried out by computer. A number of suitable programs for this purpose have now been devised. Some require trial and error insertion of approximate J and $\Delta\nu$ values; others converge on a correct solution if given line positions and intensities. The spectra in this book were calculated using a program devised by L. C. Snyder and R. L. Kornegay of the Bell Laboratories.

14. The Dependence of Coupling on Geometry and Chemical Structure

The proton–proton couplings with which we small be mainly concerned in polymer spectra are *geminal*, i.e., between protons on the same atom, usually carbon, and *vicinal*, between protons on neighboring atoms. These are often designated ^2J and ^3J, respectively, the superscript denoting the number of intervening bonds. Of one-bond or direct couplings, ^1J, only that between ^{13}C and directly bonded protons (or ^{19}F) is of importance in polymer spectroscopy (see Section 16).

Geminal couplings exhibit very wide variations. In groups with sp^3 hybridization, they are generally in the range of −12 to −15 Hz. For example, the geminal coupling in methane is −12.4 Hz and in the methylene groups of polymethyl methacrylate −14.9 Hz. The value increases algebraically, becoming less negative or actually positive as the H–C–H angle increases. Thus, the coupling in ethylene is +2.5 Hz (sp^2 hybridized) and −4 to −6 Hz in cyclopropanes (intermediate between sp^2 and sp^3). Substituent effects are often quite large and can overwhelm geometrical effects. An electronegative substituent on a vinyl group decreases the geminal coupling: in vinyl chloride, ^2J is -1.3 Hz. Other more marked substituent effects have also been observed, but they are seldom important in polymer spectra.

Vicinal couplings are markedly dependent on geometry, in both unsaturated and saturated systems, and likewise show substituent effects. They do not, however, vary over so large a range as geminal couplings and are rarely—probably never—negative. *Trans* couplings across double bonds are always about twice as large as *cis* couplings in the same structure, but both decrease with increasing electronegativity of the substituent. Thus, in ethylene, $J_{cis} = 11.5$ Hz and $J_{trans} = 19.0$ Hz, while in vinyl chloride $J_{cis} = 7.4$ Hz and $J_{trans} = 14.8$ Hz. In saturated systems, *trans* couplings (a) as shown, are substantially larger than *gauche* couplings (b). Both in ring compounds like

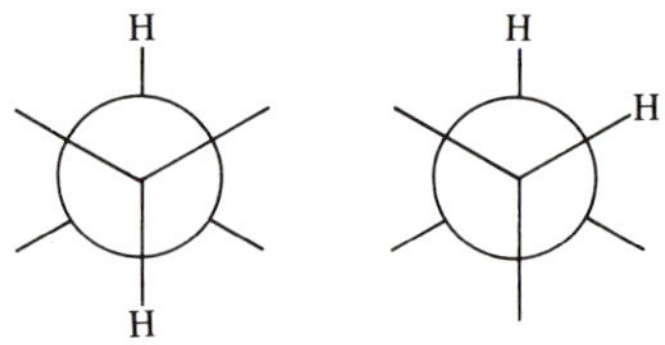

(a) (b)

cyclohexanes and in open-chain compounds, J_{gauche} (commonly denoted J_g) is typically 2–4 Hz and J_{trans} (J_t) ranges from 8 to 13 Hz. These considerations are of great importance in studying the conformation of polymer chains, and will be considered in greater detail in Chapter IX. Karplus[22,23] has given a valence-bond interpretation of these observations and concludes that vicinal coupling of protons on adjacent sp^3 carbon atoms should depend on φ, the dihedral angle, as follows:

$$J = \begin{cases} 8.5\cos^2\varphi - 0.28 \quad \text{(Hz)} & \text{for} \quad 0° \leqslant \varphi \leqslant 90° \\ 9.5\cos^2\varphi - 0.28 \quad \text{(Hz)} & \text{for} \quad 90° \leqslant \varphi \leqslant 180° \end{cases} \tag{I-52}$$

A plot of these functions (Fig. I-17) shows that J should be 1.8 Hz when $\varphi = 60°$ (J_g) and 9.2 Hz when $\varphi = 180°$ (J_t). Actual couplings tend to be somewhat larger than these values, particularly in open-chain systems.

15. Rate Phenomena: Averaging of Chemical Shifts and *J* Couplings

In the ultraviolet or infrared spectrum of a molecule having two or more coexisting subspecies—for example, two different conformations or ionized forms—one may expect to see absorption peaks corresponding to both forms, unaffected even when these are in very rapid equilibration with each other.

If δt is the lifetime of one form, then from the uncertainty principle one may expect a broadening of the absorption peaks given by

$$\delta\nu \simeq 1/(2\pi \cdot \delta t) \tag{I-53}$$

(It has been pointed out, however, by Meiboom[24] that the uncertainty principle formulation is no longer valid when the broadening peaks begin to overlap, because it would lead one to expect the broadening to increase without limit

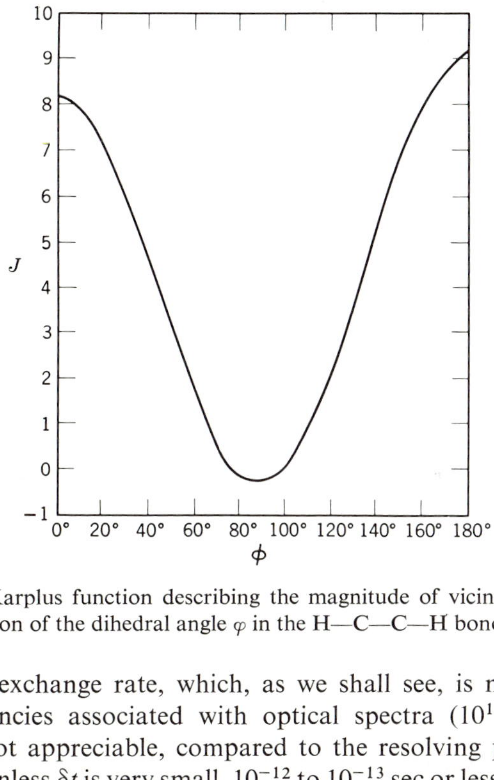

FIG. I-17. The Karplus function describing the magnitude of vicinal proton–proton J coupling as a function of the dihedral angle φ in the H—C—C—H bond system.

with increasing exchange rate, which, as we shall see, is not the case.) At the high frequencies associated with optical spectra (10^{13}–10^{15} Hz), this broadening is not appreciable, compared to the resolving power of optical spectrometers, unless δt is very small, 10^{-12} to 10^{-13} sec or less. Conformations and ionized states of organic molecules usually have lifetimes considerably longer than this. However, in NMR spectroscopy the observing frequency is one-millionth as great, spectral lines may be only 1 Hz in width, and those corresponding to interchanging subspecies may be only a few hertz apart. Broadening sufficient to merge the lines of the individual subspecies will occur even when lifetimes are as long as 10^{-1} to 10^{-3} sec. If their lifetimes are much shorter than this, the two forms will give a single narrow peak and be entirely indistinguishable. These lifetimes (say 10^{-1} to 10^{-5} sec) correspond to a range of rates of reaction and isomerization that is of great importance in chemistry. By suitable modification of the Bloch equations (Section 7), the spectral line shapes corresponding to the broadening and merging of peaks can be given analytical form, and in this way the rates of a wide variety of processes have been measured. Such measurements are not of sufficient importance in the study of polymers to warrant extended discussion here, particularly since they are amply treated elsewhere. The main point, which will be elaborated further later, is that the NMR spectra of molecules having conformational freedom—and this includes most polymers—represent an average over the attainable conformations. Considerable information concerning conformational preferences can be extracted from NMR spectra, but it is generally not presented directly to us.

Another process of importance in certain polymer spectra, particularly those of biopolymers, is *proton exchange*. The exchange rates of protons between organic molecules, in particular those of OH, CO_2H, and CONH groups, are often in the range amenable to NMR measurement and have been the object of many studies. In such systems, the rate information is contained in the collapse of spin multiplets rather than in the merging of chemically shifted peaks. For example, in pure ethanol, the hydroxyl proton is a triplet because of coupling to the two methylene protons. If we add water, and particularly if we add an acidic catalyst, this triplet will collapse to a singlet, passing through various intermediate stages (Fig. I-18) as the catalyst concentration is increased. The reason for this behavior is that the hydroxyl protons undergo intermolecular exchange, and in the presence of a sufficient quantity of catalyst, their residence time on any one molecule is so short compared to $1/J$ that the coupling is in effect "turned off." Following the departure of a hydroxyl proton, the chances are nearly even that the next proton to arrive at that site will be in the other spin state. When this happens rapidly, no appreciable coupling is possible. In Fig. I-18, τ is the residence time of the hydroxyl proton. The extent of collapse depends inversely on the product τJ, where J is the coupling of the OH and methylene protons.

16. Nuclei Other than ^1H

Of all the known isotopes, about half (or about 150) have magnetic moments. Of these only a relatively small number have been systematically observed by NMR; still fewer are of interest with respect to polymers. These are briefly discussed.

A. Deuterium

The principal utility of the deuterium nucleus in polymer structural studies, as in those on small molecules, is to simplify spectra by eliminating selected

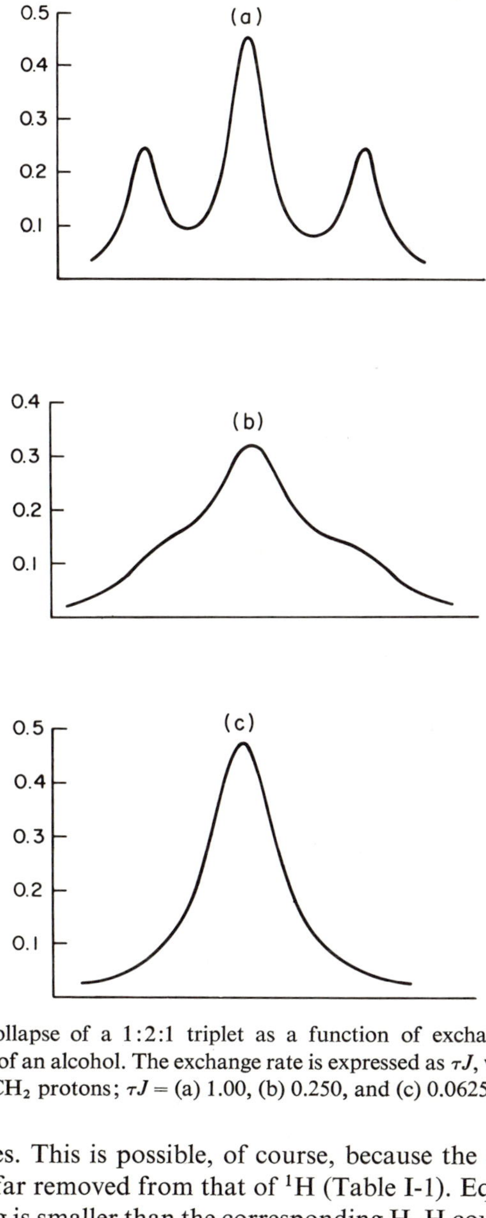

FIG. I-18. The collapse of a 1:2:1 triplet as a function of exchange rate, as in the OH—CH_2— group of an alcohol. The exchange rate is expressed as τJ, where J is the coupling of the OH and CH_2 protons; τJ = (a) 1.00, (b) 0.250, and (c) 0.0625. (From Arnold.[24a])

proton resonances. This is possible, of course, because the resonance of the deuteron is very far removed from that of ^{1}H (Table I-1). Equally important, the H–D coupling is smaller than the corresponding H–H coupling by a factor equal to the ratio $\gamma_{^1H}/\gamma_{^2H}$. Such a relationship is general for the relative couplings of isotopic nuclei, and in the present case is equal to 6.51. Since vicinal couplings in many polymer spectra are of the order of 6 Hz, this means that such couplings are virtually eliminated as well. We shall see many instances of the use of this technique in later discussion.

B. Fluorine

The magnetic moment of ^{19}F is only slightly smaller (Table I-1) than that of ^{1}H, and its relative observing sensitivity is therefore high. (At constant magnetic field strength, relative sensitivities of nuclei are approximately proportional to the cubes of the ratios of their magnetic moments.) It has a spin of $\frac{1}{2}$, and therefore experiences no complications from quadrupole relaxation. Because of the greater polarizability of its electron cloud, ^{19}F, in common with most other nuclei, exhibits a much greater range of chemical shifts than does ^{1}H—nearly 400 ppm as compared to about 10–12 ppm. This greater range often makes possible the discrimination of rather subtle environmental differences in polymer chains (Chapter V). There is no universally agreed upon scale for ^{19}F chemical shifts. Filipovich and Tiers[25] have proposed one in which CCl_3F, a volatile liquid, serves as both solvent and reference zero. Chemical shifts on this scale are expressed as Φ values (in ppm) if extrapolated to zero concentratic n or as Φ^* values if not, as is more commonly the case.

Particularly in highly fluorinated groups, ^{19}F–^{19}F couplings show unexpected behavior. Vicinal couplings in perfluoroalkyl groups are often nearly zero, but couplings through four or five bonds may be 5 to 15 Hz. Since both ^{19}F–^{19}F and ^{19}F–^{1}H couplings are transmitted to an observable degree through a larger number of bonds than ^{1}H–^{1}H couplings, ^{19}F multiplets are frequently quite complex and may appear poorly resolved and broad because of many closely spaced transitions. (This is illustrated in the model fluorochloropentane spectra in Fig. V-7.) The geometrical dependence of vicinal ^{19}F–^{19}F couplings is not as well worked out as that of ^{1}H–^{1}H couplings. Harris and Sheppard[26] found that in $CBrF_2 \cdot CBrF_2$, J_g is about 12 Hz while J_t is only 1.5 Hz ,a reversal of the relative magnitudes observed for proton couplings. However, on increasing substitution by heavier halogens, J_t increases rapidly and may overtake J_g.[27,28] Geminal ^{19}F–^{19}F couplings are much larger than vicinal couplings or geminal proton couplings. In saturated systems, they range from about 150 Hz to nearly 300 Hz; on double bonds they are generally less than 100 Hz. Vicinal ^{19}F–^{19}F couplings are negative, whereas geminal couplings are positive.

In that they tend to fall off in magnitude monotonically with the number of intervening bonds, ^{19}F–^{1}H couplings are more "normal" than ^{19}F–^{19}F couplings. Geminal couplings are in the range of about 45–55 Hz. Vicinal couplings usually have less than half the magnitude of geminal couplings. Both are positive and sensitive to substituent effects.

C. Carbon-13

The observation of ^{13}C resonance is beset by a number of difficulties which have been in large part overcome in the course of steady experimental and instrumental development. The inherent sensitivity of ^{13}C is low because its magnetic moment is relatively small (Table I-1), about one-fourth of that of the proton. In addition, it has a natural abundance of only 1.1%. Its spin–lattice relaxation time is usually rather long, and so in addition to being weak its signals are more readily saturated than those of ^1H or ^{19}F. Its spin is $\frac{1}{2}$, and it has no quadrupole moment; it therefore should in principle give narrow signals. Older methods of observation, using rapid passage in the dispersion mode to avoid saturation, broadened the signals so that for the most part only fine structure due to direct ^{13}C–^1H couplings (120–250 Hz) could be discerned, couplings through two or more bonds (about 5 Hz) being within the rather great experimental line width. More recently it has become possible by means of time-averaging devices (Section 18C) to observe ^{13}C signals in the absorption mode. Under these conditions, lines may be less than 1 Hz in width and weak couplings can be resolved.

Like ^{19}F, ^{13}C chemical shifts exhibit a range of nearly 400 ppm. "External" $^{13}CS_2$ (see Section 17B) is commonly employed as a reference standard, but this is by no means generally accepted. The most shielded ^{13}C nuclei are those bonded to heavy atoms like I or Br, whereas the least shielded are those bonded to oxygen, as in ester and keto carbonyl groups. Aliphatic hydrocarbon ^{13}C resonance is at considerably higher field than aromatic, but olefinic and aromatic are in the same region. Changes in electron density are probably primarily responsible for differing ^{13}C shieldings. Anisotropic shielding effects are no doubt comparable in absolute magnitude to those in proton spectra, but are relatively insignificant. The ^{13}C spectra of polymers are assuming major importance and many examples will be discussed in later chapters.

D. Nitrogen-14 and Nitrogen-15

Both ^{14}N and ^{15}N have magnetic moments and can give NMR spectra. However, ^{14}N has a spin of 1 (Table I-1) and therefore a quadrupole moment. The resulting rapid spin–lattice relaxation (Section 5) broadens its signals and makes their observation difficult. This difficulty does not arise with ^{15}N, which has a spin of $\frac{1}{2}$, but both its inherent sensitivity and natural abundance are substantially less than those of ^{13}C. Although ^{15}N signals have been observed in natural abundance,[29] it is customary to employ enriched samples.

The observation of the nitrogen resonance itself has not assumed any importance in polymer studies. However, the influence of ^{14}N on the attached proton in polyamides, polypeptides, and proteins is of considerable significance and has been the subject of numerous studies, which will be considered in Chapters XIII and XIV. Couplings of ^{14}N to directly bonded protons are in the range 50–65 Hz. (The corresponding ^{15}N–H couplings bear a constant proportion of $\gamma^{15}/\gamma^{14} = 1.41$ to the ^{14}N–H couplings.) We have seen that when the molecular framework has tetragonal symmetry, as in NH_4^+, the electric field within the molecule does not couple effectively to the ^{14}N quadrupole (see Section 10). The spin–lattice relaxation time of the ^{14}N nucleus is relatively long under these conditions, and the proton spectrum appears as a 1:1:1 triplet with spacing equal to $J_{^{14}N-H}$. However, in most molecules and particularly in asymmetric ones,[30] coupling of ^{14}N to the molecular electric field gradients is much stronger, and as a result the ^{14}N in effect decouples itself (see Section 18B) from the proton. The resulting proton spectrum may vary from a triplet with very broad lines, as in alkylammonium salts, to a broad singlet and even to a very narrow resonance.[30] These questions will be considered in more detail in Chapter XIII.

E. Phosphorus

The only naturally occurring isotope of phosphorus is ^{31}P. It has a spin of $\frac{1}{2}$ and its sensitivity, while much less than that of the proton, is still well within the range for convenient observation (Table I-1). It exhibits a range of chemical shifts of over 600 ppm and has been the subject of numerous studies.[31] Its chief interest in connection with polymers is in nucleic acids and oligonucleotides, which will be discussed in Chapter XV.

F. Other Nuclei

A number of other nuclei have been occasionally employed in polymer studies, particularly of proteins (Chapter XIV), but their use is at present very limited and does not warrant extended discussion here.

17. Experimental Techniques

We shall not deal with experimental methods in general, since these are fully dealt with in all of the treatises listed at the end of this chapter, but will briefly indicate the special problems presented by polymers and will describe the more recently developed techniques, where these are applicable.

A. Solution Preparation

As we have already indicated (Section 5), the rate of spin–lattice relaxation of polymer nuclei will not in general be proportional to the macroscopic solution viscosity, because long-chain polymers move segmentally. Nevertheless, the rate of local chain motion, upon which the line width depends, will depend upon the solvent viscosity. It is therefore necessary to choose solvents of moderate viscosity, even though the final solution may be highly viscous. At the same time, however, it is usually advantageous to observe polymer spectra at elevated temperatures (100–150°) to reduce dipolar broadening so far as possible. Solvents of low viscosity are also highly volatile, and even in sealed tubes will reflux, with consequent deterioration of the spectrum. This reduces the usefulness of such proton-free solvents as CCl_4, CS_2, and $CDCl_3$. A compromise is necessary. A major additional requirement of course, is that the liquid chosen be at least a reasonably good solvent for the polymer. For most vinyl polymers, aromatic solvents are suitable; *chlorobenzene* and *o-dichlorobenzene* have been found widely useful. Both give complex spectra near 2.7τ. Aromatic solvents often produce quite large shielding effects due to their pronounced shielding anisotropy (Section 11). This is sometimes useful in separating close peaks, benzene itself being particularly effective despite its relatively low boiling point. If the aromatic protons of polymers are to be observed, C_6D_6, pentachloroethane, and tetrachloroethylene can be used. Other solvents for vinyl and related polymers will be indicated in later discussion. (See Table I-2).

The choice of solvents for biopolymers is central to the entire experimental problem, for here the conformation of the chains and their dependence on

TABLE I-2

SOLVENTS FOR POLYMER NMR SPECTROSCOPY

Solvent	Boiling point, °C	τ
Benzene	80	2.73
Chlorobenzene	132	~2.7 (complex)
o-Dichlorobenzene	180	~2.7 (complex)
$CCl_3 \cdot CHCl_2$	162	3.91
Tetrachloroethylene	121	—
$CHCl_3$	61	2.73
d_6-DMSO	~100 (unstable at boiling point)	7.42[a]
d_6-acetonitrile	80	8.05[a]
Hexafluoroacetone	—	—
Trifluoroacetic acid	72	~ −2.5
Methanol	65	CH: 6.60 OH: ~5.0
Water	100	~5.2 (moves upfield with increasing temp.)

[a] Due to residual d_5 compound.

the solvent is usually the major object of investigation (Chapters XIII–XV). Dimethyl sulfoxide (DMSO), chloroform, trifluoroacetic acid (TFA), acetonitrile, hexafluoroacetone, methanol, and water are commonly employed. Interfering proton peaks can be greatly reduced by employing their deuterated derivatives, although residual peaks remain, slightly upfield (0.02–0.05 ppm) from those of the protonated solvent. For chloroform, this is a singlet but residual d_5-acetonitrile and d_5-dimethyl sulfoxide give characteristic pentuplets due to deuterium coupling to the remaining proton. In the preparation of polymer solutions, as of any others to be examined by NMR, it is important to exclude even the tiniest fragments of insoluble material, which can perturb the field homogeneity when they drift within the receiver coil. To avoid such contamination, filtration of the solution is desirable. This is most conveniently done by forcing the solution into the NMR tube through a Millipore filter membrane incorporated in a hypodermic syringe. The high viscosity of many polymer solutions can make this difficult. In earlier work, solutions as high as 10–15% (w/v) concentration were employed in order to maximize the signal-to-noise ratio. The increased sensitivity of modern instruments (see Section 18) has alleviated this problem by permitting one to obtain good spectra at concentrations of 1–2%, particularly by use of spectral accumulation.

Solvents for ^{19}F and ^{13}C spectroscopy have in general less stringent requirements. Deuterium substitution is of course unnecessary. The much wider chemical shift spacings mean large "windows" in the solvent spectra, so that interference with the solute signals is seldom a serious problem.

B. References

As is now well known, reference substances for NMR spectroscopy may be *internal*, i.e., dissolved in the solution under examination or *external*, i.e., contained in a small capillary within the NMR tube or in the annular space formed by two concentric tubes. The former has the advantage that no correction is necessary for the difference in bulk susceptibility between reference and solution (see the experimental chapters in Bovey or Becker in the suggested further reading list), but the possible disadvantage that association between reference and solution may influence the reference position somewhat. External references are seldom employed in proton spectroscopy unless made necessary by insolubility. The use of tetramethylsilane (TMS; see p. 30) is now very widespread. For most polymer work, however, it is unsuitable because at the elevated temperatures usually employed it evaporates into the vapor space in the tube and gives inconveniently weak signals. Hexamethyldisiloxane, HMDS, which has much lower volatility, is preferable; it appears at 9.94τ.

For aqueous systems, commonly employed in biopolymer studies, sodium 2,2-dimethyl-2-silapentane-5-sulfonate (DSS), has been found very useful.[32]

$$(CH_3)_3SiCH_2CH_2CH_2SO_3\,Na$$

The methyl protons give a strong singlet, but the methylene protons give appreciable multiplet signals which can be troublesome at high gain. For some purposes, *tert*-butyl alcohol (methyls: about 8.8τ), dioxane (about 6.3τ), and acetone (about 8.0τ) are sometimes employed.

The use of CCl_3F as an internal reference for ^{19}F spectroscopy has already been described (Section 16B).

External references are commonly employed in ^{13}C spectroscopy because bulk magnetic susceptibility effects are relatively insignificant. Carbon chemical shifts, δ_C, are most frequently referred to $^{13}CS_2$ as zero. Other references have been proposed, e.g., benzene, $^{13}CH_3CO_2H$, and TMS, but all suffer the disadvantage of giving multiplet ^{13}C signals due to proton coupling.

C. Spectral Integration and Peak Resolution

An important feature of polymer NMR spectroscopy is the careful measurement of relative peak intensities. As we shall see in subsequent chapters, polymer spectra commonly consist of a number of closely spaced, often partially overlapping signals the intensities of which bear no simple integral relationship, because they arise from statistical distributions of structural features such as comonomer or configurational sequences. The *areas* of peaks in NMR spectra are proportional to the numbers of nuclei contributing to them, regardless of their chemical binding. Peak *heights*, although easily measured, are not a reliable measure of relative intensities, since peak widths, being proportional to T_2^{-1}, will in general differ for different protons. Peak areas can be measured by weighing them or by counting the squares they enclose, but it is much more convenient to observe the spectrum in the integral mode. NMR spectrometers are normally equipped with electronic integrators which present the spectrum as a series of steps rather than peaks, the height of each step being proportional to the number of nuclei underneath the signal. This is illustrated in Figs. I-19a and I-19b, which show the 100 MHz spectra of two samples of polymethyl methacrylate, described in Chapter III. In each spectrum, the peaks between 8.5τ and 9.0τ represent α-CH_3 protons, and the remainder of the spectrum the β-CH_2 protons. If the integration is properly carried out, taking particular care to avoid differential saturation, the total step heights corresponding to each of these regions must be in the ratio 3:2. More significant, as we shall see in later discussion, are the relative heights of the steps in the α-CH_3 spectrum; these can be readily measured in the present spectra and provide important structural information.

In the 60 MHz spectra of these materials (shown in Fig. III-2) the peaks are less clearly separated; the methylene peaks at 8.0τ in Fig. I-19b present a similar problem. In such cases, it may be advantageous to employ an analog

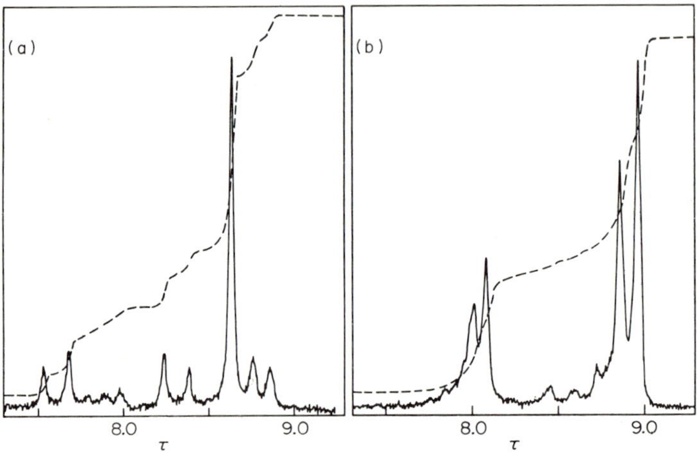

FIG. I-19. 100 MHz spectra of two samples of polymethyl methacrylate, illustrating use of integral spectra (dashed lines) to determine relative peak areas; (a) predominantly isotactic and (b) predominantly syndiotactic polymers.

computer, such as the du Pont Model 310 Curve Resolver, by means of which the experimental curves can be visually matched to computer-generated curves, the relative areas of which are given by the instrument.

18. Techniques for Spectral Simplification and Optimization

A. LARGE H_0: SUPERCONDUCTING MAGNETS

We have seen (Section 11) that as H_0 is increased there is a directly proportionate increase in the chemical shift differences between different nuclei, with no change in the magnitude of the spin–spin J coupling. This not only permits greater discrimination of details of molecular structure, but also (Section 13) can effect striking simplification of complex, second-order spectra by grouping the lines in well separated multiplets and decreasing the observable number of transitions. This simplification is particularly significant in polymer spectra because they are commonly very complex and because they exhibit dipolar broadening. This will be discussed in detail in later chapters.

A recent large step toward the goal of homogeneous fields of great strength is the development of superconducting magnets.[33, 34] Solenoids wound from niobium-alloy wire and maintained at liquid helium temperature are capable at present of producing fields of the order of 50 kG, and it appears that this figure might be doubled without major change of design. In later chapters, we shall see many examples of the advantageous use of instruments of this design in polymer studies.

B. Spin Decoupling

Another effective means of simplifying spectra is double resonance or spin decoupling. To take the simplest case, suppose that we have an AX system (two doublets), and that while observing A with the usual rf field H_1, we simultaneously provide a second, much stronger rf field H_2, at the resonant frequency of X. We find that the A doublet collapses to a singlet. The field H_2 has caused the magnetic moment of nucleus X to jump so rapidly back and forth between its spin states that it is no longer observably coupled to nucleus A.

By reversing the experiment, we can also observe the collapse of the doublet of X upon irradiating A. Experimentally, double irradiation may be carried out by either keeping field H_2 constant on the chosen resonance while sweeping H_1 through the spectrum ("frequency sweep") or sweeping H_0, which is the same as sweeping H_1 and H_2 together while maintaining a fixed frequency difference between them equal to the frequency difference $\Delta\nu$ between A and X ("field sweep"). Both methods are commonly employed. The first procedure requires very high stability in H_2, but is quite feasible and is decidedly more convenient, since one may "sit" on one resonance while exploring the spectrum to find the others to which it is coupled. (An example of the use of decoupling in elucidating polymer structure is illustrated in Fig. V-1.)

Decoupling of different nuclear species is often important. In polymer studies, a common use of this technique is the irradiation of deuterium nuclei (Section 16A) to remove broadening arising from unresolved multiplicity. Figure I-20(a) shows 60 MHz spectra of the β-methylene protons of polyisopropyl-α-*cis*-β-d_2-acrylate (Chapter VIII, Section 4B). These protons are coupled to one geminal and two vicinal deuterons. They therefore appear as

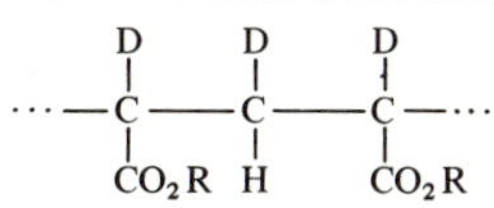

rather broad signals in Fig. I-20(a). Figure I-20(b) shows the calculated transitions and their envelopes. Note that these are rather broader than the observed resonances. This is characteristic of such multiplets for protons on polymer chains. Because of relatively slow molecular reorientation, the quadrupolar relaxation of the deuterium nuclei (recall that $I = 1$ for ^{2}H) is particularly effective in partially decoupling them from the protons. Upon irradiation of the deuterium at 9.1 MHz, the peaks become narrow [Fig. I-20(c)].

The use of heteronuclear decoupling is particularly important in observing ^{13}C spectra. The resonances of carbon nuclei are first-order multiplets, corresponding to the number of attached protons, with spacings of about 150–250 Hz (Section 16C). They are helpful in making general assignments. Once this has been done, it is advantageous to irradiate the protons in order to collapse the carbon resonances to singlets. This aids quantitative measurement and improves resolution by increasing the effective signal-to-noise ratio. In order to irradiate all the protons at once regardless of their chemical shift, the

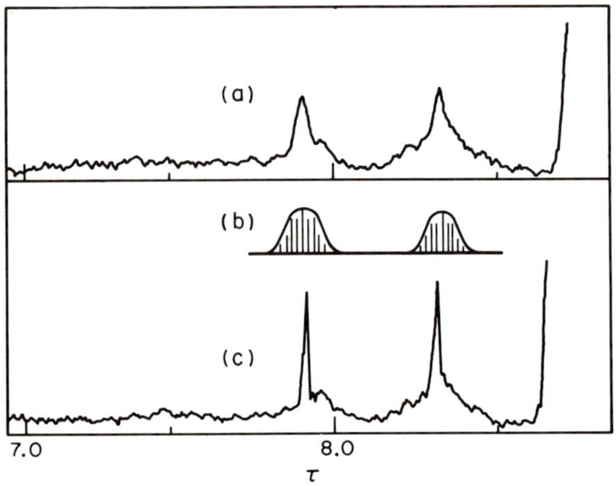

FIG. I-20. 60 MHz β-methylene proton spectra of polyisopropyl-α-*cis*-β-d_2-acrylate; (a) observed, (b) calculated from deuterium couplings, and (c) on irradiation of deuterium.

irradiating field H_2 is modulated with random noise.[35] The ^{13}C spectra shown in subsequent chapters (Figs. III-7, IV-6, VI-6, VII-8, XII-1, and XII-2) are decoupled spectra of this type.

In cases where the carbon multiplets overlap, the spectrum can be clarified by another double resonance technique, in which the protons are irradiated off-resonance.[36] For reasons which we shall not describe here, this results in diminishing the spacing of the multiplets while preserving their form, and can help in disentangling them.

A further effect which is closely related to and may accompany the double resonance experiment is the nuclear Overhauser effect (abbreviated NOE). This is a complex phenomenon, analogous to a long known electron–nucleus interaction, in which the populations of the nuclear energy levels rather than the transition energies are perturbed. The NOE does not arise from J coupling, which in fact complicates it and should be minimal, but rather from direct dipole–dipole relaxation, which for pairs of like interacting nuclei depends [see Eq. (I-19b)] upon γ^4 and on the inverse sixth power of the internuclear distance. The optimal conditions for the observation of the effect exist in structures where the observed nucleus is normally relaxed principally by interaction with the nucleus (or nuclei, in which case the contributions of each nucleus to $1/T_1$ may be summed) being irradiated in the NOE experiment. Under these conditions, an enhancement of the signal strength of the observed proton may be observed provided r^6 is not too great. For protons, it can be shown that the maximum enhancement is 50%. For ^{13}C–^{1}H interaction, as in noise-decoupled C-13 spectra, the effect will be proportional to $\gamma^2_{^1H} \cdot \gamma^2_{^{13}C}$, the maximum enhancement being given by

$$\frac{I}{I_0} = 1 + \frac{1}{2}\frac{\gamma_{^1H}}{\gamma_{^{13}C}} \simeq 3 \qquad \text{(I-54)}$$

This theoretical factor has been attained within experimental error for $H^{13}CO_2H$ by Kulhmann and Grant.[37] A similar enhancement may in general be expected in ^{13}C polymer spectra, although this may vary with the particular structures involved.

C. Signal Enhancement

Although the sensitivity of NMR increases with the strength of the magnetic field (in theory as $H_0{}^2$), even in superconducting fields the signal-to-noise ratio may not be adequate when observing dilute solutions or minor components. This is particularly true in the study of biopolymers, e.g., proteins, where the measurement of single-proton resonances (out of about 1000 or more per molecule) may be necessary. The successful observation of these systems and of nuclei of low sensitivity such as ^{13}C requires some form of *signal enhancement* to increase the signal-to-noise ratio to acceptable levels. In principle, the signal-to-noise ratio of any spectrum can be increased by extending the measuring period t. Background noise accumulates in proportion to $t^{1/2}$ whereas the strength of a coherent signal increases as t. Thus one may obtain increased sensitivity by sweeping through the spectrum over a period of several hours instead of a few minutes, as is usual. This technique is of limited utility, however, because of low-frequency noise, changes in gain, and other sources of instability.

A more practical way is to sweep through the spectrum rapidly many times in succession, beginning each new sweep at exactly the same point in the spectrum and then summing up the traces. A practical means of performing this summation is to use a multichannel pulse-height analyzer—essentially a small computer which can remember several hundred traces. A "computer of average transients" (CAT) with 1024 channels is commonly used for this purpose. In Fig. I-21 are shown spectra of the methylene quartet of a 0.1% solution of ethylbenzene in carbon tetrachloride after accumulating 1, 4, 16, 64, 225, and 1600 scans. The spectrum progresses from a meaningless jumble to a well-resolved multiplet.

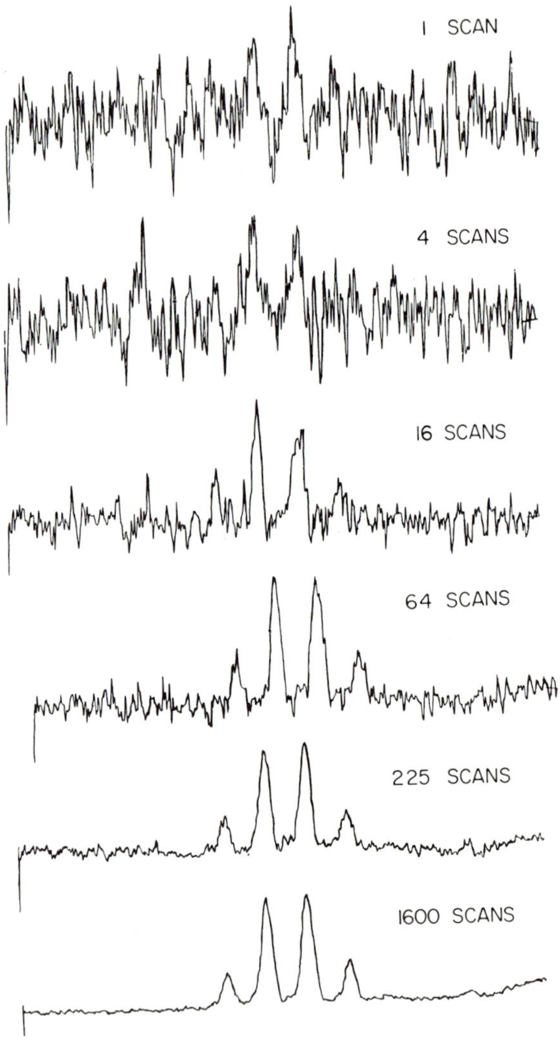

FIG. I-21. The methylene quartet of a 0.1% solution of ethylbenzene in carbon tetrachloride as a function of the number of CAT scans. (Figure courtesy of Varian Associates, Palo Alto, California.)

In some circumstances, it is necessary to accumulate several thousand scans (true of several of the ^{13}C spectra in later chapters). By conventional scanning this is possible but tedious. To obviate this difficulty, advantage can be taken of the fact that the decay tail of the free induction signal following an rf pulse (Section 9) contains all of the information of the slow passage spectrum and is in fact its Fourier transform. The decay tails can be stored in the CAT and their accumulation transformed into the high resolution spectrum with a saving of time of the order of 100- to 1000-fold.[38] No spectra in this book have been obtained in this way, but it is clear that it will be of major importance, particularly for the observation of the ^{13}C spectra of biopolymers. Further description can be found in Ernst and Anderson,[38] in Chapter 2 of Bovey, and in Chapter 3 of Becker, both in the list of suggested further reading at the end of this chapter.

REFERENCES

1. W. Pauli, *Naturwissenschaften* **12**, 741 (1924).
2. I. Rabi, S. Millman, P. Kusch, and J. R. Zacharias, *Phys. Rev.* **55**, 526 (1939); J. B. M. Kellogg, I. Rabi, N. F. Ramsey, and J. R. Zacharias, *Phys. Rev.* **56**, 728 (1939).
3. E M. Purcell, H. C. Torrey, and R. V. Pound, *Phys. Rev.* **69**, 37 (1946).
4. F. Bloch, W. W. Hansen, and M. E. Packard, *Phys. Rev.* **69**, 127 (1946).
5. J. T. Arnold, S. S. Dharmatti, and M. E. Packard, *J. Chem. Phys.* **19**, 507 (1951).
6. N. Bloembergen, E. M. Purcell, and R. V. Pound, *Phys. Rev.* **73**, 679 (1948).
7. R. Kubo and K. Tomita, *J. Phys. Soc. Jap.* **9**, 935 (1954).
8. I. Solomon, *Phys. Rev.* **99**, 559 (1955).
9. D. W. McCall and F. A. Bovey, *J. Polym. Sci.* **45**, 530 (1960).
10. W. P. Slichter and D. D. Davis, *Macromolecules* **1**, 47 (1968).
11. G. E. Pake, *J. Chem. Phys.* **16**, 327 (1948).
12. J. H. Van Vleck, *Phys. Rev.* **74**, 1168 (1948).
13. E. R. Andrew and R. Bersohn, *J. Chem. Phys.* **18**, 159 (1950).
14. D. W. McCall and E. W. Anderson, *J. Polym. Sci. Part A* **1**, 1175 (1963).
15. F. Bloch, *Phys. Rev.* **70**, 460 (1946).
16. G. Lindström, *Phys. Rev.* **78**, 817 (1950).
17. H. A. Thomas, *Phys. Rev.* **80**, 901 (1950).
18. W. D. Knight, *Phys. Rev.* **76**, 1259 (1949).
19. A. D. Buckingham, *Can. J. Chem.* **38**, 300 (1960).
20. G. V. D. Tiers, *J. Phys. Chem.* **62**, 1151 (1958).
21. J. A. Pople, W. G. Schneider, and H. J. Bernstein, "High Resolution Nuclear Magnetic Resonance," p. 98; p. 117 *et seq.* McGraw-Hill, New York, 1959.
22. M. Karplus, *J. Chem. Phys.* **30**, 11 (1959).
23. M. Karplus, *J. Amer. Chem. Soc.* **85**, 2870 (1963).
24. S. Meiboom, *Z. Elektrochem.* **64**, 50 (1960).
24a. J. T. Arnold, *Phys. Rev.* **102**, 136 (1956).
25. G. Filipovich and G. V. D. Tiers, *J. Phys. Chem.* **63**, 761 (1959).
26. R. K. Harris and N. Sheppard, *Trans. Faraday Soc.* **59**, 606 (1963).
27. D. S. Thompson, R. A. Newmark, and C. H. Sederholm, *J. Chem. Phys.* **37**, 411 (1962).

28. R. A. Newmark and C. H. Sederholm, *J. Chem. Phys.* **43**, 602 (1965).
29. A. J. R. Bourn and E. W. Randall, *Mol. Phys.* **8**, 567 (1964).
30. G. V. D. Tiers and F. A. Bovey, *J. Phys. Chem.* **63**, 302 (1959).
31. M. M. Crutchfield, C. H. Dungan, J. H. Letcher, V. Mark, and J. R. Van Wazer, "^{31}P Nuclear Magnetic Resonance." Wiley (Interscience), New York, 1967.
32. G. V. D. Tiers and R. I. Coon, *J. Org. Chem.* **26**, 2097 (1961).
33. F. A. Nelson and H. E. Weaver, *Science* **146**, 223 (1964).
34. W. D. Phillips and R. C. Ferguson, *Science* **157**, 257 (1967).
35. R. R. Ernst, *J. Chem. Phys.* **45**, 3845 (1966).
36. A. L. Bloom and J. N. Shoolery, *Phys. Rev.* **97**, 1261 (1955).
37. K. F. Kuhlmann and D. M. Grant, *J. Amer. Chem. Soc.* **90**, 7355 (1968).
38. R. R. Ernst and W. A. Anderson, *Rev. Sci. Instr.* **37**, 93 (1966).

SUGGESTED FURTHER READING

E. D. Becker, "High Resolution NMR." Academic Press, New York, 1969.

F. A. Bovey, "Nuclear Magnetic Resonance Spectroscopy." Academic Press, New York, 1969.

J. W. Emsley, J. Feeney, and L. H. Sutcliffe, "High Resolution Nuclear Magnetic Resonance," Vols. 1 and 2. Pergamon Press, Oxford, 1966.

J. A. Pople, W. G. Schneider, and H. J. Bernstein, "High Resolution NMR." McGraw-Hill, New York, 1959.

Chapter II

ISOMERISM IN POLYMER CHAINS

1. Introduction

Of the many applications of high resolution NMR to the study of polymer structure, its use for the observation and measurement of isomerism in polymer chains has without doubt so far proved to be the most significant. This emphasis stems in large part from the power of NMR to reveal the symmetry properties of organic molecules. In this chapter, we shall briefly review some of the most important types of polymer isomerism, reserving the detailed discussion of their relationship to NMR for later chapters.†

2. Vinyl Polymers

Of the various options available to a growing polymer chain of a vinyl monomer of the type shown, where A or B may also be H,‡ the principal

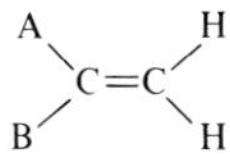

† For more extended discussions of the subject of this chapter, the reader is referred to "The Stereochemistry of Macromolecules"[1] and particularly to Goodman's chapter in "Topics in Stereochemistry."[1a]

‡ H may be replaced by F, but we shall consider only protons for the present discussion.

ones are as follows:

a. Head-to-tail (i) versus head-to-head tail-to-tail (ii) addition

(i) 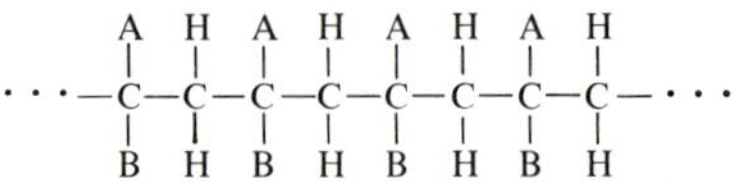

(ii) 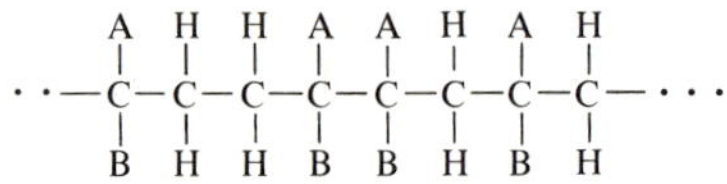

with a single monomer unit turned "backward" or with the chain continuing to propagate in the new sense.

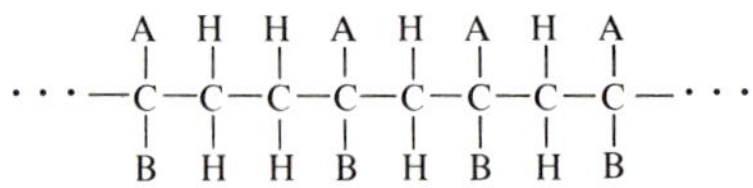

We shall see (Chapter V) that by NMR one can detect and distinguish these modes of addition.

b. The formation of branch points, either trifunctional (i) or tetrafunctional (ii), and crosslinks (iii). (Several variations in the nature and position of the

(i) 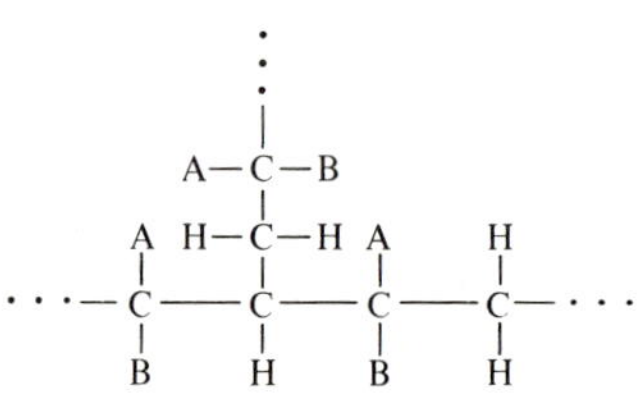

(ii) 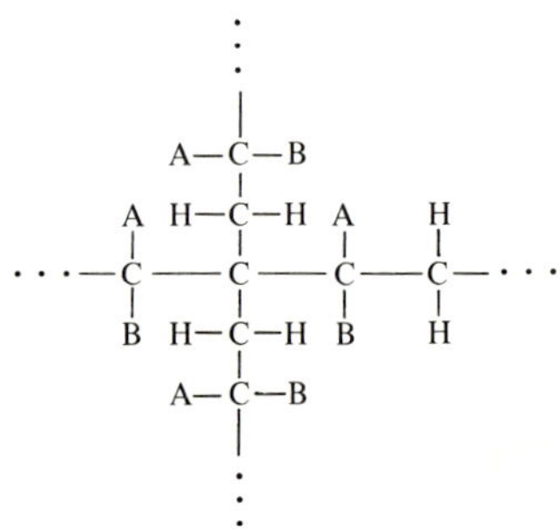

(iii) 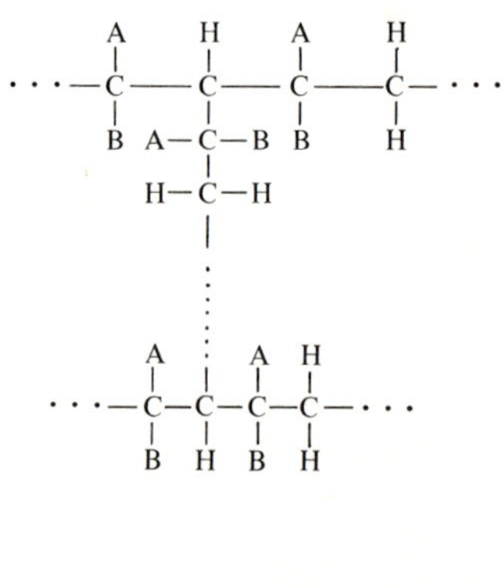

branches in these structures are clearly possible.) Although NMR has now become an analytical tool of fairly high sensitivity, it has not yet been effectively applied to the measurement of such units. There is no reason in principle why this cannot be done.

c. Relative handedness of successive monomer units, of which the simplest extreme types are now well known to be *syndiotactic* (i) and *isotactic* (ii).

(i) 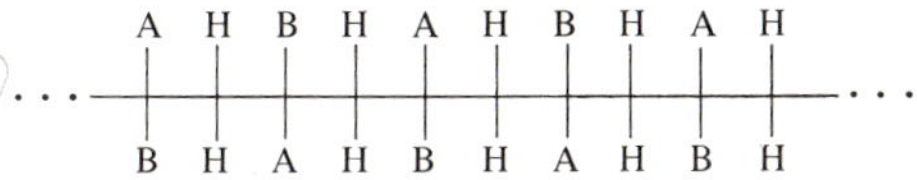

(ii) 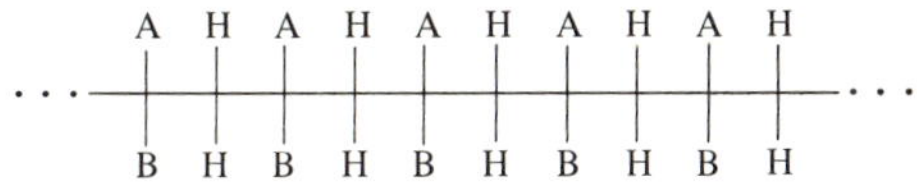

We shall see later that one can define an indefinitely large number of such "pure" (i.e., regularly repeating) configurational sequence types, the experimental significance of which depends upon one's ability to discriminate between them. The next simplest and most self-evident one is the *heterotactic*[1b] (iii).

(iii) 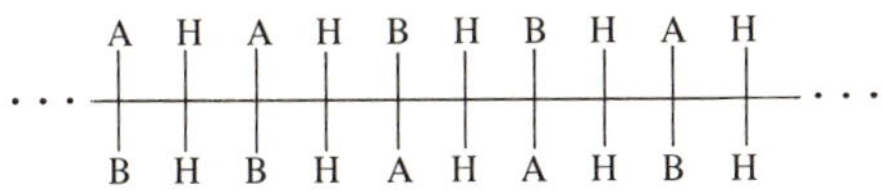

Such structures can usually be readily distinguished by NMR (Chapter III). In the succeeding discussion we shall often employ this simplified projection of a vinyl polymer, in which the chain is represented in the planar zigzag conformation, viewed in either direction along the zigzag plane. Where appropriate, the CH_2 groups are reduced to "sticks" as shown.

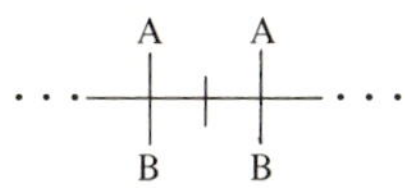

The asymmetric centers in these chains are in fact pseudoasymmetric, the chain ends being for the present disregarded, and they therefore give rise to no optical activity, either actual or potential. The term *asymmetric* is reserved for polymer structures (as for small molecules) which have no symmetry elements, even of rotation or translation. The term *dissymmetric* is used to describe structures which cannot be superimposed on their mirror images, but which may contain rotational or translational symmetry axes. An isotactic sequence is neither asymmetric nor dissymmetric, being superimposable on its mirror image; it has translational symmetry, but no symmetry axes. A syndiotactic sequence has translational symmetry, and if it contains an even number of pseudoasymmetric carbon atoms it will contain a twofold symmetry axis as well. Nevertheless, it is not superimposable on its mirror image

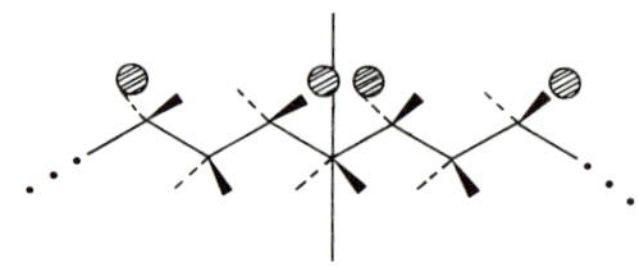

and so is dissymmetric. A sequence with an odd number of pseudoasymmetric carbon atoms has a plane of symmetry and so is neither asymmetric nor dissymmetric.

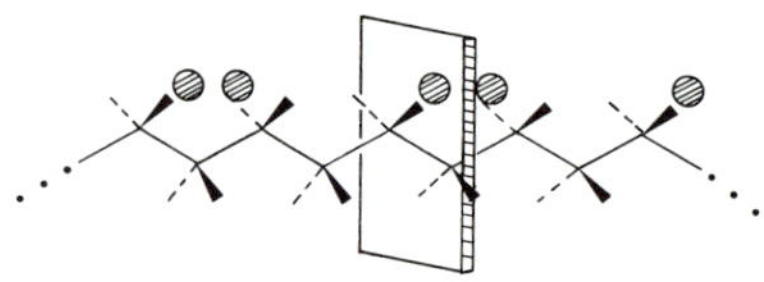

Similar generalizations can be easily made for any of the "pure" configurational sequence types discussed in Chapter III.

Since such chains thus never have true asymmetric centers, their designation in terms of *d* or *l* is open to serious objection. This will be discussed more fully in Chapter III.

Vinyl polymers may acquire true asymmetric centers by ring closure, as for example in polyvinyl formal (Chapter IV, Section 6). Questions of chirality

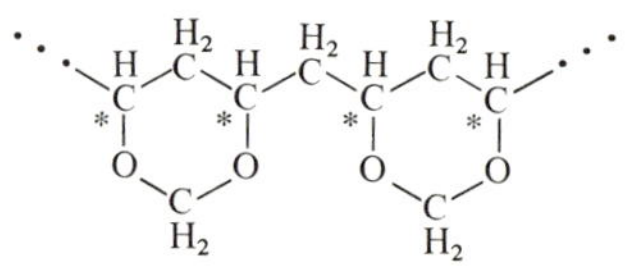

then become superimposed on (and intimately related to) questions of isotactic or syndiotactic sequence. In copolymers (Chapter X), true asymmetric centers

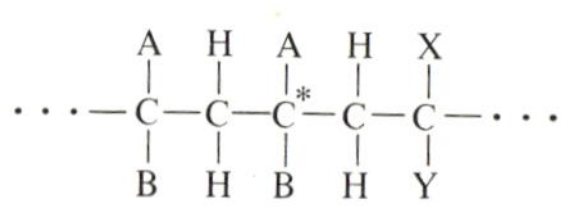

may be present but this raises no question that cannot be treated within the framework already outlined.

Vinyl monomers which are α,β-disubstituted have additional options in

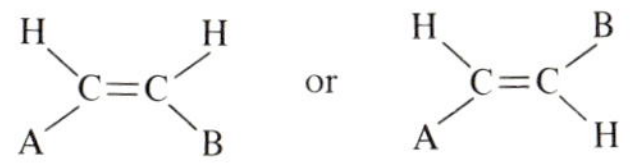

that the polymer chain now has two interleaved systems of nonidentical, pseudoasymmetric carbon atoms. Three extreme configurations can be defined: *disyndiotactic* (i); *erythrodiisotactic* (ii); and *threodiisotactic* (iii).†

(i) 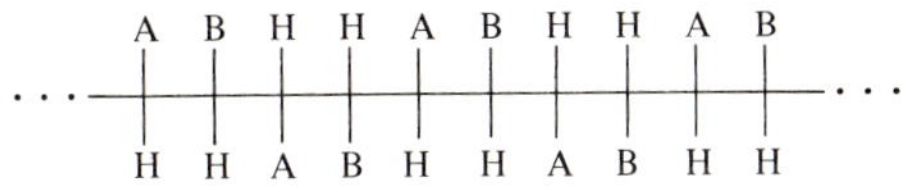

(ii) 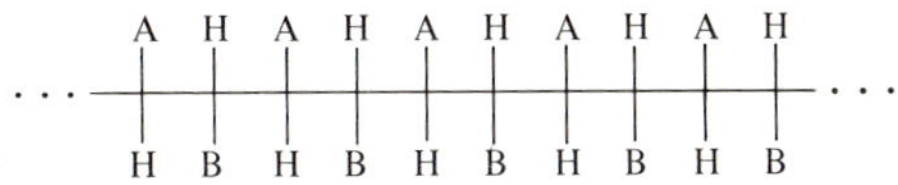

(iii) 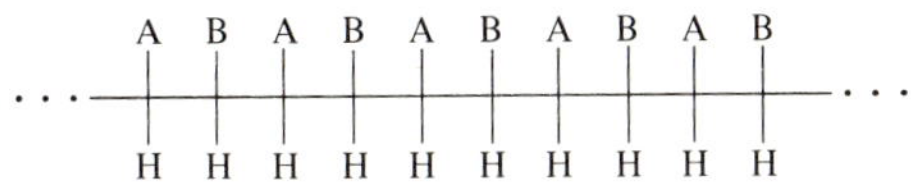

These configurations can be readily distinguished by NMR, but as we shall see (Chapters III, VII, and VIII), it may not be easy to tell which is which.

When adjacent carbon atoms are linked by rings, complications multiply. We list some of the possibilities, although most polymers of this type have not yet been investigated by NMR.

(i) Erythrodiisotactic (ring is *cis*). The rings are represented in a generalized

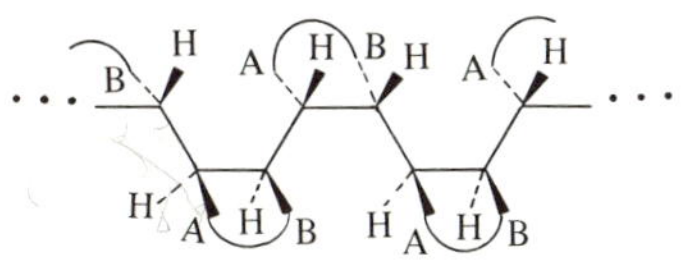

† The terms *erythro* and *threo* are of course based on the configurations of the sugars erythrose and threose. A mnemonic device that may help some readers is to recall that if A ≡ B, the *erythro* configuration is *meso* (shown in eclipsed "sawhorse" representation), whereas the *threo* configuration is *racemic.*

erythro 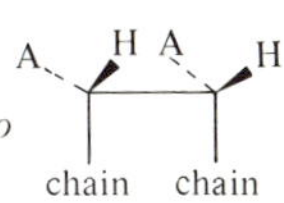

threo 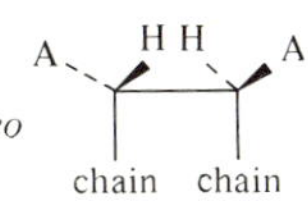

When the monomer has three unlike substituents, the definition becomes somewhat arbitrary.

manner. A projection of the planar zigzag conformation is not employed because the chains are structurally incapable of assuming this conformation.

(ii) Erythrodisyndiotactic (ring is *cis*).

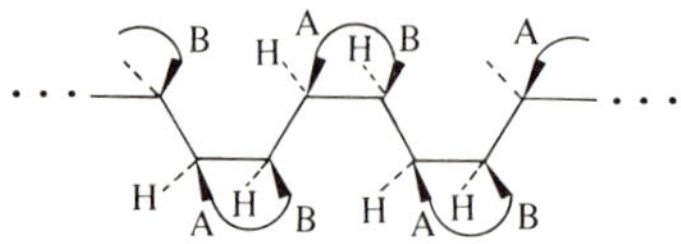

(iii) Threodiisotactic (ring is *trans*). This can also be represented as shown on the right.

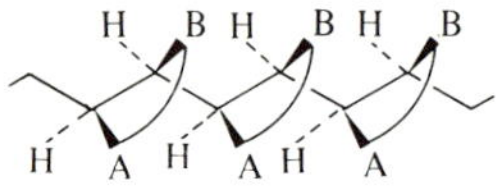

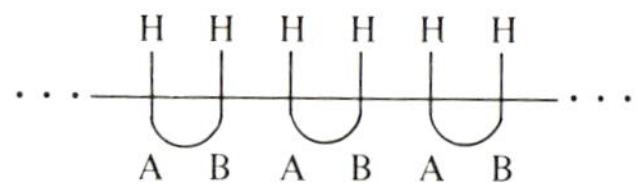

(iv) Threodisyndiotactic (ring is *trans*). This can also be represented as shown on the right.

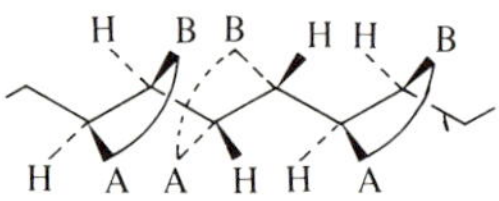

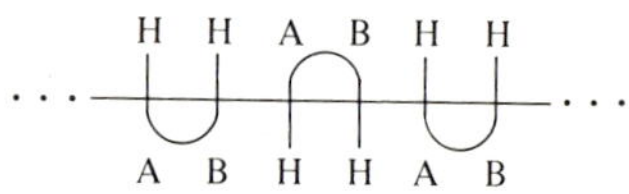

The rings may be symmetrical (A = B) as in polycyclobutene,[2,3] or unsymmetrical (A ≠ B), exemplified by polybenzofuran.[4,5]

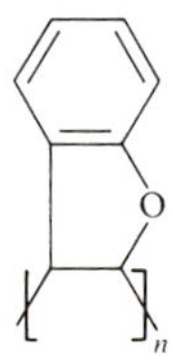

In both cases, all the main-chain carbon atoms are asymmetric.

3. Diene Polymers

The polymerization of conjugated dienes can produce structures having intriguing combinations of geometrical and configurational isomerism: 1,3-butadiene can give 1,4 or 1,2 enchainment. The 1,4 structures may be *cis* (ia) or *trans* (ib). These are distinguishable by NMR (Chapter XI).

(ia) 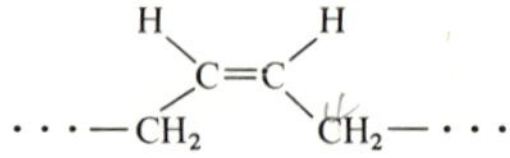 (ib) 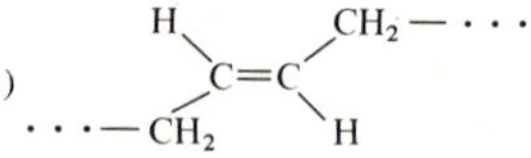

The 1,2 structures have the same configurational properties as vinyl polymer chains and occur in isotactic (iia) or syndiotactic (iib) sequences.

(iia) 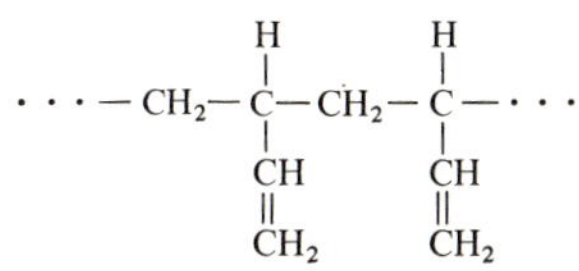

(iib) 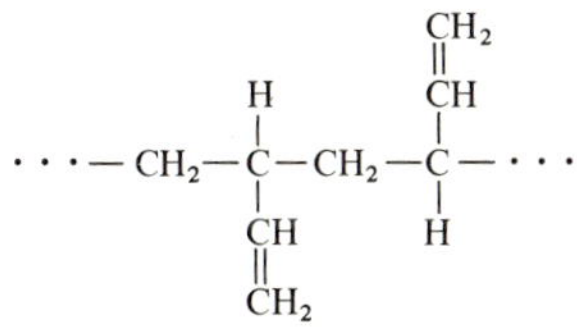

Polymers from 2-substituted butadienes such as chloroprene (A = Cl) or isoprene ($A = CH_3$) may give both 1,2 (iii) and 3,4 (iv) structures, each of which may occur in isotactic or syndiotactic sequences as shown previously.

(iii) 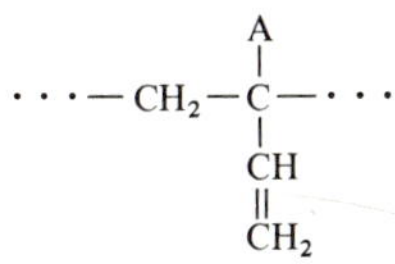

(iv) 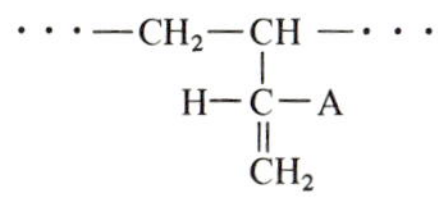

In none of these polydiene chain units is optical isomerism possible. When 1,2 (or 3,4) units and 1,4 units are juxtaposed in a chain, asymmetric carbon atoms are present in principle, but in fact have not been made to lead to observable optical activity.

The 1-substituted 1,3-butadienes such as 1,3-pentadiene can yield such mixed structures, which can be obtained in optically active form; thus, Natta *et al.*[6] prepared an optically active *cis*-1,4-poly-1,3-pentadiene. Such monomers

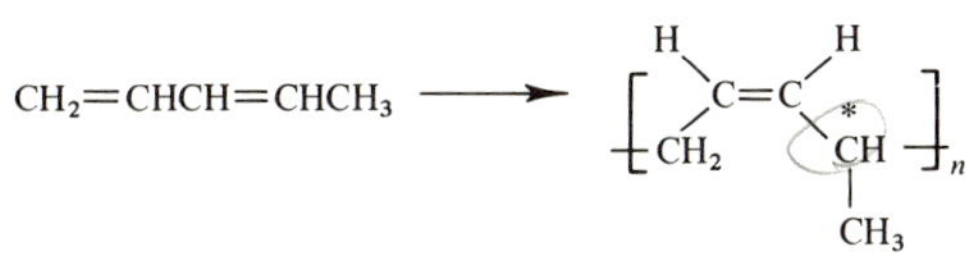

clearly have the capability of yielding *cis*-isotactic, *trans*-isotactic, *cis*-syndiotactic, and *trans*-syndiotactic structures, but as yet these have not been prepared and established.

Dienes of the type ACH=CHCH=CHA, which themselves may of course be *cis-cis*, *trans-trans*, or *cis-trans*, can give rise to the following isomeric chains by 1,4-addition:

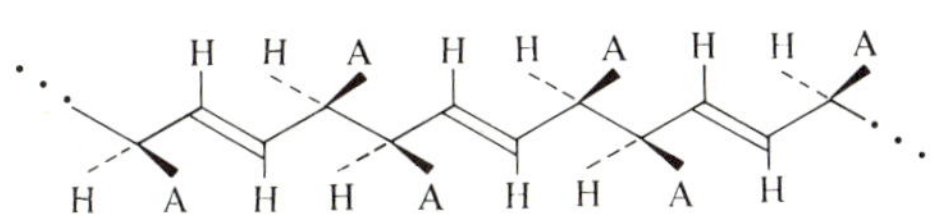

(i) *trans-dl*

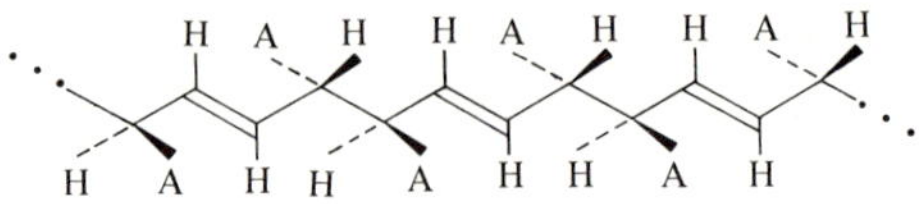

(ii) *trans-meso*

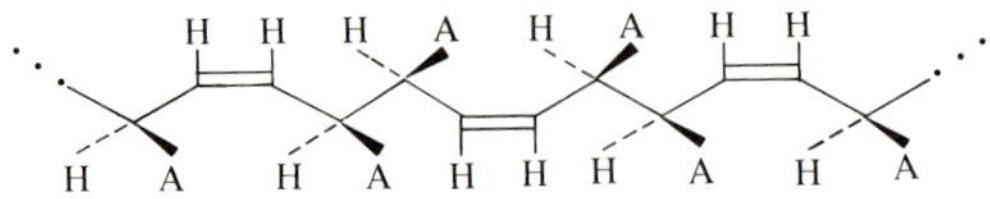

(iii) *cis-dl*

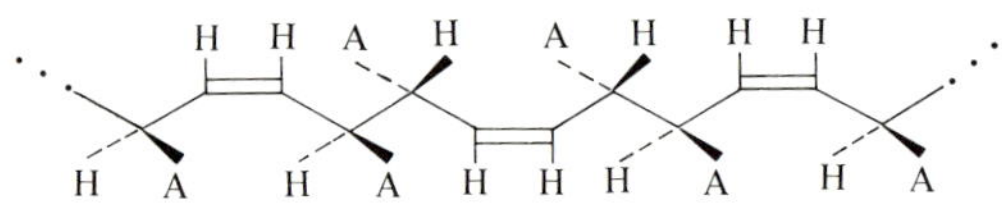

(iv) *cis-meso*

If the substituents are unlike, ACH=CHCH=CHB (*cis-cis*, *trans-trans*, *trans-cis*, and *cis-trans* geometrical isomers of the monomer are now possible), the following structures can be generated:

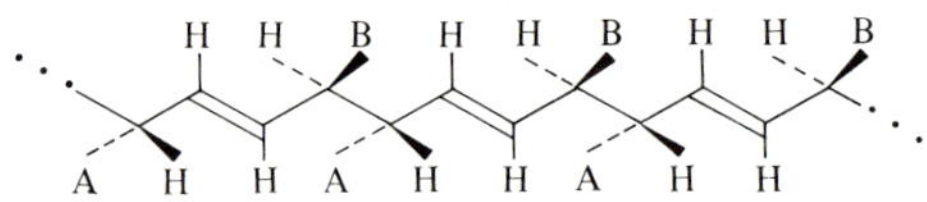

(v) *dl-trans-erythro*

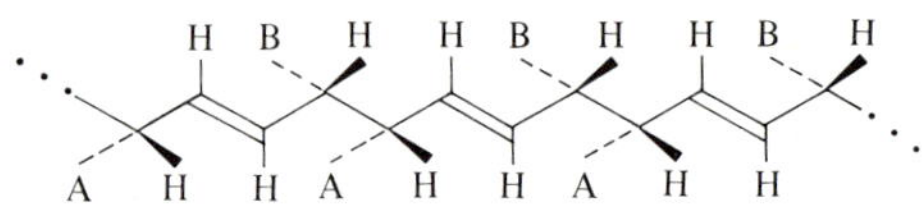

(vi) *dl-trans-threo*

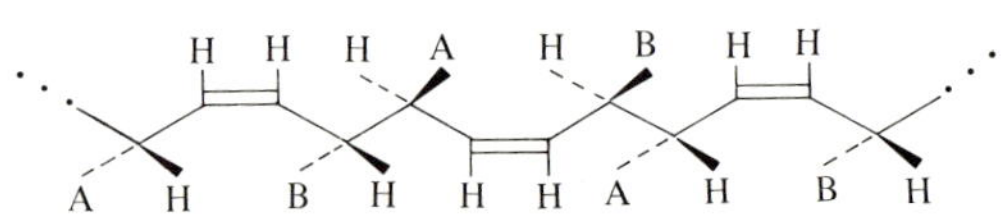

(vii) *dl-cis-erythro*

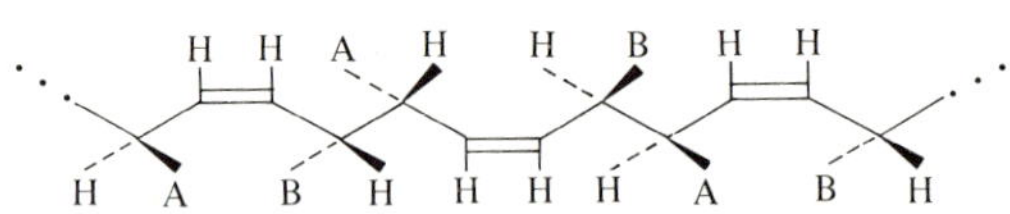

(viii) *dl-cis-threo*

These may be termed diisotactic; one can also imagine syndiotactic structures, with A or A and B alternating from side to side of the plane of the main-chain carbons.

It is evident that the symmetry properties of these chains are such that all substituent groups are equivalent within any chain, and that therefore an absolute determination of stereochemical configuration cannot be made by NMR on this basis despite the fact all the substituent carbon atoms are asymmetric. Their spectra must, however, all be different to some degree. From conformational analysis and the measurement of vicinal proton–proton couplings, they could in principle be distinguished. Natta *et al.*[2,7] have prepared optically active crystalline polymers from esters of *trans-trans* sorbic acid, and showed that they have the *trans*-1,4-*erythro* diisotactic structure (v).

4. Vinyl Polymers with Optically Active Side Chains

Vinyl polymers may contain true asymmetric centers in the side chains.† These centers are not generated during polymerization or by ring closure following polymerization, but are inherent in the monomer. The production of optically active polymers thus does not require the use of optically active catalysts unless the *dl* monomer is employed. Examples of such monomers are (a) olefins, (b) vinyl ethers, and (c) (by extension of the term "vinyl") aldehydes, when polymerization occurs by opening of the C=O double bond.

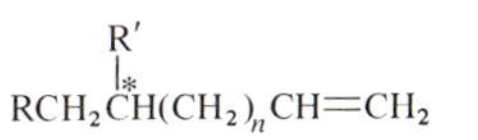

(a)

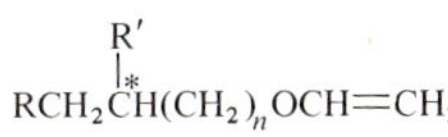

(b)

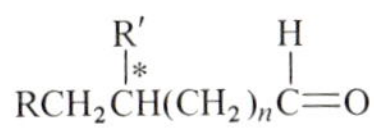

(c)

An all-*d* chain cannot be distinguished from an all-*l* chain by NMR, unless the polymers are examined in a chiral solvent, where a distinction is at least possible in principle[9–11] although not actually demonstrated. However, when *d* and *l* units occur in the same chain, the resulting diastereoisomeric sequences are distinguishable, although not always readily.

5. Polymers with Main-Chain Asymmetric Centers from Asymmetric Monomers

Ring-opening polymerizations of monomers of the general type shown,

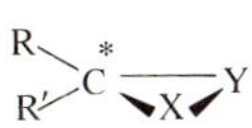

† For reviews, see Goodman[1a] and Pino.[8]

where X is commonly a hetero atom and Y may represent one or more carbon atoms or other atoms, will give polymers with true asymmetric centers in the main chain. Such systems differ from those represented by benzofuran and cyclobutene (p. 58) in that the monomer is asymmetric and so the generation of asymmetric polymer chains does not necessarily require the use of optically active catalysts.

The simplest example of such a system is propylene oxide,[12–14] which can

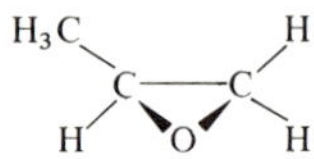

yield optically active "isotactic" chains of opposite chirality (a) and (b),† and a "syndiotactic" chain with alternating *d* and *l* units (c). (The quotation marks indicate some reserve on this author's part as to whether the terms isotactic and syndiotactic need or should be applied to such structures.)

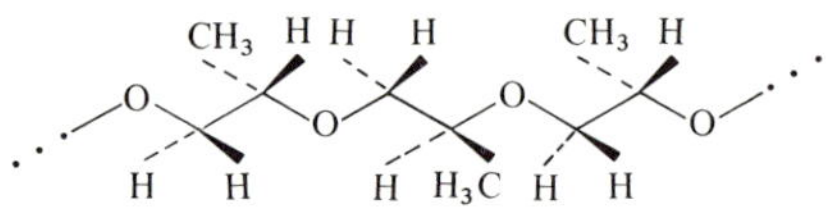

(a)

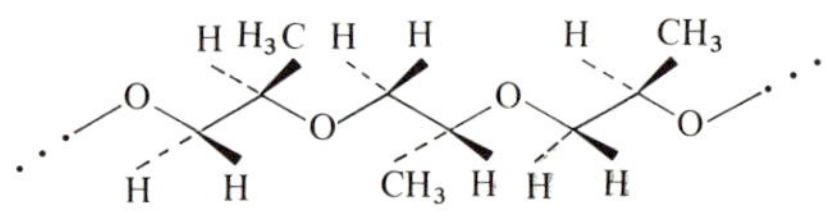

(b)

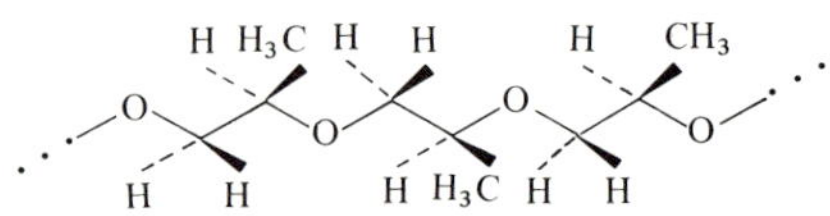

(c)

The NMR of polypropylene oxide is discussed in Chapter XII. Olefin–SO_2 copolymers, discussed in Chapter X, are closely analogous, even though not prepared by ring-opening polymerization.

Other examples are poly-α-amino acids (a), dealt with *in extenso* in Chapter XIII,‡ and polylactic acid (b).[15–18] With respect to NMR, the same remarks

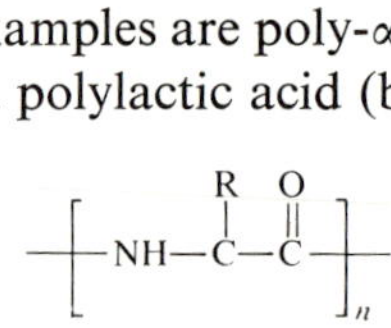

(a)

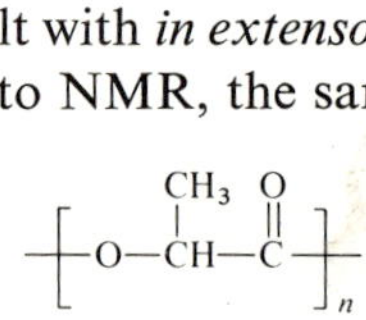

(b)

† The maintenance of optical activity in the chain is not easy to explain mechanistically, however, even when the pure *d* or *l* monomer is used. See the review of Goodman (pp. 146–148).[1a]

‡ The polymerization of *N*-carboxyanhydrides to yield poly-α-amino acids has been reviewed by Szwarc.[15]

apply to these polymers as to those in Section 4: information concerning configurational sequence and optical purity can be obtained, but none concerning absolute chirality as such.

6. Other Types of Isomerism

It perhaps goes without saying that in polymers other than those already mentioned, there are many types of structural isomerism which can usually be readily determined by NMR, as for example, the presence of *ortho*- or *meta*-linked units in aromatic polyesters, isomerism in phenol-formaldehyde resins, polyamides, polysulfones, etc.† Studies of this kind are often very useful, but for the most part involve problems in isomer identification which are no different from those of the corresponding small molecules. They do not take advantage of the unique ability of NMR to provide sequence information.

Another type of isomerism, if one may so term it, is the variation of sequence distribution in vinyl copolymers and other types of copolymers. These are discussed in Chapters X and XI.

REFERENCES

1. A. D. Ketley, ed., "The Stereochemistry of Macromolecules," 3 Vols. Dekker, New York, 1967.
1a. M. Goodman, Concepts of Polymer Stereochemistry *in* "Topics in Stereochemistry" (N. L. Allinger and E. L. Eliel, eds.), Vol. 2. Wiley, New York, 1967.
1b. F. A. Bovey and G. V. D. Tiers, *J. Polym. Sci.* **44**, 173 (1960).
2. G. Dall'Asta, G. Mazzanti, G. Natta, and L. Porri, *Makromol. Chem.* **56**, 224 (1962).
3. G. Natta, G. Dall'Asta, G. Mazzanti, and G. Motroni, *Makromol. Chem.* **69**, 163 (1963).
4. G. Natta, M. Farina, M. Peraldo, and G. Bressan, *Chim. Ind. (Milan)* **43**, 161 (1961).
5. G. Natta, *Makromol. Chem.* **43**, 68 (1961).
6. G. Natta, L. Porri, and S. Valenti, *Makromol. Chem.* **65**, 106 (1963).
7. G. Natta, M. Farina, M. Donati, and M. Peraldo, *Chim. Ind. (Milan)* **42**, 1363 (1960).
8. P. Pino, *Fortschr. Hochpolym. Forsch.* **4**, 399 (1966).
9. W. H. Pirkle, *J. Amer. Chem. Soc.* **88**, 1837 (1966).
10. T. G. Burlingame and W. H. Pirkle, *J. Amer. Chem. Soc.* **88**, 4294 (1966).
11. J. C. Jochens, G. Taigel, and A. Selinger, *Tetrahedron Lett.* **1967**, 1901 (1967).
12. M. E. Pruitt and J. M. Baggett, U.S. Pat. 2,706,181 (1955).
13. C. C. Price and M. Osgan, *J. Amer. Chem. Soc.* **78**, 4787 (1956).

† Such NMR studies are reviewed by Willis and Cudby[19] and by Woodbrey.[20]

14. M. Osgan and C. C. Price, *J. Polym. Sci.* **34**, 153 (1959).
15. M. Szwarc, *Fortschr. Hochpolym. Forsch.* **4**, 1 (1965).
16. J. Kleine and H. Kleine, *Makromol. Chem.* **30**, 23 (1959).
17. R. C. Schultz and J. Schwaab, *Makromol. Chem.* **87**, 90 (1965).
18. M. Goodman and M. D'Alagni, *J. Polymer Sci. Part B* **5**, 515 (1967).
19. H. A. Willis and M. E. A. Cudby, *App. Spectrosc. Rev.* **1** (2), 237 (1968).
20. J. C. Woodbrey, *in* "The Stereochemistry of Macromolecules" (A. D. Ketley, ed.), Vol. 3. Dekker, New York, 1968.

Chapter III

CONFIGURATIONAL SEQUENCES AND THEIR OBSERVATION: POLYMERS OF α,α'-DISUBSTITUTED VINYL MONOMERS

1. Introduction

Of the various forms of isomerism possible in polymer chains (Chapter II), the relative stereochemical configuration of pseudoasymmetric centers is of particularly great significance and particularly amenable to study by high resolution NMR. It is now of course well known that the physical, mechanical, and (to a lesser extent) chemical properties of vinyl and diene polymers are critically dependent upon their stereochemical configuration. For example, isotactic polymethyl methacrylate has a crystalline melting point of 160°, whereas the syndiotactic polymer melts at 200°; the atactic polymer does not crystallize. Isotactic polypropylene is crystalline and melts at 180°, whereas the atactic material is oily or rubbery.

During the past 15 years, polymer chemistry has undergone a major revolution, beginning with Ziegler's development of metal halide–metal alkyl initiators for ethylene polymerization and greatly extended by Natta and his colleagues in Milan, who showed how similar catalyst systems can be used to produce polymers of *controlled stereochemistry* from a wide variety of vinyl and diene monomers.[1,2] Very highly stereoregular polymers, since they can be made to crystallize, can be studied by x-ray diffraction. Polymers of lower regularity generally do not crystallize (this is not always true) yet their properties may be very dependent upon their degree of stereochemical regularity.

Earlier studies[3–5] of the relationship of chain stereoregularity to crystallizability and other physical properties were hampered by a lack of any direct experimental measurements of configurational sequences. The application of high resolution NMR spectroscopy[6–8] to such polymers made such measurements feasible, and allowed one to compare actual polymer chains to theoretical predictions.[9–12]

2. Model Compounds for Vinyl Polymer Chains

The power of high resolution NMR to determine the configurations of vinyl polymer chains arises primarily from its sensitivity to the equivalence or nonequivalence of similarly bonded magnetic nuclei. This sensitivity makes it an extraordinarily effective means for the determination of certain kinds of molecular symmetry or lack of symmetry. Before amplifying these statements, we must first define what we mean by "equivalence," a term which in our context has some meanings peculiar to nuclear magnetic resonance.

A. Equivalence in Chemical Shift

Equivalence in this sense implies no more than fortuitous equivalence in chemical shift for nuclei which may actually occupy quite different environments. Thus, methylacetylene $CH_3C{\equiv}CH$, in chloroform solution shows a single peak for all protons. This type of equivalence can usually be removed by using a higher magnetic field or a different solvent. In structural studies in general, and especially in polymer studies (where lines tend to be broad), such equivalence degrades the information available and has often proved misleading, particularly in the older literature. Nuclei having chemical shift equivalence have been termed "isochronous."[13]

B. Equivalence by Reason of Molecular Symmetry

If a molecule has an element of symmetry, such as a translation axis, a two-fold or higher rotational axis, or a mirror plane, then those magnetic nuclei which exchange their positions under the appropriate symmetry operation have *symmetry equivalence*, and must have the same chemical shift. In polymer chains formed of identical repeating units and long enough so that end effects may be ignored, hundreds or thousands of nuclei will be equivalent in this sense. On the other hand, in nonrepeating structures such as protein molecules, few of the nuclei have symmetry equivalence, although substantial groups of them may appear to have chemical shift equivalence. In our subsequent discussions, the terms *equivalence* (and *nonequivalence*) will be assumed to mean symmetry equivalence (and nonequivalence).

C. Magnetic Equivalence

A group of spins A_n possesses *magnetic equivalence* if all n spins have the same chemical shift and are equally coupled to each of the m spins in every other group of magnetically equivalent spins $B_m \cdots B_i \cdots$ in the molecule. (Chapter I, Sections 12 and 13). Thus, the protons and the fluorine nuclei in difluoromethane CF_2H_2 and in 1,1-difluoroallene (a) each form a magnetically equivalent group, and it can be designated as an A_2X_2 spin system. The protons in ethane form a single magnetically equivalent group, A_6. But in 1,1-difluoro-

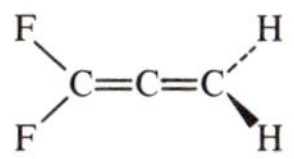

(a)

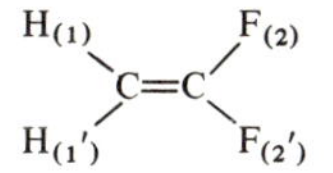

(b)

ethylene (b), the protons and fluorine nuclei do not form magnetically equivalent groups, since the *cis* couplings J_{12} and $J_{1'2'}$ are different in magnitude from the *trans* couplings $J_{12'}$ and $J_{1'2}$. Such a spin system is designated AA′XX′, or as AA′BB′ if the chemical shift difference in Hz between A and B is comparable to their coupling. As we shall see, such AA′XX′ or AA′BB′ spin systems (with certain complications) are of common occurrence in polymer chains.

The only structure which always constitutes a system of magnetically equivalent spins and which occurs frequently in polymers is the methyl (or CF_3) group; here, rotation ensures complete equivalence, regardless of the environment.

The above cases have been chosen because they are fairly obvious. We must now consider some structures which can occur in polymers and in which questions of equivalence or nonequivalence in the B sense above do not have quite such self-evident answers. Consider the type of molecule represented by (c). If the groups X and Y both have a plane or axis of symmetry (e.g., phenyl, methyl, halogen, etc.) the groups M are equivalent (or "enantiotopic"[14]). In a chiral solvent, they will be nonequivalent, at least in principle, but this does not normally concern us in polymer studies. If, however, one of the groups—say, Y—has in itself no element of symmetry, the groups M will be nonequivalent or "diastereotopic."[14] The molecule is not necessarily asymmetric as a whole, however, for Y may, for example, have the structure (d), in which case the molecule has a plane of symmetry [see (e)]. Such a

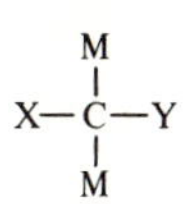

(c)

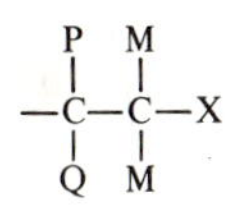

(d)

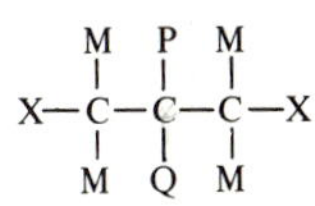

(e)

molecule nevertheless satisfies the requirements for steric differentiation of the geminal M groups in each pair. To simplify the ensuing discussion, we shall refer to all such CM_2 groups in which the M groups are equivalent as *homosteric*, and to those in which the M groups are nonequivalent as *heterosteric*.†

Let us consider some specific examples which are relevant to polymers.

1. *Y Is an Asymmetric Carbon Atom, $CR_1R_2R_3$*

For each optical isomer (not ordinarily distinguishable by NMR), there are three staggered conformers. In each the environments of M_a and M_b are nonequivalent. Most of the observed differentiation between M_a and M_b probably

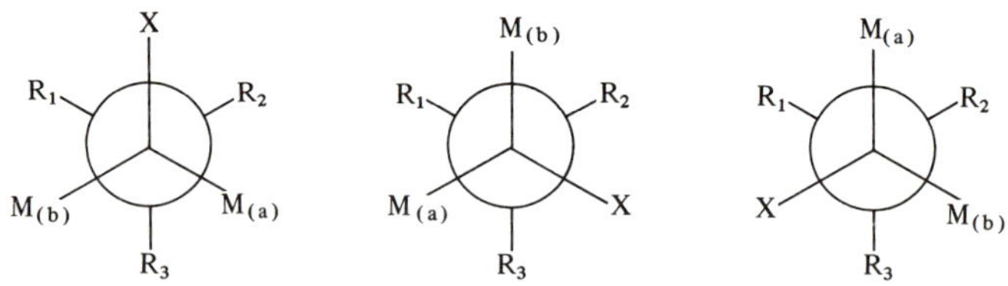

arises from an energetic preference for one of the conformers,[16] but it is obvious from elementary symmetry considerations that in principle some degree of nonequivalence must persist even if all three were equally populated and in rapid equilibration with each other,[17] for the M groups are in different environments in each conformer shown and in any other possible conformers. A great many examples of this type are now known. (For a review, see van Gorkom and Hall.[18])

2. *$X = Y = CR_1R_2R_3$; the Molecule Has Two Similar Asymmetric Centers*

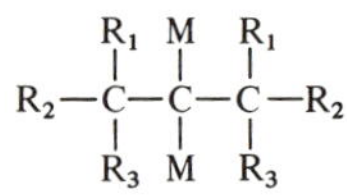

If X and Y are of the same handedness we have either the *d* or *l* diastereoisomer. Unless observed in a chiral solvent[19–21] the *racemic* mixture of *d* and *l* will be indistinguishable from either alone by NMR. The enantiomers are dissymmetric, but the CM_2 group is homosteric in each. If X and Y are of opposite handedness, we have the *meso* diastereoisomer, which of course has a plane of symmetry, but in which the CM_2 group is heterosteric. Molecules of this type are clearly very closely related to vinyl polymers. The most carefully studied examples are *racemic* and *meso*-2,4-disubstituted pentanes. Tiers and Bovey[22] showed that the fluorine nuclei of the central CF_2 group of

† The author is indebted to Professor Murray Goodman for the suggestion of this terminology.[15]

meso-$CF_2ClCFClCF_2CFClCF_2Cl$ have differing chemical shifts and couplings, while those of the *racemic* isomer are identical. A number of molecules of this type have now been studied, including the 2,4-dichloro-,[23–27] -bromo-,[26] -cyano-,[26, 28, 29, 30] hydroxy-,[26, 31, 32] acetoxy-,[24, 30] trifluoroacetoxy-,[29] carboxy-,[31] carbomethoxy-,[33, 34] phenyl,[31, 35–37] and methoxypentanes.[38] The spectra of isomeric 2,4-diphenylpentanes (Fig. III-1), analogs of poly-

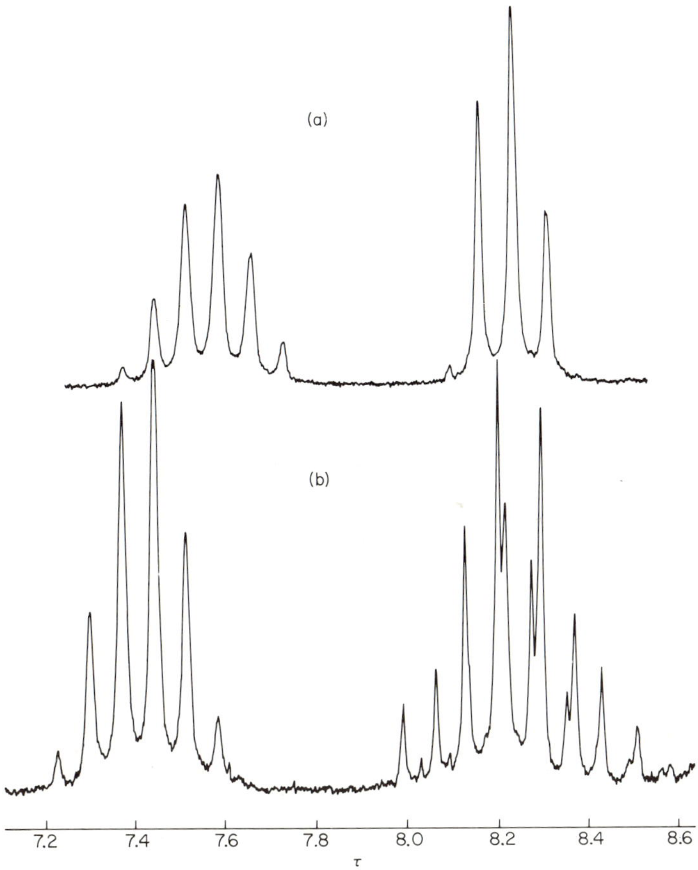

FIG. III-1. The 100 MHz spectra of (a) *racemic* and (b) *meso*-2,4-diphenylpentane, 10% (v/v) in chlorobenzene at 35°.

styrene, may serve as an example. Analysis of these spectra (Chapter IX) shows that the methylene group protons differ by 0.21 ppm in chemical shift in the *meso* isomer, but are equivalent in the *racemic* isomer; this conclusion is qualitatively evident from the greater complexity of the CH_2 portion of the spectrum, centered at about 8.2τ, in Fig. III-1(b). In both spectra, the sextuplet at lower field is that of the methine protons. (The aromatic and methyl resonances are omitted.)

Such studies have been extended to the 2,4,6-trisubstituted heptanes, including the trichloro-,[39–41] tricarboxy-,[33] tricarbomethoxy-,[33,34] tricyano-,[42,43] and triphenylpentanes.[44–46] Of these, there are one *racemic* and two *meso* diastereoisomers. They are closely related to isotactic, syndio-

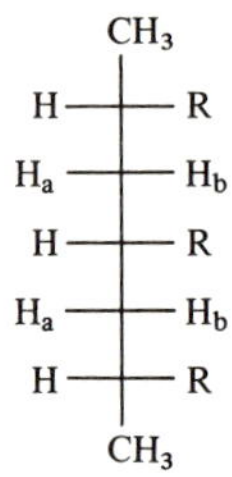

"isotactic"
(*meso*-1)

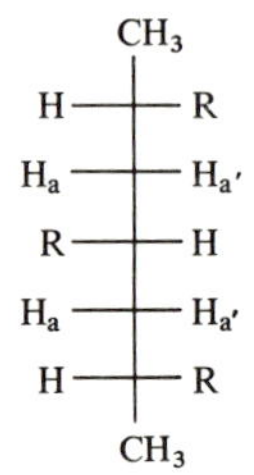

"syndiotactic"
(*meso*-2)

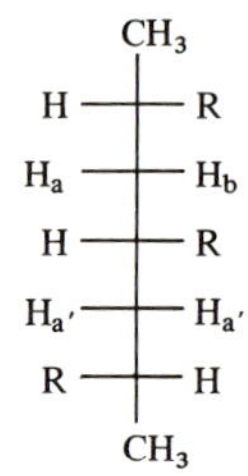

"heterotactic"
(*racemic*)

tactic, and heterotactic triad sequences (Section 3) of monomer units in polymer chains, and are so named. All the methylene groups are heterosteric. The *meso* isomers each have two like heterosteric methylene groups. The *racemic* isomer has two unlike heterosteric methylene groups so that all four of its methylene protons are nonequivalent. One might expect that the nonequivalence of the a′ and a″ protons in the "heterotactic" isomer and of the a and a′ protons in the "syndiotactic" isomer would be small, while that of the a and b protons might be quite large. This expectation appears to be borne out.[39–46]

It has been assumed in the foregoing discussion that the molecules may exist in any or all of their possible conformations and that the rate of interconversion of conformations is rapid on the NMR time scale. For large molecules (including polymers), as for the simple ethanic rotors considered earlier, the existence of heterostericity or homostericity does not in principle depend upon any particular conformer's being energetically preferred, nor upon there being a slow equilibration of the conformers.

The conformations of these model molecules and their relationship to conformations of vinyl polymer chains will be discussed in Chapter IX.

Objections to the use of *d* and *l* to designate the configurations of the α-carbon atoms in these chains have been indicated in Chapter II (Section 2). A further objection becomes evident when we consider the notation appropriate to these small molecules. Thus, a methylene group in an isotactic chain can be regarded as the center unit in a *meso* dyad of monomer units having the same configuration, and such a dyad might be designated *dd* (or *ll*), but in the analogous small molecule, the 2,4-disubstituted pentane, the two asymmetric carbon atoms are of opposite configuration and are conventionally designated *d* and *l*. Similarly, for *racemic* dyads in syndiotactic chains, the monomer units are regarded by the polymer chemist as being of opposite configuration, whereas in the analogous pentane they have the same configuration.

3. The Observation of Configuration Sequences: Polymethyl Methacrylate

It is now well recognized that the symmetry considerations discussed in the preceding section can be extended to the long chains of polymers. In a way, very long chains simplify matters. If the degree of polymerization exceeds about 100, so that end effects can be neglected, then for a purely syndiotactic chain H_a and $H_{a'}$ can be considered strictly equivalent. We shall begin by discussing polymethyl methacrylate, one of the class of α,α'-disubstituted polymers in which there can be no vicinal coupling of main-chain protons.

Polymethyl methacrylate has proved particularly amenable to NMR study.[6,8,47–77] Figure III-2 shows the 60 MHz spectra of (a) predominantly

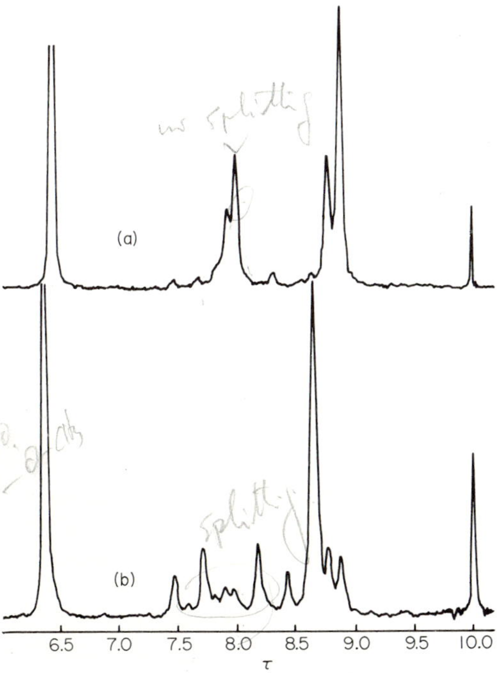

FIG. III-2. The 60 MHz spectra of 15% (w/v) solutions in chlorobenzene of polymethyl methacrylate prepared with (a) a free-radical initiator and (b) an anionic initiator (PhMgBr). The ester methyl resonance appears near 6.5τ, the β-methylene protons appear near 8.0τ and the β-methyl protons give three peaks between 8.5 and 9.0τ.

syndiotactic (called I) and (b) predominantly isotactic (called II) polymethyl methacrylate in chlorobenzene solution at 150°. The polymer shown as spectrum (a), which we shall designate as I, was prepared with a free radical initiator; the polymer in spectrum (b), designated II, was prepared in toluene at −78° with phenylmagnesium bromide initiator. The methylene resonance in (a) is approximately the expected singlet, although somewhat broadened and complicated by the residual isotactic resonance and by additional fine structure with which we shall deal in Section 5. The methylene resonance in (b) is predominantly the expected AB quartet ($J = -14.9$ Hz), with additional structure due to syndiotactic sequences. The nature of these multiplets is an absolute measure of the polymer's predominant configuration, and it would not be necessary to have recourse to x-ray diffraction, even if it were possible.

The spectrum of the α-substituents does not in itself provide an absolute indication of configuration. It must be correlated with the β-methylene resonance or some other absolute measurement. If this is done, the α-substituent spectrum may give more detailed configurational information than the β-methylene spectrum. The chemical shifts of α-substituents commonly vary appreciably with the relative configurations of the nearest neighboring monomer units, but this is not always the case. For example in chlorobenzene solution the α-methyl group resonance in polymethyl methacrylate is markedly dependent on configuration (Fig. III-2), whereas the ester methyl resonance is not (but in benzene, it is found that different configurations can be distinguished in the ester methyl region[59]). When such discrimination is possible one can observe three species of α-substituents in a polymer which is not stereochemically pure: those on the central monomer units of isotactic, syndiotactic, and heterotactic triads of monomer units (Chapter II, Section 2). The definitions of these correspond to the configurations shown previously for the 2,4,6-trisubstituted heptanes. The simplest notations for these triad sequences are *i*, *s*, and *h*, respectively.[6] A more general notation will be introduced in Section 4. In Fig. III-2, the *i*, *s*, and *h* α-methyl resonances appear at 8.67, 8.90, and 8.79τ in both spectra. From their relative intensities one can obtain a measure of the average configuration of the chains. The measurement of the areas of such overlapping peaks has been discussed in Chapter I, Section 16.

4. The Generation of Configurational Sequences

In Table III-1, the designations of monomer dyad and triad sequences are indicated. The *meso* dyad is designated *m* and the *racemic* dyad *r*. This system of nomenclature can be extended to sequences of any length. Thus, an isotactic triad is *mm*, a heterotactic triad *mr*, and a syndiotactic triad *rr*. Let us now

TABLE III-1

	α Substituent				β-CM_2		
	Designation	Projection	Bernoullian probability		Designation	Projection	Bernoullian probability
Triad	Isotactic, *mm* (*i*)		$P_m{}^2$	Dyad	*meso*, *m*	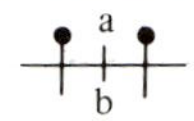	P_m
	Heterotactic, *mr* (*h*)	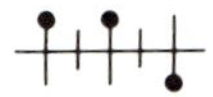	$2P_m(1-P_m)$		*racemic*, *r*	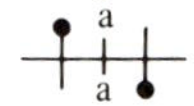	$(1-P_m)$
	Syndiotactic, *rr* (*s*)	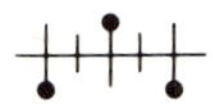	$(1-P_m)^2$	Tetrad	*mmm*	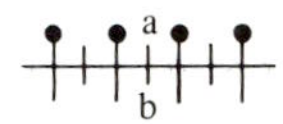	$P_m{}^3$
Pentad	*mmmm* (isotactic)	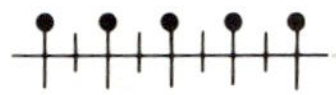	$P_m{}^4$		*mmr*	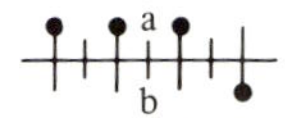	$2P_m{}^2(1-P_m)$
	mmmr	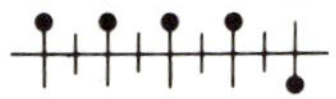	$2P_m{}^3(1-P_m)$		*rmr*	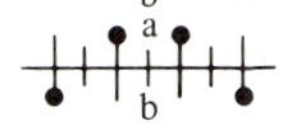	$P_m(1-P_m)^2$
	rmmr	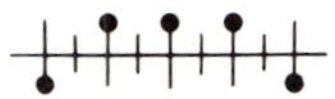	$P_m{}^2(1-P_m)^2$		*mrm*	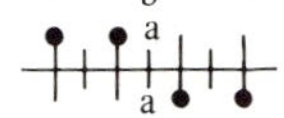	$P_m{}^2(1-P_m)$
	mmrm	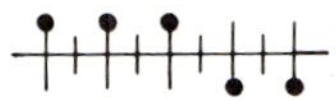	$2P_m{}^3(1-P_m)$		*rrm*	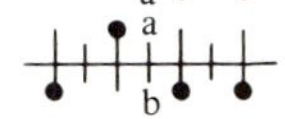	$2P_m(1-P_m)^2$
	mmrr	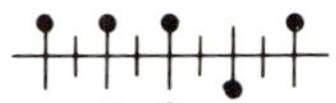	$2P_m{}^2(1-P_m)^2$		*rrr*	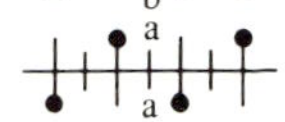	$(1-P_m)^3$
	rmrm (heterotactic)		$2P_m{}^2(1-P_m)^2$				
	rmrr		$2P_m(1-P_m)^3$				
	mrrm		$P_m{}^2(1-P_m)^2$				
	rrrm	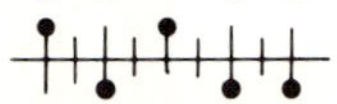	$2P_m(1-P_m)^3$				
	rrrr (syndiotactic)	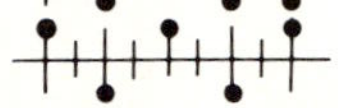	$(1-P_m)^4$				

assume that the probability of generating a *meso* sequence when a new monomer unit is formed at the end of a growing chain can be denoted by a single parameter, which we shall call P_m[62] (this probability has been previously designated as α[4] and as σ[6]). In these terms, the generation of the chain is a Bernoulli-trial process. We may liken such a process to that of reaching into a large jar of balls marked "*m*" or "*r*" and withdrawing a ball at random. The proportion of "*m*" balls in the jar is P_m. The probability of forming an *r* sequence is of course $1 - P_m$. A plot of these relations is shown in Fig. III-3. It will be noted that the proportion of *mr* units rises to a maximum at $P_m = 0.5$, corresponding to random propagation. For a random polymer, the proportion *mm*:*mr*:*rr* will be 1:2:1. (A similar plot of dyad frequencies versus P_m would obviously be two straight lines with slopes of +1 and −1 for *m* and *r*, respectively.) For any given polymer, if Bernoullian, the *mm*, *mr*, and *rr* sequence frequencies, as estimated from the relative areas of the appropriate peaks (α-methyl peaks for methyl methacrylate polymers), would lie on a single vertical line in Fig. III-3, corresponding to a single value of P_m. If this

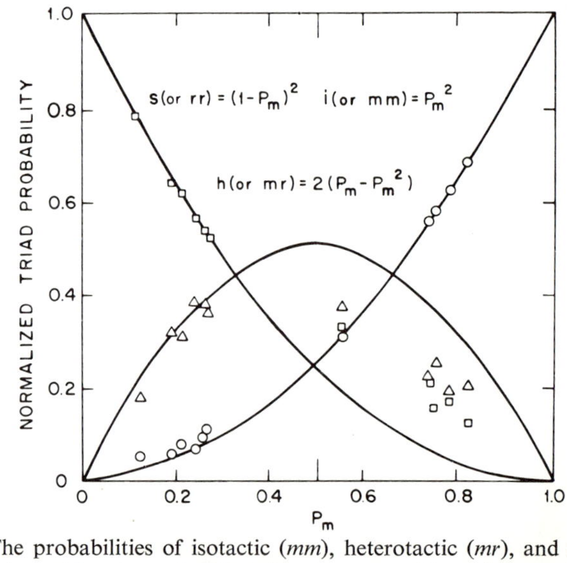

FIG. III-3. The probabilities of isotactic (*mm*), heterotactic (*mr*), and syndiotactic (*rr*) triads as a function of P_m, the probability of *m* placement. The points on the left-hand side are for methyl methacrylate polymers prepared with free radical initiators and those on the right-hand side for polymers prepared with anionic initiators.

is not the case, then the polymer's configurational sequence deviates from Bernoullian. The polymer whose spectrum is shown in Fig. III-2(a) obeys these simple statistics within experimental error, P_m being 0.24 ± 0.01. Analysis of the spectrum in Fig. III-2(b) shows that this polymer does not obey Bernoullian statistics. As indicated in the caption to Fig. III-2, methyl methacrylate polymers produced by free-radical initiators usually follow Bernoullian statistics within experimental error, while those produced using anionic initiators do not.

The mechanism of propagation, as deduced from NMR measurements, is discussed in detail in Chapter VIII.

5. The Observation of Longer Configurational Sequences

We have seen (p. 72) that the β-methylene resonance of polymethyl methacrylate cannot be fully explained in terms of dyads even when observed at 60 MHz. We must now consider the observation of sequences longer than dyads and triads, for this has become possible through improved techniques and advances in instrumental design, particularly with increasing strength of the magnetic field of the spectrometer (Chapter I, Section 17). With respect to β-methylene groups (or, more generally, β-CM_2 groups) one may expect to resolve tetrad sequences of monomer units, appearing as a fine structure on the *m* and *r* resonances. As indicated in Table III-1, the *m* resonance

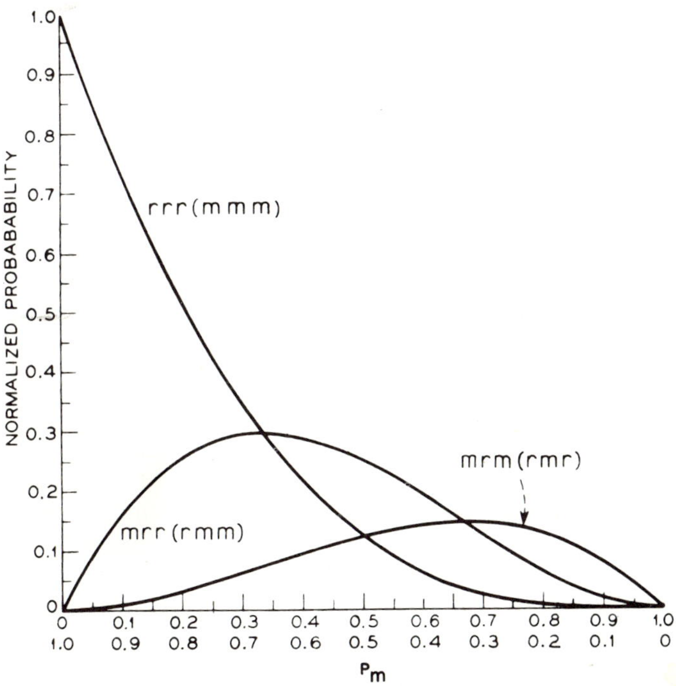

FIG. III-4. Tetrad sequence probabilities as a function of P_m. For *rrr*, *mrr*(*rrm*), and *mrm*, the upper P_m scale is used; for *mmm*, *rmm*(*mmr*), and *rmr* the lower P_m scale is used.

should be resolved into three tetrad resonances, all heterosteric, giving six different chemical shifts. The *r* resonance should be split into two homosteric resonances and one heterosteric resonance, giving a total of ten observable β-CM_2 chemical shifts in a polymer which is not too highly stereoregular. One may also expect the α-substituent resonance to be resolved into ten peaks, corresponding to ten different pentad configurational sequences. If the polymer has been generated by a Bernoulli-trial propagation, the sequence frequencies will be as shown in the third and last columns of Table III-1.

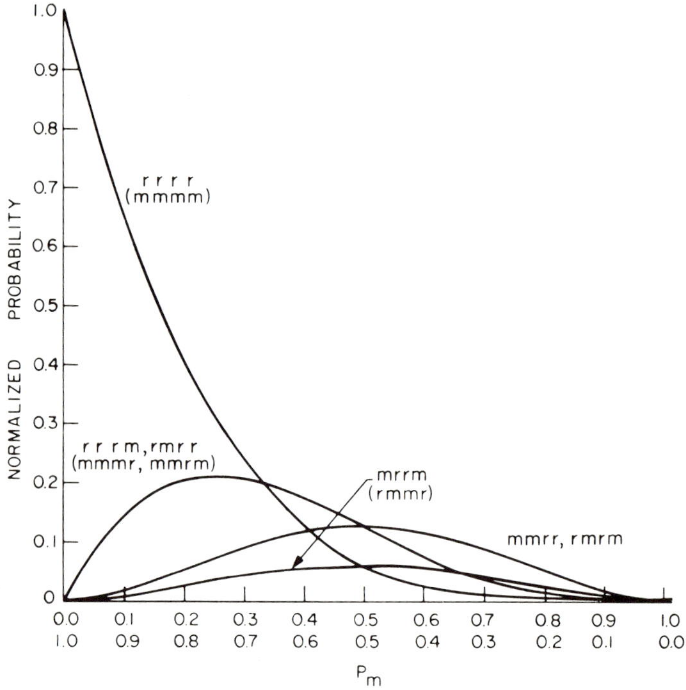

FIG. III-5. Pentad sequence probabilities as a function of P_m. The lower scale should be used for sequences in parentheses.

These relationships are plotted for tetrad sequences in Fig. III-4 and for pentad sequences in Fig. III-5. Since *m* and *r* are conjugate terms in these relationships, the frequency plots are symmetrical, as implied in the legends to these figures: three curves serve for six tetrad species and four curves for ten pentad species, since there are eight pairs of sequences of which the members of each pair have the same probability.

It should be pointed out that the written *direction* of the sequences is entirely arbitrary and therefore immaterial, except perhaps at the ends of the chain, which we ignore. Thus, for example, *mmr* cannot be observationally distinguished from *rmm*, nor can *rmrr* be distinguished from *rrmr*. Thus unsymmetrical sequences of this sort must be counted twice, i.e., in both directions, and this is the reason that the factor 2 appears in the Bernoullian probability expressions (Table III-1). In polymers which have true asymmetric centers in the main chain, such as polypeptides or polypropylene oxide, a chain direction can be defined, and therefore, for example, the sequence *ddld* can (in principle, at least) be observationally distinguished from *dldd*.

The methylene spectrum of the predominantly syndiotactic polymethyl methacrylate [Fig. III-2(a)] shows a clear splitting (alluded to previously) into two peaks with a third weaker one as a shoulder on the low-field side. These same resonances can be seen, in altered relative intensity, in the spectrum of the predominantly isotactic polymer [Fig. III-2(b)]. It is also significant that in the residual *meso* quartet in Fig. III-2(a), the upfield doublet is at

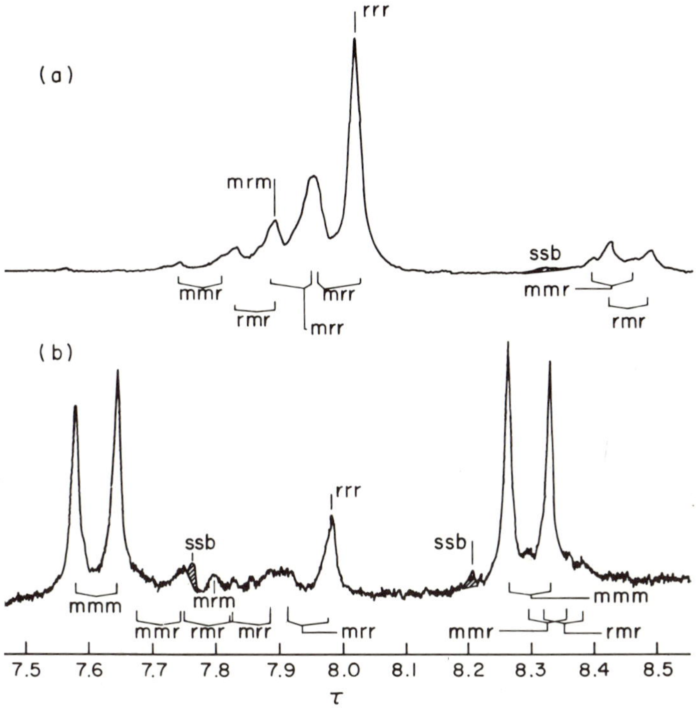

FIG. III-6. (a) β-methylene proton spectrum of predominantly syndiotactic polymethyl methacrylate (polymer I) observed at 220 MHz (about 10% w/v solutions in chlorobenzene at 135°); (b) β-methylene proton spectrum of predominantly isotactic polymethyl methacrylate (polymer II), same conditions as (a). In these spectra, the vertical markers for the doublet components of *mmm*, *mmr*, *rmr*, and *mrr* quartets are joined by slanting lines which intersect at the chemical shift; "ssb" designates spinning side band.

markedly higher field than the corresponding doublet in Fig. III-2(b). These features clearly represent a discrimination of tetrad sequences.

The 100 MHz spectra of these polymers show an appreciable improvement over the 60 MHz results (see Fig. I-19). A much more dramatic improvement

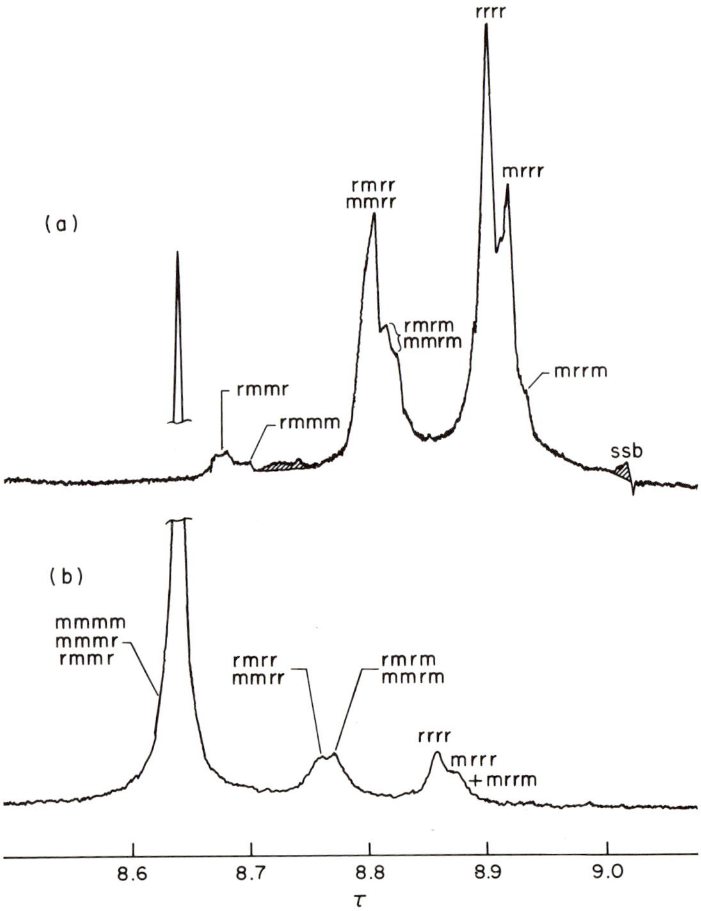

FIG. III-7. (a) α-methyl proton spectrum of predominantly syndiotactic polymethyl methacrylate (polymer I) observed at 220 MHz (about 10% w/v solutions in chlorobenzene at 135°); (b) α-methyl proton spectrum of predominantly isotactic polymethyl methacrylate (polymer II), same conditions as (a); "ssb" designates spinning side band.

in discrimination, however, is achieved with a 220 MHz instrument (Chapter I, Section 18A). In Figs. III-6(a) and III-7(a) are shown the 220 MHz spectra of the β-methylene and α-methyl protons of the predominantly syndiotactic polymethyl methacrylate, polymer I. It is assumed that the propagation is a Bernoulli-trial process, and that the triad, tetrad, and pentad probabilities are therefore given by the curves of Figs. III-3, III-4, and III-5. This enables the peaks of Fig. III-6(a) to be assigned to tetrad sequences on the basis of their relative areas. The assignments are as shown in the figure. The *meso* tetrads appear as approximately AX quartets; the centers of the doublets may be taken as the chemical shifts without significant error. It is believed (see Chapter VIII, Section 4D) that the less shielded proton in the *meso* methylene groups is *syn* (or *erythro*) and that the more shielded proton is *anti* (or *threo*).

In Figs. III-6(b) and III-7(b) are shown the β-methylene and α-methyl proton spectra, respectively, of the predominantly isotactic polymethyl methacrylate, polymer II. We assume that the proton shieldings, while they may be (and in fact are) slightly altered when the predominant stereochemistry is altered,† will remain in the same order for tetrad sequences. (It is obvious that they do so for triad sequences.) The assignments in Fig. III-6(b) are then essentially the same as in Fig. III-6(a). The *mmm* tetrad resonance, not discernible in Fig. III-6(a), is now the dominant feature of the methylene spectrum in Fig. III-6(b). The much weaker AX spectra of the *mmr* and *rmr* sequences are difficult to see in Fig. III-6(b); although clearer in the original spectra, relative intensities cannot be accurately measured.

Turning to the α-methyl spectra [Figs. III-7(a) and III-7(b)], we see the *mm*, *mr*, and *rr* peaks clearly split into further peaks, which represent pentad sequences. The assignments are again made in polymer I on the basis of Bernoullian intensities and assumed to be the same in II. The pentad resonances are slightly shifted upfield in the isotactic polymer.

It should be pointed out again that there are many types (in fact, an infinite number) of regular configurational sequences other than isotactic and syndiotactic. One of these, the heterotactic, has already been indicated (Chapter II, Section 2). Such higher sequence types can be generated only by non-Bernoullian propagation processes, which are discussed in Chapter VIII.

6. Numbers and Necessary Relationships of *n*(ads)

The number of observationally distinguishable *n*(ads), i.e., the number of types of sequences containing *n* monomer units, is designated as $N(n)$ and obeys the following relationship[62]:

n	2	3	4	5	6	7	8
$N(n)$	2	3	6	10	20	36	72

or, in general

$$N(n) = 2^{n-2} + 2^{m-1}$$

† The arguments advanced by Flory and Baldeschwieler,[78] are probably not applicable to α,α'-disubstituted polymer chains.

where $m = n/2$ if n is even and $m = (n-1)/2$ if n is odd. It is evident that, regardless of the generating statistics, the number of possible types of sequences increases rapidly with their length (asymptotically as 2^{n-2} as $N \to \infty$) and that their discrimination beyond pentads, or hexads at most, will be very difficult.

The probabilities of occurrence of sequences are governed by certain necessary relationships which are based on elementary statistical considerations and are entirely independent of the configurational statistics generated by any particular mechanism. Some of these necessary relationships are summarized in Table III-2. Since these relationships are independent of mechanism,

TABLE III-2

SOME NECESSARY RELATIONSHIPS AMONG SEQUENCE FREQUENCIES

Dyad	$(m) + (r) = 1$
Triad	$(mm) + (mr) + (rr) = 1$
Dyad–Triad	$(m) = (mm) + \frac{1}{2}(mr)$ $(r) = (rr) + \frac{1}{2}(mr)$
Triad–Tetrad	$(mm) = (mmm) + \frac{1}{2}(mmr)$ $(mr) = (mmr) + 2(rmr) = (mrr) + 2(mrm)$ $(rr) = (rrr) + \frac{1}{2}(mrr)$
Tetrad–Tetrad	sum $= 1$ $(mmr) + 2(rmr) = 2(mrm) + (mrr)$
Pentad–Pentad	sum $= 1$ $(mmmr) + 2(rmmr) = (mmrm) + (mmrr)$ $(mrrr) + 2(mrrm) = (rrmr) + (rrmm)$
Tetrad–Pentad	$(mmm) = (mmmm) + \frac{1}{2}(mmmr)$ $(mmr) = (mmmr) + 2(rmmr) = (mmrm) + (mmrr)$ $(rmr) = \frac{1}{2}(mrmr) + \frac{1}{2}(rmrr)$ $(mrm) = \frac{1}{2}(mrmr) + \frac{1}{2}(mmrm)$ $(rrm) = 2(mrrm) + (mrrr) = (mmrr) + (rmrr)$ $(rrr) = (rrrr) + \frac{1}{2}(mrrr)$

they serve as useful tests of peak assignments. Any set of assignments requiring peak intensity ratios in conflict with the appropriate relationships in Table III-2 cannot be correct. It must be understood that the intensities are normalized over each species of n(ad).

7. ^{13}C Spectrum of Polymethyl Methacrylate

The ^{13}C spectrum (see Chapter I, Section 16C) of a predominantly syndiotactic polymethyl methacrylate (similar to polymer I, discussed in Sections 4 and 5) has been reported.[79, 80] Because of the low sensitivity and natural abundance of ^{13}C, a 50% solution in chlorobenzene was employed; the temperature of observation was about 60°. These are not conditions for optimum resolution. Nevertheless, because of the smaller magnetic moment of ^{13}C, line widths are not excessive and useful results can be obtained. Figure III-8(a) is the spectrum of all the carbon atoms except that of the carbonyl group; Fig. III-8(b) is the carbonyl carbon spectrum. The spectra were

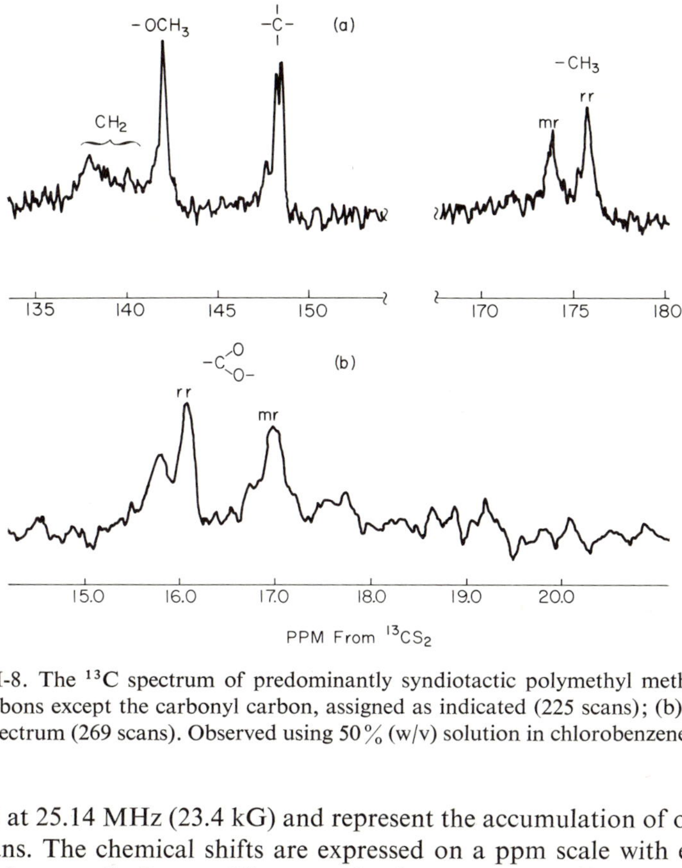

FIG. III-8. The ^{13}C spectrum of predominantly syndiotactic polymethyl methacrylate; (a) all carbons except the carbonyl carbon, assigned as indicated (225 scans); (b) carbonyl carbon spectrum (269 scans). Observed using 50% (w/v) solution in chlorobenzene at about 60°.

observed at 25.14 MHz (23.4 kG) and represent the accumulation of over 200 CAT scans. The chemical shifts are expressed on a ppm scale with external $^{13}CS_2$ as the zero of reference (Chapter I, Section 17B). Direct ^{13}C–H couplings, which are of the order of 125 Hz, would cause splitting of some of the ^{13}C resonances: $^{13}CH_2$ would be a triplet and $^{13}CH_3$ a quadruplet, while $^{13}C{=}O$ and the quaternary carbon are singlets. These splittings furnish a means of assigning the peaks in a general way. In Fig. III-8, multiplicity has been removed by double resonance in order to improve the effective signal-to-noise ratio. This has been done by "noise decoupling," as described in Chapter I, Section 18B.

The pattern of peaks in Fig. III-8(a) is reminiscent of that of the proton spectrum [Figs. III-6(a) and III-7(a)] except that the spectrum contains no absolute configurational information, since the homosteric or heterosteric nature of a β-CH_2 group is not evident in its ^{13}C resonance. Somewhat more informative is the carbonyl spectrum [Fig. III-8(b)]. Although the noise level is rather high, a splitting into pentad resonances is evident. In fact, the spectrum has the same pattern as the 220 MHz α-methyl proton spectrum [Fig. III-7(a)] but is reversed from left to right, differing in this respect from all the other ^{13}C peaks in this polymer and in others to be described later. It is also noteworthy that the carbonyl spectrum is spread out over a tenfold greater range of chemical shifts than the α-CH_3 proton spectra, both being expressed in ppm.

The time required to accumulate such spectra can be substantially reduced by employing Fourier transform spectroscopy (Chapter I, Section 18C). This will undoubtedly be the method of choice in future studies.

8. Other Methacrylates and Related Polymers

The NMR study of ethyl methacrylate[74] will be described in Chapter VIII. Sobue *et al.*[81] observed that the polymerization of menthyl methacrylate with free radical and anionic initiators was similar to that of the methyl ester except that the free radical polymers were characterized by higher values of P_m, about 0.4. Wiles and Brownstein[82] reported that the anionic polymerization of allyl methacrylate proceeds entirely via the vinyl rather than the allyl double bond and that the effect of solvents on the chain configuration is much like that described for ethyl methacrylate. Yokota and Ishii[83] studied benzyl, β-phenylethyl, and γ-phenylpropyl methacrylate polymers. Tsurata *et al.*[67] observed the configuration of free radical and anionic polymers of a number of methacrylate monomers with ester groups of varying degrees of branching, viz., ethyl, isopropyl, *tert*-butyl, *n*-butyl, *d*-bornyl, *dl*-isobornyl, and 2,4,6-triphenylbenzyl. They converted all the polymers to polymethyl methacrylate by ester exchange prior to NMR measurements. The free radical polymers of the more highly branched monomers appeared to deviate from Bernoullian statistics, although the deviations were not large. Yuki *et al.*[84] studied the tacticity of anionically initiated D-α-methylbenzyl methacrylate, again by conversion to polymethyl methacrylate. Liu *et al.*[85] observed this polymer directly and were able to use the α-CH_3 peaks for configurational studies. The response of this monomer to anionic initiators and to ultraviolet initiation over a wide temperature range is much like that of methyl methacrylate.

Polymethacrylonitrile has also been studied by conversion to polymethyl methacrylate.[86,87] Its direct observation will be described in Section 11 of this chapter. Polymers of methacrylamide have been observed, but the chief objectives were to determine to what extent the polymerization proceeded by hydrogen migration, yielding the chain units shown (predominant with

$$\begin{array}{c} \quad CH_3 \\ \quad | \\ -CH_2-CH-CONH- \end{array}$$

alkali metal alkoxide initiators[88]), and to measure the rates of rotation and proton exchange of the amide group in aqueous solution.[89]

The configuration of polymethacrylic acid polymerized in aqueous solution was measured as a function of the *p*H of the polymerization medium.[90] For this purpose, the polymer was converted to polymethyl methacrylate by methylation with diazomethane. It was found that the polymer was distinctly more syndiotactic as the *p*H was increased, presumably because of coulombic repulsion between charged monomer and growing chain end. Tsuruta *et al.*[67] have observed polymethacrylic acid directly.

The configuration of polymethacrylic anhydride has been measured by saponification and methylation of polymethyl methacrylate.[63,91–93] In the polymerization of 1,6-heptadienes such as this, propagation occurs by alternating intramolecular and intermolecular steps which presumably must differ to some degree in configurational preference, and will therefore have different values of P_m for each step (assuming the propagation to be Bernoullian for each type of addition). Calling these P_{m_A} and P_{m_B}, we then have for the triad frequencies in the resulting polymer:

$$mm = P_{m_A} P_{m_B} \tag{III-1}$$

$$rr = (1 - P_{m_A})(1 - P_{m_B}) \tag{III-2}$$

$$mr = P_{m_A} + P_{m_B} - 2P_{m_A} P_{m_B} \tag{III-3}$$

These parallel the triad frequency relations given in Table III-1. In fact, it was found in the first study of this system[91] that for the free radical polymerization at 100° both steps had the same value of P_m within experimental error, and that P_m was somewhat higher, about 0.31, than for normally prepared free radical polymethyl methacrylate. Reinmöller and Fox[63] confirmed this finding for polymers prepared in the temperature range 5–135°, but reported that those prepared at –30° and –55° fitted first-order Markov (Chapter VIII) rather than Bernoullian statistics. They concluded that the intramolecular propagation was Bernoullian at all temperatures, but that the intermolecular steps became non-Bernoullian at low temperature.

9. Polyalkyl α-Chloroacrylates

The stereoregularity of polymethyl α-chloroacrylate has been examined at 100 MHz in d_6-DMSO by Matsuzaki *et al.*[94] A polymer prepared with phenyl magnesium bromide at 0° showed a spectrum in which the methylene resonance was principally a *meso* quartet. The methoxyl resonance was best suited to triad analysis. These authors' analysis gave a value of 0.68 for P_m although the methylene resonance suggests a larger value. The Grignard-initiated polymers prepared at −40° and −78° were much less stereoregular, ranging from nearly random to predominantly syndiotactic. Free radical polymers were also predominantly syndiotactic, as expected.

Similar results were obtained from polyethyl α-chloroacrylate by Wesslén and Lenz,[95] except that under no conditions was the polymer so highly isotactic as that prepared with the Grignard reagent at 0° by Matsuzaki *et al.*[94] The 220 MHz spectra of ethyl α-chloroacrylate polymers were interpreted in terms of tetrad frequencies by Wesslén *et al.*[96]

10. Poly-α-Methylstyrene

Poly-α-methylsytrene has been studied by a number of workers,[97–103] but the interpretation of the spectra is still in some doubt. Variations in regularity with polymerization conditions can be clearly seen. In the earliest work (at 40 MHz),[97] one could not distinguish triads in the α-methyl spectrum, but the striking upfield shift of all the methyl groups, caused by the phenyl group ring currents, was evident. Brownstein *et al.*[98] found that a polymer prepared with $SnCl_4$ in toluene at −78° showed apparent singlets for both methyl and methylene protons, whereas a polymer prepared with sodium naphthalenide in THF showed two additional α-methyl peaks at lower field. The three α-methyl peaks were assigned to *mm*, *mr*, and *rr*, sequences in order of increasing shielding because the first polymer was believed to be highly syndiotactic. Sakurada *et al.*[99] found the most shielded peak to predominate in polymers prepared with a variety of initiators, being most dominant for Ziegler and cationic initiators. They reversed the assignment of Brownstein *et al.* because they believed from x-ray evidence that the most stereoregular polymers were isotactic. Ramey and Statton[100] observed some structure in the methylene resonance at 100 MHz and interpreted the spectrum as an *m* quartet and *r* singlet; they attributed the α-methyl peaks to *mr*, *mm*, and *rr* in order of increasing shielding, an assignment neither reasonable nor required by the methylene analysis. Fujii *et al.*[101] appear to have refuted the interpretation of Ramey and Statton but were not able to draw any definite conclusions of their own. Observations at 220 MHz[102, 103] confirm the original interpretation of Brownstein *et al.*[98] Figure III-9 shows the spectrum of a

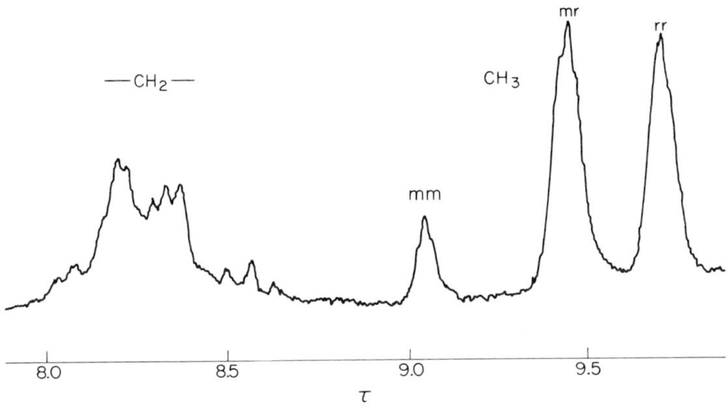

FIG. III-9. 220 MHz spectrum of poly-α-methylstyrene observed in 10% (w/v) solution in *o*-dichlorobenzene at 140° (Heatley *et al.*[102, 103]).

predominantly syndiotactic polymer prepared with *n*-butyllithium in THF at −78°.[102, 103] The triad assignments and intensities are consistent with a P_m of about 0.31, although it is not yet established that this polymer is Bernoullian. The β-methylene region resembles that of predominantly syndiotactic polymethyl methacrylate [Fig. III-6(a)] except that additional peaks appear in the 8.3τ region. Assignments have not been established in detail.

11. Polymethacrylonitrile

Methacrylonitrile can be polymerized by free radical initiators and by anionic coordination initiators such as diethylzinc and diethylmagnesium.[104, 105] The NMR spectra of polymers prepared with the latter have been observed in trifluoroacetic acid.[106, 107] They resemble spectra of predominantly isotactic polymethyl methacrylate in that the β-CH_2 region exhibits a quartet for *m* dyads, in the center of which is a singlet for the *r* dyads. Peak area measurements indicated a proportion of *m* as high as 80% for the most highly crystalline fraction. The spectra of these polymers differ in that only one α-CH_3 peak could be resolved for polymethacrylonitrile even at 100 MHz. Polymers prepared with γ radiation as initiator showed the *r* peak to be predominant, as in free radical polymethyl methacrylate. Burnett *et al.*[108] reported that the stereoregularity of polymethacrylonitrile is unaffected by using isobutraldehyde as solvent for free radical polymerization. Alterations in physical properties were found to be due to reduced molecular weight. It has been claimed that this solvent encourages regular placement in vinyl chloride polymerization; we shall discuss this claim in Chapter VIII, Section 2.

12. Poly-α-methylvinyl Methyl Ether

Goodman and Fan[109] reported an NMR study of poly-α-methylvinyl methyl ether prepared with cationic initiators at low temperatures. All polymers gave very simple spectra, consisting of three singlets at 6.86, 8.04, and 8.57τ (in chlorobenzene at 140°) for the methoxyl, β-methylene, and α-methyl protons, respectively. The singlet β-methylene resonance was interpreted as indicating that the configuration was highly syndiotactic, a conclusion supported by its infrared spectrum. The evidence is admittedly somewhat negative, because there is always a possibility of accidental equivalence of *meso* methylene protons. However, since all isotactic or stereoirregular polymer chains with α-methyl substituents have so far been observed to have markedly heterosteric *meso* methylene protons, their conclusion is a very reasonable one. It is confirmed by measurements at 220 MHz, which also show the β-methylene resonance to be a singlet,[110] but has been questioned by Matsuzaki *et al.*[111] who reported triplet α-methyl signals and believed the polymer to be stereoirregular. This report has been conclusively refuted, and their observations shown to be due to decomposition of the polymer.[112,113] More recently, energy minimization and x-ray diffraction[114] studies have demonstrated the syndiotactic configuration beyond any question.

REFERENCES

1. G. Natta, *Chem. Ind.* (*Milan*) **46**, 397 (1964) (Nobel Prize Address).
2. K. Ziegler, *Angew. Chem.* **76**, 545 (1964) (Nobel Prize Address).
3. H. L. Frisch, C. Schuerch, and M. Szwarc, *J. Polym. Sci.* **11**, 559 (1953).
4. B. D. Coleman, *J. Polym. Sci.* **31**, 155 (1958).
5. J. W. L. Fordham, *J. Polym. Sci.* **39**, 321 (1959).
6. F. A. Bovey and G. V. D. Tiers, *J. Polym. Sci.* **44**, 173 (1960).
7. F. A. Bovey and G. V. D. Tiers, *Fortschr. Hochpolymer. Forsch.* **3**, 139 (1963).
8. F. A. Bovey, *Accounts Chem. Res.* **1**, 175 (1968).
9. R. L. Miller and L. E. Nielsen, *J. Polym. Sci.* **46**, 303 (1960).
10. R. L. Miller, *J. Polym. Sci.* **56**, 375 (1962).
11. R. L. Miller, *Soc. Plast. Eng. Trans.* **1963**, 123 (1963).
12. F. P. Price, *J. Chem. Phys.* **36**, 209 (1962).
13. A. Abragam, "The Principles of Nuclear Magnetism." Oxford Univ. Press, London and New York, 1961.
14. K. Mislow and M. Raban, *in* "Topics in Stereochemistry" (N. L. Allinger and E. L. Eliel, eds.), Vol. I, p. 1. Wiley, New York, 1967.

15. M. Goodman *in* "Topics in Stereochemistry" (N. L. Allinger and E. L. Eliel, eds.), Vol. II, p. 73. Wiley, New York, 1967.
16. H. S. Gutowsky, *J. Chem. Phys.* **37**, 2916 (1962), and references cited therein.
17. J. S. Waugh and F. A. Cotton, *J. Phys. Chem.* **65**, 562 (1961).
18. M. van Gorkom and G. E. Hall, *Quart. Rev. Chem. Soc.* **22**, 14 (1968).
19. W. H. Pirkle, *J. Amer. Chem. Soc.* **88**, 1837 (1966).
20. T. G. Burlingame and W. H. Pirkle, *J. Amer. Chem. Soc.* **88**, 4294 (1966).
21. J. C. Jochens, G. Taigel, and A. Selinger, *Tetrahedron Lett.* **1967**, 1901 (1967).
22. G. V. D. Tiers and F. A. Bovey, *J. Polym. Sci. Part A* **1**, 833 (1963).
23. D. Doskocilova, *J. Polym. Sci. Part B* **2**, 421 (1964).
24. D. Doskocilova and B. Schneider, *Coll. Czech. Chem. Commun.* **29**, 2290 (1964).
25. S. Satoh, *J. Polym. Sci. Part A* **2**, 5221 (1964).
26. P. E. McMahon and W. C. Tincher, *J. Mol. Spectrosc.* **15**, 180 (1965).
27. B. Schneider, J. Stokr, D. Doskocilova, S. Sykora, J. Jakes, and M. Kolinsky, *J. Polym. Sci. Part C* **22**, 1073 (1969).
28. H. G. Clark, *Makromol. Chem.* **63**, 69 (1963).
29. R. Yamadera and M. Murano, *J. Polym. Sci. Part A-1* **5**, 1855 (1967).
30. Y. Fujiwara and S. Fujiwara, *Bull. Chem. Soc. Jap.* **37**, 1005 (1964).
31. D. Doskocilova and B. Schneider, *J. Polym. Sci. Part B* **3**, 209 (1965).
32. K. Fujii, S. Brownstein, and A. M. Eastham, *J. Polym. Sci. Part A-1* **6**, 2377 (1968).
33. H. G. Clark, *Makromol. Chem.* **86**, 107 (1965).
34. D. Lim, B. Obereigner, and D. Doskocilova, *J. Polym. Sci. Part B* **3**, 893 (1965).
35. F. A. Bovey, F. P. Hood, E. W. Anderson, and L. C. Snyder, *J. Chem. Phys.* **42**, 3900 (1965).
36. F. Heatley and F. A. Bovey, *Macromolecules* **1**, 301 (1968).
37. A. D. Williams, J. I. Brauman, N. J. Nelson, and P. J. Flory, *J. Amer. Chem. Soc.* **89**, 4807 (1967).
38. K. Matsuzaki, K. Sakota, and M. Okada, *J. Polym. Sci. Part A-1* **7**, 1444 (1969).
39. T. Shimanouchi, M. Tasumi, and Y. Abe, *Makromol. Chem.* **86**, 43 (1965).
40. Y. Abe, M. Tasumi, T. Shimanouchi, S. Satoh, and R. Chujo, *J. Polym. Sci. Part A-1* **4**, 1413 (1966).
41. D. Doskocilova, J. Stokr, B. Schneider, H. Pivcova, M. Kolinsky, J. Petranek, and D. Lim, *J. Polym. Sci. Part C* **16**, 215 (1967).
42. M. Murano and R. Yamadera, *J. Polym. Sci. Part B* **5**, 483 (1967).
43. M. Murano and R. Yamadera, *J. Polym. Sci. Part A-1* **6**, 843 (1968).
44. D. Lim, M. Kolinsky, J. Petranek, and D. Doskocilova, *J. Polym. Sci. Part B* **4**, 645 (1966).
45. H. Pivcova, M. Kolinsky, D. Lim, and B. Schneider, *J. Polym. Sci. Part C* **22**, 1093 (1969).
46. A. D. Williams and P. J. Flory, *J. Amer. Chem. Soc.* **91**, 539 (1969).
47. F. A. Bovey, *J. Polym. Sci.* **47**, 480 (1960).
48. F. A. Bovey, *J. Polym. Sci.* **46**, 59 (1960).
49. A. Nishioka, H. Watanabe, K. Abe, and Y. Sono, *J. Polym. Sci.* **48**, 241 (1960).
50. D. Braun, M. Herner, U. Johnsen, and W. Kern, *Makromol. Chem.* **51**, 15 (1962).
51. B. D. Coleman and T. G. Fox, *J. Polym. Sci. Part A* **1**, 3183 (1963).
52. Y. Kato, H. Watanabe, and A. Nishioka, *Bull. Chem. Soc. Jap.* **37**, 1762 (1964).
53. Y. Kato and A. Nishioka, *Bull. Chem. Soc. Jap.* **37**, 1614 (1964).
54. F. A. Bovey, *Chem. Eng. News* **43**, Aug. 30, p. 113 (1965).
55. H. Abe, K. Imai, and M. Matsumoto, *J. Polym. Sci. Part B* **3**, 1503 (1965).
56. K. K. Zasshi, *J. Chem. Soc. Jap. Ind. Chem. Sect.* **68**, 113, 852 (1965).

57. M. Yamamoto and G. Oster, *J. Polym. Sci. Part A-1* **4**, 1683 (1966).
58. H. Abe, K. Imai, and M. Matsumoto, *J. Polym. Sci. Part B* **4**, 589 (1966).
59. K. C. Ramey and J. Messick, *J. Polym. Sci. Part A-2*, 155 (1966).
60. F. A. Bovey, *Pure Appl. Chem.* **12**, 525 (1966).
61. F. A. Bovey, *Pure Appl. Chem.* **15**, 349 (1967).
62. H. L. Frisch, C. L. Mallows, and F. A. Bovey, *J. Chem. Phys.* **45**, 1565 (1966).
63. M. Reinmöller and T. G. Fox, paper presented to Div. of Polymer Chem., September 1966 Meeting of Amer. Chem. Soc. (preprints p. 987); see also *Polymer Preprints* 7(2), 999, 1005 (1966).
64. B. D. Coleman, T. G. Fox, and M. Reinmöller, *J. Polym. Sci. Part B* **4**, 1029 (1966).
65. D. Lim, J. Coupek, K. Juzl, J. Baca, S. Sykora, and B. Schneider, paper presented at *I.U.P.A.C. Int. Symp. on Macromol. Chem., Tokyo–Kyoto*, 1966.
66. T. Higashimura, Y. Ohsumi, and S. Ikamura, paper presented at *I.U.P.A.C. Int. Symp. on Macromol. Chem., Tokyo–Kyoto*, 1966.
67. T. Tsuruta, T. Makimoto, and H. Kanai, *J. Macromol. Chem.* **1**, 31 (1966).
68. R. C. Ferguson and W. D. Phillips, *Science* **157**, 257 (1967).
69. R. C. Ferguson, *Trans. N.Y. Acad. Sci.* **29**(4), 495 (1967).
70. R. C. Ferguson, *Macromolecules* **2**, 237 (1969).
71. K. C. Ramey, *J. Polym. Sci. Part B* **5**, 958 (1967).
72. K. Hatada, K. Ota, and H. Yuki, *J. Polym. Sci. Part B* **5**, 225 (1967).
73. K.-J. Liu, *J. Polym. Sci. Part A-1* **5**, 1199 (1967).
74. W. Fowells, C. Schuerch, F. A. Bovey, and F. P. Hood, *J. Amer. Chem. Soc.* **89**, 1396 (1967).
75. H. L. Frisch, C. L. Mallows, F. Heatley, and F. A. Bovey, *Macromolecules* **1**, 533 (1968).
76. S. Okuzawa, H. Hirai, and S. Makishima, *J. Polym. Sci. Part A-1* **7**, 1039 (1969).
77. H. Hirai, T. Ikegami, and S. Makishima, *J. Polym. Sci. Part A-1* **7**, 2059 (1969).
78. P. J. Flory and J. D. Baldeschwieler, *J. Amer. Chem. Soc.* **88**, 2873 (1966).
79. F. A. Bovey, paper presented to *Div. Polymer Chem., Amer. Chem. Soc. Meeting, 157th, Minneapolis, April* 1969.
80. L. F. Johnson, F. Heatley, and F. A. Bovey, *Macromolecules* **3**, 175 (1970).
81. H. Sobue, K. Matsuzaki, and S. Nakano, *J. Polym. Sci. Part A* **2**, 3339 (1964).
82. D. M. Wiles and S. Brownstein, *J. Polym. Sci. Part B* **3**, 951 (1965).
83. K. Yokota and Y. Ishii, *J. Polym. Sci. Part B* **3**, 771 (1965).
84. H. Yuki, K. Ohta, K. Uno, and S. Murashashi, *J. Polym. Sci. Part A-1* **6**, 829 (1968).
85. K.-J. Liu, S. J. Lignowski, and R. Ullman, *Makromol. Chem.* **115**, 119 (1968).
86. H. Sobue, T. Uryu, K. Matsuzaki, and Y. Tabata, *J. Polym. Sci. Part A* **2**, 3333 (1964).
87. K. Matsuzaki and T. Uryu, *J. Polym. Sci. Part B* **4**, 255 (1966).
88. K. Yokota, Y. Sakai, and Y. Ishii, *J. Polym. Sci. Part B* **3**, 839 (1965).
89. F. A. Bovey and G. V. D. Tiers, *J. Polymer. Sci. Part A* **1**, 849 (1963).
90. F. A. Bovey, *J. Polym. Sci. Part A* **1**, 843 (1963).
91. G. V. D. Tiers and F. A. Bovey, *J. Polym. Sci.* **47**, 479 (1960).
92. W. L. Miller, W. S. Brey, Jr., and G. B. Butler, *J. Polym. Sci.* **54**, 329 (1961).
93. J. C. H. Hwa, *J. Polym. Sci.* **60**, S12 (1962).
94. K. Matsuzaki, T. Uryu, and K. Ito, *Makromol. Chem.* **126**, 292 (1969).
95. B. Wesslén and R. W. Lenz, *Preprints of Polymer Div., Amer. Chem. Soc. Meeting, Houston, February 1970*, p. 105.
96. B. Wesslén, R. W. Lenz, and F. A. Bovey, in press.
97. F. A. Bovey, G. V. D. Tiers, and G. Filipovich, *J. Polym. Sci.* **38**, 73 (1959).
98. S. Brownstein, S. Bywater, and D. J. Worsfold, *Makromol. Chem.* **48**, 127 (1961).

99. Y. Sakurada, M. Matsumoto, K. Imai, A. Nishioka, and Y. Kato, *J. Polym. Sci. Part B* **1**, 633 (1963).
100. K. C. Ramey and G. L. Statton, *Makromol. Chem.* **85**, 287 (1965).
101. F. Fujii, D. J. Worsfold, and S. Bywater, *Makromol. Chem.* **117**, 275 (1968).
102. F. Heatley, D. J. Worsfold, and F. A. Bovey, unpublished results.
103. F. Heatley, S. Bywater, and D. J. Worsfold, paper presented to *Div. of Org. Coatings and Plastics Chem., Joint Amer. Chem. Soc.–Can. Inst. Chem. Meeting, Toronto, May 1970.*
104. G. Natta and G. Dall'Asta, *Chim. Ind.* (*Milan*) **46**, 1429 (1964).
105. Y. Joh, T. Yoshihara, Y. Kotake, F. Ide, and K. Nakatsuka, *J. Polym. Sci. Part B* **3**, 933 (1965); **4**, 673 (1966).
106. A. L. Segre, F. Ciampelli, and G. Dall'Asta, *J. Polym. Sci. Part B* **4**, 633 (1966).
107. K. Ishigure, Y. Tabata, and K. Oshima, *J. Polym. Sci. Part B* **4**, 669 (1966).
108. G. M. Burnett, F. L. Ross, and J. N. Hay, *Macromol. Chem.* **105**, 1 (1967).
109. M. Goodman and Y.-L. Fan, *J. Amer. Chem. Soc.* **86**, 4922 (1964).
110. M. Goodman, Y.-L. Fan, J. J. Ryan, and F. A. Bovey, unpublished observations.
111. K. Matsuzaki, M. Hamada, and K. Arita, *J. Polym. Sci. Part A-1* **5**, 1233 (1967).
112. K.-J. Liu and S. J. Lignowski, *J. Polym. Sci. Part B* **6**, 191 (1968).
113. M. Goodman and Y.-L. Fan, *Macromolecules* **1**, 163 (1968).
114. V. Y. Chen, G. Allegra, P. Corradini, and M. Goodman, *Macromolecules*, **3**, 274 (1970).

Chapter IV

POLYACRYLATES, POLYACRYLONITRILE, AND POLYVINYL ALCOHOL AND ITS DERIVATIVES

1. Introduction

In the previous chapter, we have considered polymers in which vicinal coupling of main-chain protons is not possible. Such spectra are simpler to interpret but contain no direct information concerning chain conformation (Chapter IX). In this and the succeeding three chapters we discuss the more complex spectra of the large class of polymers of α-monosubstituted monomers in which vicinal coupling of α- and β-protons must be dealt with.

2. Polyalkyl Acrylates: The Calculation of Main-Chain Spectra

The structure and conformation of polyalkyl acrylates have received considerable study by NMR,[1-9] particularly with respect to the direction of addition to the double bond during propagation (Chapter VIII). This class of polymers is well suited to such studies, the *m* methylene protons differing widely in chemical shift, as in methacrylates. Figure IV-1(a) shows the 100 MHz spectrum of the backbone protons of isotactic polyisopropyl acrylate prepared with a Grignard initiator.[2,3] (The side-chain protons give a doublet at 8.76τ for the methyl protons and a septet at 4.96τ for the methinyl proton which are not particularly useful analytically and are not shown.) Figure

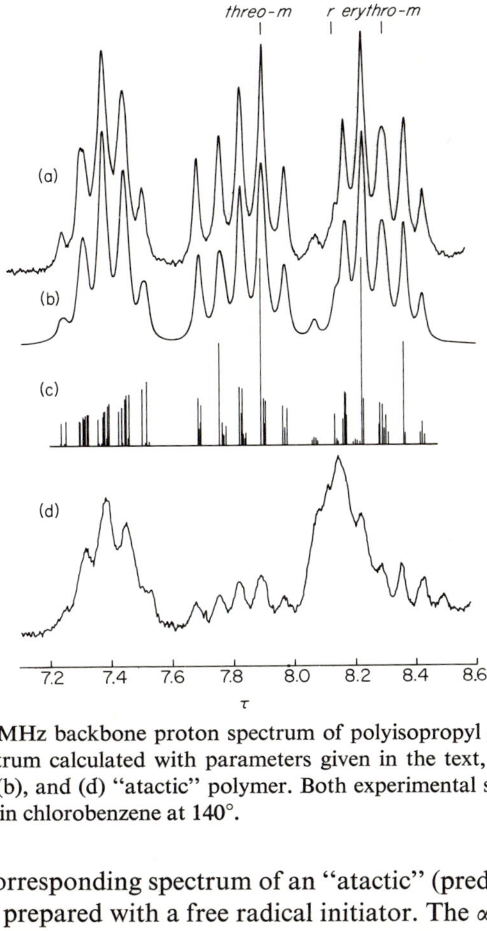

FIG. IV-1. 100 MHz backbone proton spectrum of polyisopropyl acrylate: (a) isotactic polymer, (b) spectrum calculated with parameters given in the text, (c) "stick" spectrum corresponding to (b), and (d) "atactic" polymer. Both experimental spectra were obtained on 10% solutions in chlorobenzene at 140°.

IV-1(d) is the corresponding spectrum of an "atactic" (predominantly syndiotactic) polymer prepared with a free radical initiator. The α-proton resonance is centered near 7.42τ in both spectra, being insensitive to configuration. It is the pentuplet expected for approximately equal couplings (see below) to four neighboring β-protons. The β-proton multiplets at 7.86 and 8.31τ have been identified by Yoshino *et al.*[5] as *syn* and *anti* (or *erythro* and *threo*), respectively, (Chapter II, Section 2) from an ingenious study of model compounds.

The calculation of such polymer spectra poses the problem of representing the main-chain protons by a limited but adequate spin system, since it is obviously impossible (and unnecessary) to include *all* the spins in a regularly repeating structure such as this.

It is possible to reproduce such spectra with reasonable accuracy by treating the α- and β-protons as five-spin and four-spin systems respectively. A more satisfactory approach however, is to treat the polymer chain as a six-spin "dimer."[10, 11] This representation should not be taken too literally in the structural sense since it is primarily a spin bookkeeping system. It is realistic because J couplings through more than three intervening bonds are very small

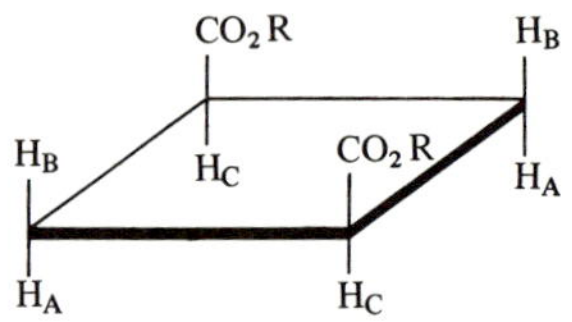

in such polymer chains (but see p. 140). Spectrum (b) in Fig. IV-1 was calculated using this model and the following parameters:

$$\tau_A = 8.31 \qquad \tau_B = 7.86 \qquad \tau_C = 7.42$$

$$J_{AB} = -14.0 \text{ Hz} \qquad J_{AC} = 6.0 \text{ Hz} \qquad J_{BC} = 7.5 \text{ Hz}$$

Line width at half-height: 2.4 Hz

In spectrum (d) the r methylene resonance appears as a poorly resolved multiplet centered at 8.16τ. A residual resonance occurs in nearly this same position in spectrum (a). This multiplet can be satisfactorily reproduced with the following spin model parameters:

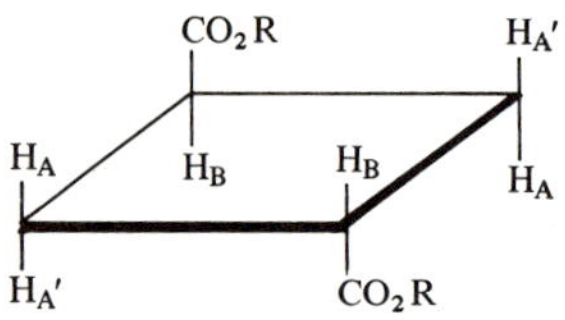

$$\tau_A = 8.16 \qquad \tau_B = 7.42$$

$$\tfrac{1}{2}(J_{AB} + J'_{AB}) = 7.0 \text{ Hz}$$

Line width at half-height: 2.4 Hz

In an earlier calculation on the same polymer, based on its 60 MHz spectrum,[3] slightly different parameters were used, and no r units could be detected. (The computed spectrum was also a significantly less satisfactory match to the experimental, mainly because the lower observing frequency made the six-spin model a less satisfactory approximation.) The good match shown in Fig. IV-1 could be achieved only by assuming 5% r units, i.e., that

$$\frac{(m)}{(m)+(r)} = 0.95$$

The polymer is probably not strictly Bernoullian in configurational statistics, and so cannot be designated in terms of P_m (Chapter III, Section 4). The free radical polymer, which is very probably Bernoullian, had a P_m of approximately 0.30.

The study of the mechanism of addition in acrylate polymerization using β-labelled monomers will be discussed in Chapter VIII. In Section 3B of Chapter IX, the interpretation of the above and additional NMR data are discussed in relation to chain conformation in solution.

3. Polyacrylamides

Polyacrylamide[12] and poly-*N*,*N*′-dimethylacrylamide[13] have been observed by NMR not with the purpose of determining their stereochemical configuration (which is no doubt similar to that of free radical polyacrylates) but in order to study the rotational mobility of the side chains. It is well known† that in the amide group the C—N bond has partial double bond character.

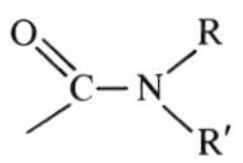

The substituents R and R′, whether protons or alkyl groups, occupy different magnetic environments which are exchanged by rotation about the C—N bond. Because the activation energy for this process is about 20 kcal, the exchange is slow enough that the R groups give separate resonances at room temperature. As the temperature is increased, the peaks broaden and then coalesce; since most amides exhibit about the same rotational barrier, the coalescence temperature depends primarily on the separation of the R-group resonances (in Hz).

The NH_2 spectrum of polyacrylamide in aqueous solution exhibits a doublet resonance in contrast to the broad singlet shown by some small molecule amides (see Fig. XIII-9). This is attributed to the relatively slow reorientation rate of the polymer chain, permitting more effective coupling of the ^{14}N nucleus to the motions of the molecular framework and thus nearly completely abolishing its coupling to the protons (see Chapter I, Section 10 and Chapter XIII, Section 3). Measurement of the collapse of this doublet as a function of temperature gave a rotational activation energy of 10.5 ± 1 kcal at *p*H 4.5, probably an underestimate due to spurious peak broadening, but at least no higher than for amide groups not attached to polymer chains. At this *p*H, the rotation is more rapid than proton exchange between the amide group and water. Rotation was found not to be catalyzed by acid or base, whereas exchange is catalyzed by both and becomes more rapid than rotation at high and low *p*H. The two amide protons exchange at equal rates.

† For a review, see Robin *et al.*[14] and also Chapter XIII, Section 3.

Comparison of the 60 MHz spectra of the *N*-methyl groups of poly-*N*,*N'*-dimethylacrylamide[13] (a) with that of the model compound *N*,*N'*-dimethylisobutyramide (b) showed that the polymer spectrum is severely perturbed by

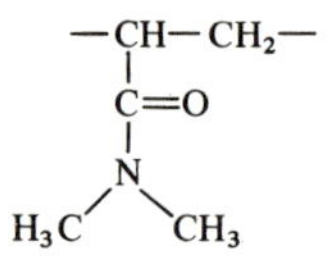

(a)

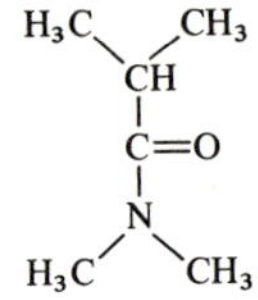

(b)

effects of stereochemical configuration. Instead of a doublet, the *N*-methyl resonance is a singlet with a doublet appearing as two shoulders on the downfield side. The spectrum of a 90:10 acrylic acid:*N*,*N'*-dimethylacrylamide copolymer was similar but with a greater separation of singlet and doublet. Evidently, in one of the triad sequences (it is not clear which) the methyl group resonances happen to coincide in chemical shift.

Measurement of the collapse of the doublets gave an activation free energy of about 19 kcal, equal within experimental error to that for the model compound. There was thus no evidence for interlocking of the groups or for cooperative rotation in the polymer chain.

4. Polyacrylonitrile

The NMR spectrum of polyacrylonitrile[15–22] has proved particularly susceptible to interpretive errors, despite study of dimer[23–26] and trimer[26,27] model compounds (Chapter III, Section 2; Chapter IX, Section 2). The difficulty is in the understanding of the β-methylene proton spectrum, aggravated in this instance by a peculiar sensitivity to the choice of observing solvent.[20] These difficulties have been thoroughly discussed,[21] and will not be reviewed here. Correct analyses in terms of configuration appear to have been reported by Svegliado *et al.*[19] and by Yoshino *et al.*[21] who find that the *m* dyads are observably heterosteric at 100 MHz. The β-methylene spectra of polymers prepared with free radical initiators at 120, 35, and −78° are shown in Fig. IV-2.[19] The multiplet can be analyzed into a *racemic* methylene triplet and the AB part of an ABX_2 multiplet corresponding to the *meso* protons. This analysis is supported convincingly by Yoshino's[21] spectrum of the polymer prepared from the α,β-d_2 monomer, in which all J coupling is effectively eliminated. The *meso* multiplet is reduced to two equal singlets (Fig. IV-3) corresponding to equal probability of *cis* and *trans* double bond

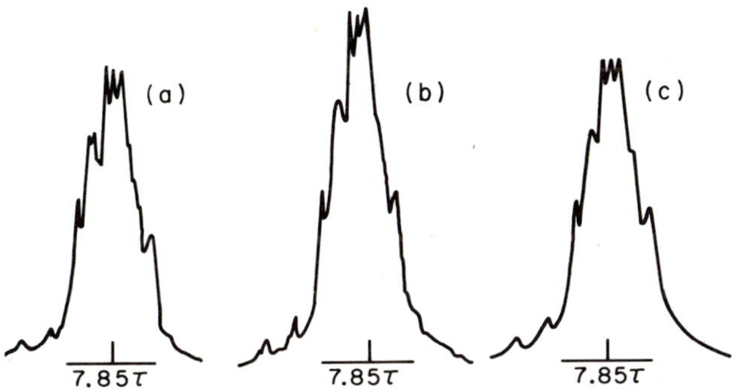

FIG. IV-2. 100 MHz spectra of the β-methylene protons of polyacrylonitrile prepared with free radical initiators at (a) 120°, (b) 35°, and (c) −78°. Spectra observed using 10% (w/v) solutions in d_6-DMSO (Svegliado *et al.* [19]).

opening in free-radical propagation, as we shall see in Chapter VIII. The *racemic* resonance is the expected singlet, tetrad sequences not being resolved in these spectra; the $(m)/(r)$ ratio is about 1:1, and so $P_m \simeq 0.5$, corresponding to random propagation.

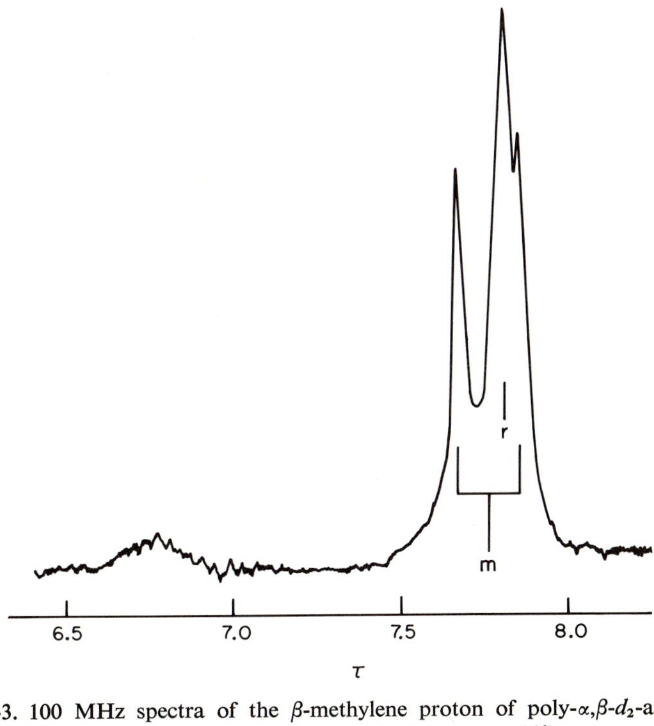

FIG. IV-3. 100 MHz spectra of the β-methylene proton of poly-α,β-d_2-acrylonitrile, observed in 5% solution in aqueous perchloric acid (Yoshino *et al.*[21]).

Bracke *et al.*[22] have reported the spectrum of a head-to-head oligomer of polyacrylonitrile with hydroxy end groups.

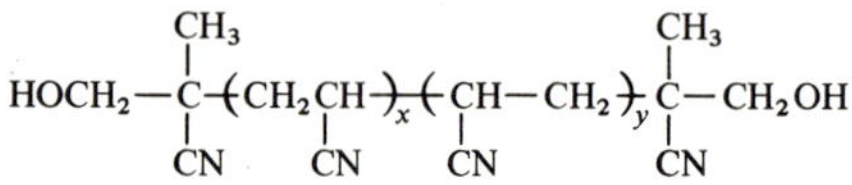

The effect of temperature on the NMR spectrum is discussed in Chapter VIII, Section 2D.

5. Polyvinyl Alcohol and Its Esters

The NMR spectra of polyvinyl alcohol[28–32] and its acetate,[33, 34, 37, 40] trifluoroacetate,[29, 34, 38, 39] and formate[40] esters are all rather similar. In the 100 MHz spectrum of relatively high molecular weight polyvinyl alcohol at 37°, and α- and β-proton multiplets are unresolved; irradiation of the β-protons reveals three peaks in the α-proton spectrum. Upon raising the

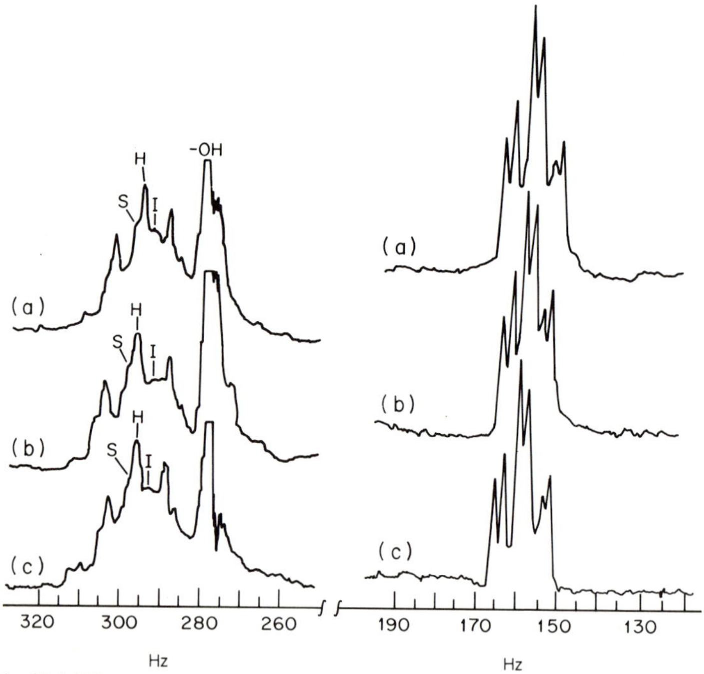

FIG. IV-4. 60 MHz spectra of the α-protons (left) and β-protons (right) of polyvinyl alcohols derived from polyvinyl acetate prepared (a) in bulk, (b) in *n*-butyraldehyde, and (c) in methyl ethyl ketone. Spectra observed in D_2O at 175°; peak positions in the downfield from an unspecified reference (Friedlander *et al.*[31]).

observing temperature and particularly upon cleaving the polymer chains with periodic acid,[31, 32] much better resolved spectra are obtained, as shown in Fig. IV-4. (Periodic acid cleaves the main chain at the anomalous head-to-head 1,2-glycol units, which occur to the extent of about 1–2%.[41]) The α-proton spectrum appears as three overlapping pentuplets,[31] and the β-protons as two triplets[30, 31] which, in polymers prepared from polyvinyl acetate made with a free radical initiator, are of nearly equal intensity. The appearance of triplets for both *r* and *m* dyads indicates that the latter is not observably heterosteric at frequencies as high as 100 MHz. Absolute configuration determinations are therefore based on the configuration of the polymer (acetate, trifluoroacetate, or *tert*-butyl ether) from which the polyvinyl alcohol was prepared, which may be known from NMR[29, 31, 35] or x-ray[36] measurements. In this way, the α-proton multiplets are assigned to *rr*, *mr*, and *mm* triads and β-methylene triplets to *m* and *r* dyads in order of increasing shielding.

From the close similarity of the multiplets in Fig. IV-4, Friedlander *et al.*[31] concluded that the solvent had little or no influence on the stereochemistry of the free radical polyvinyl acetate precursor; in particular isobutyraldehyde did not produce the previously claimed increase in stereoregularity.[42] They also observed little or no effect of polymerization temperature, the polymer being very nearly atactic at all temperatures (see Chapter VIII, Section 2E).

At lower frequencies, the spectrum of polyvinyl acetate shows unresolved α- and β-proton multiplets.[33, 34] Upon irradiation of the β-protons, the α-proton resonance collapses to three singlets which are assigned to *mm*, *mr*, and *rr* in order of increasing shielding.[34] The methyl resonance is clearly split into triad peaks.[33–35] Because the β-methylene spectrum is again not very informative, primary reliance has been placed on x-ray[38] and infrared[35, 38] measurements both on the polymer itself and on others to which it could be securely related chemically without racemization. It is concluded that, as shown in Fig. IV-5, the methyl shielding in $CDCl_3$ solution increases in the order *mm*, *mr*, *rr*, the same as for the α-protons. It seems very probable that future work at higher frequencies will permit the absolute determination of configuration without reference to other measurements.

The α-proton[34] and ^{19}F spectra[38, 39] of polyvinyl trifluoroacetate have been reported. The α-protons appear to increase in shielding in the same order as in polyvinyl acetate: *mm*, *mr*, *rr*; and the ^{19}F was reported by Pritchard *et al.*[39] to behave in the same way. However, Fujii *et al.*[38] observed more than three peaks in the ^{19}F spectrum and were unable to reach a definite conclusion. It seems probable, however, that the assignment of Pritchard *et al.*[39] is correct in general, although there may be complications from longer sequences.

Ramey *et al.*[40] reported that the 100 MHz spectrum of polyvinyl formate in d_6-acetone revealed no configurational fine structure in the α- or β-proton region even upon decoupling. However, the formyl proton at 2τ proved to be

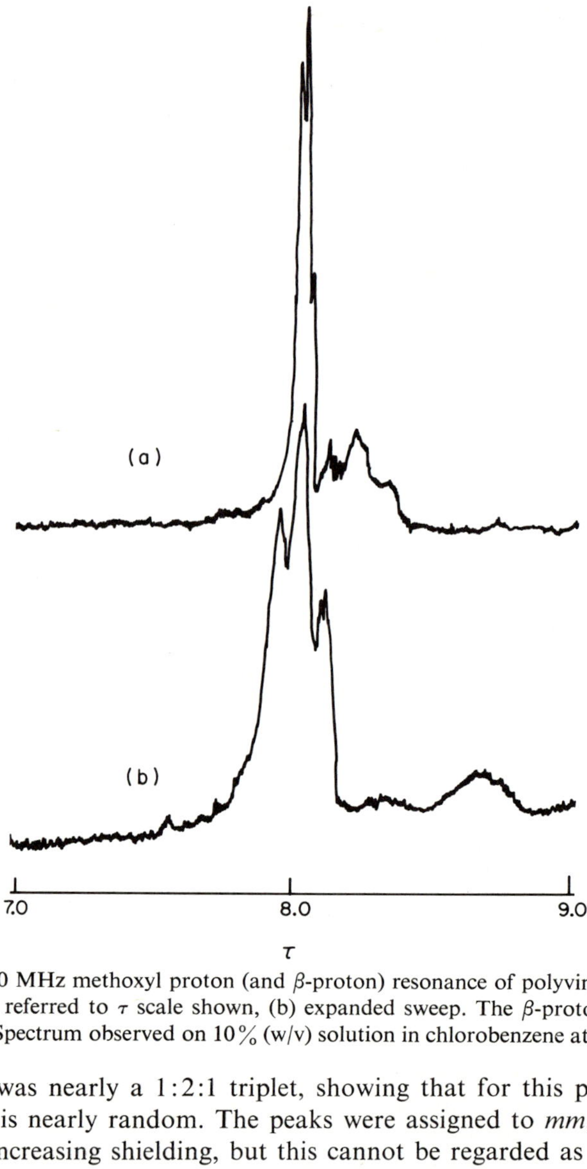

FIG. IV-5. 60 MHz methoxyl proton (and β-proton) resonance of polyvinyl acetate: (a) normal sweep, referred to τ scale shown, (b) expanded sweep. The β-proton multiplet is at about 8.2τ. Spectrum observed on 10% (w/v) solution in chlorobenzene at about 50°.

sensitive. It was nearly a 1:2:1 triplet, showing that for this polymer also propagation is nearly random. The peaks were assigned to *mm*, *mr*, and *rr* in order of increasing shielding, but this cannot be regarded as established.

6. Polyvinyl Methyl Ether and Other Vinyl Ethers

Polyvinyl methyl ether shows a typical quartet-singlet β-methylene spectrum when the α-protons are irradiated,[33] and hence its stereoregularity can be readily determined. The methoxyl resonance is not well resolved in some solvents, particularly aromatic solvents,[33,43,44] but in nitromethane[45] and chlorinated aliphatic solvents[44,45] three signals can be seen which correspond to *rr*, *mr*, and *mm* triads in order of increasing shielding.

It has long been known that polyvinyl methyl ether and its homologs can be polymerized with cationic initiators such as BF_3·etherate to yield either soft and rubbery products or hard, crystalline materials, depending upon the conditions of reaction. It was early recognized that these differences reflected chain stereochemistry, the crystalline materials being, as judged from x-ray diffraction results, predominantly isotactic.[46,47] More recently, stereoregular polymers have been prepared with Ziegler initiators.[48] NMR studies confirmed the earlier conclusions concerning the isomeric forms of these polymers. Brownstein and Wiles[45] found that in the crystalline materials, the *mm* methoxyl peak is strongly predominant; more rubbery materials have a more irregular structure, but *mm* predominates to some extent in all the materials reported.

Ohsumi and co-workers[49,50] employed the methoxyl resonance to elucidate the mechanism of propagation in the cationic polymerization of vinyl methyl ether. They found that with BF_3·etherate and $SnCl_4 \cdot CCl_3CO_2H$ initiators, the propagation was not Bernoullian, whereas the polymer prepared with a sulfuric acid-aluminium sulfate initiator apparently was. These results were interpreted in terms of penultimate effects versus an enantiomorphic catalyst site model, respectively; such conclusions should be treated with reserve until longer sequences (see Chapter VIII) can be examined, for it may be that the apparently Bernoullian statistics are actually more complex.[51]

The ^{13}C spectrum of polyvinyl methyl ether has been observed.[52,53] The spectrum of a predominantly isotactic polymer is shown in Fig. IV-6. In spectrum (a), the ^{13}C–H couplings have been completely removed by noise decoupling, as explained in Chapter I, Section 18B. The peaks appear as singlets on the scale shown. In (b), the protons are partially decoupled by irradiation off resonance. This, as we have seen in Chapter I, has the effect of collapsing the spacing of the multiplets (by a factor of about five in the present case) while preserving their form and enables one to identify the peaks by their multiplicity without any confusion from overlapping. Complete decoupling increases the effective signal-to-noise ratio and is normally employed once the chemical shifts have been assigned in a general way. In Fig. IV-6(a), the $^{13}CH_2$ resonance appears as two peaks at 152 and 154 ppm (upfield from $^{13}CS_2$), assigned on the basis of their relative intensities and the proton spectrum[33] to *r* and *m* dyads, respectively. The methoxyl peak at 137 ppm appears to be the most sensitive to stereochemical configuration. This is not evident in Fig. IV-6(a) but is clear from the expanded methoxyl spectrum shown in Fig. IV-7(b). At least eight peaks can be distinguished, evidently

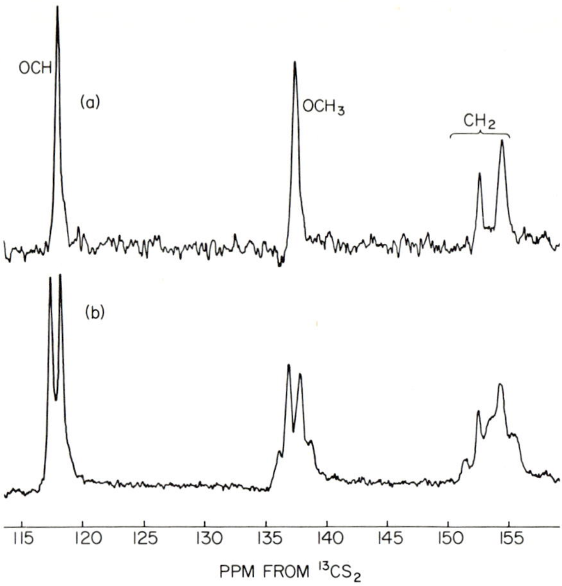

FIG. IV-6. Natural abundance ^{13}C spectrum (25 MHz) of polyvinyl methyl ether: (a) protons completely decoupled (50 scans); (b) decoupling "off resonance" (1038 scans). Spectra observed on 50% (w/v) solution in chlorobenzene at about 50°.

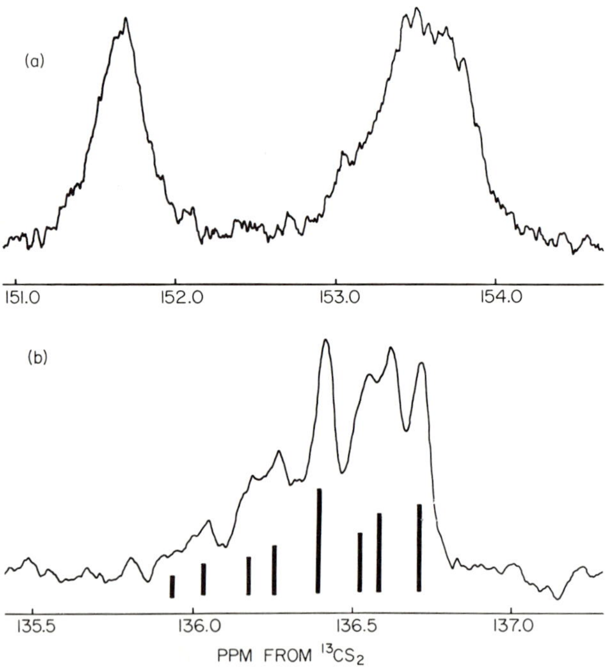

FIG. IV-7. Expanded ^{13}C spectrum of (a) the β-methylene carbons (2128 scans) and (b) the methoxyl carbons of polyvinyl methyl ether (200 scans), same conditions as in Fig. IV-6(a). The noise level in (b) is relatively high; bars mark features believed to be real peaks.

associated with pentad sequences. Here, then, is one means for the more searching statistical study suggested previously. Figure IV-7(a) shows the β-methylene ^{13}C spectrum similarly expanded. Substructure is observable in the *m* peak.

A number of higher homologs of polyvinyl methyl ether have been studied, including ethyl,[54] isopropyl,[54] isobutyl,[43] and *tert*-butyl.[43,54] In general, their spectra have similar interpretations but there are complications from overlapping side-chain resonances.

7. Polyvinyl Formal

Polyvinyl formal is one of the simplest examples of vinyl polymers in which the α-carbons are connected by rings.

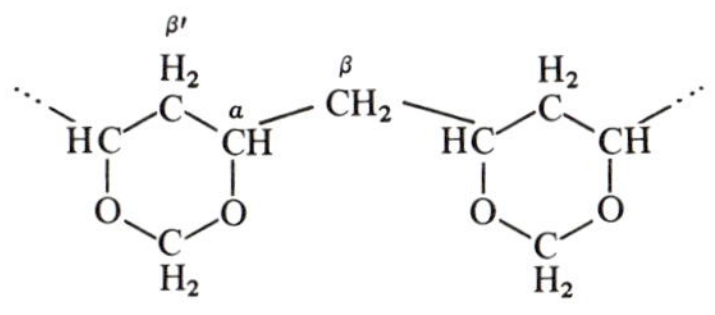

We have already seen in Chapter II (Section 2) that with the formation of rings the α-carbons become true asymmetric centers. Each *racemic* (syndiotactic) pair of neighboring hydroxyl groups, upon reacting with formaldehyde, gives a *trans*-1,3-dioxane ring. This isomerizes rapidly between two equal-energy forms ("identomers"; see Chapter IX, Section 2), making both the formal and β'-methylene groups homosteric (a). (In rings formed at the center

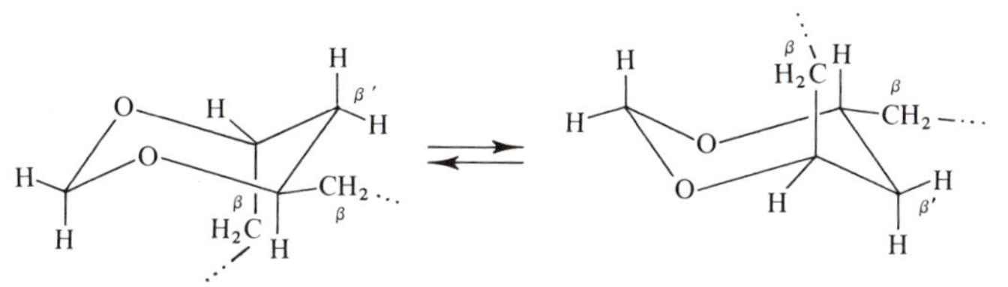

(a)

units of *rrr* tetrads or longer *r* sequences, the β-methylenes will also be effectively homosteric). Each *meso* (isotactic) pair of hydroxyls, however, will form a *cis*-dioxane, which will strongly prefer the conformation in which the β-CH_2 groups are *eq-eq* (both equatorial) (b). The formal and β'-methylene

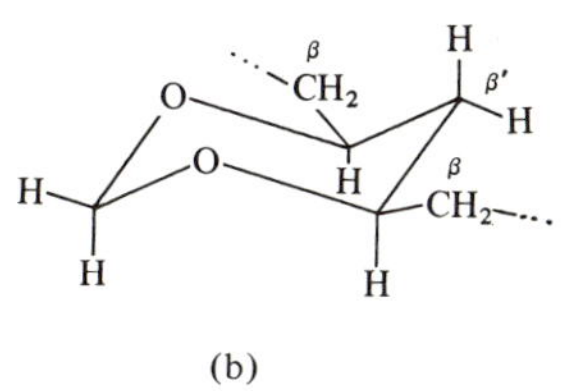

(b)

groups are now heterosteric [and in fact would be so for any ratio of *eq-eq* and *ax-ax* (both axial) conformers]. The formal group is the most clearly distinguished and measured;[55,56] in Fig. IV-8 is shown the 100 MHz spectrum

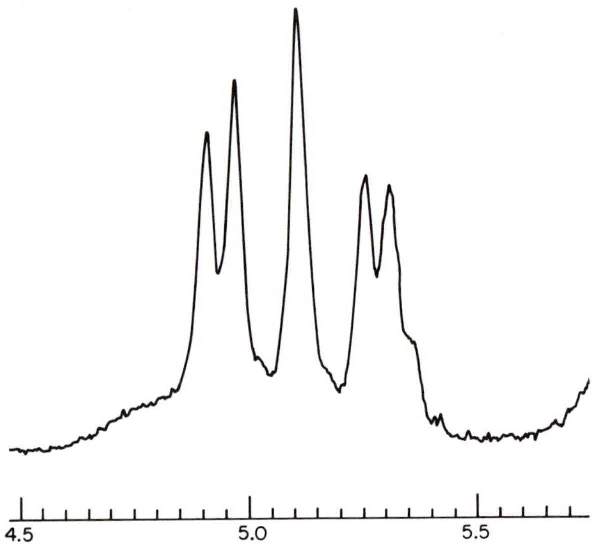

FIG. IV-8. 100 MHz formal proton spectrum of polyvinyl formal, observed in 10% (w/v) solution in $CDCl_3$ at 80° (Bovey and Ryan[56]).

of polyvinyl formal in $CDCl_3$ solution.[56] The formal protons exhibit the expected quartet and singlet, the ratio of which is the $m:r$ ratio in the original polyvinyl alcohol.

In the above representation of the *racemic* or *trans*-dioxane rings, we have shown only one of the two possible enantiomers. In an achiral reaction environment, these will be formed with equal probability, but (depending upon the details of the reaction mechanism) with varying frequencies of *ddd*, *ddl*, and *ldl* sequences (and their antipodes). The occurrence of these diastereoisomeric sequences may not prove readily detectable, but may broaden the peaks. There are indeed hints of additional fine structure in Fig. IV-8. At least a part of this, however, may arise from incomplete formalization of the hydroxyls.

8. Polymers of α,β-Disubstituted Vinyl Ethers (Vinylene Ethers)

The configurational isomerism possible in polymers of α,β-disubstituted vinyl monomers (termed vinylene monomers when neither substituent is D) have been presented in Chapter II (Section 2). When the β-substituent is D, we can still obtain absolute configurational information (and mechanistic information as well; see Chapter VIII). But in all other polymers of this class we cannot. There is now no fundamental distinction in the symmetry properties of the α- and β-protons or substituents, and in fact the labeling as "α" or "β" is arbitrary. Only odd sequences have observational significance. Their numbers are given by (see p. 79)

n	1	3	5	7	···
$N(n)$	2	6	20	72	···

or

$$N(n) = 2^{n-1} + 2m \tag{IV-1}$$

where $n = 2m + 1$. Thus for triads, we have the arrangement shown. An

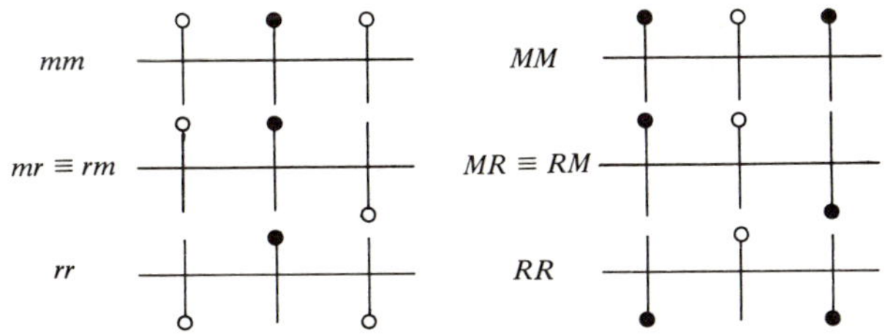

alternative scheme of nomenclature as well as a more extensive statistical treatment have been offered.[51] In terms of spin models, one can readily represent *erythro*-diisotactic, *threo*-diisotactic, and disyndiotactic chains in terms of four-spin "dimer" models (Section 2), giving AA′BB′ spectra. Less stereoregular chains will require for their representation the summation of these and two additional ABCD spectra corresponding to "dimer" models with lower symmetry. Since no computer simulations of any α,β-disubstituted polymer spectra have yet been reported, we shall not enlarge further on these matters here.

From J couplings and reasonable assumptions concerning conformational preferences (Chapter IX) one can in principle distinguish disyndiotactic from diisotactic configurations, but of the latter one cannot *a priori* tell *erythro* from *threo*, information which is essential to determining the direction of addition to the double bond (Chapter VIII). Heavy reliance on infrared and (particularly) x-ray diffraction results is therefore characteristic of NMR studies of these polymers; NMR supplies quantitative detail which the other methods cannot.

A. Propenyl Ethers

The chain unit has the structure shown. Natta *et al.*[57,58] have determined

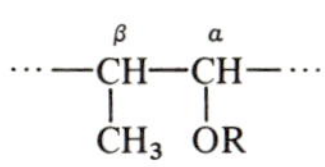

by x-ray diffraction that the cationic polymerization of *trans*-propenyl isobutyl ether yields the *threo*-diisotactic polymer, double bond opening therefore being *cis* (Chapter VIII). This furnishes an important clue to the interpretation of the spectra of propenyl methyl ether polymers observed by Ohsumi *et al.*[59, 60] They found that the crystalline polymers obtained from the *trans*-rich monomer by homogeneous cationic initiation gave a predominantly singlet methoxyl and doublet methyl resonance, and are presumably *threo*-diisotactic, whereas the crystalline polymer obtained from the *cis* monomer by heterogeneous catalysis was *erythro*-diisotactic. The further conclusion was drawn that in the amorphous polymer obtained from the *cis* monomer by homogeneous catalysis, the α-carbons were isotactic and the β-carbons atactic, corresponding to a complex pattern of bond opening. It is rather difficult, however, to be sure of conclusions of this latter sort because of the lack of absolute configurational clues by which one can tell whether it is the α- or the β-carbons which are stereoregular.

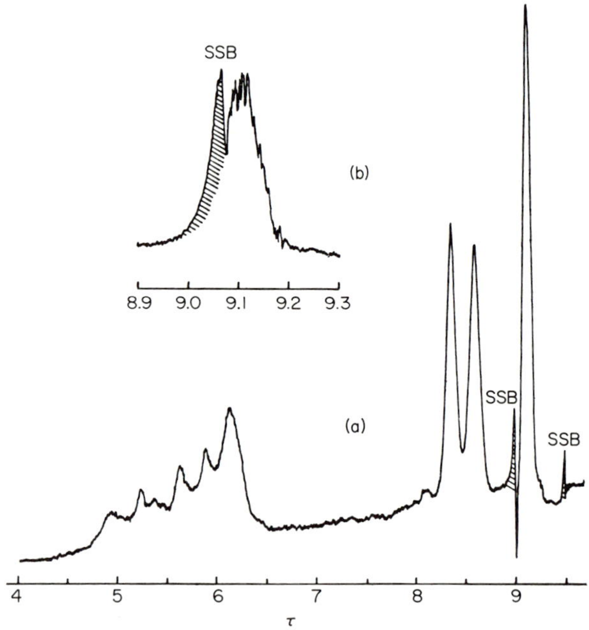

Fig. IV-9. 220 MHz spectra of poly-*cis*-β-chlorovinyl *n*-butyl ether, observed at 125° in 10% (w/v) chlorobenzene solution. The main-chain protons are the least shielded; the side-chain protons increase in shielding from α′-(next to oxygen) to δ- (methyl). The expanded spectrum (b) of the δ-protons is shown above; "ssb" is a spinning sideband (Goodman *et al.*[62]).

B. β-CHLOROVINYL ETHERS

Natta *et al.*[61] reported that *cis*- and *trans*-β-chlorovinyl *n*-butyl ether and *trans*-β-chlorovinyl isobutyl ether polymerize with cationic initiators at low temperature to yield crystalline polymers. They found the *cis*-β-chlorovinyl isobutyl ether and both the *cis*- and *trans*-β-chlorovinyl ethyl ethers gave polymers which could not be crystallized. The configurations of the crystalline polymers corresponded to *cis* double bond openings, as with the propenyl ethers.

NMR study of the noncrystallizable polymers has shown them to be stereoirregular but a quantitative interpretation is not yet possible.[62] The 220 MHz spectrum of the *cis*-β-chlorovinyl *n*-butyl ether polymer is shown in Fig. IV-9. The side-chain methylene protons give apparently broad multiplets representing the overlap of many chemical shifts corresponding to triad and higher sequences. Only in the terminal methyl group can structure be resolved, because here sequences longer than triads do not broaden the multiplet peaks sufficiently to obscure them.

REFERENCES

1. T. Yoshino, Y. Kikuchi, and J. Komiyama, *J. Phys. Chem.* **70**, 1059 (1966).
2. F. Heatley and F. A. Bovey, *Macromolecules* **2**, 303 (1968).
3. C. Schuerch, W. Fowells, A. Yamada, F. A. Bovey, and F. P. Hood, *J. Amer. Chem. Soc.* **86**, 4481 (1964).
4. T. Yoshino, J. Komiyama, and M. Shinomiya, *J. Amer. Chem. Soc.* **86**, 4482 (1964).
5. T. Yoshino, M. Shinomiya, and J. Komiyama, *J. Amer. Chem. Soc.* **87**, 387 (1965).
6. T. Yoshino and K. Kuno, *J. Amer. Chem. Soc.* **87**, 4404 (1965).
7. T. Yoshino and J. Komiyama, *J. Amer. Chem. Soc.* **88**, 176 (1966).
8. T. Yoshino and J. Komiyama, *J. Polym. Sci. Part B* **4**, 991 (1966).
9. K. Matsuzaki, T. Uryu, A. Ishida, T. Ohki, and M. Takeuchi, *J. Polym. Sci. Part A-1* **5**, 2167 (1967).
10. W. C. Tincher, *J. Polym. Sci.* **62**, S148 (1962).
11. W. C. Tincher, *Makromol. Chem.* **85**, 20 (1965).
12. F. A. Bovey and G. V. D. Tiers, *J. Polym. Sci. Part A* **1**, 849 (1963).
13. Y. Miron and H. Morawetz, *Macromolecules* **2**, 162 (1969).
14. M. B. Robin, F. A. Bovey, and H. Basch, *in* "The Chemistry of Amides" (J. Zabicky, ed.). Wiley (Interscience), New York, 1970.
15. K. Matsuzaki, T. Uryu, K. Ishigure, and M. Takeuchi, *J. Polym. Sci. Part B* **4**, 93 (1966).
16. J. Bargon, K. H. Hellwege, and U. Johnsen, *Kolloid Z.* **212**, 51 (1966)
17. M. Murano and R. Yamadera, *J. Polym. Sci. Part B* **5**, 333 (1967).
18. R. Yamadera and M. Murano, *J. Polym. Sci. Part A-1* **5**, 1059 (1967).
19. G. Svegliado, G. Talamini, and G. Vidotto, *J. Polym. Sci. Part A-1* **5**, 2875 (1967).
20. M. Murano, *Makromol. Chem.* **112**, 281 (1968).
21. T. Yoshino, H. Kenjo, and K. Kuno, *J. Polym. Sci. Part B* **5**, 703 (1967).
22. W. Bracke, J. A. Empen, and C. S. Marvel, *Macromolecules* **1**, 465 (1968).

23. P. E. McMahon and W. C. Tincher, *J. Mol. Spectrosc.* **15**, 180 (1965).
24. H. G. Clark, *Makromol. Chem.* **63**, 69 (1963).
25. R. Yamadera and M. Murano, *J. Polym. Sci. Part A-1* **5**, 1855 (1967).
26. M. Murano and R. Yamadera, *J. Polym. Sci. Part B* **5**, 483 (1967).
27. M. Murano and R. Yamadera, *J. Polym. Sci. Part A-1* **6**, 843 (1968).
28. A. Danno and N. Hayakawa, *Bull. Chem. Soc. Jap.* **35**, 1748 (1962).
29. K. C. Ramey and N. D. Field, *J. Polym. Sci. Part B* **3**, 63 (1965).
30. J. Bargon, K. H. Hellwege, and U. Johnsen, *Makromol. Chem.* **85**, 291 (1965).
31. H. N. Friedlander, H. E. Harris, and J. G. Pritchard, *J. Polym. Sci. Part A-1* **4**, 649 (1966).
32. W. C. Tincher, *Makromol. Chem.* **85**, 46 (1965).
33. F. A. Bovey, E. W. Anderson, D. C Douglass, and J. A. Manson, *J. Chem. Phys.* **39**, 1199 (1963).
34. K. C. Ramey and N. D. Field, *J. Polym. Sci. Part B* **3**, 69 (1965).
35. S. Murahashi, S. Nozakura, M. Sumi, H. Yuki, and K. Hatada, *J. Polym. Sci. Part B* **4**, 65 (1966).
36. K. Fujii, Y. Fujiwara, and S. Fujiwara, *Makromol. Chem.* **89**, 278 (1965).
37. K. C. Ramey and D. C. Lini, *J. Polym. Sci. Part B* **5**, 39 (1967).
38. K. Fujii, S. Brownstein, and A. M. Eastham, *J. Polymer Sci. Part A-1* **6**, 2387 (1968).
39. J. G. Pritchard, R. L. Vollmer, W. C. Lawrence, and W. B. Black, *J. Polym. Sci. Part A-1* **4**, 707 (1966).
40. K. C. Ramey, D. C. Lini, and G. L. Statton, *J. Polym. Sci. Part A-1* **5**, 257 (1967).
41. P. J. Flory and F. S. Leutner, *J. Polym. Sci.* **3**, 880 (1948); **5**, 267 (1950).
42. I. Rosen, P. H. Burleigh, and J. F. Gillespie, *J. Polym. Sci.* **54**, 31 (1961).
43. R. J. Kern, J. J. Hawkins, and J. D. Calfee, **66**, 126 (1963).
44. K. C. Ramey, N. D. Field, and I. Hasegawa, *J. Polym. Sci. Part B* **2**, 865 (1964).
45. S. Brownstein and D. M. Wiles, *J. Polym. Sci. Part A* **2**, 1901 (1964).
46. C. E. Schildknecht, S. T. Gross, and A. O. Zoss, *Ind. Eng. Chem.* **41**, 1998 (1949).
47. C. E. Schildknecht, A. O. Zoss, and F. Grosser, *Ind. Eng. Chem.* **41**, 2891 (1949).
48. E. J. Vandenberg, R. F. Heck, and D. S. Breslow, *J. Polym. Sci.* **41**, 519 (1959).
49. Y. Ohsumi, T. Higashimura, and S. Okamura, *J. Polym. Sci. Part A-1* **5**, 849 (1967).
50. T. Higashimura, Y. Ohsumi, K. Kuroda, and S. Okamura, *J. Polym. Sci. Part A-1* **5**, 863 (1967).
51. H. L. Frisch, C. L. Mallows, and F. A. Bovey, *J. Chem. Phys.* **45**, 1565 (1966).
52. F. A. Bovey, *Polymer Preprints* (*Amer. Chem. Soc.*) **10**, 9 (1969).
53. L. F. Johnson, F. Heatley, and F. A. Bovey, *Macromolecules* **3**, 175 (1970).
54. K. C. Ramey, N. D. Field, and A. E. Borchert, *J. Polym. Sci. Part A* **3**, 2885 (1965).
55. K. Fujii, K. Shibatani, Y. Fujiwara, Y. Ohyanagi, J. Ukida, and M. Matsumoto, *J. Polym. Sci. Part B* **4**, 787 (1966).
56. F. A. Bovey and J. J. Ryan, unpublished observations.
57. G. Natta, *J. Polym. Sci.* **48**, 219 (1960).
58. G. Natta, M. Farina, and M. Peraldo, *Chim. Ind.* (*Milan*) **42**, 255 (1960).
59. Y. Ohsumi, T. Higashimura, S. Okamura, R. Chujo, and T. Kuroda, *J. Polym. Sci. Part A-1* **5**, 3009 (1967).
60. Y. Ohsumi, T. Higashimura, S. Okamura, R. Chujo, and T. Kuroda, *J. Polym. Sci. Part A-1* **6**, 3015 (1968).
61. G. Natta, M. Peraldo, M. Farina, and G. Bressan, *Makromol. Chem.* **55**, 139 (1962).
62. M. Goodman, V. Y. Chen, J. J. Ryan, F. Heatley, and F. A. Bovey, unpublished observations.

Chapter V

HALOETHYLENE POLYMERS

1. Polyvinyl Chloride

Interpretation of the NMR data for polyvinyl chloride has been protracted and controversial, in part because of ambiguities in the spectra and in part because of conflicting conclusions drawn from other measurements, particularly from vibrational spectroscopy (infrared and Raman). These ambiguities and disagreements appear to have been largely resolved by NMR observations at higher magnetic fields. Study of model 2,4-dichloropentanes and 2,4,6-trichloroheptanes (Chapter III, Section 2; Chapter IX, Section 2) has been very helpful in deducing the conformation of the polymer chain, but at one time caused some confusion with regard to its stereochemical configuration. This concerned mainly the β-methylene spectrum, in which the chemical shift difference between the protons decreases with the number of neighboring *m* dyads. We shall not discuss here all the contributions to this problem, which have been extensive,[1–24] but will instead emphasize the earliest and most recent results.

As we have seen in the previous chapter, one would expect on the simplest basis that the α-proton spectrum would be a pentuplet from coupling to the four β-protons and the β-protons a triplet from coupling to the two α-protons. In fact, this is not observed even at 60 MHz. Johnsen[8] reported that at this frequency the α-proton spectrum of polyvinyl chloride in chlorobenzene at 160° gave an apparent "pentuplet" at 5.53τ, as expected, but that the β-protons appeared as five peaks near 8.8τ. These were interpreted as two overlapping triplets centered at 7.78 and 7.96τ, corresponding to *meso* and *racemic* methylene groups, respectively. This view was questioned,[1, 11, 13] because it interpreted the *meso* methylene groups as homosteric, in contradiction to basic principles. However, it received confirmation from double resonance experiments,[3, 12] (Chapter I, Section 18B) as shown in Fig. V-1. (The polymers

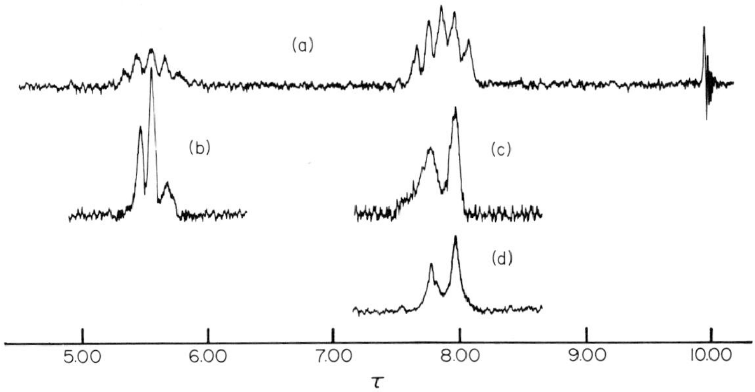

FIG. V-1. Normal and decoupled spectra of polyvinyl chloride, observed in 15% (w/v) solution in chlorobenzene. (a) normal spectrum (at 170°); (b) α-protons upon irradiation of β-protons (150°); (c) β-protons upon irradiation of α-protons (150°); (d) poly-α-d_1-vinyl chloride (156°) (Bovey *et al.*[12]).

illustrated were prepared with a free radical initiator at 50°.) The normal spectrum is shown in (a). When the β-protons are decoupled from the α-protons two peaks are observed, separated by 0.20 ppm [spectrum (c)]. A very similar spectrum is shown by the polymer of α-deuterovinyl chloride, $ClDC{=}CH_2$ [spectrum (d)]. When the β-protons are irradiated the α-proton resonance shows peaks at 5.48, 5.59, and 5.71τ [spectrum (b)]. The dependence of the decoupled α-proton spectrum upon the decoupling difference frequency $\Delta\nu$ was employed to demonstrate that these peaks correspond to *rr*, *mr*, and *mm* triads, respectively, assuming the correctness of the dyad assignments, which are compatible with model compound measurements. Relative peak areas in the α- and β-proton spectra indicated the polymer to be nearly atactic with a small preponderance of *r* sequences.

Careful examination[20] of spectrum (d) in Fig. V-1 shows that a dyad analysis is inadequate, because, as previously observed,[11, 13] there are small peaks which cannot be accounted for in this way. Yoshino and Komiyama[14] supplied the explanation for these by showing that in the 100 MHz spectrum of poly-α,β-d_2-vinyl chloride, peaks corresponding to all ten chemical shifts of the six β-methylene tetrads can be distinguished. Such discrimination is easier in the α,β-d_2-polymer than in the α-d_1-polymer because in the latter the heterosteric tetrad resonances (*mmm*, *mmr*, *rmr*, and *mrr*; see Table III-1) appear as AB quartets, with the resonance intensity concentrated at the center (see Chapter I, Section 13), rather than as more widely spaced pairs of singlets.

The 220 MHz spectrum of poly-α-d_1-vinyl chloride has been shown[22] to be consistent with the chemical shift assignments of Yoshino and Komiyama.[14] At these higher frequencies, many more features can be seen (Fig. V-2); the chance of fitting the spectrum with a self-consistent but erroneous set of assignments becomes very small. [Figure V-2 should be compared with Fig.

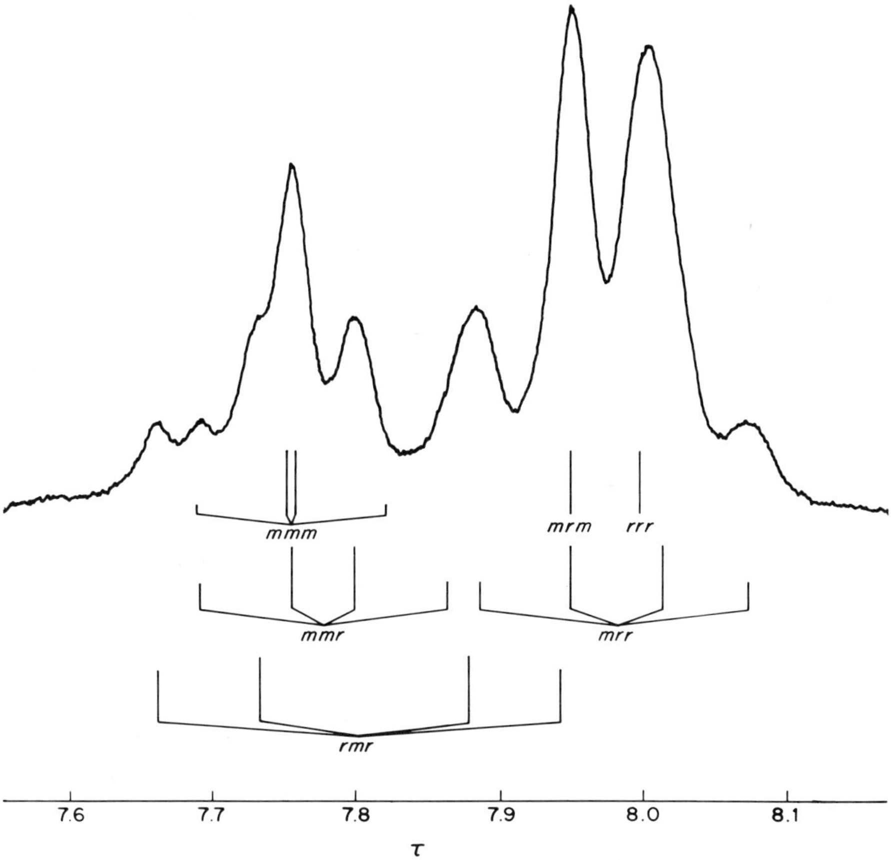

FIG. V-2. 220 MHz proton spectrum of poly-α-d_1-vinyl chloride, observed in 15% (w/v) solution in chlorobenzene at 140° (Heatley and Bovey[22]).

V-1(d), the same polymer at 60 MHz.] The analysis of the spectrum into four AB quartets ($J_{AB} = -15$ Hz) for the heterosteric tetrads and two singlets for the homosteric tetrads is indicated. Their intensities correspond to Bernoullian statistics with a P_m of 0.43. Comparing the chemical shifts in the series *mmm*, *mmr*, *rmr*, it is seen that the configuration of the second nearest asymmetric center has as large an effect on the differentiation of the chemical shifts in each methylene group as the configuration of the adjacent α-carbons. The fact that the least heterosteric character is found in the *mmm* tetrad indicates that chemical shift differentiation from the nearest and next nearest α-carbons of the same configuration must be similar in magnitude but opposite in sign. There is thus a much less rapid attenuation of the influence of the α-carbon configuration along the chain than in polymethyl methacrylate.

A second consistent trend is the upfield movement of the mean tetrad chemical shift when neighboring *m* dyads are replaced by *r*. We have seen the same trend in polymethyl methacrylate (Chapter III, Section 2). These chemical shift correlations are no doubt intimately related to chain conformation as well as configuration. At present, no consistent effort has been made to understand them in polar chains such as these. It is more than probable that the present theoretical framework for the interpretation and prediction of chemical shifts is inadequate to the task.

The undeuterated polymer spectrum has also yielded to a more complete analysis. Johnsen and Kolbe[21] have employed decoupling to distinguish pentad resonances in the α-proton spectrum at 100 MHz. Cavalli *et al.*[23] obtained very similar results for poly-β-d_2-vinyl chloride at 100 MHz and

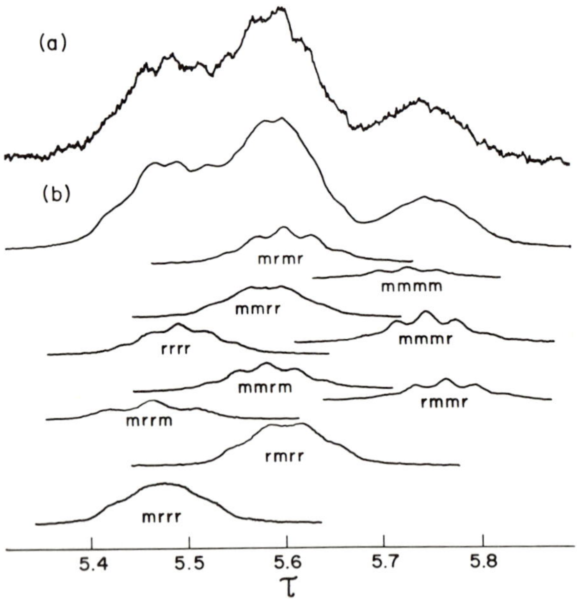

FIG. V-3. (a) 220 MHz α-proton spectrum of polyvinyl chloride, observed in 5% solution in chlorobenzene at 140°; (b) spectrum calculated using the parameters in Table V-1, with the component pentad subspectra (Heatley and Bovey[22]).

even at 60 MHz. Using pentad chemical shifts established by these authors and the tetrad chemical shifts just discussed, Heatley and Bovey.[22] were able to carry out a computer simulation of the entire polyvinyl chloride spectrum. Figure V-3 shows the α-proton spectrum and Fig. V-4 the β-proton spectrum, as observed at 220 MHz in chlorobenzene at 140°. Spectrum (b) in each figure is the simulated spectrum. Below these are the calculated pentad (Fig. V-3) and tetrad (Fig. V-4) subspectra. The pentad subspectra each represent a single type of proton, the resonance of which is split by coupling to the β-

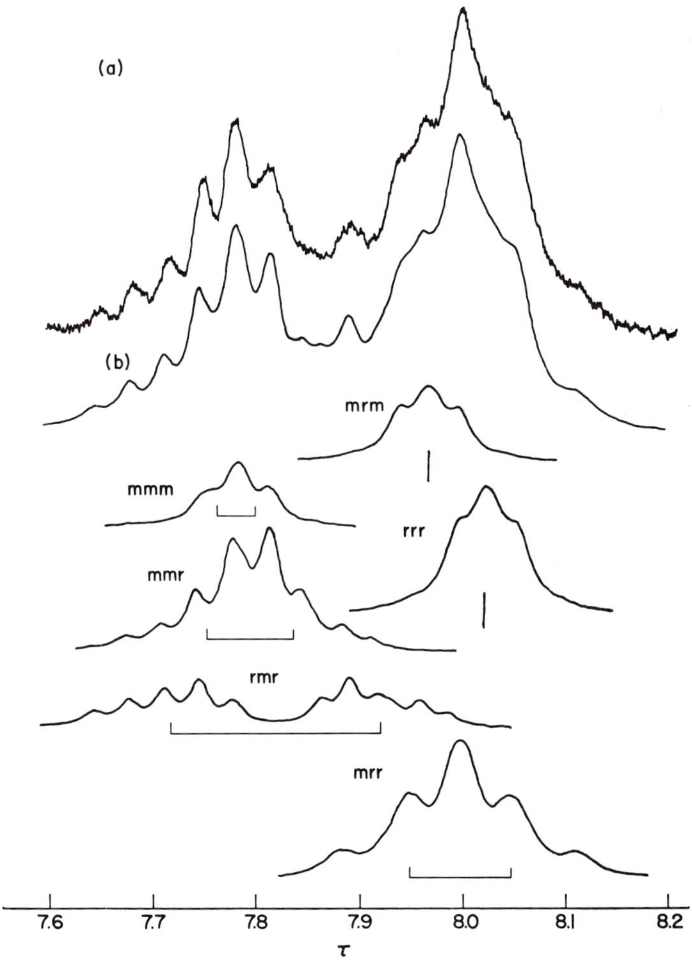

FIG. V-4. (a) 220 MHz β-proton spectrum of polyvinyl chloride, observed in 5% solution in chlorobenzene at 140°; (b) spectrum calculated using the parameters in Table V-1, with the component tetrad subspectra (Heatley and Bovey[22]).

TABLE V-1

CHEMICAL SHIFTS (SEE TEXT) AND J COUPLINGS (Hz) FOR POLYVINYL CHLORIDE (220 MHz)

	Solvent			
	Chlorobenzene	*o*-Dichloro-benzene	Penta-chloroethane	Relative intensity
m centered tetrads				
mmm ν_A	−44.5	−40.5	−48.5	0.080
ν_B	−46.0	−32.5	−42.5	
mrm ν_A	−46.0	−42.5	−48.5	0.208
ν_B	−28.5	−28.5	−32.5	
rmr ν_A	−55.0	−49	−59.5	0.141
ν_B	−10.5	−10.5	−16.5	
J_{AB}	−15	−15	−15	
J_{AX}	7.5	7.5	7.5	
J_{BX}	6.5	6.5	6.5	
Line width	5.0	5.0	5.0	
r centered tetrads				
mrm	0	0	0	0.105
mrm, τ	7.96	7.90	7.87	
mrr ν_A	−4.0	−3.5	−4.5	0.281
ν_B	17.5	17.0	9.5	
rrr	12.5	12.5	1.0	0.185
$J_{AA'}$	−15	−15	−15	
$J_{AX'}$	11.0	11.0	10.0	
$J_{AX'}$	2.0	2.0	3.0	
Line width	7.0	6.0	5.0	
Pentads				
mrrm	−543		−527	0.060
mrrr	−541		−525	0.160
rrrr	−537		−526	0.105
mmrm	−517		−513.5	0.091
mmrr	−517		−513.5	0.119
mrmr	−513		−507.5	0.119
rmrr	−513		−505.5	0.160
mmmm	−485		−492	0.035
rmmr	−477		−479	0.060
Line width	5.0		5.0	

protons. The tetrad spectra correspond to the same chemical shifts (indicated by vertical bars, which are connected by horizontal bars for the heterosteric methylenes) shown in Fig. V-2, but now complicated by vicinal couplings. These subspectra are those which would be expected for polymers each of which contained only the repeating sequence in question. The subspectra are summed by the computer assuming the same Bernoullian intensities ($P_m = 0.43$) as in Fig. V-2. The spectra in three solvents (*o*-dichlorobenzene and pentachloroethane in addition to chlorobenzene) are calculated on the assumption that the polymer chain may be satisfactorily represented as a cyclic "dimer," as described in Chapter IV. Agreement within experimental error (there is clearly more leeway in Fig. V-3 than in Fig. V-4) is obtained using the parameters shown in Table V-1; chemical shifts of tetrad and pentad protons are expressed in Hz with respect to *mrm* as zero in each solvent (actual τ values are given for *mrm* in each solvent). The interpretation of the vicinal couplings will be discussed in Chapter IX.

The effect of temperature and solvent on the stereoregularity of free radical vinyl chloride polymers is discussed in Chapter VIII, Section 2.

2. Polyvinyl Bromide

Polyvinyl bromide has been observed by Ramey *et al.*[25] The spectra of 60 and 100 MHz are very similar to those of polyvinyl chloride. The interpretation of these authors was not entirely conclusive. It appears that analysis along very much the same lines as described in the previous section should prove effective. For the present, it is probably safe to conclude that, like polyvinyl chloride, polyvinyl bromide is predominantly but not strongly syndiotactic.

3. Polyvinyl Fluoride

The proton spectrum of polyvinyl fluoride, observed in benzaldehyde solution at 60 MHz,[12] shows a doublet of quintets near 5τ and an unsymmetrical triplet near 8τ. The low-field resonance corresponds to the α-protons; the large doublet splitting arises from geminal $^1H_\alpha$–^{19}F coupling (47.0 Hz) and the smaller splitting from vicinal H_α–H_β coupling. The spacing of the β-proton triplets corresponds to the $^1H_\beta$–^{19}F coupling of 20 Hz, but there was no further resolution of the members of the triplet into the multiplicity expected from vicinal proton coupling. Upon irradiation of the β-protons, the α-proton resonance became two sets of doublets, the more shielded being the weaker set. Reversed decoupling revealed nothing helpful in the β-proton region. From these results, no definite conclusion could be drawn except that the polymer clearly was not stereoregular.

The ^{19}F resonance of polyvinyl fluoride has been reported by Naylor and Lasoski[26] and considered in detail by Wilson and Santee.[27] The 56.4 MHz spectrum (in a solvent composed of equal volumes *N,N'*-dimethylacetamide and hexamethyl phosphoramide) is shown in Fig. V-5. The chemical shifts

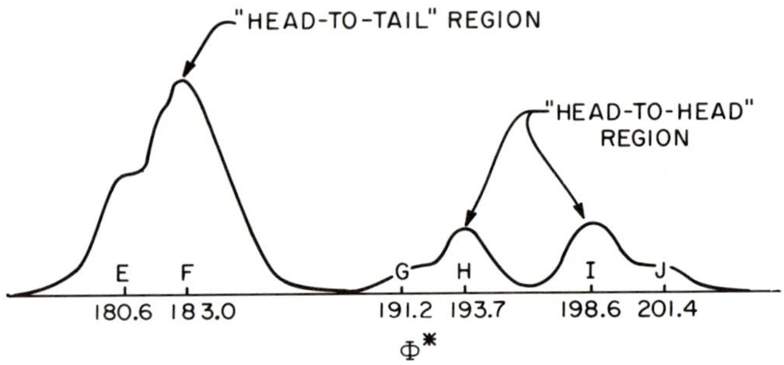

FIG. V-5. 56.4 MHz ^{19}F spectrum of polyvinyl fluoride in a solvent composed of equal volumes of *N,N'*-dimethylacetamide and hexamethyl phosphoramide, observed at 89°. For a discussion of possible assignments of the peaks lettered E through J see Wilson and Santee.[27] For a discussion of the chemical shift scale, see Chapter I, Section 16B.

are expressed on the Φ scale,[28] on which internal $CFCl_3$ is taken as zero (Chapter I, Section 16B). At least eight peaks can be seen. It is unlikely that these are due primarily to different configurational sequences, although this is no doubt an added complication, because the range of chemical shifts is quite large and the pattern similar to that shown by polyvinylidene fluoride (Section 4) to which such an explanation could not apply. It was concluded that, as in the latter, the more shielded fluorines are associated with head-to-head units

$$\cdots\text{—}CFHCH_2\cdot CH_2CFH\cdot CFHCH_2\text{—}\cdots$$

but a detailed assignment was not made. About 32% of the ^{19}F intensity is in the upfield peaks; this corresponds to one out of six of the monomer units being inserted "backward" (as shown above), a surprisingly high ratio.

4. Polyvinylidene Fluoride

Naylor and Lasoski[26] reported a small peak upfield from the main ^{19}F peak in the polyvinylidene fluoride spectrum, and attributed it tentatively to head-to-head units. Wilson[27,29] obtained better resolved spectra showing a number of additional peaks (Fig. V-6). In the absence of any possible stereochemical complications, the only logical explanation for these peaks is that head-to-head, tail-to-tail monomer units must occur with considerable

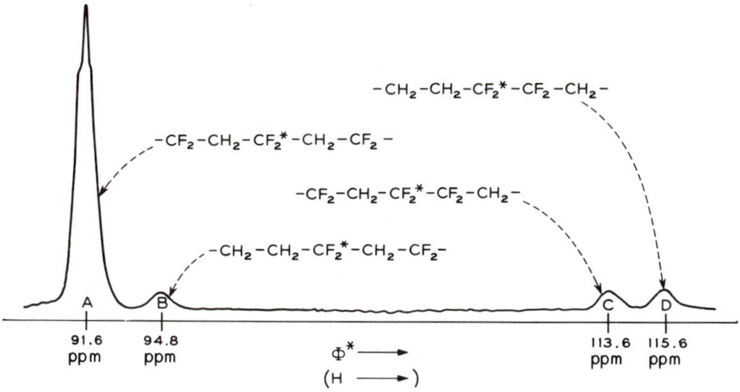

FIG. V-6. 56.4 MHz ^{19}F spectrum of polyvinylidene fluoride in 25% solution in *N,N'*-dimethylacetamide, observed at room temperature (Wilson and Santee[27]). For a discussion of the chemical shift scale, see Chapter I, Section 16B.

frequency, although not so often as in polyvinyl fluoride (Section 3). In this case, reference to model compounds enabled the peaks to be assigned with assurance. The chain units are identified in the representative segment shown.

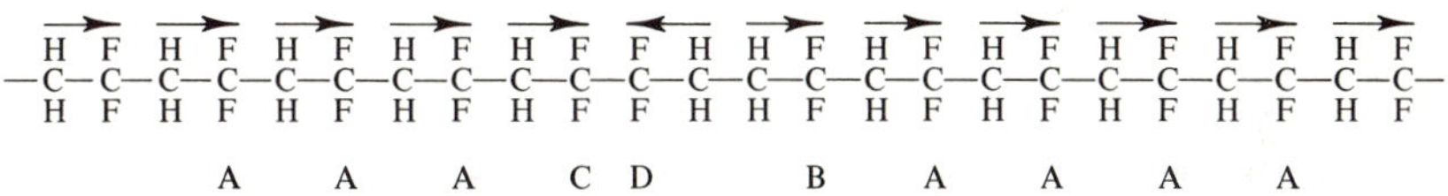

The letters refer to the corresponding peaks in Fig. V-6. It can be seen that a single reversed unit leads to three distinguishable fluorine resonances, all upfield from the "normal" peak. Since peaks B, C, and D have about equal intensity, it is evident that this reversed unit has little or no tendency to propagate further units of its kind. Quantitative estimation indicates that 5–6% of the monomer units are reversed.

5. Polytrifluorochloroethylene

Naylor and Lasoski[26] examined the ^{19}F spectrum of polytrifluorochloroethylene at relatively low resolution and were able to distinguish only one Cl–C–^{19}F resonance. Tiers and Bovey[30] examined polymers of relatively low molecular weight prepared by free radical initiation over a 110° temperature range, and compared their spectra to those of the model compounds *meso*- and *dl*-$CF_2Cl \cdot CFCl \cdot CF_2 \cdot CFCl \cdot CF_2Cl$, the central three carbons of which should be a close facsimile of the environment in the isotactic and syndiotactic polymer chains, respectively. The spectra of the polymer and of a mixture of the *meso*- and *dl*-pentanes, observed at 40 MHz in 3,3′-bistrifluoromethylbiphenyl, are shown in Fig. V-7. The most significant portion of the model

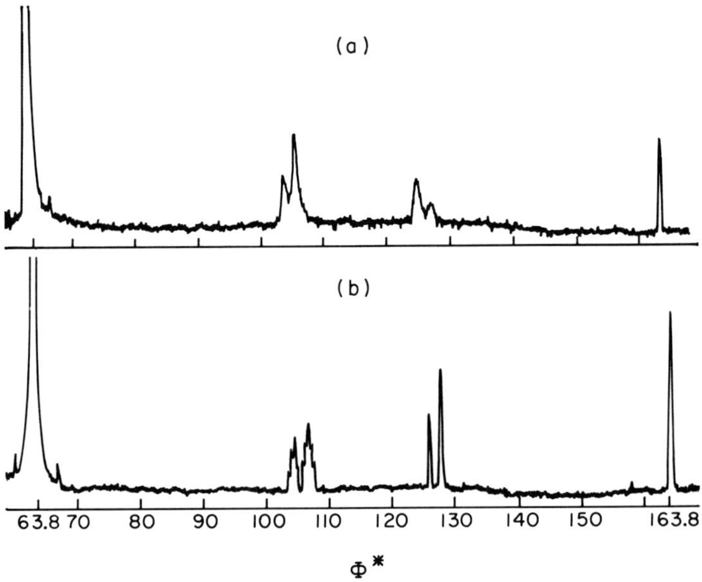

FIG. V-7. (a) 40 MHz spectrum of polytrifluorochloroethylene, about 20% (w/v) in 3,3′-bistrifluoromethylbiphenyl, observed at 150°; (b) spectra of mixed *meso*- and *racemic* $CF_2ClCFClCF_2CFClCF_2Cl$, 10% in the same solvent, same temperature. For a discussion of the chemical shift scale, see Chapter I, Section 16B.

compound spectrum is the central CF_2 resonance near 106Φ. Detailed analysis shows that the more shielded multiplet must correspond to the *dl* compound. In the polymer spectrum, two unresolved multiplets are observed in this position. (Peaks corresponding to the CFCl groups also appear in closely corresponding positions near 128Φ in both spectra.) The close correspondence of chemical shifts makes it highly probable that the polymer peaks correspond to *meso* and *racemic* CF_2 groups in order of increasing shielding. The *racemic* peak has about twice the intensity of the *meso*, and so the polymer appears to be about two-third syndiotactic on a dyad basis. The polymer spectra are independent of the temperature of polymerization within experimental error. The stereochemically irregular structure of polyvinyltrifluorochloroethylene is somewhat surprising in view of the relatively high melting point (218°) and high attainable degree of crystallinity of this polymer. But such behavior is by no means unprecedented (see Chapter VIII, Section 2).

REFERENCES

1. D. Doskocilova, *J. Polym. Sci. Part B* **2**, 421 (1964).
2. D. Doskocilova and B. Schneider, *Coll. Czech. Chem. Commun.* **29**, 2290 (1964).
3. S. Satoh, *J. Polym. Sci. Part A* **2**, 5221 (1964).
4. H. G. Clark, *Makromol. Chem.* **63**, 69 (1963).
5. T. Shimanouchi, M. Tasumi, and Y. Abe, *Makromol. Chem.* **86**, 43 (1965).
6. Y. Abe, M. Tasumi, T. Shimanouchi, S. Satoh, and R. Chujo, *J. Polym. Sci. Part A-1* **4**, 1413 (1966).
7. D. Doskocilova, J. Stokr, B. Schneider, H. Pivcova, M. Kolinsky, J. Petranek, and D. Lim, *J. Polym. Sci. Part C* **16**, 215 (1967).
8. U. Johnsen, *J. Polym. Sci.* **54**, 56 (1961).
9. F. A. Bovey and G. V. D. Tiers, *Chem. Ind. (London)* **1962**, 1826 (1962).
10. R. Chujo, S. Satoh, T. Ozeki, and E. Nagai, *J. Polym. Sci.* **61**, S12 (1962).
11. W. C. Tincher, *J. Polym. Sci.* **62**, S148 (1962).
12. F. A. Bovey, E. W. Anderson, D. C. Douglass, and J. A. Manson, *J. Chem. Phys.* **39**, 1199 (1963).
13. W. C. Tincher, *Makromol. Chem.* **85**, 20 (1965).
14. T. Yoshino and J. Komiyama, *J. Polym. Sci. Part B* **3**, 311 (1965).
15. O. C. Böckman, *J. Polym. Sci. Part A* **3**, 3399 (1965).
16. B. Schneider, J. Stokr, D. Doskocilova, M. Kolinsky, S. Sykora, and D. Lim, *IUPAC Symp. Macromol. Chem., Prague, 1965* preprints, p. 599.
17. J. Bargon, K. H. Hellwege, and U. Johnsen, *Makromol. Chem.* **95**, 187 (1966).
18. K. C. Ramey, *J. Phys. Chem.* **70**, 25 (1966).
19. S. Enemoto, M. Asahina, and S. Satoh, *J. Polym. Sci. Part A-1* **4**, 1373 (1966).
20. F. A. Bovey, F. P. Hood, E. W. Anderson, and R. L. Kornegay, *J. Phys. Chem.* **71**, 312 (1967).
21. U. Johnsen and K. Kolbe, *Kolloid Z.* **221**, 64 (1967).
22. F. Heatley and F. A. Bovey, *Macromolecules* **2**, 241 (1969).
23. L. Cavalli, G. C. Borsini, G. Carraro, and G. Confalonieri, *J. Polym. Sci. Part A-1* **8**, 801 (1970).
24. P.-Q. Tho, *J. Polym. Sci. Part B* **7**, 103 (1969).
25. K. C. Ramey, D. C. Lini, and W. B. Wise, *J. Polym. Sci. Part B* **6**, 523 (1968).
26. R. E. Naylor and S. W. Lasoski, Jr., *J. Polym. Sci.* **44**, 1 (1960).
27. C. W. Wilson III and E. R. Santee, Jr., *J. Polym. Sci. Part C* **8**, 97 (1965).
28. G. Filipovich and G. V. D. Tiers, *J. Phys. Chem.* **63**, 761 (1959).
29. C. W. Wilson III, *J. Polym. Sci. Part A* **1**, 1305 (1963).
30. G. V. D. Tiers and F. A. Bovey, *J. Polym. Sci. Part A* **1**, 833 (1963).

Chapter VI

POLYSTYRENE AND RELATED POLYMERS

1. Polystyrene

The earliest reported NMR spectra of polymers were those of polystyrene.[1-3] While not in every case correctly interpreted,[2] these spectra clearly indicated the potential of NMR for the study of polymer structure and laid to rest the assumption that the resonances would be too broad to be useful. In Fig. VI-1 are shown 100 MHz spectra of (a) atactic and (b) isotactic polystyrene (the syndiotactic polymer does not appear to exist). It will be observed that the aromatic resonance near 3.5τ is divided into a three-proton and a two-proton part; substitution experiments[3] show that these correspond to *meta–para* and *ortho* protons, respectively. (This spectral feature is evident even at 40 MHz.[1-3]) All the aromatic protons are abnormally shielded compared to those of simple alkylbenzenes and diphenylpentanes, the *ortho* protons being particularly affected. This is illustrated in Fig. VI-2, where the aromatic proton chemical shifts are schematically compared. The source of this shielding lies in the ring currents of the benzene rings (Chapter I, Section 11). Inspection of molecular models makes it very plausible that the *ortho* protons should show the observed excess shielding of about 0.7 ppm. This question is explored more fully in Chapter IX, where the conformations of polystyrene chains are discussed.

The α- and β-protons give readily distinguishable resonances, particularly in the isotactic polymer.[4-9] In Fig. VI-1 they appear as multiplets centered at 7.6τ and 8.3τ, respectively. Quite well-resolved multiplets can be obtained for the isotactic polymer even at 60 MHz[6] in both the aliphatic and aromatic regions. However, at 60 MHz and at 100 MHz [Fig. VI-1(b)] it is not possible

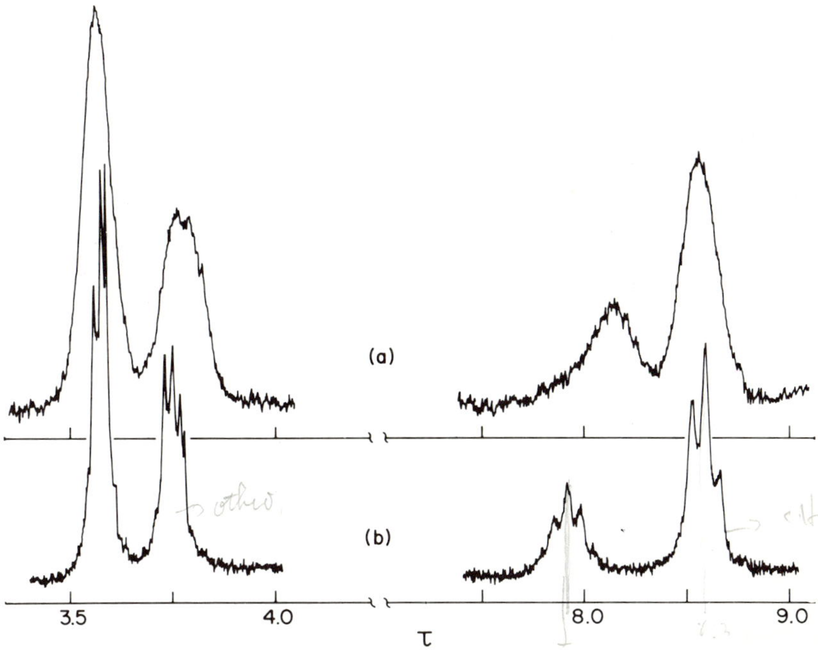

FIG. VI-1. 100 MHz spectra of (a) "atactic" and (b) isotactic polystyrene, observed in 5% solution in tetrachloroethylene at 140° (A. I. Brewster and F. A. Bovey, unpublished observations).

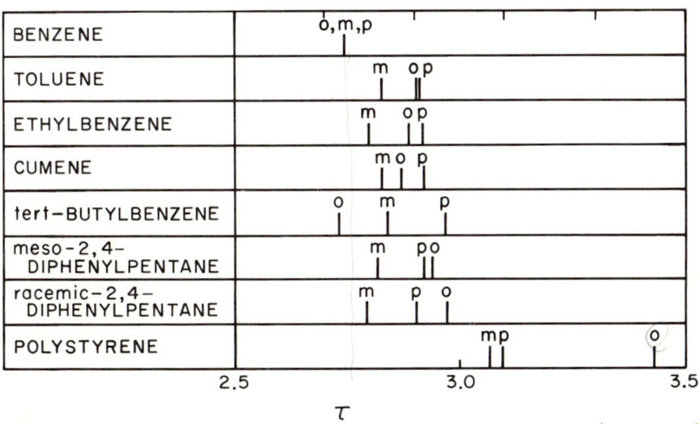

FIG. VI-2. A schematic comparison of aromatic proton chemical shifts in polystyrene (isotactic or atactic) and model aromatic compounds.

to distinguish the two different chemical shifts of the *meso* methylene groups, and hence the stereochemical configuration of the polymer cannot be unequivocally demonstrated by NMR. This is achieved at 220 MHz (see below). It will be noted that resolution is apparently much poorer in the spectrum of the atactic polymer [Fig. VI-1(a)]. As we shall see in later discussion, this is because the spectrum, like that of polyvinyl chloride (Chapter V, Section 1), is actually the superposition of at least ten subspectra.

The β-methylene proton ambiguity is resolved at 220 MHz.[7] In Fig. VI-3 spectra (a) and (d) are those of the backbone protons, observed in *o*-dichlorobenzene at 130°. The multiplet centered at 7.8τ is that of the α-proton and

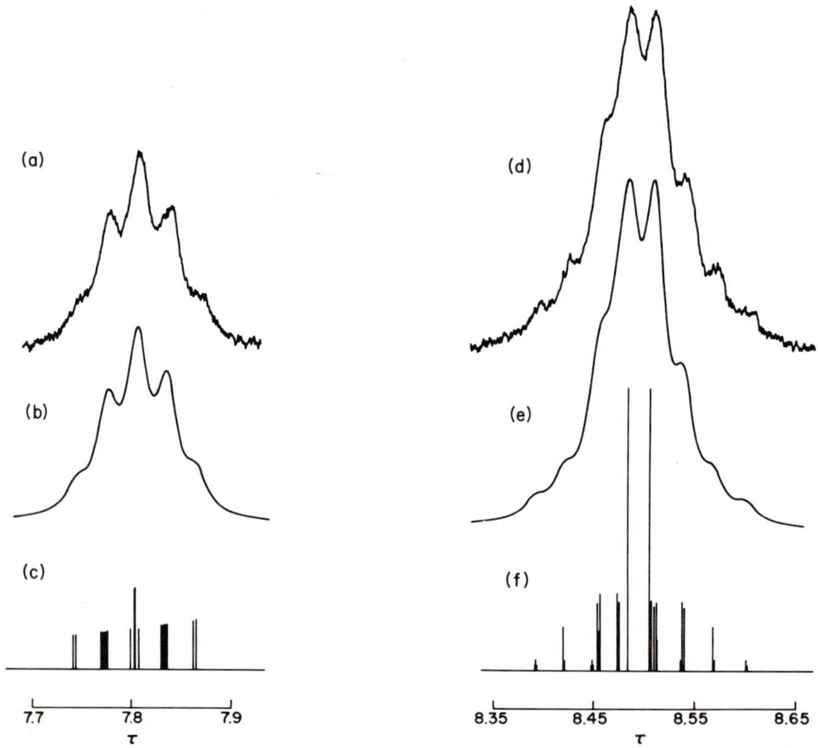

FIG. VI-3. 220 MHz backbone proton spectrum of isotactic polystyrene, 2% in *o*-dichlorobenzene at 130°. (a) and (d) are the experimental spectra of the α- and β-protons, respectively; (b) and (e) are spectra calculated with the parameters shown in Table VI-1 (for 130°); (c) and (f) are "stick" spectra corresponding to (b) and (e) (Heatley and Bovey[7]).

that at about 8.5τ corresponds to the β-methylene protons. Spectra (b) and (e) are computer simulations of the observed spectrum, using the same six-spin model described in Chapter IV and V. [Spectra (c) and (f) are the same with zero line width—so-called "stick" spectra.] The computed spectrum was obtained using the chemical shifts shown in Table VI-1 (for 130°). The

TABLE VI-1

CHEMICAL SHIFTS AND COUPLING CONSTANTS FOR ISOTACTIC POLYSTYRENE (2% IN *o*-DICHLOROBENZENE)[a]

Temp, °C	τ_A	τ_B	τ_C	J_{AB}, Hz	J_{AC}, Hz	J_{BC}, Hz	Line width, Hz
55	8.527	8.433	7.806	−14.5	7.5	6.0	8.0
80	8.518	8.436	7.795	−14.5	7.5	6.0	6.0
100	8.499	8.426	7.771	−14.5	7.25	6.25	6.0
130	8.462	8.403	7.746	−14.5	7.25	6.25	5.5
160	8.413	8.369	7.682	−14.5	7.25	6.25	4.0

[a] A, high-field β-proton; B, low-field β-proton; C, α-proton.

methylene portion of the spectrum cannot be matched by assuming that $\tau_A = \tau_B$, but only by assuming a difference of 0.059 ppm. Note that this difference decreases with increasing observing temperature. The interpretation of the couplings, which are nearly invariant with temperature, will be discussed in Chapter IX.

Segre *et al.*[8] observed the spectrum of isotactic polystyrene of 97.6% steric purity prepared from d_7-styrene $C_6D_5CH{=}CD_2$, using an $AlEt_3$–$TiCl_3$ catalyst. The deuterium nuclei were irradiated to remove vicinal and longer range D–H couplings, including the weak couplings of the phenyl groups (see Chapter VII, Section 2). The single narrow line of the α-protons was accompanied only by weak multiplets of the residual β-protons and a satellite α-doublet arising from their vicinal coupling. No other peaks were observable, so the polymer was at least 97.6% isotactic.

Analysis[6] of the spectrum of the aromatic protons of the isotactic polymer [Fig. VI-1(b)] shows that it can be well represented as AA′BB′C, i.e., as a spin system symmetrical about a plane bisecting the ring through the C-1 and C-4 carbon atoms, with normal *ortho*, *meta*, and *para* spin–spin couplings. (The simulated spectrum is shown in Bovey *et al.*[6]) This would not be expected if the phenyl group is so restricted as to rotate slowly on the NMR time scale, as is probably true for poly-*N*-vinylcarbazole (Section 4). It may therefore be concluded that the barrier to rotation of the phenyl group is less than approximately 15 kcal.

The interpretation of the proton spectrum of "atactic" polystyrene, usually prepared by free radical initiators, is not clear in detail. Brownstein *et al.*[4] concluded that in the spectrum of stereoirregular poly-β,β′-d_2-styrene, resonances corresponding to isotactic, heterotactic, and syndiotactic triads could be distinguished, and that the last of these dominates. Bovey *et al.* supported this interpretation.[6] Segre *et al.*[8] reported the 100 MHz spectrum of poly-d_7-styrene and an 80:20 copolymer of d_8-styrene and d_7-styrene, both prepared with a free radical initiator. Irradiation of the deuterium nuclei produced slightly better resolution in this copolymer than in the homopolymer. Heatley and Bovey[10] obtained the 220 MHz spectrum of poly-β,β'-d_2-styrene (without deuterium irradiation). These spectra[6,8-10] are compared in Fig.

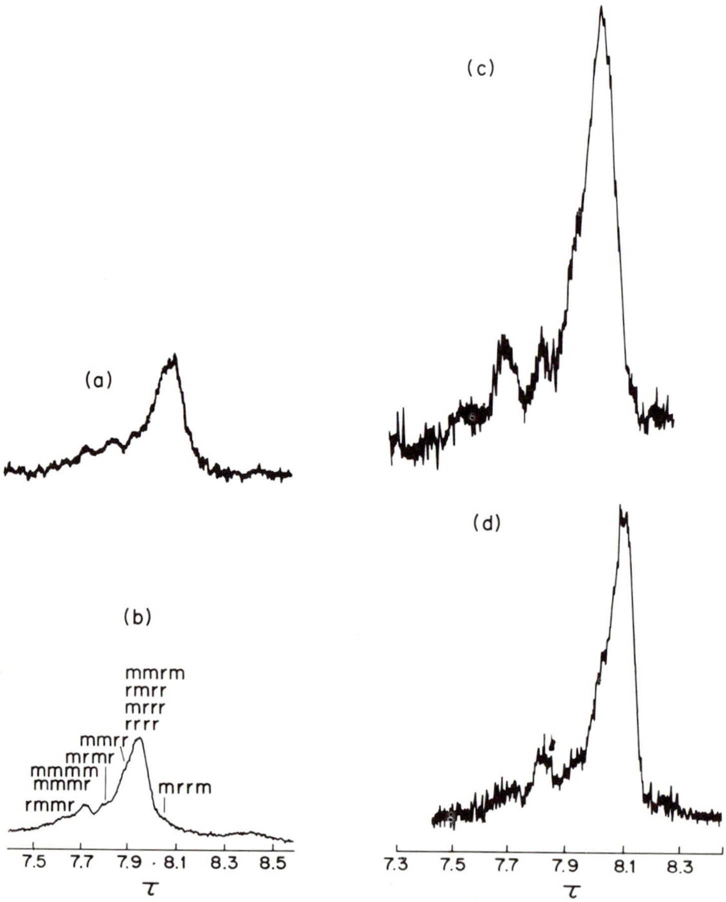

FIG. VI-4. Spectra of "atactic" poly-β,β'-d_2-styrene: (a) at 60 MHz in *o*-dichlorobenzene at 200° with deuterium irradiation (Bovey *et al.*[6]); (b) at 220 MHz in *o*-dichlorobenzene at 150° without deuterium irradiation (Heatley and Bovey[10]); (c) 100 MHz spectrum of poly-d_7-styrene in *o*-dichlorobenzene at 160° with deuterium irradiation (Segre *et al.*[8]); (d) 100 MHz spectrum of an 80:20 copolymer (mole ratio) of d_8-styrene and d_7-styrene, same conditions as (c) (Segre *et al.*[8]).

VI-4. It will be noted that with the exception of a small splitting of the major peak in (d), the spectra are all very similar; spectrum (d) reveals little or no additional information over spectrum (a) despite a nearly fourfold increase in chemical shift discrimination. The small chemical shift differences between pentad and longer "odd"(ad) sequences are not clearly resolved and broaden the resonances so that no additional structural information can be obtained. Such behavior depends upon the relative magnitudes of the chemical shift influences of neighboring asymmetric centers and how rapidly they are attenuated with the number of intervening bonds. These in turn depend upon the chain conformation (Chapter IX) as well as on the type of substituent group. Heatley and Bovey[10] tentatively assigned the pentad chemical shifts in the manner shown in Fig. VI-4(d), based on Bernoullian intensities and an analysis of the spectrum with a curve resolver; the position of the *mmmm* pentad was taken from the experimental data of Heatley and Bovey.[7]

Fujiwara and Flory[11] have calculated the spectrum of the α-protons of polystyrene from considerations of chain conformation and ring-current shielding. This will be discussed in Chapter IX.

TABLE VI-2

Sample	Mole ratio, CCl_4–monomer	$\bar{M}_n$, calc. from transfer constant of 0.012[a]	$\bar{M}_n$ from terminal group intensity
(a)	5	16.5	16.0
(b)	10	8.4	7.9
(c)	20	4.1	3.7
(d)	40	2.1	2.4

[a] See Walling and Huyser.[12a]

The spectra of styrene oligomers prepared with carbon tetrachloride as solvent and reference have been observed.[3,12] The 100 MHz spectra of four such oligomers are shown in Fig. VI-5. The spectra (a), (b), (c), and (d) correspond to the designations of the samples in Table VI-2. These materials are presumed to have the general structure shown, where n decreases with increas-

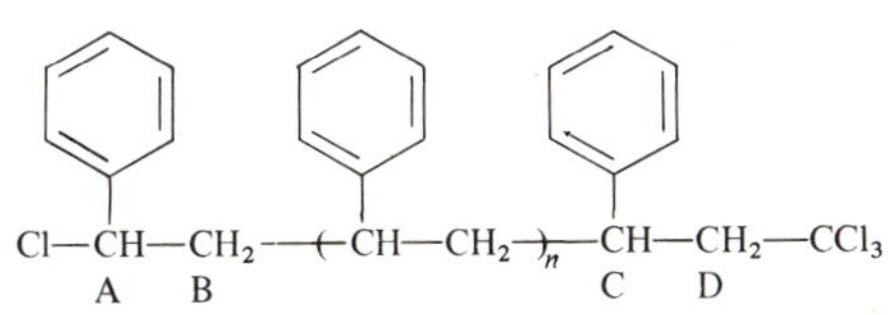

ing solvent–monomer ratio. The resonance at about 5.7τ is that of the α-proton at the Cl terminal, marked A in the structural formula. The values of $\bar{M}_n$

in the fourth column of Table VI-2 are calculated from the intensity of this peak, and agree well with the values in the third column based on the chain-transfer constant of polystyryl radicals with carbon tetrachloride. The complex

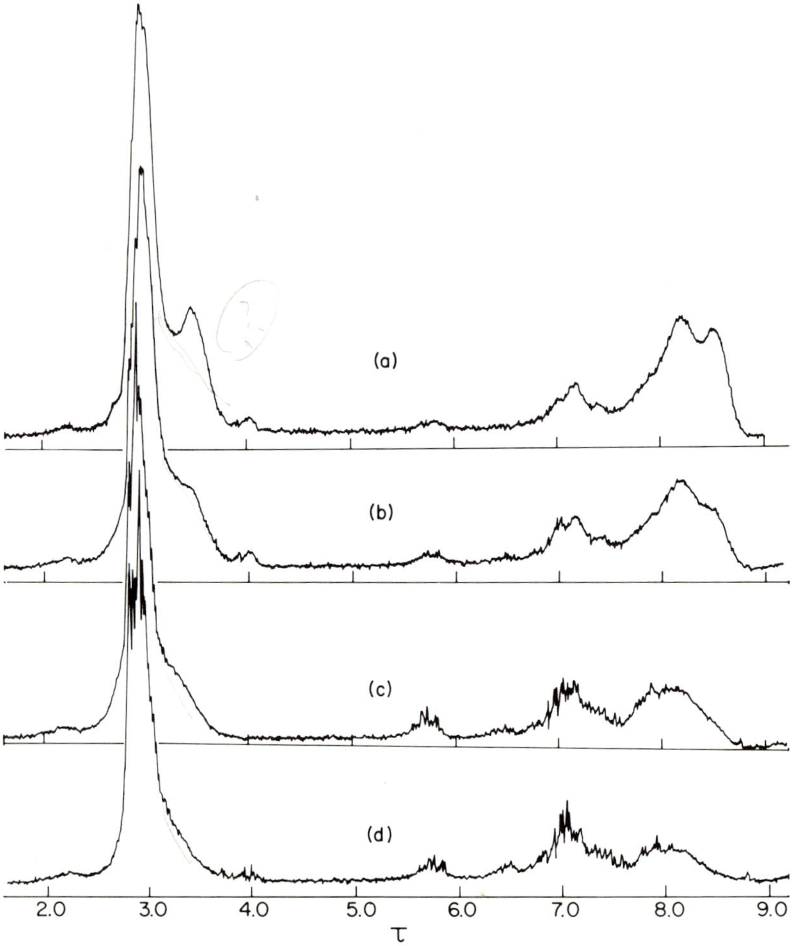

FIG. VI-5. Spectra of styrene oligomers, observed in chlorobenzene (5% w/v) at 125°. Letters on spectra correspond to those in Table VI-1 (Bovey *et al.*[12]).

resonance near 7τ probably corresponds to protons B, C, and D. The α- and β-protons of nonterminal monomer units appear at about 7.9τ and 8.2τ, respectively.

The point of principal interest in these spectra is that the resonance of the *ortho* protons of the phenyl groups at about 3.4τ decreases in intensity as $\bar{M}_n$ is decreased, being virtually absent in spectrum (d). This furnishes a measure of how many phenyl groups have neighboring phenyl groups on both sides. This peak thus allows an estimate of the lengths of styrene block sequences in copolymers.[3,13] Kinstle and Harwood[14] employed styrene deuterated at the *meta* and *para* positions in order to reduce signal broadening arising from coupling of the ring protons. Three triad *ortho* proton signals were clearly resolved in the homopolymer. The measurement of styrene sequences in copolymers by *ortho* proton observations will be discussed further in Chapter X.

The ^{13}C spectrum of stereoirregular polystyrene has been observed.[14a] The most informative peak is that of the C_1 carbon atom of the phenyl ring, i.e., the carbon bonded to C_α, appearing about 47.5 ppm upfield from $^{13}CS_2$ (Fig. VI-6). This might be expected to experience shielding influences somewhat similar to those of the α-proton, and indeed the spectrum [Fig. VI-7(a)] is strongly reminiscent of those of Fig. VI-4, although the spread of chemical shifts is over twice as great. It is clear that pentad sequences are at least partially resolved. In Fig. VI-7(b) are shown the spectra of "atactic" and isotactic polystyrene, obtained from a solution containing both. In this way, the position of the *mmmm* pentad carbon resonance was established. Most of the area of the "atactic" spectrum is upfield from the *mmmm* peak, and so the polymer must be predominantly syndiotactic. Both the carbon and proton spectra are consistent with a P_m of about 0.3 for free radical polystyrene, which thus, like all free radical polymers so far examined, is predominantly syndiotactic.

2. Ring-Substituted Polystyrenes and Related Polymers

Bovey *et al.*[3] reported the 40 MHz spectra of poly-*o*-chlorostyrene, poly-*m*-chlorostyrene, and poly-*p*-chlorostyrene, all prepared with free radical initiators. The aromatic spectrum of the first showed a single resonance, while the spectra of the other two showed two peaks. The spectra were so poorly resolved, however, that no quantitative structural conclusions could be drawn. Kobayashi *et al.*[15] reported somewhat more informative spectra of isotactic and atactic poly-*p*-chlorostyrene at 100 MHz; they were closely similar to the corresponding spectra of polystyrene. Brady and Salovey[16] have reported the spectrum of iodinated isotactic polystyrene. The phenyl resonance was predominantly a symmetrical AA′BB′ "quartet," showing that iodine substituted nearly exclusively at the *para* position in the polymer.

Ohsumi *et al.*[17] have made a thorough study of poly-*o*-methoxystyrene, initiated by both free radical and cationic initiators. The 60 MHz aromatic and main-chain spectra were not readily interpretable, but the methoxyl resonance was split into three peaks (at 6.64τ, 6.76τ and 7.93τ in $CDCl_3$) which were assigned to *rr*, *mr*, and *mm* triads in order of increasing shielding.

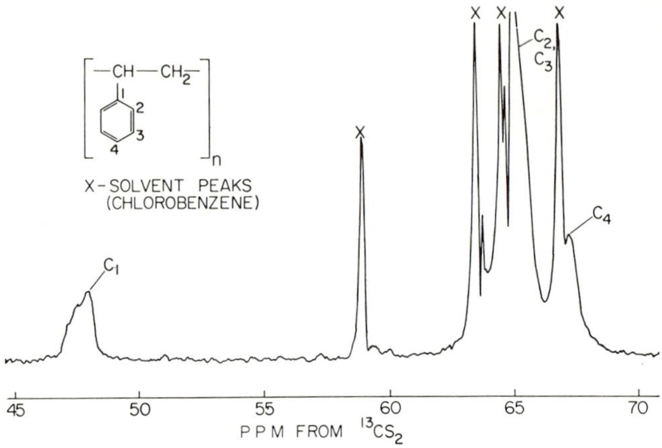

FIG. VI-6. ^{13}C spectrum (25 MHz) of the aromatic carbons of "atactic" polystyrene, 50% (w/v) solution in chlorobenzene–chloroform solvent at about 50°, protons decoupled. The peaks marked "X" are due to chlorobenzene; 1639 scans (Johnson *et al.*[14a]).

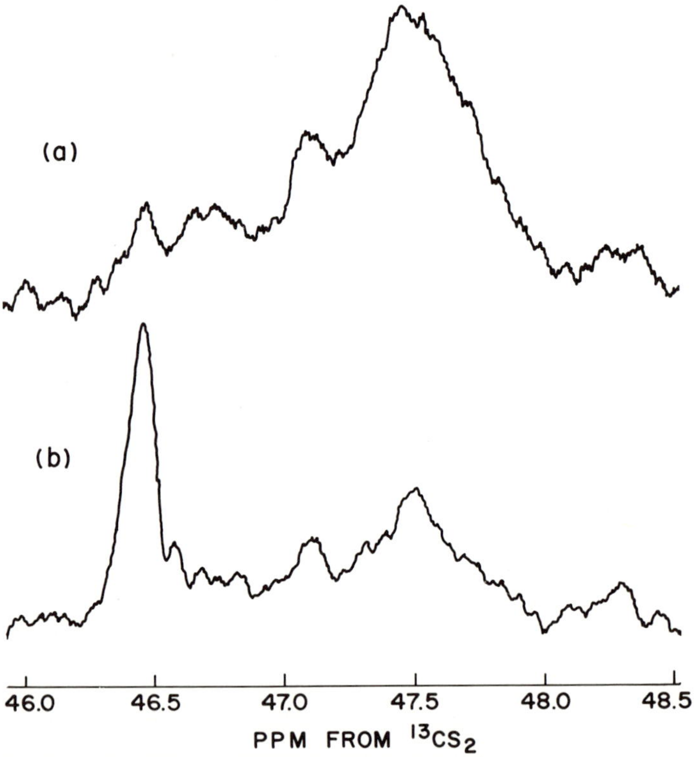

FIG. VI-7. (a) Expanded ^{13}C spectrum (25 MHz) of the aromatic C_1 spectrum of "atactic" polystyrene, same conditions as in Figure VI-6 (3125 scans); (b) C_1 spectrum of mixed atactic and isotactic polystyrene; 593 scans (Johnson *et al.*[14a]).

It was found that the sequence statistics were Bernoullian with both classes of initiator. For free radical propagation, P_m was about 0.4, decreasing slightly at lower temperatures. The cationic polymers were characterized by somewhat smaller values of P_m, which was found to be independent of the polarity of the solvent.

3. Poly-2-Vinylpyridine

The NMR spectrum of poly-2-vinylpyridine might be expected to be closely

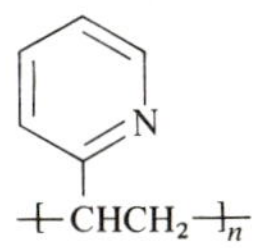

similar to that of polystyrene. Although initial attempts to analyze the "atactic" polymer were inconclusive,[18] the isotactic polymer,[19,20] prepared with a Grignard initiator, gives a well-resolved spectrum at 100 MHz which is less ambiguous than that of isotactic polystyrene because the β-protons are well separated in chemical shift. The corresponding polymer of the *trans*-α,β-d_2 monomer showed two singlets of which the smaller was the more shielded. Because the *erythro–threo* assignment is not established, it is not known from this observation whether *cis* or *trans* addition predominates (see Chapter VIII, Section 4).

The isotactic polymer showed a vicinal coupling pattern which is very similar to that of isotactic polystyrene, and leads to similar conclusions with respect to its solution conformation[19] (Chapter IX). Polymers prepared with *n*-butyllithium were less stereoregular and those prepared with diethylmagnesium in tetrahydrofuran were apparently nearly atactic, although the spectra did not lend themselves to precise analysis.[19]

4. Poly-*N*-Vinylcarbazole

The NMR spectrum of poly-*N*-vinylcarbazole was reported by Heller

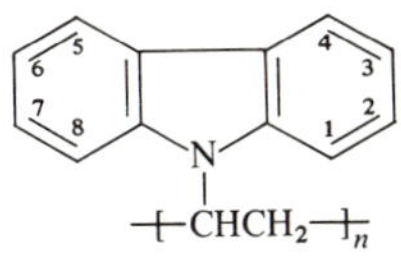

et al.[21] but specific assignments of the peaks were not made. Yoshimoto *et al.*[22] reported chemical shifts (unfortunately without showing the spectra or specifying the solvent) of the polymer and its β,β'-d_2 derivative prepared by thermal (i.e., free radical) and cationic polymerization. The methylene protons appear at about 8.5τ and the methine protons as two peaks at 6.4τ and 7.3τ. Of the aromatic protons, seven appear near 3τ, while *one* shows very abnormal shielding, appearing at 5.0τ. Thus, the two sides of each carbazyl group are apparently nonequivalent. This was attributed, perhaps correctly, to very restricted rotation of the side chain about the C—N bond; inspection of molecular models shows that this is plausible, for the structure is a severely crowded one in both *meso* and *racemic* dyads. The rather wide separation (0.9τ) of the two observable methine resonances were also attributed to this restriction, but this does not seem logical. The temperature dependence of the aromatic proton spectrum has been similarly interpreted.[23] A more complete analysis of poly-*N*-vinylcarbazole spectra, particularly with respect to stereochemical configuration, is clearly needed.

5. Poly-*N*-Vinylindoles

Gipstein and Hewett[24] attempted to obtain the 100 MHz spectra of homopolymers and copolymers of *N*-vinylindole, *N*-vinyl-2-methylindole, and *N*-vinyl-3-methylindole in d_6-benzene, but reported resolution to be so poor that only analytical information concerning the aliphatic–aromatic proton ratio was forthcoming.

6. Poly-1-Vinylnaphthalene, Poly-2-Vinylnaphthalene, and Poly-4-Vinylbiphenyl

The 100 MHz spectra of poly-1-vinylnaphthalene (a), poly-2-vinylnaphthalene (b), and poly-4-vinylbiphenyl (c), have been reported by Heller and

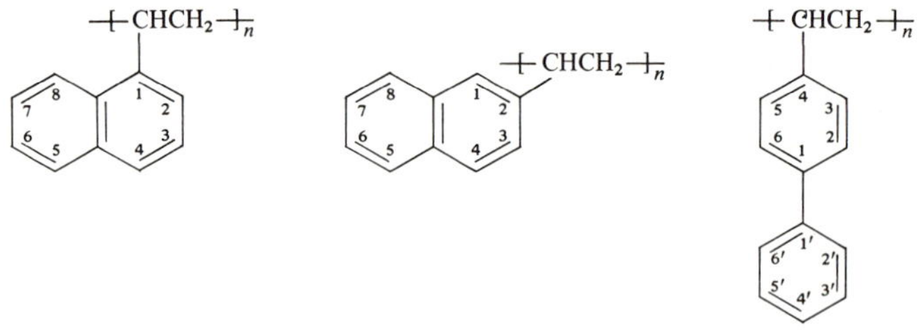

(a) (b) (c)

Miller.[25] The polymers were prepared in "atactic" form using Et_3Al–$TiCl_4$, Et_2AlCl–$TiCl_3$, Et_3Al–$TiCl_3$ catalysts. The last of these yielded polymers which were believed to be at least 90% isotactic. Butanone extraction provided an atactic fraction. The spectra were obtained in $CDCl_3$ at 25°, except for those of the poly-4-vinylbiphenyl, which were observed in d_8-naphthalene at 130°. The latter were somewhat better resolved. The spectra of the aliphatic protons closely resembled those of polystyrene (Section 1), including the greater separation of the α- and β-proton resonances in the isotactic polymer. However, in no case was the resolution sufficient to yield any information concerning configuration. The aromatic region showed the expected increased complexity over that of polystyrene. These materials deserve to be reexamined under better experimental conditions, preferably at 220 MHz.

A conformational transformation in solutions of poly-2-vinylnaphthalene has been indicated by NMR studies of Eskin and Serdjuk.[26] This is probably similar to the transformation observed in polystyrene solutions and discussed in Chapter IX.

REFERENCES

1. M. Saunders and A. Wishnia, *Ann. N.Y. Acad. Sci.* **70**, 870 (1958).
2. A. Odajima, *J. Phys. Soc. Jap.* **14**, 777 (1959).
3. F. A. Bovey, G. V. D. Tiers, and G. Filipovich, *J. Polym. Sci.* **38**, 73 (1959).
4. S. Brownstein, S. Bywater, and D. J. Worsfold, *Makromol. Chem.* **48**, 127 (1961).
5. T. Yoshino, H. Kyogoku, J. Komiyama, and Y. Manabe, *J. Chem. Phys.* **38**, 1026 (1963).
6. F. A. Bovey, F. P. Hood, E. W. Anderson, and L. C. Snyder, *J. Chem. Phys.* **42**, 3900 (1965).
7. F. Heatley and F. A. Bovey, *Macromolecules* **1**, 301 (1968).
8. A. L. Segre, P. Ferruti, E. Toja, and F. Danusso, *Macromolecules* **2**, 35 (1969).
9. A. D. Williams and P. J. Flory, *J. Amer. Chem. Soc.* **91**, 3111 (1969).
10. F. Heatley and F. A. Bovey, unpublished observations.
11. Y. Fujiwara and P. J. Flory, *Macromolecules* **3**, 43 (1970).
12. F. A. Bovey, A. I. Brewster, and F. P. Hood, unpublished observations.

12a. C. Walling and E. S. Huyser, "Organic Reactions" (A. C. Cope, ed.), Vol. 13. Wiley, New York, 1963.

13. V. D. Mochel, *Macromolecules* **2**, 537 (1969).
14. J. F. Kinstle and H. J. Harwood, *Polymer Preprints* **10**(2), 1389 (1969).

14a. L. F. Johnson, F. Heatley, and F. A. Bovey, *Macromolecules* **3**, 175 (1970).

15. M. Kobayashi, K. Tsumura, and H. Tadokoro, *J. Polym. Sci. Part A-2* **6**, 1493 (1968).
16. G. W. Brady and R. Salovey, *J. Amer. Chem. Soc.* **86**, 3499 (1964).
17. Y. Ohsumi, S. Tani, T. Higashimura, and S. Okamura, *J. Macromol. Sci.* **A2**, 727 (1968).
18. G. Geuskens, J. C. Lubikulu, and C. David, *Polymer* **7**, 63 (1966).
19. G. Weill and G. Hermann, *J. Polym. Sci. Part A-2* **5**, 1293 (1967).
20. K. Matsuzaki and T. Sugimoto, *J. Polym. Sci. Part A-2* **5**, 1320 (1967).
21. J. Heller, D. O. Tieszen, and D. B. Parkinson, *J. Polym. Sci. Part A* **1**, 125 (1963).
22. S. Yoshimoto, Y. Akana, A. Kimura, H. Hirata, S. Kusabayashi, and H. Mikawa, *Chem. Commun.* **1969**, 987 (1969).
23. D. R. Williams, *Macromolecules* **3**, 602 (1970).
24. E. Gipstein and W. A. Hewett, *Macromolecules* **2**, 82 (1969).
25. J. Heller and D. B. Miller, *J. Polym. Sci. Part A-1* **5**, 2323 (1967).
26. V. E. Eskin and I. M. Serdjuk, paper presented at IUPAC Symp. on Macromol. Chem., Tokyo–Kyoto, Japan, 1966; *Preprints* **6**, 17 (1966).

Chapter VII

POLYOLEFINS AND RELATED POLYMERS

1. Polyethylene

The structural variables in polyethylene—long and short branches and olefinic unsaturation—occur in relatively small proportions (of the order of 1–2% or less) but are important in relation to physical properties. The traditional method for their determination has been infrared spectroscopy,[1] but NMR appears to be at least as powerful and has the advantage that it gives the concentrations directly without the need for establishing extinction coefficients. Kato and Nishioka[2] showed that methyl groups (i.e., branch terminals) can be detected in low density polyethylene by employing up to 400 "CAT" scans (Chapter I, Section 18C). As expected, the observation of these features is greatly facilitated by observing at 220 MHz. Ferguson[3] and Ferguson and Phillips[4] have reported the high frequency spectrum of low density polyethylene, run in the melt at 140°. The features (Fig. VII-1) observed are identified as A and C: vinyl $CH_2{=}CH{-}$; B: vinylene, $-CH{=}CH-$; D: vinylidene, $=CH_2$; E: $(CH_2)_n$; and F: CH_3.

Liu[5, 6] has shown that in aromatic solvents long-chain alkanes, i.e., polyethylene oligomers, exhibit a splitting of the methylene resonance. This behavior is illustrated in Fig. VII-2, which shows 60 MHz spectra of a number of normal alkanes from n-C_6H_{14} to n-$C_{28}H_{58}$, observed in α-chloronaphthalene at 35°. (The sharp peak at 0 Hz is that of the external dioxane reference.) The more shielded of the two hydrocarbon resonances is that of the terminal methyl groups; it of course decreases in intensity as the chain is lengthened. At C_{18}, the methylene line becomes observably split into two peaks. (This splitting would probably be resolved earlier in the series at higher observing frequency.) As the chain is lengthened the larger of the two peaks moves to lower field and increases in intensity relative to the other, the position of which does not change with increasing chain length beyond C_{18}. For polyethylene itself, the lower-field peak dominates so strongly that the other cannot be seen, and the spectrum becomes the same as peak E in Fig. VII-1. In nonaromatic solvents, this splitting of the alkane methylene resonance is not observed. Even dilution of α-chloronaphthalene with as little as 20% (by volume) of carbon tetrachloride greatly reduces it.[6] The effect is also reduced at higher temperatures.

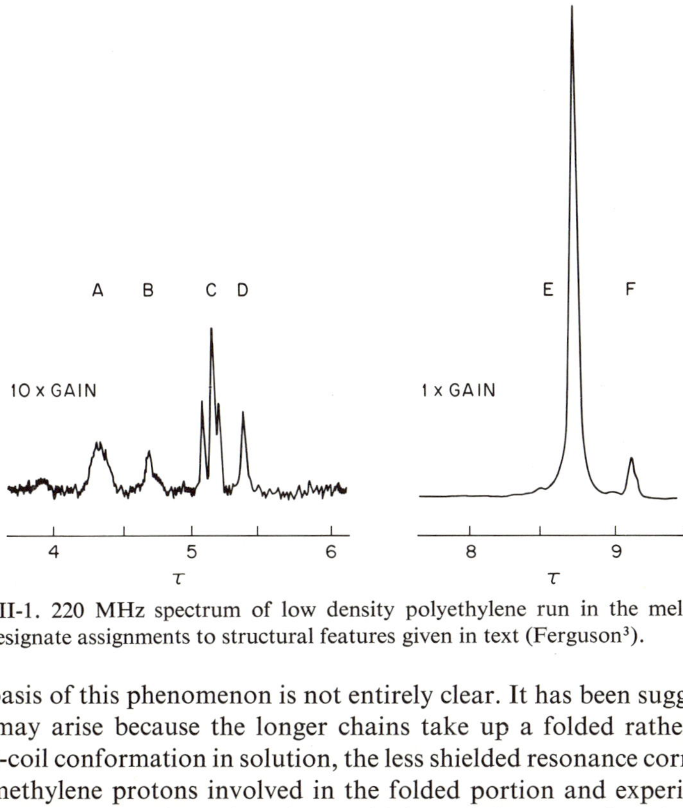

FIG. VII-1. 220 MHz spectrum of low density polyethylene run in the melt at 140°. Letters designate assignments to structural features given in text (Ferguson[3]).

The basis of this phenomenon is not entirely clear. It has been suggested[5–7] that it may arise because the longer chains take up a folded rather than a random-coil conformation in solution, the less shielded resonance corresponding to methylene protons involved in the folded portion and experiencing a different environment from the others, which are fully exposed to the solvent. It is also not clear whether the aromatic solvent allows the folding to occur, or whether it also occurs in nonaromatic solvents but that in such solvents the chemical shift difference is not sufficiently large to be resolved. Upon adding carbon tetrachloride to α-chloronaphthalene, it is observed that the upfield peak appears to coalesce with the lower-field peak without any obvious change in intensity, both moving toward a common midpoint. This suggests that the second alternative may be more nearly correct.

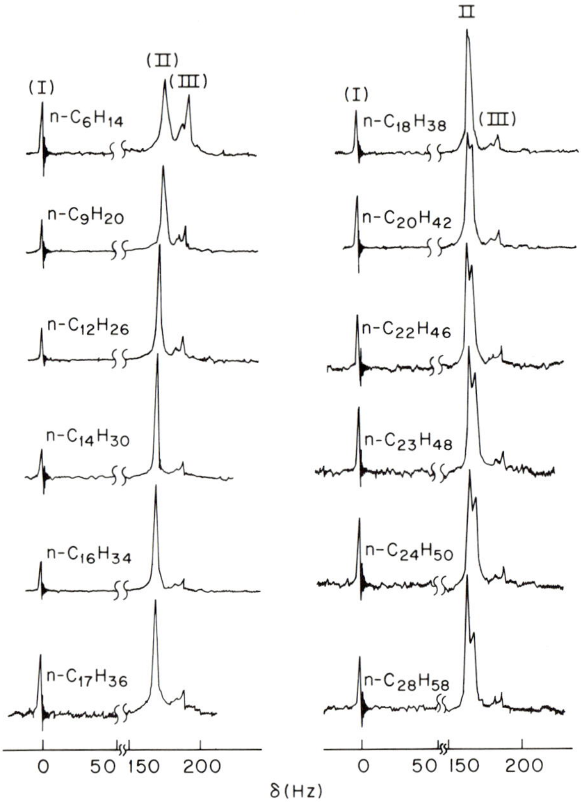

FIG. VII-2. 20 MHz spectra of *n*-alkanes, 6% in α-chloronaphthalene at 35°. Chemical shifts are expressed in Hz upfield from external dioxane (peak I). Peaks II and III are methylene and methyl absorptions, respectively (Liu[5]).

2. Polypropylene

Polypropylene is of major technological significance and also presents quite challenging problems in spectral interpretation. For these reasons, the NMR literature is quite large.[8–30] Despite apparent structural simplicity, the NMR spectra of propylene polymers are complex because the methyl protons couple strongly to the α-protons, and because the chemical shift differences between the α-, β-, and methyl protons are relatively small. The NMR of polypropylene has been thoroughly reviewed by Woodbrey[31] through approximately mid-1967, and we shall therefore emphasize the most recent results—particularly data obtained at 220 MHz.

The increased spectral simplicity attained in high magnetic fields is particularly striking in the case of polypropylene. In Fig. VII-3 are shown the 60 MHz spectra of (a) isotactic and (b) syndiotactic polymer observed in *o*-dichlorobenzene solution at 150°.[12] Near 8.4τ is the octet of the α-protons, the peaks broadened by the multiplicity of transitions in this strongly coupled system.

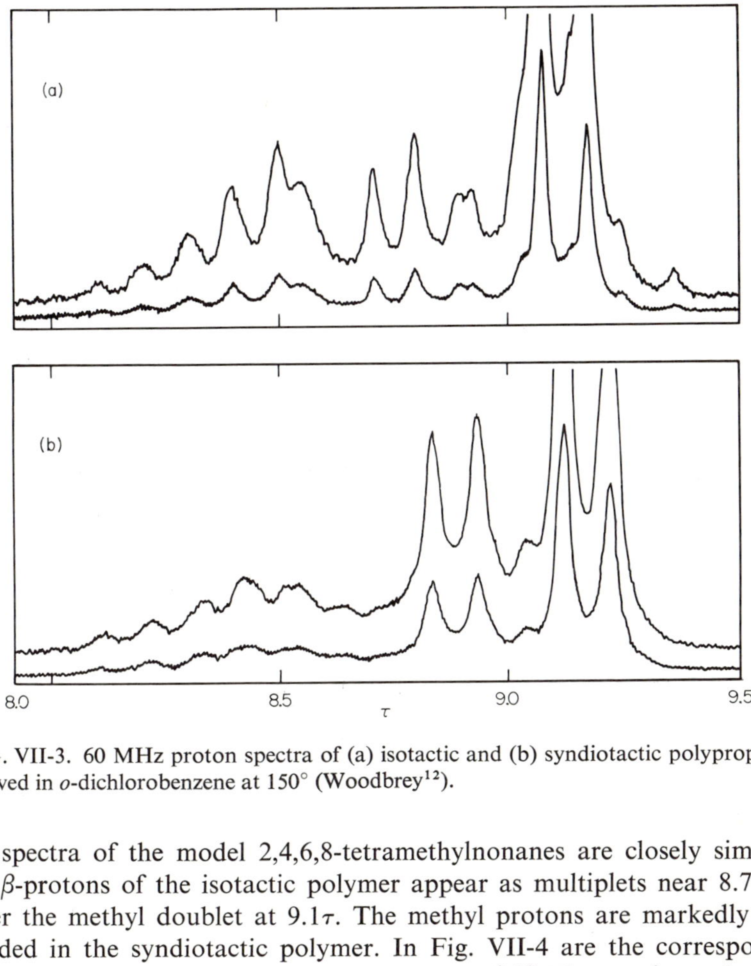

FIG. VII-3. 60 MHz proton spectra of (a) isotactic and (b) syndiotactic polypropylenes observed in *o*-dichlorobenzene at 150° (Woodbrey[12]).

The spectra of the model 2,4,6,8-tetramethylnonanes are closely similar.[16] The β-protons of the isotactic polymer appear as multiplets near 8.7τ and under the methyl doublet at 9.1τ. The methyl protons are markedly more shielded in the syndiotactic polymer. In Fig. VII-4 are the corresponding spectra at 220 MHz as observed by Ferguson.[3,22] The α-, β-, and methyl resonances are now clearly separated, the spectrum being nearly first order.

The spectrum in Fig. VII-4(a) fully confirms earlier conclusions from infrared and x-ray studies (see Woodbrey's review for detailed references), as well as from NMR studies of deuterated polymers,[9,17,20] that in polymers prepared with initiators of the $TiCl_3$–aluminum alkyl class,[32] the least soluble

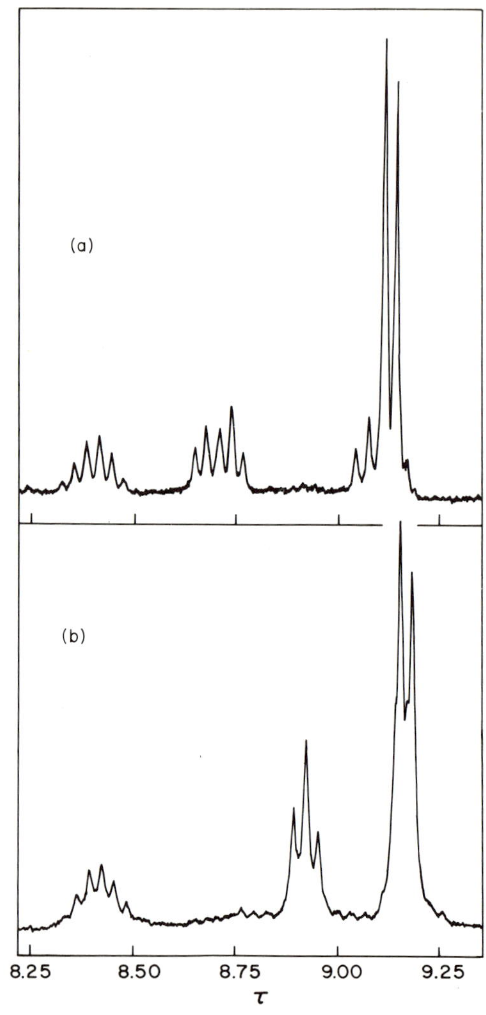

FIG. VII-4. 220 MHz proton spectra of (a) isotactic and (b) syndiotactic polypropylenes, observed in *o*-dichlorobenzene at 165° (Ferguson[3,22]).

and most readily crystallizable fraction is isotactic. The following parameters are obtained[20,27] (in *o*-dichlorobenzene at 110°):

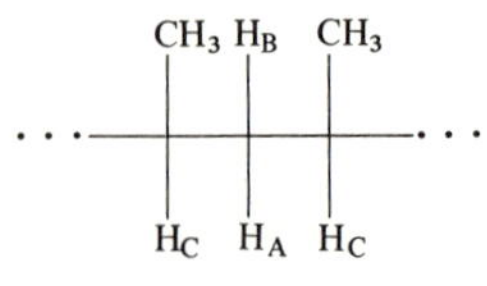

H_A (anti, or *threo* in β-d_1 polymers): $8.73_9\tau$ J_{AB}: -13.5 Hz

H_B (syn, or *erythro*): $9.13_5\tau$ J_{AC}: 6.0 Hz

H_C: $8.42_5\tau$ J_{BC}: 7.0 Hz

CH_3: $9.14_4\tau$ J_{CH_3}–H_C: 6.5 Hz

Spectrum VII-4(b) confirms earlier conclusions that polypropylenes prepared with homogeneous initiators such as VCl_4–anisole–$Al(C_2H_5)_2Cl$ in hydrocarbon solvents at low temperatures are predominantly syndiotactic. The parameters are [20, 24, 26] (in *o*-dichlorobenzene at 110°):

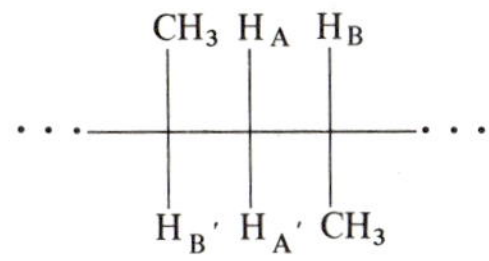

$H_A(H_{A'})$: $8.97_3\tau$ $J_{AA'}$: -13.5 Hz

$H_B(H_{B'})$: $8.46_3\tau$ $J_{AB}, J_{AB'}$: 4.8, 8.3 Hz

CH_3: $9.19_9\tau$ J_{CH_3}–$H_B(H_{B'})$: 6.5 Hz

The interpretation of these parameters in terms of chain conformation is discussed in Chapter IX.

The 220 MHz spectrum of "atactic" polypropylene, prepared with VCl_4–$Al(C_2H_5)_3$ at −78°,[24, 26] is shown in Fig. VII-5. The α- and *racemic* β-proton resonances appear relatively well resolved because the chemical shift differences among the triad and tetrad sequences, respectively, are very small for

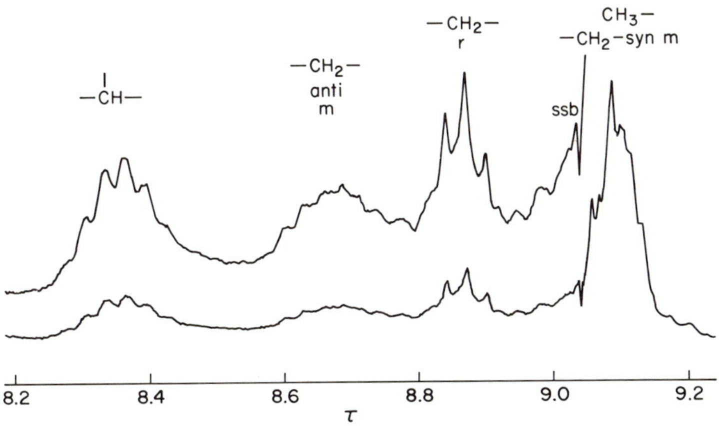

FIG. VII-5. 220 MHz proton spectrum of "atactic" polypropylene observed in *o*-dichlorobenzene at 150° (Zambelli and Segre[24]; Heatley and Zambelli[26]).

these protons. The *anti-meso-β*-protons give the more complex and seemingly more poorly resolved resonance at about 8.7τ.

Complete tetrad assignments have been made[24,28] using polymers prepared from *cis*- and *trans*-1,2,3,3,3-d_5 monomers,† i.e., with one remaining β-proton.

cis *trans*

Since propagation is known to proceed exclusively by *cis* addition to the double bond (Chapter VIII, Section 4),[19,25,33] only the *syn*- and *anti-meso*-methylene protons appear for the *cis* and *trans* monomers, respectively. Figure VII-6 shows the 220 MHz spectra of the polymers from (a) the *cis* monomer and (b) the *trans* monomer. Equivalent results can be obtained at

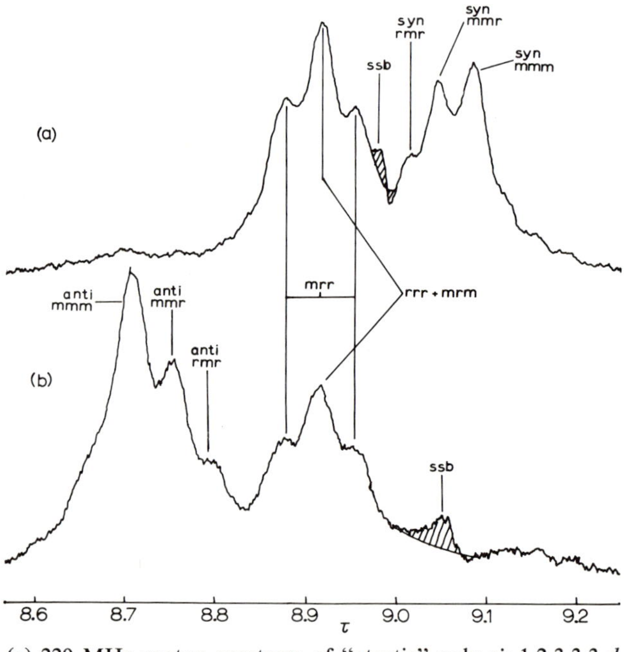

FIG. VII-6. (a) 220 MHz proton spectrum of "atactic" poly-*cis*-1,2,3,3,3-d_5-propylene, observed in 10% (w/v) solution in *o*-dichlorobenzene at 145°; (b) spectrum of poly-*trans*-1,2,3,3,3-d_5-propylene run under the same conditions (Heatley and Zambelli[28]).

† In some of the literature on polymers of β-deutero propylenes, the nomenclature is inconsistent with that employed for other β-deutero monomers (Chapter VIII, Section 4). The definitions adopted here are reversed from those of Zambelli *et al.*[19] and Heatley and Zambelli,[28] and are based on the relationship of the β-deuterium to the α-substituent (or on the relationship of the α-deuterium to the β-proton).

100 MHz by irradiating the deuterium nuclei. The *mmm* and *rrr* assignments were made by comparison with the corresponding stereoregular polymers. In contrast to the *m* centered tetrads, both chemical shifts of the *mrr* tetrad, which appear as a doublet straddling the central singlet, *must* be observed with equal intensity regardless of the mechanism of propagation. One of the *mrr* peaks corresponds to those formed in the direction $\overrightarrow{mrr}$ while the other corresponds to those formed in the direction $\overrightarrow{rrm}$. The *mmr* and *rmr* peaks were assigned on the basis of an expected regular progression in chemical shift of the central *m* dyad as the flanking *m* dyads are replaced by *r* dyads; this has been observed for polymethyl methacrylate (Chapter III, Section 2) and for polyvinyl chloride (Chapter V, Section 1), for which assignments can be rigorously made on the basis of Bernoullian statistics for the free radical polymers. The *mrm* was assigned the same chemical shift as *rrr* on the basis of intensity measurements and necessary statistical relationships among sequence probabilities which hold regardless of mechanism (Chapter III, Section 6).

An important aspect of the relationship between chain configuration and conformation—and one which is also intimately related to the mechanism of propagation—is the observation of the number and nature of configurational defects in highly stereoregular chains. In Fig. VII-7 are shown observed (a)

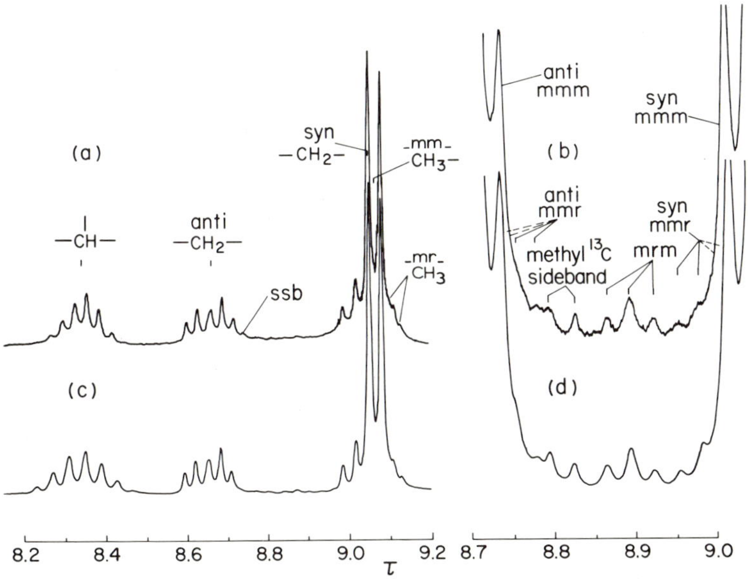

FIG. VII-7. (a) 220 MHz proton spectrum of isotactic polypropylene, observed in 10% (w/v) solution in *o*-dichlorobenzene at 140°; (b) expanded portion of (a); (c) and (d): calculated spectra corresponding to (a) and (b), using parameters in text (Heatley *et al.*[27]).

and calculated (c) spectra of a highly isotactic polypropylene, prepared with a $TiCl_3$–$Al(C_2H_5)_2Cl$ catalyst.[27] In (a) a weak but visible methyl doublet can be seen just to the high-field side of the *mm* doublet. Its position and intensity are consistent with the assignment to the *mr* or heterotactic triad. The calculated spectra were generated using the parameters given in Table VII-1. In (b), the region near 8.9τ, in which residual *racemic* resonances are expected to occur, is expanded horizontally (2×) and also vertically. A number of small peaks are observed which cannot be seen in (a). The triplet centered on 8.8τ is assigned to the *mrm* tetrad since it lies at the same chemical shift as the line assigned to this position in the d_5-atactic polymers, discussed previously. Although the *rrr* chemical shift is the same as that of *mrm*, the *rrr* tetrad is excluded because there are unlikely to be any observable syndiotactic sequences of three or more dyads in length in such a highly isotactic polymer. (It should be noted that Flory[30] has questioned this assignment, and has advanced reasons for assigning this resonance to *rrr* rather than *mrm*. This question is considered further in Chapter VIII, Section 3D and Chapter IX, Section 3C). The *syn* and *anti mmr* tetrad resonances are assigned as shown. The *mrr* tetrad is excluded on the basis of trial calculations of its chemical shift, which do not match observed features. The remaining resonance in this region is the lower ^{13}C sideband of the *mm* methyl doublet, identified from measurement of the spacing of the upper sideband, the assignment of which is unambiguous. Since the intensity of this doublet is known from the known abundance of ^{13}C (1.1%), it serves as a convenient intensity reference.[23] It is found that the proportion of *r* units is 2%.

In (d) is shown the spectrum in this region calculated using essentially the same chemical shift parameters (Table VII-1) as obtained for the atactic d_5-polymers, which are very close to the chemical shifts obtained from the isotactic and syndiotactic polymer spectra, the differences being probably chiefly due to the higher observing temperature. It must be emphasized that there are thus no arbitrary chemical shift adjustments in these calculations. The vicinal couplings of the α- and β-protons in *mmr* and *mrm* are likewise very similar to those obtained for the isotactic and syndiotactic polymers, respectively.

The identification of such configurational defects is of some significance for the interpretation of catalyst mechanism. This question is discussed in Section 3D of Chapter VIII.

Precise knowledge of the number and nature of *r* dyads in predominantly isotactic chains, combined with a measurement of the end-to-end distance of the polymer chain in solution, also permits conclusions to be drawn concerning the potential energies of various conformations of the chain. We shall therefore reserve discussion of these questions for Chapter IX. Flory and Fujiwara[29] have presented a careful theoretical study of the proton chemical

TABLE VII-1

NMR PARAMETERS USED IN THE CALCULATION OF THE SPECTRUM OF PREDOMINANTLY ISOTACTIC POLYPROPYLENE (IN *o*-DICHLOROBENZENE AT 140°)

Spin system notation:

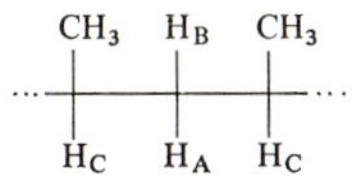

m

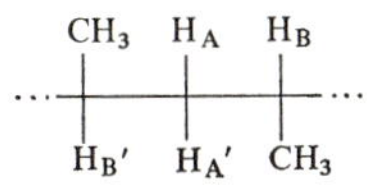

r

Resonance	Chemical shift, τ	Coupling constants, Hz	Line width, Hz	Intensity[a]
mmm	ν_A 8.67	$J_{AB} = -13.5$	3.0	0.94
	ν_B 9.05	$J_{AC} = 9.0$		
	ν_C 8.34	$J_{BC} = 7.0$		
mmr	ν_A 8.71	$J_{AB} = -13.5$	3.0	0.04
	ν_B 9.00	$J_{AC} = 6.0$		
	ν_C 8.34	$J_{BC} = 6.0$		
mrm	ν_A 8.88	$J_{AB} = 8.0$	3.0	0.02
	ν_B 8.34	$J_{AB'} = 5.0$		
		$J_{AA'} = -13.5$		
mm	ν_{CH_3} 9.07	$J_{\alpha-CH_3} = 6.5$	2.5	0.97
		$J_{^{13}C-H} = 124$		
mr	ν_{CH_3} 9.12	$J_{\alpha-CH_3} = 6.5$	2.5	0.03

[a] Normalized for each *n*(ad) species.

shifts in stereoirregular polypropylene. This is based primarily on consideration of the statistical probabilities of conformations of tetrad sequences, and is discussed in Chapter IX, Section 3C.

The high specificity of addition to the double bond in isotactic propagation demonstrated by NMR does not in itself enable one to assign the *erythro* and *threo* β-proton resonances. This assignment, assumed in the preceding discussion, is actually based in part on infrared studies. The determination of the direction of addition in syndiotactic placements requires more elaborate experiments. These questions are discussed in Chapter VIII, Section 4.

Segre[23] has employed highly isotactic poly-2,3,3,3-d_4-propene to demonstrate that the line widths of polymer spectra are not necessarily always as great as may be supposed (see Chapter I, Section 6). Upon irradiation of the deuterium nuclei, the lines of the AB quartet of the β-protons (Fig. VII-8) had a width of less than 0.5 Hz, similar to that of small molecules. It was concluded that in this polymer, and by extension most stereoregular polymers with comparable segmental freedom, the apparent line width results largely

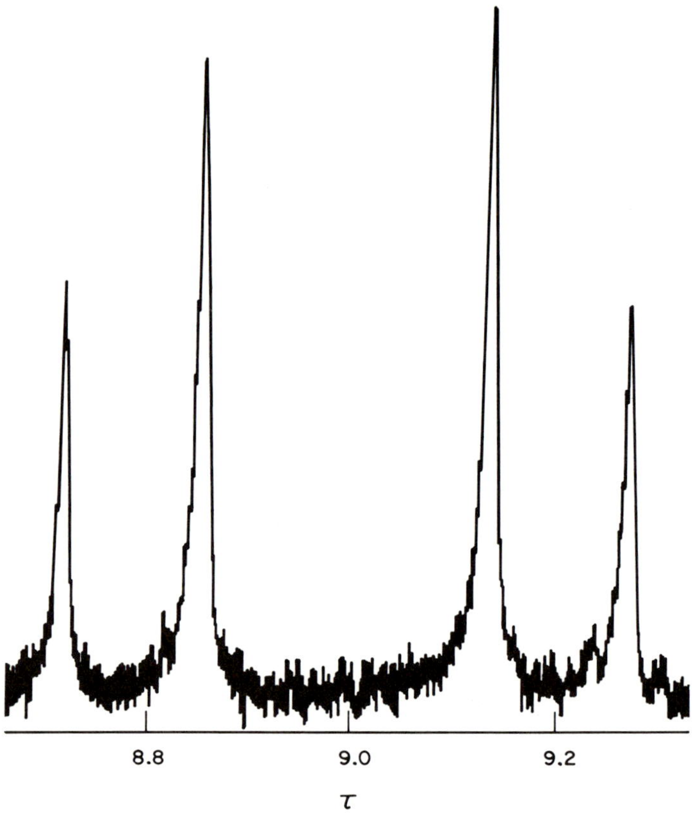

FIG. VII-8. 100 MHz spectrum of isotactic poly-2,3,3,3-d_4-propylene observed in *o*-dichlorobenzene solution at 150° with deuterium decoupling (Segre[23]).

from unresolved long-range couplings. It is not at present clear, however, that the line widths of isotactic polyisopropyl acrylate (Chapter IV, Section 2) and polystyrene (Chapter VI, Section 1) spectra, which seem to be nearly an order of magnitude greater than this, can be entirely explained on this basis.

The ^{13}C spectra of isotactic and atactic polypropylene were reported by Johnson *et al.*[34, 35] In Fig. VII-9, spectrum (a) is that of the isotactic polymer and (b) that of the atactic polymer—the latter a semiliquid material of low molecular weight. In both spectra, the protons are noise-decoupled (Chapter I, Section 18B). In (a), three singlets are observed, as expected. The chemical shifts are in reasonably good agreement with those calculated from the additive parameters obtained by Grant and Paul[36] from the ^{13}C chemical shifts in small linear and branched alkanes. Thus, the calculated shifts of the methylene, methine, and methyl carbons (referred to external $^{13}CS_2$) are 146, 166, and 173 ppm, respectively, compared to the experimental values of 147, 165, and 172 ppm.

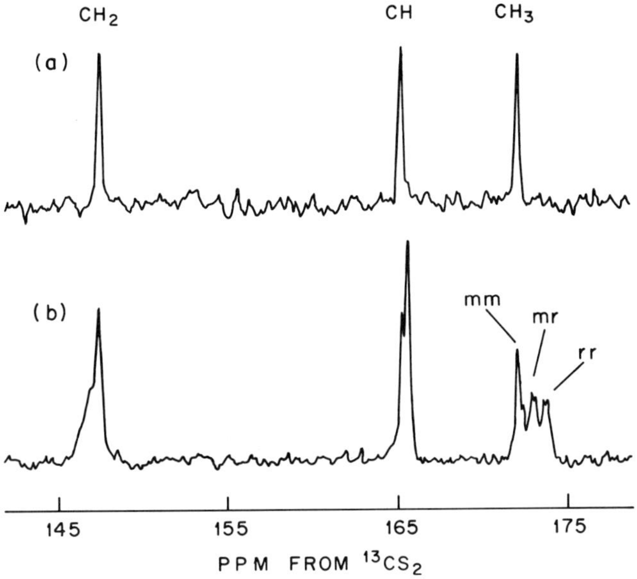

FIG. VII-9. ^{13}C spectrum (25 Hz) of polypropylenes; (a) isotactic, about 5% (w/v) in *o*-dichlorobenzene at 60°; 100 scans; (b) atactic, about 30% (w/v) in *o*-dichlorobenzene at 60°; 100 scans (Bovey[34]; Johnson *et al.*[35]).

In the spectrum of atactic polypropylene, Fig. VII-9(b), all three carbon resonances show stereochemical fine structure. The methyl resonances fall into three distinct groups, which were assigned as shown to isotactic, heterotactic, and syndiotactic triads on the basis of their chemical shifts relative to that of the isotactic polymer. Pentad resonances may also be distinguished as fine structure on the triad peaks, but resolution was not sufficient to permit assignments to be made, particularly since the propagation statistics are unknown. The triad intensities indicate that this "atactic" polymer is essentially of random configuration.

The methine carbon chemical shift is little affected by the polymer stereochemistry. However, the methylene carbon is observably split, and this represents a partial resolution of tetrad sequences.

3. Polymers of Higher Olefins

The NMR of higher olefin polymers has received less study than their technological significance justifies, mainly because at lower observing frequencies the side-chain resonances tend to obscure those of the main chain. It seems likely that measurements at higher frequencies will prove useful, but at present spectral interpretation is uncertain. In Fig. VII-10 the 220 MHz spectra of isotactic (a) and atactic (b) polybutene-1, observed in *o*-dichlorobenzene at 100°, are compared.[37] The expected apparently poorer resolution

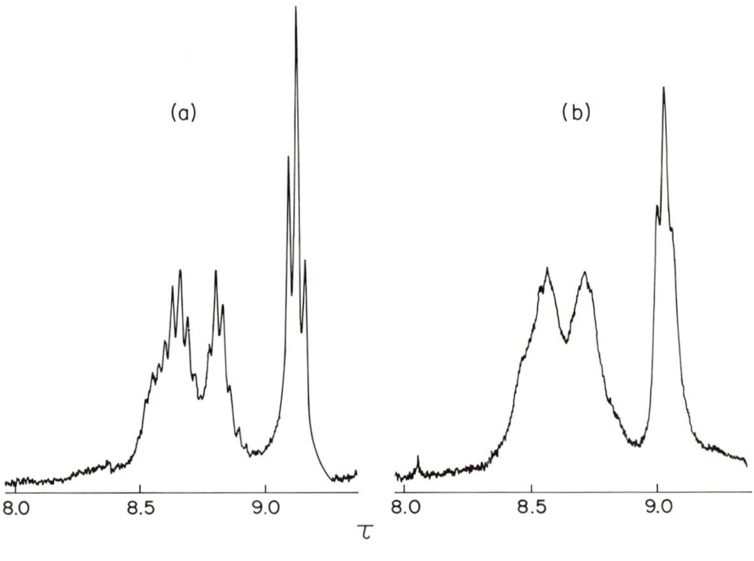

FIG. VII-10. 220 MHz proton spectra of (a) isotactic polybutene-1 and (b) "atactic" polybutene-1, observed in 5% (w/v) solutions in *o*-dichlorobenzene at 100° (Bovey and Brewster[37]).

is seen in (b), but the differences in the spectra are much less than for the corresponding polypropylenes. The methyl peak near 9τ is slightly less shielded in (b), with no clear indication of a residual *mm* peak, suggesting that this polymer is predominantly syndiotactic. It appears that the differentiation in *m* dyad proton chemical shifts must be much smaller than in polypropylene. Use of specifically deuterated monomers may be necessary for further analysis.

Isomerization during cationic polymerization has been demonstrated.[38–44]

For example, Edwards and Chamberlain[38] showed that 4-methylpentene-1 rearranges during polymerization:

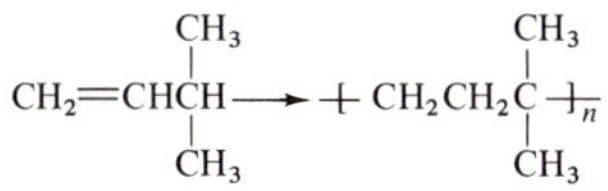

Similarly, Hudson[44] found that 3-methylbutene-1, 4-methylpentene-1, and 5-methylhexene-1 give predominantly polymers of structure (a) where $n = 2$, 3, and 4, respectively, while 3,3-dimethylbutene-1 gives (b).

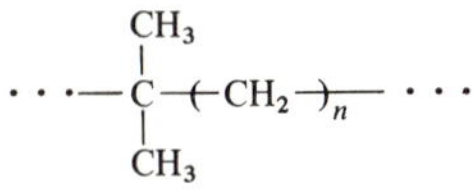

(a)

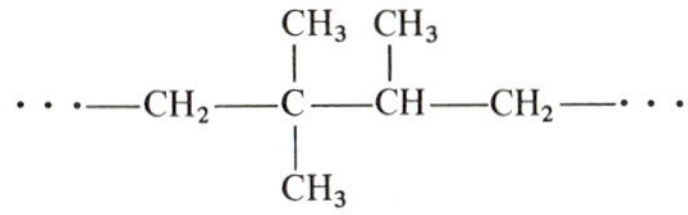

(b)

4. Polymers of Vinylcyclopropane and Substituted Vinylcyclopropanes

Ketley *et al.*[45] have examined the products of the cationic polymerization of vinylcyclopropane (a) and substituted vinylcyclopropanes (b), (c), (d), and (e). Spectra were obtained at 60 MHz and 190° using solutions in hexa-

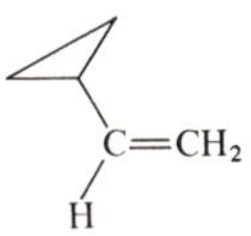

(a)

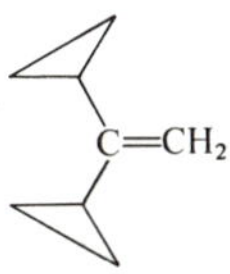

(b)

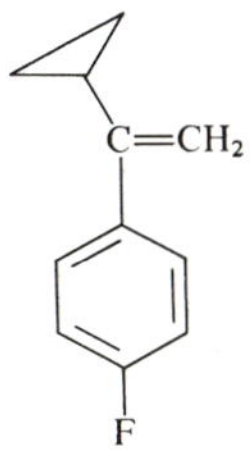

(c)

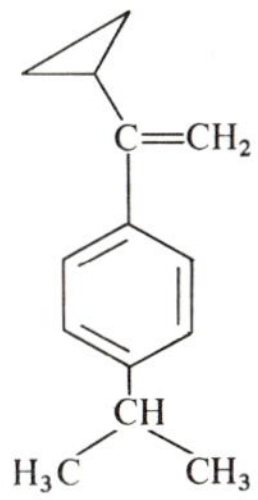

(d)

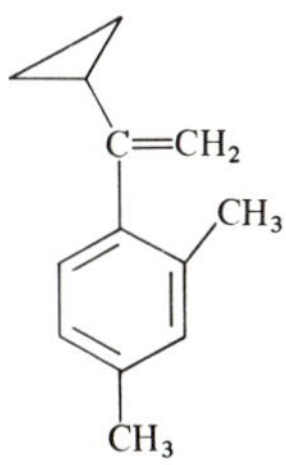

(e)

chlorobutadiene. The polymerization proceeds partially by simple opening of the vinyl double bond, yielding structure I, and partially by opening of the cyclopropane ring to yield II and III. The cyclopropyl ring protons are well known to appear at unusually high fields, usually about 9.5–9.8τ, probably

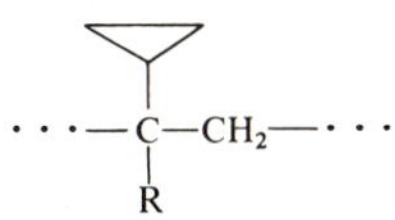

(I)

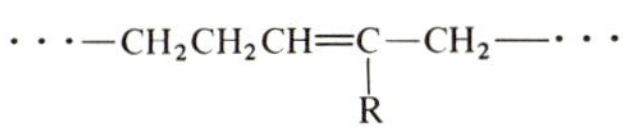

(II)

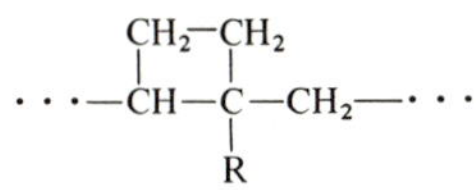

(III)

because of ring-current effects,[46–48] and could thus readily be identified. Table VII-2 shows their analysis of the polymers, based on NMR peak position

TABLE VII-2

Substituent R in 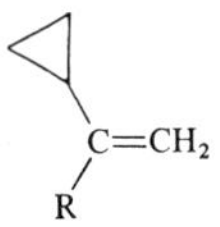	H		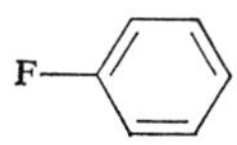	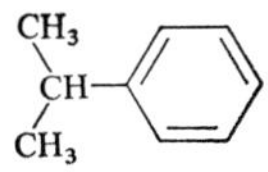	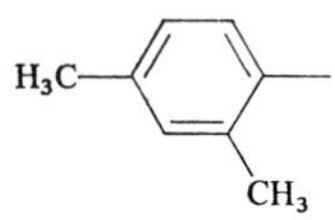
Chain structure I	60	73	20	0	0
II	10	27	50	0	60
III	30	0	30	100	30

and intensities. When R is a bulky substituent, the vinyl polymer chain is highly hindered sterically. The strain is relieved by ring opening.

REFERENCES

1. F. W. Billmeyer, Jr., "Textbook of Polymer Science," Chapter 13. Wiley, New York 1964, and references therein.
2. Y. Kato and A. Nishioka, *Bull. Chem. Soc. Jap.* **37**, 1614 (1964).
3. R. C. Ferguson, *Polymer Preprints* **8**(2), 1026 (1967).
4. R. C. Ferguson and W. D. Phillips, *Science* **157**, 257 (1967).
5. K.-J. Liu, *J. Polym. Sci. Part A-2* **5**, 1209 (1967).
6. K.-J. Liu, *J. Polym. Sci. Part A-2* **6**, 947 (1968).
7. K.-J. Liu and R. Ullman, *J. Polym. Sci. Part A-2* **6**, 451 (1968).
8. S. Satoh, R. Chujo, T. Ozeki, and E. Nagai, *J. Polym. Sci.* **62**, S101 (1962).
9. S. Stehling, *J. Polym. Sci. Part A* **2**, 1815 (1964).
10. W. C. Tincher, *Polymer Preprints* **3**, 142 (1962).
11. Y. Kato, N. Ashikari, and A. Nishioka, *Bull. Chem. Soc. Jap.* **37**, 1630 (1964).
12. J. C. Woodbrey, *J. Polym. Sci. Part B* **2**, 315 (1964).
13. J. C. Woodbrey and Q. A. Tremontozzi, *J. Polym. Sci. Part C* **8**, 113 (1965).
14. W. C. Tincher, *Makromol. Chem.* **85**, 34 (1965).
15. S. Onishi and K. Nukada, *J. Polym. Sci. Part B* **3**, 179, 1001 (1965).
16. P. Pino, S. Pucci, E. Benedetti, and P. Bucci, *J. Amer. Chem. Soc.* **87**, 3263 (1965).
17. G. Natta, E. Lombardi, A. L. Segre, A. Zambelli, and A. Marinangeli, *Chim. Ind.* **47**, 378 (1965).
18. E. Benedetti, S. Pucci, P. Pino, V. Schettino, and S. Califano, *Spectrochim. Acta* **A23**, 2371 (1967).
19. A. Zambelli, A. L. Segre, M. Farina, and G. Natta, *Makromol. Chem.* **110**, 1 (1967).
20. E. Lombardi, A. L. Segre, A. Zambelli, and A. Marinangeli, *J. Polym. Sci. Part C* **16**, 2539 (1967).

References

21. A. Zambelli, *Chem. Commun.* **1967**, 1252 (1967).
22. R. C. Ferguson, *Trans. N.Y. Acad. Sci.* **29**, 495 (1967).
23. A. L. Segre, *Macromolecules* **1**, 93 (1968).
24. A. Zambelli and A. L. Segre, *J. Polym. Sci. Part B* **6**, 473 (1968).
25. A. Zambelli, M. G. Giongo, and G. Natta, *Makromol. Chem*, **112**, 183 (1968).
26. F. Heatley and A. Zambelli, private communication (1968).
27. F. Heatley, R. Salovey, and F. A. Bovey, *Macromolecules* **2**, 619 (1969).
28. F. Heatley and A. Zambelli, *Macromolecules* **2**, 618 (1969).
29. P. J. Flory and Y. Fujiwara, *Macromolecules* **2**, 327 (1969).
30. P. J. Flory, *Macromolecules* **3**, 613 (1970).
31. J. C. Woodbrey *in* "The Stereochemistry of Macromolecules" (A. D. Ketley, ed.), Vol. 3. Dekker, New York, 1968.
32. G. Natta, I. Pasquon, A. Zambelli, and G. Gatti, *J. Polym. Sci.* **51**, 387 (1961).
33. T. Miyazawa and T. Ideguchi, *J. Polym. Sci. Part B* **1**, 389 (1963).
34. F. A. Bovey, *Polymer Preprints* **10**(1), 9 (1969).
35. L. F. Johnson, F. Heatley, and F. A. Bovey, *Macromolecules* **3**, 175 (1970).
36. D. M. Grant and E. G. Paul, *J. Amer. Chem. Soc.* **86**, 2984 (1964).
37. F. A. Bovey and A. I. Brewster, unpublished observations.
38. W. R. Edwards and N. F. Chamberlain, *J. Polym. Sci. Part A* **1**, 2299 (1963).
39. J. P. Kennedy, L. S. Minckler, Jr., G. G. Wanless, and R. M. Thomas, *J. Polym. Sci. Part A* **2**, 2093 (1964).
40. J. P. Kennedy and A. W. Langer, Jr., *Fortschr. Hochpolym. Forsch.* **3**, 508 (1964).
41. J. P. Kennedy, P. Borzel, W. Naegele, and R. G. Squires, *Makromol. Chem.* **93**, 191 (1966).
42. J. P. Kennedy, J. J. Elliott, and B. E. Hudson, *Makromol. Chem.* **79**, 109 (1964).
43. J. P. Kennedy, W. Naegele, and J. J. Elliott, *J. Polym. Sci. Part B* **3**, 729 (1965).
44. B. E. Hudson, *Makromol. Chem.* **94**, 172 (1966).
45. A. D. Ketley, A. J. Berlin, and L. P. Fisher, *J. Polym. Sci. Part A-1* **5**, 227 (1967).
46. J. R. Lacher, J. W. Pollock, and J. D. Park, *J. Chem. Phys.* **20**, 1047 (1952).
47. D. J. Patel, M. E. H. Howden, and J. D. Roberts, *J. Amer. Chem. Soc.* **85**, 3218 (1963).
48. J. J. Burke and P. C. Lauterbur, *J. Amer. Chem. Soc.* **86**, 1870 (1964).

Chapter VIII

NMR INVESTIGATIONS OF THE MECHANISM OF PROPAGATION IN VINYL POLYMERIZATION

1. Introduction

High resolution NMR is capable of supplying important and in some cases quite subtle information concerning the mechanism of vinyl polymerization. One instance of this is the determination of the structure of styrene telomers (Chapter VI, Section 1), which confirms the accepted ideas concerning initiation and chain transfer in these systems. This sort of information, however, can also be obtained by classical methods of organic structure determination. In this chapter we shall be concerned with the measurement of configurational sequences and of *erythro* and *syn* proton intensities, which for the most part can be carried out only by NMR.

2. The Effect of Polymerization Temperature on Configuration

The dependence of stereochemical configuration on the temperature of polymerization has a significant bearing on our theoretical understanding of the mechanism of propagation and is clearly amenable to measurement by NMR.

If we express the propagation steps in terms of absolute reaction rate theory, we may express the rate constant for isotactic or m placement of monomer units as

$$k_m = (kT/h)\exp[(\Delta S^{\ddagger}_{m}/R) - (\Delta H^{\ddagger}_{m}/RT)] \qquad \text{(VIII-1)}$$

and for syndiotactic or r replacement

$$k_r = (kT/h)\exp[(\Delta S_r^{\ddagger}/R) - (\Delta H_r^{\ddagger}/RT)] \tag{VIII-2}$$

from which we have for the relative frequency of m placement (Chapter III, Section 4):

$$P_m = \frac{k_m}{k_m + k_r} \tag{VIII-3a}$$

$$= \frac{\exp[-(\Delta F_m^{\ddagger} - \Delta F_r^{\ddagger})/RT]}{1 + \exp[-(\Delta F_m^{\ddagger} - \Delta F_r^{\ddagger})/RT]} \tag{VIII-3b}$$

and

$$k_m/k_r = P_m/(1 - P_m)$$

$$= \exp\{[(\Delta S_m^{\ddagger} - \Delta S_r^{\ddagger})/R] - [(\Delta H_m^{\ddagger} - \Delta H_r^{\ddagger})/RT]\} \tag{VIII-4}$$

The differences in activation enthalpies and entropies for isotactic and syndiotactic propagation (the measurable quantities) are given by

$$\Delta H_m^{\ddagger} - \Delta H_r^{\ddagger} = \Delta(\Delta H_p^{\ddagger}) = -R\,\partial \ln[P_m/(1 - P_m)]/\partial(1/T) \tag{VIII-5}$$

and

$$\Delta S_m^{\ddagger} - \Delta S_r^{\ddagger} = \Delta(\Delta S_p^{\ddagger}) = R\ln[P_m/(1 - P_m)] + \Delta(\Delta H_p^{\ddagger})/T \tag{VIII-6}$$

This treatment of course implies that the configurational statistics are Bernoullian, i.e., can be described by a single probability of isotactic placement, as is almost always the case for free radical propagation (Chapter III, Section 4).

A. Methyl Methacrylate and Related Monomers

Measurements of P_m for free radical methyl methacrylate propagation have been obtained over a 178° polymerization temperature range by Bovey,[1] Fox and Schnecko,[2] and Otsu *et al.*[3] The values range from 0.13 at –78°[1] to 0.36 at 250°.[3] Figure VIII-1 shows an Arrhenius plot of these data according to Eq. (VIII-5), the low temperature points of Fox and Schnecko[2] being favored as very probably the most accurate. From this plot, values of 1 kcal for $\Delta(\Delta H_p^{\ddagger})$ and 1 eu for $\Delta(\Delta S_p^{\ddagger})$ are obtained. Thus, in this system, syndiotactic placement appears to be favored by the additional enthalpy of activation required for isotactic placement, which itself is favored by entropy.

For 1-menthyl methacrylate, Sobue *et al.*[4] observed P_m for free radical propagation to vary from 0.37 at –78° to 0.44 at 80°. The bulky side chain appears to encourage isotactic placement; $\Delta(\Delta H_p^{\ddagger})$ is about 0.5 kcal, only about half the value for methyl methacrylate.

The temperature dependence of the configuration of polycyclohexyl methacrylate is reported[5] to be slightly smaller than for polymethyl methacrylate. Loebl and O'Neill[5a] found that for the polymerization of methacrylic acid in methanol, $\Delta(\Delta H_p^{\ddagger}) = 1.2$ kcal and $\Delta(\Delta S_p^{\ddagger}) = 1.5$ kcal.

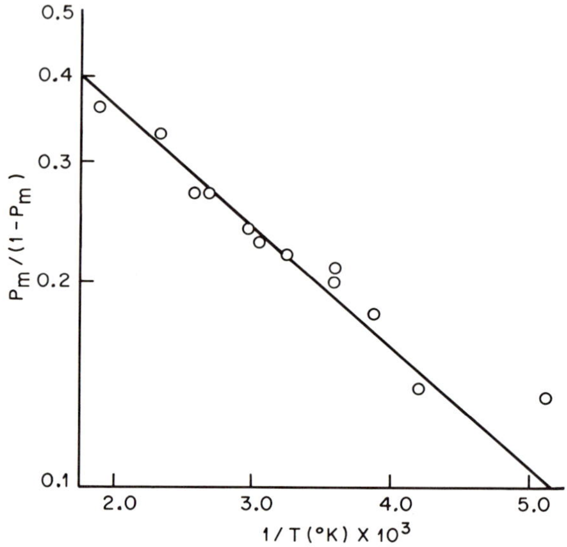

FIG. VIII-1. Arrhenius plot of the stereoregularity of free radical polymethyl methacrylate, plotted according to Eq. (VIII-5) (data from Bovey[1]; Fox and Schnecko[2]; Otsu *et al.*[3]).

B. METHACRYLONITRILE

Contradictory reports have been published (Chapter III, Section 11) concerning the stereochemistry of polymethacrylonitrile and its dependence on temperature.[6,7] It was first believed[6] that free radical initiators produce a polymer containing no *rr* units, but it was later[7] stated that the polymer is nearly atactic (P_m about 0.55) and that its configuration is independent of the polymerization temperature. This latter conclusion is clearly more reasonable than the first, but contrasts unexpectedly with the observations for methyl methacrylate. This polymer would repay further study.

C. VINYL CHLORIDE

The effect of temperature on the stereochemical configuration of polyvinyl chloride has been the subject of much study and dispute, centering principally on contradictory interpretations of NMR and infrared findings. Fordham *et al.*[8] estimated a value of 600 cal for $\Delta(\Delta H_p^\ddagger)$, while Bovey and Tiers[9] suggested a value of only about 200 cal. Neither group had a substantial experimental basis for their conclusions, and it now appears that they bracket the correct value. Germar *et al.*[10] reported NMR and infrared data pointing to a much larger value of $\Delta(\Delta H_p^\ddagger)$ than either of these earlier estimates, but there were clearly discrepancies in samples prepared at the lower temperatures.

Bargon *et al.*[11] reported 350 cal and 1200 cal for polymers prepared in emulsion systems with water and methanol, respectively, as the continuous phase. Such a difference is very hard to rationalize and is probably not real. Bovey *et al.*[12] deduced a value of 310 cal for solution-polymerized poly-α-d_1-vinyl chloride, using computer simulation of 60 MHz spectra, and essentially the same chemical shift analysis already presented in Chapter 5, Section 1. Cavalli *et al.*[13] reported 630 cal for bulk-polymerized polymer. In most studies positive values of $\Delta(\Delta S_p^{\ddagger})$ of the order of 1 eu were obtained. It appears probable that the discrepancies in activation enthalpy are not true effects and that the larger values represent interpretive errors.

There is, however, no doubt that polyvinyl chloride prepared at low temperature differs in its physical properties, particularly in crystallizability, from that prepared at normal temperatures.[8,14] Talamini and Vidotto[15] suggested that *r* blocks of four or five units might form crystallites, but in a Bernoullian structure with a P_m of about 0.4 such blocks would constitute at most only about 0.1–0.2 of the monomer units. There is now ample evidence that vinyl polymer chains may crystallize despite a high degree of configurational irregularity. It appears likely that the enhanced crystallizability is due primarily to decreased branching in the low-temperature polymers[13,14,16–18] rather than to the relatively small increase in syndiotactic sequences (e.g., P_m of 0.46 and 0.37 for polymers prepared at 100 and $-78°$, respectively[12]).

Rosen *et al.*[19] concluded from x-ray diffraction studies that polyvinyl chloride prepared by free radical initiation in solvents such as butyraldehyde is more stereoregular (presumably more syndiotactic) than when prepared in emulsion or in ketones. A number of similar claims have been made by other authors (references given in Böckman[14]) for polyvinyl chloride and other vinyl monomers. However, Bovey and Tiers[9] were not able to detect any difference in the stereoregularity of polyvinyl chloride so prepared, and this finding was confirmed by Böckman,[14] who suggested that decreased molecular weight led to the enhanced crystallinity reported.

D. Acrylonitrile

Polyacrylonitrile is semicrystalline, suggesting a high degree of stereoregularity. This indeed has been claimed on the basis of infrared observations[20,21] although other infrared studies[22] have led to conclusions in agreement with the NMR findings (Chapter IV, Section 4) that polyacrylonitrile is in fact nearly random. Differences in crystallization behavior[23,24] and solution dimensions[25] with polymerization temperature have led to suggestions of greater syndiotactic regularity at lower temperatures. This conclusion is similar to those advanced for polyvinyl chloride (Section 2C) and has even less foundation in experimental fact. The β-methylene spectra of polymers prepared with free radical initiators over a nearly 200° temperature range (Fig. IV-2) are unchanged within experimental error. It is clear that both $\Delta(\Delta H_p^{\ddagger})$ and $\Delta(\Delta S_p^{\ddagger})$ are nearly zero, in agreement with the potential energy calculations of Krigbaum and Tokita,[26] the relatively compact cyano groups, like chlorine atoms, being accepted into the crystal lattice whether the configuration is regular or not. The effects of low polymerization temperature are due very probably primarily to decreased molecular weight and branching, resulting in enhanced crystallizability as in polyvinyl chloride.

E. Vinyl Esters and Polyvinyl Alcohol

We have seen in Chapter IV, Section 5, that the polymerization solvent has little or no influence on the stereoregularity of polyvinyl acetate or, of course, on that of the derived polyvinyl alcohol; in particular, isobutyraldehyde did not produce the claimed increase [19] in syndiotactic structure. The polymer is nearly atactic at all polymerization temperatures; therefore, as for acrylonitrile, $\Delta(\Delta H_p^{\ddagger})$ and $\Delta(\Delta S_p^{\ddagger})$ must be nearly zero. Despite this, polyvinyl alcohol can attain a substantial degree of crystallinity, this being no doubt aided by hydrogen bonding. Polyvinyl trifluoroacetate derived from polyvinyl acetate becomes increasingly crystalline as the polymerization temperature of the vinyl acetate is decreased.[27] The most probable explanation is that already indicated in previous sections.

F. Trifluorochloroethylene

The case of polytrifluorochloroethylene is particularly striking, for this polymer can attain a high degree of crystallinity (mp 218°), yet is highly irregular in configuration (Chapter V, Section 5). The approximately 2:1 m/r ratio is independent of the polymerization temperature over a 110° range, and we therefore have

$$\Delta(\Delta H_p^{\ddagger}) \simeq 0.0 \text{ cal}; \qquad \Delta(\Delta S_p^{\ddagger}) \simeq -1.4 \text{ eu}$$

G. Alkyl Acrylates

Polymethyl and isopropyl acrylates, prepared with free radical initiators, are about 2:1 in m/r ratio, and this is substantially independent of polymerization temperature.[28] The values of $\Delta(\Delta H_p^{\ddagger})$ and $\Delta(\Delta S_p^{\ddagger})$ in Section 2F therefore apply.

The general conclusions that one may tentatively draw from the limited data at present available is that for α,α'-disubstituted monomers (methyl methacrylate is actually the only well established example) free radical m placement requires about 1 kcal more activation enthalpy than r placement, but is somewhat favored by entropy. In α-monosubstituted monomers, *r* placement tends to be favored by both entropy (but when the substituent is CN or OAc there is little or no difference) and enthalpy, although $\Delta(\Delta H_p^{\ddagger})$ does not exceed about 300 cal and is often nearly zero. Solvation of the free radical is not a significant effect, since chain configuration appears to be independent of the choice of polymerization solvent. (There remains the possibility, however, that strong hydrogen bonding between monomer and solvent might have a significant effect. This has not been tested.) In short, the only configurational control of free radical vinyl polymerization open to the experimentalist is temperature, and this is at best a weak one.

3. Configurational Statistics and Propagation Mechanism

In Chapter III we described the observation of longer configurational sequences (up to pentad) with particular reference to methyl methacrylate polymerization and discussed the generation of such sequences by Bernoulli trial processes, the simplest type of sequence statistics. We now wish to consider this problem in a more general sense.

A. Bernoulli and First-Order Propagation Models

In Fig. VIII-2 are shown the building up of a chain by (a) Bernoulli trial and (b) first-order Markov steps. It is perhaps generally understood what each of these types of sequences implies concerning the mechanism producing it, but we shall nevertheless consider these distributions in some detail here, because one sometimes finds statements in the literature concerning their interpretation that are inconsistent or misleading. In Bernoulli trial propagation, the chain end is not represented as having any particular stereochemistry, i.e., not only is it not *d* or *l* (a terminology generally to be avoided; see Chapter II, Section 2; Chapter III, Section 2), but it is also unimportant whether it is *m* or *r*. As we have seen (Chapter III, Section 4), the process is thus like reaching at random into a large jar containing balls marked *m* and *r*, the fraction of *m* balls being P_m. The outcome of the choice does not depend on the previous choice, the jar being assumed large enough so that neither species is appreciably depleted. It is sometimes said of this mechanism that the addition "is influenced only by the end unit of the growing chain." Such statements should not be understood to imply that it is influenced by the *stereochemistry* of the end of the growing chain, for *one* monomer chain unit considered alone as in (a) in Fig. VIII-2 has no stereochemistry.†

† It is probable that in free radical propagation (and possibly anionic and cationic as well) the growing radical chain end is actually trigonal, i.e., "flat," and that its stereochemistry is decided as the next monomer adds. These considerations do not affect the formal statistical arguments.

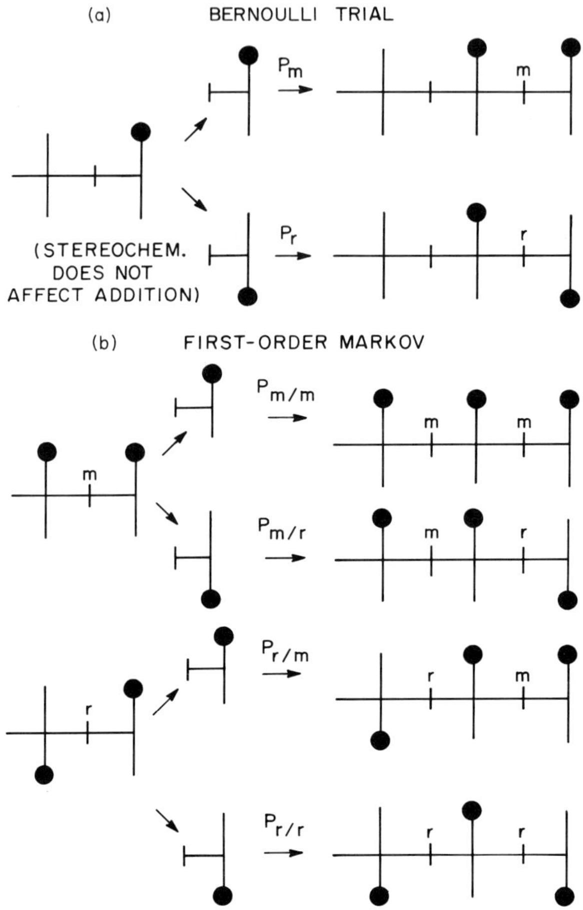

FIG. VIII-2. Schematic representation of Bernoulli trial and first-order Markov propagation steps.

The first-order Markov sequence is generated by propagation steps in which the adding monomer *is* influenced by the stereochemistry of the growing chain end, which may be m or r. We now have four probabilities characterizing the addition process: $P_{m/m}$, $P_{r/m}$, $P_{m/r}$, and $P_{r/r}$. (The designation $P_{r/m}$ means here the probability that the monomer adds in m fashion to an r chain end, etc.) For the triad frequencies we have:

$$(mm) = \frac{P_{m/m} P_{r/m}}{P_{m/r} + P_{r/m}} \tag{VIII-7}$$

$$(mr) = \frac{2P_{m/r} P_{r/m}}{P_{m/r} + P_{r/m}} \tag{VIII-8}$$

$$(rr) = \frac{P_{r/r}P_{m/r}}{P_{m/r} + P_{r/m}} \tag{VIII-9}$$

We also have the conservative relationships

$$P_{m/m} + P_{m/r} = 1 \tag{VIII-10}$$

and

$$P_{r/r} + P_{r/m} = 1 \tag{VIII-11}$$

There are thus really only two independent probabilities, which we shall take as $P_{r/m}$ and $P_{m/r}$. The propagation may be characterized if we can determine values of these conditional probabilities. Of course, one can also easily imagine second-order Markov processes, characterized by four conditional probabilities, and implying an influence of an additional chain unit, and still higher-order Markov propagation steps. One can also imagine non-Markov processes, one of which we shall discuss later.

Concerning the fitting of data to postulated mechanisms, we can say the following[29]:

(i) From dyad information alone, i.e., (m) and (r), any mechanism can be fitted but none can be tested. By "tested," we mean that a polymer can be shown to be consistent or inconsistent with a given model at a given level of sequence discrimination. But there is always the possibility that examination with higher discrimination may reveal inconsistencies with the proposed model.

(ii) From triad information (which implies dyad as well), i.e., (mm), (mr), and (rr), a Bernoulli model can be tested and Markov models of any order can be fitted. In addition, certain non-Markov models can be tested, as we shall see.

(iii) From tetrad information, a first-order Markov model can be tested, and higher orders fitted.

(iv) From *pentad* information, a second-order Markov model can be *tested*, and higher orders fitted.

The extension of such statements to higher sequences is obvious.

From Fig. III-3 and Table III-1 we find that:

$$(mm) = P_m{}^2 = (m)^2 \tag{VIII-12}$$

$$(mr) = 2P_m(1 - P_m) = 2(m)(r) \tag{VIII-13}$$

$$(rr) = (1 - P_m)^2 = (r)^2 \tag{VIII-14}$$

These relationships may be used for testing peak intensities for conformity to Bernoullian statistics; this is equivalent to fitting the observed points to the parabolic curves of Fig. III-3. Dyad data provide useful support through the dyad–triad necessary relationships of Table III-2, but are not necessary.

One point should be emphasized in connection with such fitting to supposed Bernoullian schemes. Triad intensities will fit Bernoullian frequencies so long as the placement of the adding monomer unit is independent of whether the *end* is *m* or *r*; it can be influenced to an arbitrarily high degree by the stereochemistry of the next unit from the end and still appear Bernoullian by this test. Thus, for example, we may imagine that by a second-order Markov process, we may generate chains with repeating *mmrr* units:

$$\cdots mmrrmmrrmmrr \cdots$$

By observation of only dyad and triad intensities, such a polymer would appear to be random or "atactic," since $(m) = (r)$ and $(mm) = (rr) = 0.5(mr)$. From the observation of tetrad frequencies, however, it would become obvious that this is not the case, since from Fig. III-3 and Table III-2 we find that for a random polymer,

$$(mmm) = (rrr) = (rmr) = (mrm) = \tfrac{1}{8}$$

and

$$(mmr) = (rrm) = \tfrac{1}{4}$$

while for the above polymer we have only

$$(mmr) = (rrm) = \tfrac{1}{2}$$

since the other tetrad sequences are absent.

The 60 MHz spectrum of free radical polymethyl methacrylate shown in Fig. III-2(a), polymer I, may be tested for conformity to Bernoullian statistics by application of Eqs. (VIII-12)–(VIII-14), or in a slightly different way, by using Eqs. (VIII-15) and (VIII-16), obtained from Eqs. (VIII-7)–(VIII-11).

$$P_{m/r} = \frac{(mr)}{2(mm) + (mr)} \quad [= (1 - P_m)] \tag{VIII-15}$$

$$P_{r/m} = \frac{(mr)}{2(rr) + (mr)} \quad [= P_m] \tag{VIII-16}$$

The relationships in brackets are true if the statistics are indeed Bernoullian, so that in this case $P_{m/r} + P_{r/m} = 1$. (We shall designate $P_{m/r} + P_{r/m}$ as P_Σ in some contexts.) From measurements (Chapter III, Section 3) of α-methyl resonance intensities we find (Table VIII-1):

$$(mm) = 0.04 \qquad P_{m/r} = \frac{0.36}{0.08 + 0.36} = 0.82$$

$$(mr) = 0.36 \qquad P_{r/m} = \frac{0.36}{1.20 + 0.36} = 0.23 = P_m$$

$$(rr) = 0.60 \qquad P_\Sigma = 1.05$$

The polymer is thus Bernoullian within the probable experimental error.

When this test is applied to the spectrum of polymer II [Fig. III-2(b)], we find (see triad data in Table VIII-2):

$$P_{m/r} = \frac{0.14}{1.50 + 0.14} = 0.08_5$$

$$P_{r/m} = \frac{0.14}{0.22 + 0.14} = 0.39$$

$$P_\Sigma = 0.47_5$$

Thus, as already indicated, this polymer clearly does not conform to Bernoullian statistics.

The 220 MHz spectra of these polymers (Figs. III-6 and III-7) permit a more searching test of the configurational statistics. In Table VIII-1 the tetrad and pentad intensities for polymer I, assigned as shown in these figures, are compared to the predicted intensities based on a Bernoullian model. They agree within experimental error, assuming a value of P_m of 0.24. (This value is slightly higher than that obtained from the preceding triad analysis and should be regarded as more nearly correct, since the frequencies of higher sequences are of course more sensitive measures of P_m.)

In order to explore further the statistics of polymer II, we must examine the fitting of tetrad and pentad sequence frequencies to the first-order Markov model. We shall find that we must also consider more complex propagation models: (i) second-order Markov, and (ii) possible non-Markovian models of which we select the Coleman–Fox model.

B. Fitting of Longer Sequences to the First-Order Markov Model

We first recast the triad relations of Eqs. (VIII-7) and (VIII-9) in terms of $P_{m/r}$ and $P_{r/m}$.

$$(mm) = \frac{(1 - P_{m/r})P_{r/m}}{P_{m/r} + P_{r/m}} \tag{VIII-17}$$

$$(mr) = \frac{2P_{m/r}P_{r/m}}{P_{m/r} + P_{r/m}} \tag{VIII-8}$$

$$(rr) = \frac{(1 - P_{r/m})P_{r/m}}{P_{m/r} + P_{r/m}} \tag{VIII-18}$$

For tetrad sequences:

$$(mmm) = \frac{(1 - P_{m/r})^2 P_{r/m}}{P_{m/r} + P_{r/m}} \quad \left[= \frac{(mm)^2}{(m)}\right] \tag{VIII-19}$$

TABLE VIII-1

BERNOULLI MODEL FITTING FOR POLYMER I

Dyad
(m): 0.24
(r): 0.76

		Observed	Bernoulli trial, $P_m = 0.24$
Triad			
	(mm)	0.04	0.06
	(mr)	0.36	0.36
	(rr)	0.60	0.58
Tetrad			
	(mmm)	~0.00	0.01
	(mmr)	0.07	0.09
	(rmr)	0.19	0.20
	(mrm)	0.04	0.04
	(mrr)	0.23	0.23
	(rrr)	0.43	0.44
Pentad			
	($mmmm$)	~0.00	0.003
Peak 1	($rmmr$)	0.05	0.03
Peak 2	($mmmr$)	0.02	0.02
Peak 3	($rmrr$), ($mmrr$)	0.25	0.27
Peak 4	($rmrm$)	0.06	0.07
Peak 5	($mmrm$)	0.02	0.02
Peak 6	($rrrr$)	0.40	0.39
Peak 7	($mrrr$)	0.14	0.16
Peak 8	($mrrm$)	0.05	0.03

$$(mmr) = \frac{2P_{m/r}(1-P_{m/r})P_{r/m}}{P_{m/r}+P_{r/m}} \quad \left[=\frac{(mm)(mr)}{(m)}\right] \tag{VIII-20}$$

$$(mrm) = \frac{P_{m/r}P_{r/m}^2}{P_{m/r}+P_{r/m}} \quad \left[=\frac{(mr)^2}{4(r)}\right] \tag{VIII-21}$$

$$(mrr) = \frac{2P_{m/r}P_{r/m}(1-P_{r/m})}{P_{m/r}+P_{r/m}} \quad \left[=\frac{(mr)(rr)}{(m)}\right] \tag{VIII-22}$$

$$(rmr) = \frac{P_{m/r}^2P_{r/m}}{P_{m/r}+P_{r/m}} \quad \left[=\frac{(mr)^2}{4(m)}\right] \tag{VIII-23}$$

$$(rrr) = \frac{P_{m/r}(1 - P_{r/m})^2}{P_{m/r} + P_{r/m}} \quad \left[= \frac{(rr)^2}{(r)} \right] \tag{VIII-24}$$

It is desirable and sufficient to show that *four* (not five) of these six relationships hold, but this is not always experimentally feasible because of overlapping and masking of resonances. The relationships in brackets hold for Bernoullian statistics and for first-order Markov statistics. They can therefore be useful for testing deviations from first-order Markov statistics. Thus for polymer II we find (see Table VIII-2, p. 163) that $(mm) = 0.75$, $(m) = 0.82$, and so $(mm)^2/(m) = 0.69$; the observed value of (mmm) is 0.70, in agreement with Eq. (VIII-19) within experimental error. As we shall see shortly, more stringent tests can be applied.

For pentad sequences:

$$(mmmm) = \frac{P_{r/m}(1 - P_{m/r})^3}{P_{m/r}(1 - P_{r/m}) + 2P_{m/r}P_{r/m} + P_{r/m}(1 - P_{m/r})} \quad \left[= \frac{(mmm)^2}{(mm)} \right] \tag{VIII-25}$$

$$(mmmr) = \frac{2P_{m/r}P_{r/m}(1 - P_{m/r})^2}{P_{m/r}(1 - P_{r/m}) + 2P_{m/r}P_{r/m} + P_{r/m}(1 - P_{m/r})} \quad \left[= \frac{(mmm)(mmr)}{(mm)} \right] \tag{VIII-26}$$

$$(mmrm) = \frac{2P_{m/r}P_{r/m}^2(1 - P_{m/r})}{P_{m/r}(1 - P_{r/m}) + 2P_{m/r}P_{r/m} + P_{r/m}(1 - P_{m/r})} \quad \left[= \frac{2(mmr)(mrm)}{(mr)} \right] \tag{VIII-27}$$

$$(mmrr) = \frac{2P_{m/r}P_{r/m}(1 - P_{m/r})(1 - P_{r/m})}{P_{m/r}(1 - P_{r/m}) + 2P_{m/r}P_{r/m} + P_{r/m}(1 - P_{m/r})} \quad \left[= \frac{(mmr)(mrr)}{(mr)} \right] \tag{VIII-28}$$

$$(mrmr) = \frac{2P_{m/r}^2P_{r/m}^2}{P_{m/r}(1 - P_{r/m}) + 2P_{m/r}P_{r/m} + P_{r/m}(1 - P_{m/r})} \quad \left[= \frac{4(mrm)(rmr)}{(mr)} \right] \tag{VIII-29}$$

$$(mrrm) = \frac{P_{m/r}P_{r/m}^2(1 - P_{r/m})}{P_{m/r}(1 - P_{r/m}) + 2P_{m/r}P_{r/m} + P_{r/m}(1 - P_{m/r})} \quad \left[= \frac{(mrr)^2}{4(rr)} \right] \tag{VIII-30}$$

$$(mrrr) = \frac{2P_{m/r}P_{r/m}(1 - P_{r/m})^2}{P_{m/r}(1 - P_{r/m}) + 2P_{m/r}P_{r/m} + P_{r/m}(1 - P_{m/r})} \quad \left[= \frac{(mrr)(rrr)}{(rr)} \right] \tag{VIII-31}$$

$$(rrr) = \frac{P_{m/r}(1 - P_{r/m})^3}{P_{m/r}(1 - P_{r/m}) + 2P_{m/r}P_{r/m} + P_{r/m}(1 - P_{m/r})} \quad \left[= \frac{(rrr)^2}{(rr)} \right] \tag{VIII-32}$$

Again, the bracketed relationships hold for Bernoullian and first-order Markov but not for higher Markovian (or non-Markovian) statistics. Because of necessary relations (Chapter III, Section 6) among sequence frequencies, it is necessary to show that *six* of the unbracketed relationships hold in order to establish consistency with first-order Markov statistics. This is at present even less likely to be experimentally possible than for the tetrad sequences, usually because of multiple overlaps.

It might be appropriate to summarize the four possible types of chains that can be described in first-order Markov terms; polymers resembling the last two can of course be produced by Bernoullian propagation, the limits being the corresponding pure polymer types:

1. "Block" $P_{m/r} < 0.5$, $P_{r/m} < 0.5$, $P_{\Sigma} < 1$

2. Heterotacticlike ("Alternating") $P_{m/r} > 0.5$, $P_{r/m} > 0.5$, $P_{\Sigma} > 1$

3. Isotacticlike $P_{m/r} < 0.5$, $P_{r/m} > 0.5$

4. Syndiotacticlike $P_{m/r} > 0.5$, $P_{r/m} < 0.5$

Note that for a "pure" heterotactic polymer, $P_{m/r} = P_{r/m} = 1.0$.

Coleman and Fox[30,31] have defined a useful quantity, the *persistence ratio*, as

$$\rho = \frac{2[m]\,[r]}{[mr]}$$

For first-order Markov statistics, ρ is equal to $(P_{m/r} + P_{r/m})^{-1}$. For polymer II, we find that ρ is equal to 2.1, the deviation in excess of unity expressing the tendency of the r and m species to propagate themselves preferentially, i.e., in a nonrandom manner.

Another criterion for Bernoullian statistics, which employs only triad data is

$$4\,\frac{(mm)\,(rr)}{(mr)^2}$$

This quantity is unity for a Bernoullian chain. This is a very sensitive test, and should be applied only to very accurate data, especially if (mm) or (rr) are small.

C. More Complex Propagation Models

In order to explore further the statistics of polymer II, we must examine complex propagation models. We shall discuss two: (i) second-order Markov and (ii) Coleman–Fox (one of many possible non-Markovian models).

1. *Second-Order Markov Model*

This model requires the specification of eight conditional probabilities, for we must consider the influence of the relative configurations of the last three pseudoasymmetric centers of the growing chain end. For convenience, we designate these as[32]

$$\begin{aligned} P(mm/m) &= \alpha & P(mm/r) &= \bar{\alpha} \\ P(mr/m) &= \beta & P(mr/r) &= \bar{\beta} \\ P(rm/m) &= \gamma & P(rm/r) &= \bar{\gamma} \\ P(rr/m) &= \delta & P(rr/r) &= \bar{\delta} \end{aligned} \qquad \text{(VIII-33)}$$

where $P(mm/m)$ is the probability of a monomer adding in m fashion to a chain ending in mm, and so on. There are actually four independent probabilities since

$$\begin{aligned} \alpha + \bar{\alpha} &= 1 & \gamma + \bar{\gamma} &= 1 \\ \beta + \bar{\beta} &= 1 & \delta + \bar{\delta} &= 1 \end{aligned} \qquad \text{(VIII-34)}$$

This model reduces to the first-order Markov model when $\alpha = \gamma$ and $\beta = \delta$. Putting

$$s = \bar{\alpha}\bar{\beta} + 2\bar{\alpha}\delta + \gamma\delta \qquad \text{(VIII-35)}$$

we find that the observable dyad through pentad frequencies are

$$\begin{aligned} (m) &= (\bar{\alpha} + \gamma)\delta/s \\ (r) &= (\bar{\beta} + \delta)\alpha/s \end{aligned} \qquad \text{(VIII-36)}$$

$$\begin{aligned} (mm) &= \gamma\delta/s \\ (mr) &= 2\bar{\alpha}\delta/s \\ (rr) &= \bar{\alpha}\bar{\beta}/s \end{aligned} \qquad \text{(VIII-37)}$$

$$\begin{aligned} (mmm) &= \alpha\gamma\delta/s \\ (mmr) &= 2\bar{\alpha}\gamma\delta/s \\ (mrm) &= \bar{\alpha}\beta\delta/s \\ (mrr) &= 2\bar{\alpha}\bar{\beta}\delta/s \\ (rmr) &= \bar{\alpha}\bar{\gamma}\delta/s \\ (rrr) &= \bar{\alpha}\bar{\beta}\bar{\delta}/s \end{aligned} \qquad \text{(VIII-38)}$$

$$(mmmm) = \alpha^2\gamma\delta/s \quad [= (mmm)^2/(mm)] \tag{VIII-39}$$

$$(mmmr) = 2\alpha\bar{\alpha}\gamma\delta/s \; [= (mmm)(mmr)/(mm)] \tag{VIII-40}$$

$$(mmrm) = 2\bar{\alpha}\beta\gamma\delta/s \; [= 2(mmr)(mrm)/(mr)] \tag{VIII-41}$$

$$(mmrr) = 2\bar{\alpha}\bar{\beta}\gamma\delta/s \; [= (mmr)(mrr)/(mr)] \tag{VIII-42}$$

$$(mrmr) = 2\bar{\alpha}\beta\bar{\gamma}\delta/s \; [= 4(mrm)(rmr)/(mr)] \tag{VIII-43}$$

$$(mrrm) = \bar{\alpha}\bar{\beta}\delta^2/s \quad [= (mrr)^2/4(rr)] \tag{VIII-44}$$

$$(mrrr) = 2\bar{\alpha}\bar{\beta}\delta\bar{\delta}/s \; [= (mrr)(rrr)/(rr)] \tag{VIII-45}$$

$$(rmmr) = \bar{\alpha}^2\gamma\delta/s \quad [= (mmr)^2/4(mm)] \tag{VIII-46}$$

$$(rmrr) = 2\bar{\alpha}\bar{\beta}\bar{\gamma}\delta/s \; [= 2(mrr)(rmr)/(mr)] \tag{VIII-47}$$

$$(rrr) = \bar{\alpha}\bar{\beta}\bar{\delta}^2/s \quad [= (rrr)^2/(rr)] \tag{VIII-48}$$

The bracketed relationships hold for Bernoullian, first-order Markov, and second-order Markov statistics, but not for higher-order Markov models. The following expressions are also useful:

$$\begin{aligned} \alpha &= \frac{(mmm)}{(mm)} & \bar{\alpha} &= \frac{(mmr)}{2(mm)} \\ \beta &= \frac{2(mrm)}{(mr)} & \bar{\beta} &= \frac{(mrr)}{(mr)} \\ \gamma &= \frac{(mmr)}{(mr)} & \bar{\gamma} &= \frac{2(rmr)}{(mr)} \\ \delta &= \frac{(mrr)}{2(rr)} & \bar{\delta} &= \frac{(rrr)}{(rr)} \end{aligned} \tag{VIII-49}$$

2. *Coleman–Fox Model*

Another significant mechanistic scheme, which does not contemplate any influence of the chain-end stereochemistry on the mode of addition of the next monomer unit, is that put forward by Coleman and Fox,[30,31,33] primarily to explain the "blocklike" configuration exhibited in varying degrees by most if not all propagating species which deviate from Bernoullian statistics. (We do not consider here Ziegler–Natta catalysis.) Mechanistically, one can plausibly account for the tendency of an *m* sequence to propagate itself, for one may easily imagine, as many authors have suggested, that in Grignard- or metal-alkyl-initiated polymerizations the counterion chelates with the incoming monomer and the chain end in such a way as to favor *m* placements (see Section 4). It is not quite so easy to see why *r* placements should tend to propagate nonrandomly, although "penultimate" effects or complexing can again be imagined to account for this. Instead, Coleman and Fox have proposed that "block" configurations are generated because the propagating chain end may exist in two (or possibly more) states, corresponding to chelation by counterion and the interruption of this chelation by solvation. The intervals between arrival and departure of the solvating species (usually an added ether such as tetrahydrofuran) are imagined to be longer than that corresponding to an average propagation step, but not necessarily very much longer since block lengths are often quite short. It is assumed that polymerization takes place in a homogeneous phase, which is probably reasonable for the metal-alkyl-initiated polymerizations to which this treatment has been applied. In each state, the propagation is assumed to be Bernoullian, but the probability of m placement is taken to be different in each state.

To express this scheme quantitatively, the following notation is employed: The frequency of m placement is

$$(m) = \frac{(\lambda_a k_{1i} + \lambda_b k_{2i})}{(\lambda_a k_1 + \lambda_b k_2)}$$

where

$\lambda_a =$ rate constant for State 2 → State 1 of the chain end

$\lambda_b =$ rate constant for State 1 → State 2 of the chain end

$k_{1i} =$ rate constant for m placement in State 1

$k_{2i} =$ rate constant for m placement in State 2

$k_1 = k_{1i} + k_{1s}$, where k_{1s} is the rate constant for r placement in State 1

$k_2 = k_{2i} + k_{2s}$, where k_{2s} is the rate constant for r placement in State 2

$$a = \frac{\lambda_a \lambda_b k_1 k_2}{(\lambda_a k_1 + \lambda_b k_2)^2} \left(\frac{k_{1i}}{k_1} - \frac{k_{2i}}{k_2} \right)^2$$

$b = k_{1i} k_{2i} / k_1 k_2$

$c = k_{1s} k_{2s} / k_1 k_2$

$d = (\lambda_a k_1 + \lambda_b k_2) / k_1 k_2$

$x = [M]/([M] + d)$, where $[M]$ is the monomer concentration in the polymerizing system.

Coleman and Fox[30] have described how the rate constants could be determined individually, assuming that λ_a and λ_b, but not the k's, may depend on the concentration of the initiator.

Such a polymerizing system will *in general* generate chains which cannot be described by first-order Markov statistics or by Markov statistics of any finite order. Sequence frequencies are expressed by the following relationships.[29]

For triads:

$$(mm) = (m)^2 + ax \tag{VIII-50}$$

$$(mr) = 2(m)(r) - 2ax \tag{VIII-51}$$

$$(rr) = (r)^2 + ax \tag{VIII-52}$$

These equations should be compared to Eqs. (VIII-12)–(VIII-14).

For tetrads:

$$(mmm) = \frac{(mm)^2 + abx^2}{(m)} \tag{VIII-53}$$

$$(mmr) = \frac{(mm)(mr) - 2abx^2}{(m)} \tag{VIII-54}$$

$$(mrm) = \frac{(mr)^2 + 4acx^2}{4(r)} \tag{VIII-55}$$

$$(mrr) = \frac{(mr)(rr) - 2acx^2}{(r)} \tag{VIII-56}$$

$$(rmr) = \frac{(mr)^2 + 4abx^2}{4(m)} \tag{VIII-57}$$

$$(rrr) = \frac{(rr)^2 + acx^2}{(r)} \tag{VIII-58}$$

Comparison with Eqs. (VIII-19)–(VIII-24) (bracketed relationships) show clearly the deviation from expectation for Bernoullian and first-order Markov propagation.

For pentads:

$$(mmmm) = \frac{(mmm)^2 + AM^2}{(mm)} \tag{VIII-59}$$

$$(mmmr) = \frac{(mmm)(mmr) - 2AM^2}{(mm)} \tag{VIII-60}$$

$$(mmrm) = \frac{2(mmr)(mrm) + 4AMR}{(mm)} \tag{VIII-61}$$

$$(mmrr) = \frac{(mmr)(mrr) - 4AMR}{(mr)} \tag{VIII-62}$$

$$(mrmr) = \frac{4(rmr)(mrm) - 4AMR}{(mr)} \tag{VIII-63}$$

$$(mrrm) = \frac{(mrr)^2 + 4AR^2}{4(rr)} \tag{VIII-64}$$

$$(mrrr) = \frac{(rrr)(mrr) - 2AR^2}{(rr)} \tag{VIII-65}$$

$$(rmmr) = \frac{(mmr)^2 + 4AM^2}{4(mm)} \tag{VIII-66}$$

$$(rmrr) = \frac{2(mrr)(rmr) + 4AMR}{(mr)} \tag{VIII-67}$$

$$(rrrr) = \frac{(rrr)^2 + AR^2}{(rr)} \tag{VIII-68}$$

Here, $A = ax$, $M = bx$, and $R = cx$, as defined previously. These relationships should be compared to Eqs. (VIII-39)–(VIII-48) to exhibit deviations from second-order Markov statistics (as well as Bernoullian and first-order Markov).

In Table VIII-2, the observed dyad, triad, tetrad, and pentad intensities (in the last case, largely summations of superimposed peaks, unfortunately)

TABLE VIII-2

MODEL FITTING FOR POLYMER II

Dyad				
(m):	0.82			
(r):	0.18			
Triad				
(mm):	0.75			
(mr):	0.14			
(rr):	0.11			
	Observed	First-order Markov $P_{(m/r)} = 0.085$ $P_{(r/m)} = 0.39$	Second-order Markov	Coleman–Fox
Tetrad				
(mmm)	0.70	0.69	0.70	0.70
(mmr)	0.09	0.13	0.09	0.09
(rmr)	0.03	0.006	0.03	0.02
(mrm)	0.04	0.03	0.04	0.04
(mrr)	0.06	0.09	0.06	0.06
(rrr)	0.07	0.07	0.07	0.08
Pentad				
Peak 1 $\begin{Bmatrix}(mmmm)\\(mmmr)\\(rmmr)\end{Bmatrix}$	0.75	0.75	0.77	0.74
Peak 2 $\begin{Bmatrix}(rmrr)\\(mmrr)\end{Bmatrix}$	0.07	0.09	0.06	0.07
Peak 3 $\begin{Bmatrix}(rmrm)\\(mmrm)\end{Bmatrix}$	0.07	0.05	0.08	0.08
Peak 4 $(rrrr)$	0.07	0.04	0.05	0.07
Peak 5 $\begin{Bmatrix}(mrrr)\\(mrrm)\end{Bmatrix}$	0.04	0.07	0.05	0.05

of polymer II are compared to the predictions of first-order and second-order Markov mechanisms, and of a Coleman–Fox two-state mechanism. For the first-order Markov mechanism, the sequence frequencies are calculated from the values of the conditional probabilities obtained from the triad analysis. To test for conformity to second-order Markov statistics, the eight conditional probabilities are calculated from the observed triad and tetrad intensities using Eq. (VIII-49).

$$\alpha = 0.94 \qquad \bar{\alpha} = 0.06$$

$$\beta = 0.56 \qquad \bar{\beta} = 0.44$$

$$\gamma = 0.63 \qquad \bar{\gamma} = 0.37$$

$$\delta = 0.28 \qquad \bar{\delta} = 0.64$$

These values constitute also a test of first-order Markov statistics since, as we have seen, $\alpha = \gamma$ and $\beta = \delta$ if these apply. It is clear that this is not the case. From the method of calculation, it is necessary that $\beta + \bar{\beta} = 1$ and $\gamma + \bar{\gamma} = 1$, so this agreement does not constitute a test of second-order Markov statistics. The fact that $\alpha + \bar{\alpha} = 1$ and $\delta + \bar{\delta} = 1$ does constitute a valid test, however. A further test is indicated in the fourth column of Table VIII-2, in which pentad intensities, calculated from Eqs. (VIII-29)–(VIII-48), are listed. The agreement with experimental values is quite good except that the predicted ratio of *rrrr* (peak 4) to the *mrrr-mrrm* (peak 5) is somewhat lower than observed.

The calculation of a consistent set of constants for the Coleman–Fox mechanism is discussed in Frisch *et al.*,[32] to which the reader is referred for details. The quantities a/b, b/c, and (m) are obtained from experiment by an appropriate choice among Eqs. (VIII-50) through (VIII-68) by comparison with the corresponding Markov relations. If we let $P_{m_1} = k_{1i}/k_1$ be the probability of isotactic placement in state 1, $P_{m_2} = k_{2i}/k_2$, the probability of isotactic placement in state 2, and $W_1 = \lambda_a k_1/(\lambda_a k_1 + \lambda_b k_2)$, the weight fraction of polymer produced in state 1, we find for polymer II that $P_{m_1} = 0.96$, $P_{m_2} = 0.21$, and $W_1 = 0.81$, state 1 being arbitrarily taken as that in which isotactic placement predominates. In the last column of Table VIII-2, the tetrad and pentad frequencies calculated from these parameters are shown. The agreement with experimental observation is at least as good as that of the second-order Markov model, and is in fact rather better for peaks 4 and 5, not correctly predicted by the latter. Although the close similarity in the predictions of the two models means that this result does not provide conclusive evidence for the two-state model, the parameters obtained are very reasonable. They imply that about 80% of the polymer is produced under conditions such that isotactic placement is nearly exclusively preferred; the rest of the time, isotactic placement has the probability characteristic of a free radical or uncomplexed anionic chain end.

It should be noted that the two-state model, while it reduces to first-order Markov statistics when only m placements are possible in one state and only r placements in the other, cannot reduce identically to second-order Markov statistics. It may, however, approach very closely, as in the present case. Unless experimental measurements are made with the greatest care, particularly on longer sequences, one may readily arrive at "proof" of a "penultimate" stereochemical effect which is in fact quite ambiguous.

Coleman *et al.*[34] have defined the quantity Ω as a measure of the deviation from first-order Markov statistics.

$$\Omega = \frac{(r)(rrr)}{(rr)^2}$$

[This should be compared to Eqs. (VIII-24) and (VIII-58).] They observed that polymethyl methacrylate prepared with 9-fluorenyllithium initiator in 95:5 toluene:tetrahydrofuran (by volume) at $-60°$ showed a value of Ω of 1.09; when the tetrahydrofuran concentration was increased to 10 vol %, Ω decreased to unity within experimental error, corresponding to first-order Markov (or Bernoullian) propagation. Polymer II has a value of Ω of 1.04, deviating less from second-order Markov by this criterion.

In Fig. VIII-3 is shown the spectrum of a polymer with somewhat unusual configurational sequence characteristics, but which conforms closely to second-order Markov, Ω being 1.02. It was prepared with phenylmagnesium bromide initiator in toluene at $-78°$ in the presence of diethyl ether. The assignments of the m tetrads, which appear as singlets, conform to those given in Section 2 of Chapter III, and their intensities obey the triad–tetrad necessary relationship (Table III-2),

$$(mr) = (mmr) + 2(rmr) = 0.31$$

The intensity of mmr is zero within experimental error, although at first sight it would appear to be much more probable than rmr. The generation of this polymer is consistent with the following conditional probabilities[32]:

$$\begin{array}{llll} \alpha = 0.97 & \bar{\alpha} = 0.03 & \gamma = 0.06 & \bar{\gamma} = 0.94 \\ \beta = 0.25 & \bar{\beta} = 0.75 & \delta = 0.32 & \bar{\delta} = 0.68 \end{array}$$

Note the small value of γ and the large value of $\bar{\gamma}$: a chain ending in rm has a small probability of adding m but a large probability of adding r. It is not at present possible to determine the mechanistic basis of this behavior.

Zambelli[35] has presented a study of predominantly isotactic, predominantly syndiotactic, and atactic polypropylenes. The experimental aspects of this work are discussed in Chapter VII and in Section 4 of this chapter. We note

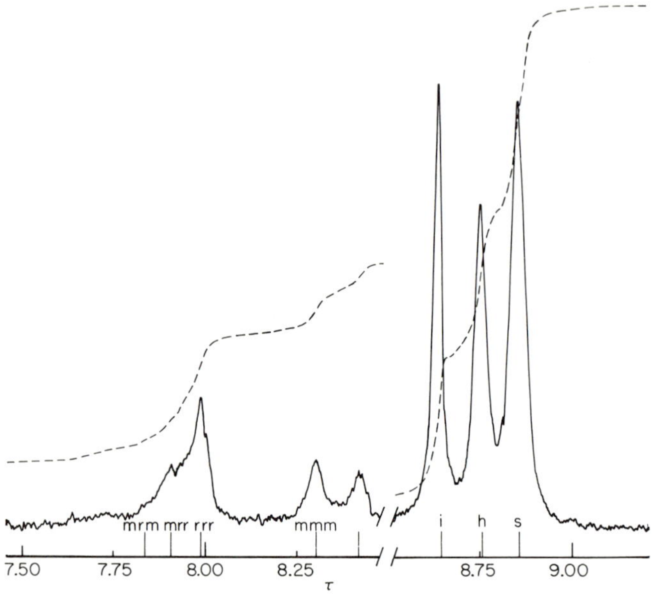

FIG. VIII-3. 60 MHz spectrum of polymethyl-*cis*-d_1-methacrylate, prepared with phenylmagnesium bromide initiator in toluene at −78° in presence of diethyl ether. Observed in chlorobenzene at 150° (broken line is integral spectrum).

here that the configurational sequences were analyzed in terms of a second-order Markov model. However, because only tetrad information was available, conformity to second-order Markov actually could not be tested. It is especially interesting that the isotactic polymer sequence was consistent with $\bar{\beta} > \beta$, i.e., a chain ending in *mr* prefers to add *r* rather than *m* even though the catalyst system is strongly *m*-directing.

D. BLOCK STATISTICS

A simpler statistical approach to the analysis of configurational sequences, and one which is often useful, is implicit in the preceding discussion. It usually requires only dyad and triad analysis.[36,37] Consider the following "block" chain:

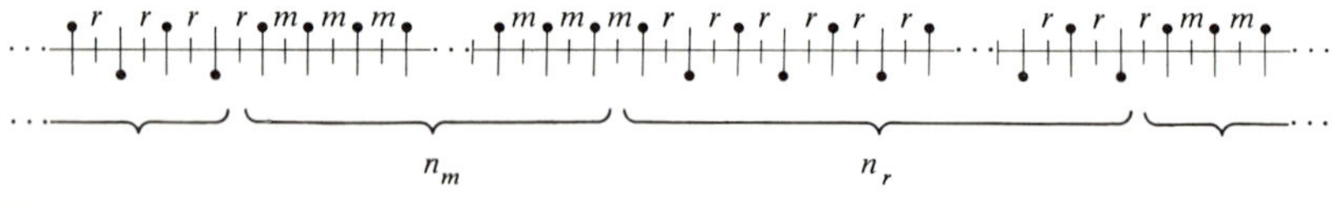

Defining the configurational blocks as shown, the isotactic sequence of n_m monomer units contains $n_m - 1$ *mm* triads and n_m *m* dyads, while the syndiotactic sequence contains $n_r - 1$ rr triads and n_r r dyads, assuming (as always) semi-infinite chains. If $N(n_m)$ is the number of isotactic sequences of length n_m, then there are $\sum N(n_m)(n_m - 1)$ mm triads. Similarly, there are $\sum N(n_r)(n_r - 1)$ rr triads in the syndiotactic sequence of length n_r. (The summations run from one to infinity over n_m and n_r.) Since the numbers of isotactic and syndiotactic sequences must be equal, $\sum N(n_m) = \sum N(n_r)$. Therefore

$$\frac{(mm)}{(rr)} = \frac{\sum N(n_m)(n_m - 1)}{\sum N(n_r)(n_r - 1)} = \frac{\sum N(n_m)n_m - \sum N(n_m)}{\sum N(n_r)n_r - \sum N(n_r)}$$

$$= \frac{(\bar{n}_m - 1)}{(\bar{n}_r - 1)}$$

since

$$\bar{n}_m = \frac{\sum N_m N(n_m)}{\sum N(n_m)}$$

and

$$\bar{n}_r = \frac{\sum n_r N(n_r)}{\sum N(n_r)}$$

where $\bar{n}_m$ and $\bar{n}_r$ are the number average lengths of the isotactic and syndiotactic sequences, respectively. It further follows, as is indeed obvious, that

$$\frac{(m)}{(r)} = \frac{(\bar{n}_m)}{(\bar{n}_r)} \tag{VIII-69}$$

From these relations it is found that

$$\bar{n}_i = \frac{[1 - (mm)/(rr)]}{[1 - (mm)(r)/(rr)(m)]} \tag{VIII-70}$$

$$\bar{n}_r = \frac{[1 - (mm)/(rr)]}{[(m)/(r) - (mm)/(rr)]} \tag{VIII-71}$$

The intensity of the heterotactic triad resonance is an inverse measure of block lengths, since there is one heterotactic triad between isotactic and syndiotactic blocks and the number of such units is thus equal to the number of blocks:

$$\sum N(n_m) + \sum N(n_r) = 2\sum N(n_m) = 2\sum N(n_r)$$

$$(mr) = \frac{2\sum N(n_m)}{\sum N(n_m)(n_m - 1) + \sum N(n_r)(n_r - 1) + 2\sum N(n_m)}$$

$$= \frac{2\sum N(n_m)}{\sum N(n_i)n_i + \sum N(n_r)n_r}$$

so that

$$(mr)^{-1} = \tfrac{1}{2}(\bar{n}_m + \bar{n}_r) = \bar{n} \tag{VIII-72}$$

where $\bar{n}$ is the number average length of all blocks. Thus a prominent *mr* resonance implies short blocks. Since

$$\bar{n}_m/(\bar{n}_m + \bar{n}_r) = (m) \qquad \bar{n}_r/(\bar{n}_m + n_r) = (r) \tag{VIII-73}$$

it follows from Eqs. (VIII-12)–(VIII-14) that

$$\bar{n}_m = \frac{2(m)}{(mr)} = 1 + \frac{2(mm)}{(mr)} \tag{VIII-74}$$

$$\bar{n}_r = \frac{2(r)}{(mr)} = 1 + \frac{2(rr)}{(mr)} \tag{VIII-75}$$

For Bernoullian chains, it follows that

$$\bar{n}_m = 1/(1 - P_m) \tag{VIII-76}$$

$$\bar{n}_r = 1/P_m \tag{VIII-77}$$

For any statistics, the persistence length (p. 158) is given by

$$\rho = \bar{n}_m(r) = \bar{n}_r(m) \tag{VIII-78}$$

For first-order Markov statistics

$$\bar{n}_m = [(r)(P_{m/r} + P_{r/m})]^{-1} \tag{VIII-79}$$

$$\bar{n}_r = [(m)(P_{m/r} + P_{r/m})]^{-1} \tag{VIII-80}$$

Extension to higher-order Markov statistics is obvious, but such relations are not particularly useful since block lengths can be obtained directly from triad (or dyad–triad) frequencies from Eqs. (VIII-74) and (VIII-75).

For the polymer shown in Fig. VIII-3, we find

$$\bar{n}_m = \frac{2 \times 0.47}{0.31} = 3.0, \qquad \bar{n}_r = \frac{2 \times 0.53}{0.31} = 3.4$$

A more conspicuously block-like structure is indicated in Fig. VIII-4, which shows the 100 MHz spectrum of a polypropylene reported by Woodbrey.[38] Intermediate in chemical shift between the *mm* and *rr* methyl doublets near 9τ (see Chapter VII, Section 2) is another weaker doublet arising from the *mr* triads. From its intensity, we find [Eq. (VIII-72)] that $\bar{n}$ is about 16.

E. Propagation Errors

A problem of some significance in relation to propagation mechanism is the nature of the configurational defects introduced into highly stereoregular chains. Even the most efficient catalyst systems can be expected to be something less than perfect in their stereoregulating abilities, and the nature of the errors introduced, if it can be determined, has a bearing on the interpretation of catalyst action.

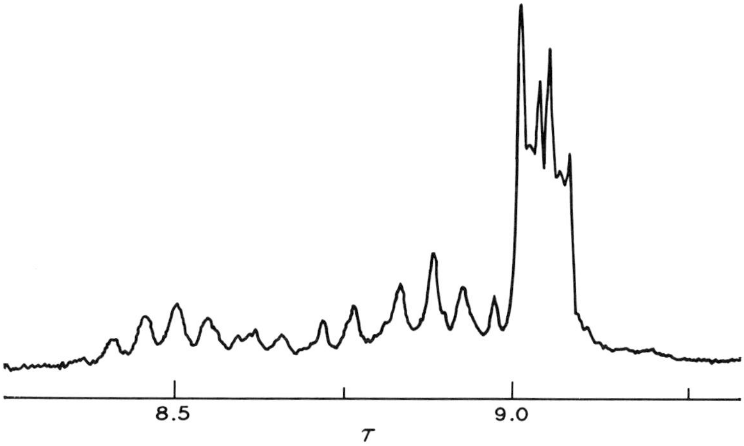

FIG. VIII-4. 100 MHz spectrum of stereoblock polypropylene in *o*-dichlorobenzene at 150° (from Woodbrey [38]).

One may imagine two ways in which a minor configurational "error" may be introduced into an isotactic chain:

(1) "Template" propagation error

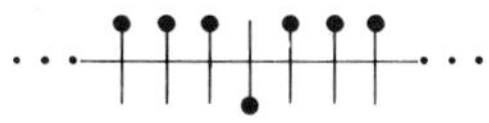

for which the (*mmr*):(*mrr*) ratio is 1:1.

(2) "Steric" propagation error

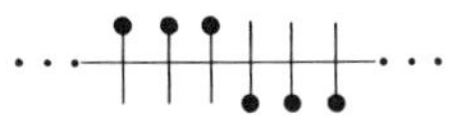

for which one will observe a 2:1 (*mmr*):(*mrm*) ratio and no *mrr*. No *rr* resonance will be observed.

The first of these implies, but of course does not rigorously prove, that the propagation is under the control of an asymmetric catalyst site which is sensitive to the symmetry of the growing chain end and corrects an *r* propagation error by restoring the absolute relationship of chain to site. The second structure implies that the principal correcting force is the steric nature of the chain end itself, possibly under the influence of a coordinating counter-ion; once a new *m* unit is added beyond the *r* error, propagation proceeds as before.

We have seen (Chapter VII, Section 2) that in a highly isotactic polypropylene prepared with a $TiCl_3$–$Al(C_2H_5)_2Cl$ catalyst, the occasional *r* units which are inserted at random are *mrm* and that the (*mmr*):(*mrm*) ratio is 1:1.† There is no *rr* resonance. This catalyst system thus introduces "steric" propagation errors in the above sense. The detailed implications of this finding for propagation mechanism have yet to be worked out. The main point is that by NMR this rather subtle type of information can be obtained.

Zambelli[35] demonstrated in an ingenious way that template control probably prevails in some isotactic propylene polymerizations. This was accomplished by copolymerizing ^{13}C-enriched ethylene in an amount small enough to ensure that virtually all of the ethylene units are inserted between propylene units. If the structure produced is exclusively of type (a) or is exclusively of type (b), the ethylene resonance will consist of two equal peaks corresponding

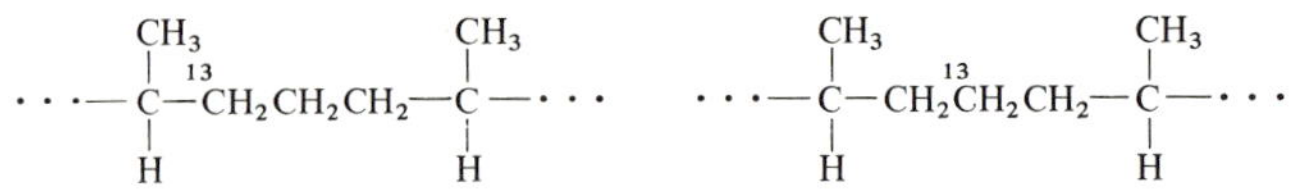

(a)

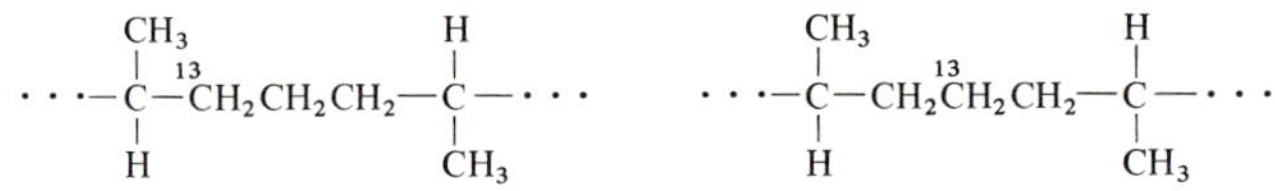

(b)

to the two equally probable positions of the carbon-13. If, on the other hand, there is no requirement that the propylene unit added beyond the ethylene unit add as in (a), i.e., no "template" control, both (a) and (b) will be produced, probably with nearly equal frequency since one can scarcely imagine that chain-end stereochemical influences will be able to bridge the intervening ethylidene unit. In this case, four ethylene peaks should be observed. In fact, two equal peaks were observed. This, in the absence of model studies, does not rigorously prove the generation of (a) rather than (b) but clearly makes it highly probable that "template" control in fact occurs in this case. Thus, the nature of the propagation error evidently depends on the particular catalyst system employed.

F. "Pure" Configurational Sequence Types

It has been indicated in the foregoing discussion that there are other configurational sequence types beside isotactic and syndiotactic. In fact, an infinite number of them exist in principle. The simplest of these higher order

† Note, however, that Flory (see Chapter VII, Flory[30]) has questioned this interpretation and has suggested that the resonance at 8.8τ in the polypropylene spectrum may actually be *rrr* (see Chapter IX, Section 3).

sequences is the heterotactic $\cdots mrmrmr \cdots$, which, as we have seen (p. 158), can be generated by a first-order Markov propagation scheme, and is the only one besides isotactic and syndiotactic that can be expressed by triad repeat units, although it requires two symbols in dyad notation. The question of such sequences is discussed in detail in Frisch *et al.*[29] No polymers of this sort are actually known to exist, although in Section 4 evidence is presented for the possible existence of one having an *mrrr* repeat unit.

4. The Direction of Addition to the Double Bond in Vinyl Polymerization

We have mentioned in Chapter VII (Section 2) that by the use of stereospecifically β-deuterated monomers, information can be obtained as to whether the mode of addition of the growing chain end to the double bond of the monomer is *cis* or *trans*. In terms of simple "ball-and-stick" concepts, the steric relationships shown must hold for β-methylene units,[39] the definitions of the diisotactic polymers conforming to those given in Chapter II, Section 2. (As indicated in the footnote on p. 136, these definitions hold when

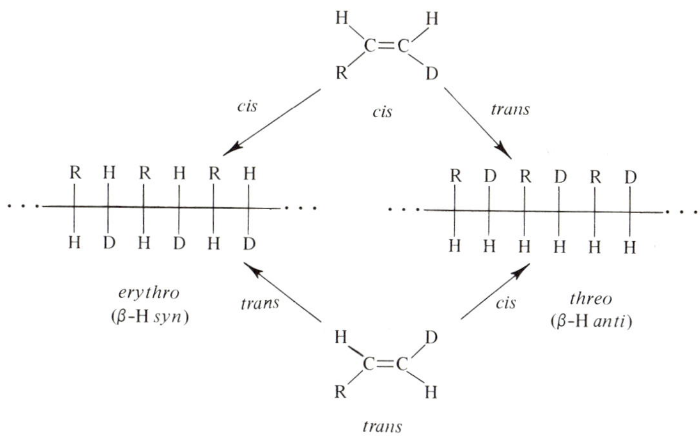

α-H is replaced by D. Of α,α'-disubstituted monomers, only methacrylates need to be considered so far. Then R is taken as the ester group, with α-CH_3 replacing α-H.)

It is clear from previous discussions that one can in general distinguish *syn* β-protons from *anti* β-protons, so that one may measure under different conditions of initiator, temperature, and solvent the extent to which growing chain ends undergo *cis*- or *trans-meso* addition to monomer. (The direction of addition in *r* placement clearly raises more difficult problems and will be dealt with a little later.)

A. Propylene

Natta *et al.*[40] and Peraldo and Farina[41] observed that the polarized infrared spectra of polymers of *cis*-1-d_1-propene and of *trans*-1-d_1-propene are quite different. Miyazawa and Ideguchi[42] showed by infrared analysis that *cis*-1-d_1-propene yields the *erythro*-diisotactic polymer (β-proton *syn*) while the *trans*-1-d_1-propene yields the *threo*-diisotactic polymer (β-proton *anti*), indicating *cis* opening of the double bond with the Ziegler–Natta catalyst system.

The high specificity of addition has also been demonstrated by NMR, as can be seen in Figure VII-6 and the accompanying discussion. Such results do not in themselves, however, tell us the mode of addition, since we cannot tell from such information alone which β-proton is *syn* and which is *anti*. So far this problem has not been convincingly solved by NMR for propylene, although a possible approach to such an assignment has been given by Zambelli.[35] This assignment has been made for isotactic polyacrylates (Section 4B).

Chemical shift information cannot tell us the mode of addition for *r* dyads. This requires somewhat more elaborate experiments employing coupling information. Zambelli[43] and Zambelli *et al.*[44] have shown that poly-*cis*-1,3,3,3-d_4-propene (the *trans* monomer would have served as well) gives, under conditions known to give a polymer syndiotactic with respect to the α-CH_3 groups, a product the NMR spectrum of which is of the AA′BB′ type. Such a symmetrical spectrum indicates that the addition must be stereospecifically *cis* or *trans*, but does not say which. To determine this, copolymers of these monomers were prepared.

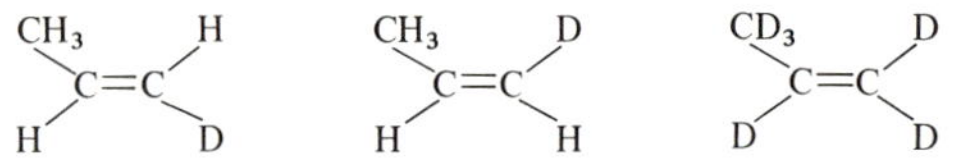

(1) (2) (3)

It was assumed that the reactivity ratios were all unity. The mode of opening being constant, pairs of monomers (1) and (3) and (2) and (3) [using an excess of (3) in both cases] will yield two different copolymers.

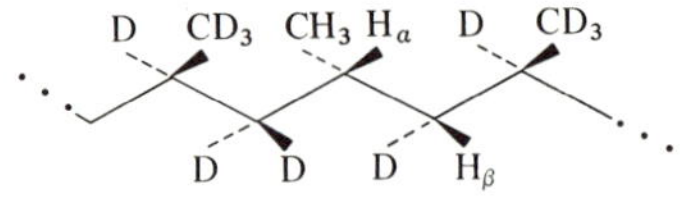

(a)

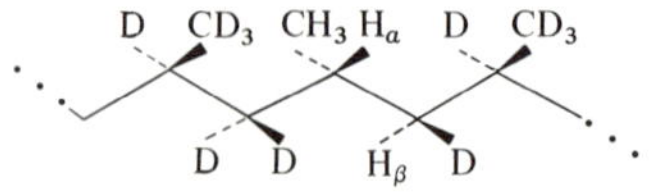

(b)

It is clear that if the *trans–trans* (planar zigzag) conformation indicated is preferred (as is the case; see Chapter IX), copolymer (a) will show the smaller (4.8 Hz) and copolymer (b) the larger (8.3 Hz) of the two vicinal couplings characterizing the unlabeled syndiotactic polymer (Chapter VII, Section 2). Copolymer (a) will be obtained from monomers (1) and (3) and copolymer (b) from monomers (2) and (3) if double bond opening is *cis*, whereas the reverse will be true if the opening is *trans*. In fact, the copolymer from (1) and (3) shows the smaller vicinal spacing, showing that syndiotactic placement, like isotactic placement, proceeds by *cis* addition.

B. Polyacrylates

Figure VIII-5 shows 60 MHz spectra[45,46] of isopropyl-α-*cis*-β-d_2-acrylate,

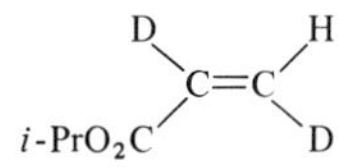

prepared with phenyl magnesium bromide in toluene at −78°, but as the figure legend indicates, varying proportions of ether. The *syn* and *anti* protons appear as singlets, the relative intensities of which vary with the proportion of ether. The polymers all appear to be well over 90% *meso*, since no *racemic*

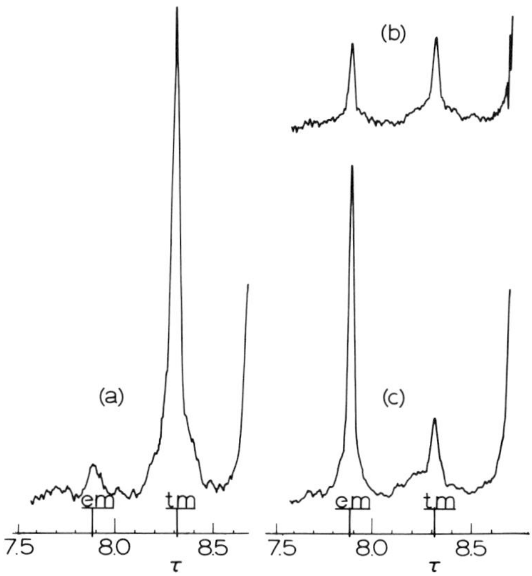

Fig. VIII-5. 60 MHz β-methylene proton spectra of polyisopropyl-α-*cis*-β-d_2-acrylate prepared with phenylmagnesium bromide and (a) large molar excess of diethyl ether; (b) partial ether removal; (c) rigorous ether removal. The notation is *em* for *erythro-meso* (*syn*) and *tm* for *threo-meso* (*anti*). Observed in chlorobenzene at 150°.

resonance is evident (see Chapter IV, Section 2). Thus, one can vary the β-carbon stereochemistry without affecting the α-carbon stereochemistry, i.e., we can control via the ether content the direction of addition to the double bond. Yoshino *et al.*[47] have assigned the *syn* and *anti* protons as shown in Fig. VIII-5 by an elegant model compound study. Note that they are reversed in relative chemical shift from the β-protons of polypropylene. These authors have also carried out mechanistic studies[48–50] related to those of Bovey[45] and Schuerch *et al.*[46] They also found a marked effect of temperature history on the behavior of Grignard initiators. Of particular interest is the study[50] of the *syn* and *anti* protons in polymers similar to that represented in Fig. VIII-5(b). They employed methyl-*trans*-β-d_1-acrylate and found that the vicinal

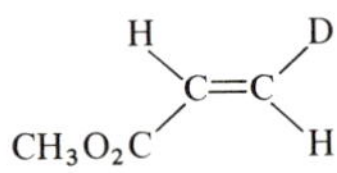

coupling pattern in the polymer spectrum was not consistent with a random arrangement of the *syn* and *anti* protons along the chain, but rather that each type occurs in long runs, possibly as whole polymer molecules. This suggests that the Grignard initiator, at least under some conditions, has sites at which *cis* addition occurs and others at which *trans* addition occurs, or perhaps a given site changes its nature periodically.

Yoshino *et al.*[48] observed that in β-d_1-labeled acrylate polymers prepared with free radical initiators, the *syn* and *anti m* β-protons appeared with equal frequency, and concluded that free radical addition proceeds with equal probability by *cis* and *trans* opening of the double bond.

C. Polyvinyl Chloride

Yoshino and Komiyama[51] observed (see Chapter V, Section 1) that the *syn* and *anti m* β-protons of poly-α-*cis*-β-d_2-vinyl chloride appeared with equal frequency in the 100 MHz spectrum of the free radical polymer, indicating that here, as in polyacrylates, free radical addition proceeds by random *cis* and *trans* addition. It appears highly probable that this conclusion holds true for the free radical propagation of most if not all vinyl monomers.

D. Polymethacrylates

It is evident that analogous information can be obtained concerning the mechanism of propagation of methacrylate monomers, since, as we have seen (Chapter III, Section 3), the *syn* and *anti* protons show a large difference in chemical shift. In this case, *syn* and *anti* assignments have not been made, but are assumed to be the same as in polyacrylates, on the grounds (not particularly strong) that the ester carbonyl group anisotropy is the dominant effect in both. The most extensive work has been done with ethyl-*cis*-d_1-methacrylate[52]:

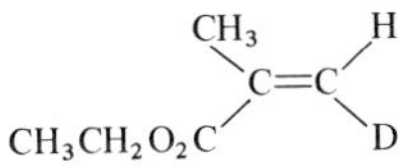

A selection of the data is presented in Table VIII-3 and three of the spectra are given in Fig. VIII-6. In toluene at −78° with 9-fluorenyllithium initiator, this monomer yields a predominantly isotactic polymer, in the same manner

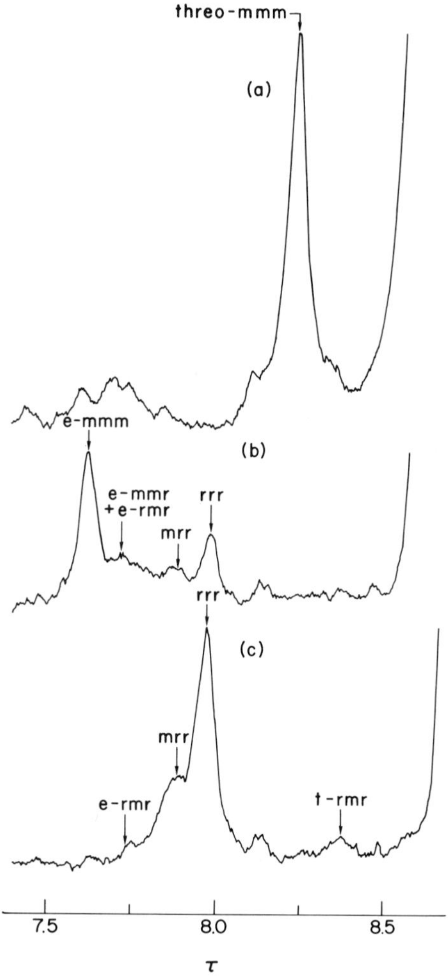

FIG. VIII-6. 60 MHz spectra showing effect of tetrahydrofuran (THF) on polymerization of ethyl-*cis*-d_1-methacrylate in toluene at −78° with 9-fluorenyllithium initiator: (a) no THF; (b) THF: initiator mole ratio 7.5; (c) THF: initiator mole ratio 2.2×10^4 (no toluene). Observed at 150°.

TABLE VIII-3

ANIONIC POLYMERIZATION OF ETHYL *cis*-β-d_1-METHACRYLATE[a]

Initiator	Initiator conc. *M*	THF mole ratio to initiator	Solvent	Temp. (°C)	(*m*)	$\frac{(tm)^b}{(m)}$	(*mm*)	(*mr*)	(*rr*)	Configuration statistics[c]
(a) Fl–Li	0.00074	0	toluene	−78	0.87	0.88	0.79	0.16	0.04	P_m 0.87
(b) Fl–Li	0.047	7.5	toluene	−78	0.64	0.11	0.55	0.18	0.27	P_Σ 0.39
(c) Fl–Li	0.00055	2.2×10^4	THF	−78	0.15	0.51	0.05	0.21	0.74	P_m 0.15
(d) Fl–Li	0.047	7.5	toluene	+30	0.80	0.09	0.72	0.17	0.11	P_Σ 0.5
(e) Fl–Cs	0.021	520	THF	−78	0.28	0.70	<0.05	0.48	0.48	—

[a] See Fowells *et al.*[52]

[b] *tm* = *threo-meso*.

[c] If polymer is Bernoullian, P_m is given; if polymer is first-order Markov (or nearly so), $P_\Sigma = (P_{m/r} + P_{r/m})$ is given (see Section 3).

as has long been known for methyl methacrylate.[53, 54] The results are shown as (a) in Fig. VIII-6 and Table VIII-3; (m) is 0.87. The m protons are 88% *threo*, indicating that the *trans* addition is very strongly preferred. In this "living" polymer system[55–57] there is a large body of evidence indicating that contact ion pairs, not specifically solvated, are the chain carriers.

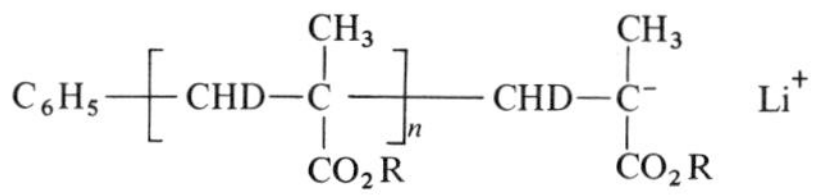

Under these conditions, a single species is responsible for chain growth, and it seems reasonable that the chain statistics should be Bernoullian or first-order Markov. Triad analysis is consistent with the former. (The α-methyl spectrum, not shown in Fig. VIII-6, is complicated by the overlapping ester methyl triplet but can still be analyzed.) In the presence of tetrahydrofuran at 7.5 mole ratio to the 9-fluorenyllithium [(b) in Fig. VIII-6 and Table VIII-3], the polymer remains predominantly isotactic [$(m) = 0.64$], but the m protons are now 98% *erythro*. The polymer departs markedly from Bernoullian (last column of Table VIII-3); a two-state propagation scheme is the most logical explanation. [When the polymerization is carried out at +30° in toluene, (d) in Table VIII-3, the result is much the same except that the m stereoblocks are longer.]

When the polymerization is carried out in tetrahydrofuran at −78°, [(c) in Fig. VIII-6 and Table VIII-3], the propagation is highly syndiotactic [$(m) = 0.15$]; the few m placements appear to occur by *cis* or *trans* addition with almost equal probability, and the configurational sequence is Bernoullian. The propagation thus resembles free radical propagation, there being no steric control. In tetrahydrofuran, the propagating species is an ion pair as in toluene but with the important difference that the ions, although not free to move independently, are solvated and separated by the solvent.

The most interesting questions are raised by the intermediate system (b), in which a relatively small concentration of tetrahydrofuran is present. The simplest conclusion would appear to be that both contact ion pairs and solvated ion pairs are present, these species interchanging in the manner contemplated in the Coleman–Fox mechanism (previous section). This is too simple an explanation, however, for if this were the case, the *meso* protons would be mostly *threo*, whereas in fact they are mostly *erythro*. The mechanism is thus more complex than can be explained by two sites or two species, and it is necessary to invoke a third species (for which there is as yet no spectroscopic evidence), which has been termed a peripherally solvated ion pair.[52] This species propagates by m placement, as does the contact ion pair, but does so by *cis* addition. It appears that this species may exist in lesser proportion at

−78° than at 30°, being partially converted to solvent-separated ion pairs when the temperature is lowered, as indicated by an increase in (*rr*).

It should be realized that the terms *cis* and *trans* do not have precisely the same meanings in terms of transition-state structures when describing vinyl monomer propagation as they have for a simple four-center reaction. It is helpful to distinguish between the true mode of addition, i.e., the actual direction in space along which the active chain end and monomer approach each other, and the apparent mode of addition, i.e., the direction of addition as conventionally judged by the stereochemical configuration about the resulting bond. We shall assume that, primarily for steric reasons, the true mode of addition is always *trans*, as represented at the top of Fig. VIII-7.

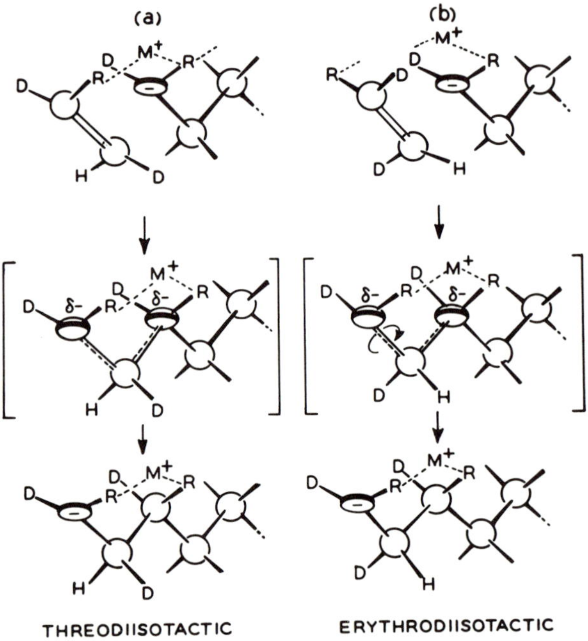

FIG. VIII-7. Monomer approach modes in isotactic placement.

Then, in the formation of the next *meso* dyad, the β-carbon configuration will depend on the mode of presentation or approach of the monomer to the growing chain end. (If a *racemic* dyad is formed, its β-configuration is of course indeterminate.) The growing chain end is assumed to be effectively planar but to have a definite and fixed stereochemical relationship to the next unit (not specifically represented in Fig. VIII-7), i.e., the end of the chain is either *m* or *r* and remains so during placement of the next monomer unit. The α-carbon configuration may of course also be determined by the monomer approach mode. The monomer might approach in a syndiotactic-like or *racemic* manner as in (b) and add on in this fashion without further change, giving an *r* dyad. But in the presence of an effectively chelating counterion (represented as M^+), rotation of the newly formed chain end could occur, giving an *erythro m* placement. Thus the α-carbon configuration depends upon chain end rotation as well as approach mode. On the other hand, if the mode of approach is "isotactic-like," as in (a) in Fig. VIII-7, the *meso* protons in the resulting dyad will be *threo*. Thus, whether the addition is apparently *cis* or *trans* depends on the mode of presentation of the monomer.

This description of events during monomer addition is verbally and pictorially elaborated in Fowells *et al.*[52]

The tendency to form solvent-separated ion pairs varies inversely as the ionic radius,[58] i.e.,

$$\text{Li} > \text{Na} > \text{K} > \text{Cs}$$

Fluorenylcesium exists in tetrahydrofuran largely as a contact ion pair. This feature of its behavior might lead one to expect isotactic propagation even in tetrahydrofuran. However, the spectrum [Fig. VIII-8; (e) in Table VIII-3]

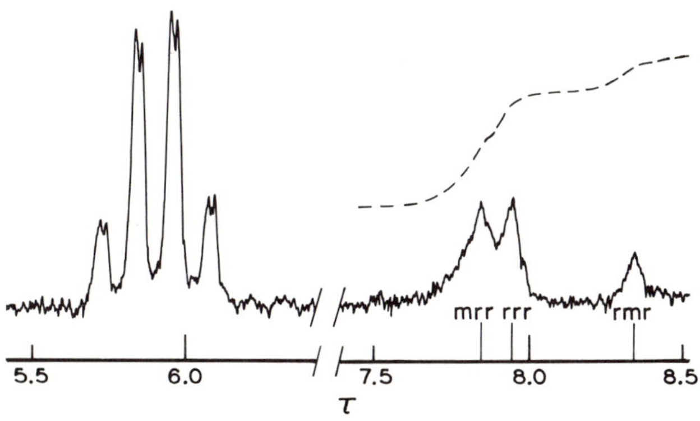

FIG. VIII-8. 60 MHz spectrum of polyethyl-*cis*-β-d_1-methacrylate prepared with fluorenylcesium initiator in tetrahydrofuran at −78°. Spectrum observed in chlorobenzene at 150° (broken line is integral spectrum).

and its analysis show that this is by no means the case, probably because the Cs^+ counter ion coordinates only weakly with chain end carbonyl groups. The sequence statistics of this polymer are interesting. Only *mr* and *rr* can be observed, *mm* being vanishingly small; this is consistent with the pair of ester methylene quartets of equal intensity near 5.9τ in Fig. VIII-8. The relative intensities of the *mrr*, *rrr*, and *rmr* peaks are 2:1:1. This result appears to indicate a structure produced by regular repetition of the unit *mrrr*, i.e., a higher "pure" polymer type in the sense described in Section 3F. Proof of this could be obtained by careful examination of the *rr* ester or α-methyl peak at high frequency, in which pentads can be discriminated (Chapter III, Section 5). If only *mrrr* can be detected, the polymer must have this repeating structure.

REFERENCES

1. F. A. Bovey, *J. Polym. Sci.* **46**, 59 (1960).
2. T. G. Fox and H. W. Schnecko, *Polymer* **3**, 575 (1963).
3. T. Otsu, B. Yamada, and M. Imoto, *J. Macromol. Chem.* **1**, 61 (1966).
4. H. Sobue, K. Matsuzaki, and S. Nakano, *J. Polym. Sci. Part A* **2**, 3339 (1964).
5. K. Matsuzaki, A. Ishida, and N. Tateno, *J. Polym. Sci. Part C* **16**, 2111 (1967).

5a. E. M. Loebl and J. J. O'Neill, *Polym. Lett.* **1**, 27 (1963).

6. H. Sobue, T. Uryu, K. Matsuzaki, and Y. Tabata, *J. Polym. Sci. Part A* **2**, 3333 (1964).
7. K. Matsuzaki and T. Uryu, *J. Polym. Sci. Part B* **4**, 255 (1966).
8. J. W. L. Fordham, P. H. Burleigh, and C. L. Sturm, *J. Polym. Sci.* **20**, 251 (1956).
9. F. A. Bovey and G. V. D. Tiers, *Chem. Ind.* (*London*) **1962**, 1826 (1962).
10. H. Germar, K. H. Hellwege, and U. Johnsen, *Makromol. Chem.* **60**, 106 (1963).
11. J. Bargon, K. H. Hellwege, and U. Johnsen, *Makromol. Chem.* **95**, 187 (1966).
12. F. A. Bovey, F. P. Hood, E. W. Anderson, and R. L. Kornegay, *J. Phys. Chem.* **71**, 312, (1967).
13. L. Cavalli, G. C. Borsini, G. Carraro, and G. Confalonieri, *J. Polym. Sci. Part A-1* **8**, 801 (1970).
14. O. C. Böckman, *J. Polym. Sci. Part A* **3**, 3399 (1965).
15. G. Talamini and G. Vidotto, *Polymer Previews* **2**, 227 (1966).
16. J. D. Cotman, Jr., *Ann. N.Y. Acad. Sci.* **57**, 417 (1953).
17. M. H. George, R. J. Grisenthwaite, and R. F. Hunter, *Chem. Ind.* (*London*) **1958**, 1114 (1958).
18. G. Boccato, A. Rigo, G. Talamini, and F. Zilio-Grandi, *Makromol. Chem.* **108**, 218 (1967).
19. I. Rosen, P. H. Burleigh, and J. F. Gillespie, *J. Polym. Sci.* **54**, 31 (1961).
20. R. Stefani, M. Chevreton, and C. Eyraud, *C. R. Acad. Sci. Paris* **251**, 2174 (1960).
21. R. Yamadera, H. Tadokoro, and S. Marahashi, *J. Chem. Phys.* **41**, 1233 (1964).
22. C. Y. Liang and S. Krimm, *J. Polym. Sci.* **31**, 531 (1958).
23. R. Chiang, *J. Polym. Sci. Part A* **3**, 2019 (1965).
24. G. Talamini and G. Vidotto, *Chim. Ind.* (*Milan*) **46**, 371 (1964).
25. H. Inagaki, K. Hayashi, and S. Marahashi, *J. Chem. Phys.* **41**, 1233 (1964).
26. W. R. Krigbaum and S. Tokita, *J. Polym. Sci.* **43**, 467 (1960).
27. W. Cooper, F. R. Johnston, and G. Vaughan, *J. Polym. Sci. Part A* **1**, 1509 (1963).
28. C. Schuerch, W. Fowells, F. P. Hood, and F. A. Bovey, unpublished observations.
29. H. L. Frisch, C. L. Mallows, and F. A. Bovey, *J. Chem. Phys.* **45**, 1565 (1966).
30. B. D. Coleman and T. G. Fox, *J. Chem. Phys.* **38**, 1065 (1963).
31. B. D. Coleman and T. G. Fox, *J. Polym. Sci. Part A* **1**, 3183 (1963).
32. H. L. Frisch, C. L. Mallows, F. Heatley, and F. A. Bovey, *Macromolecules* **1**, 533 (1968).
33. B. D. Coleman and T. G. Fox, *J. Amer. Chem. Soc.* **85**, 1241 (1963).
34. B. D. Coleman, T. G. Fox, and M. Reinmöller, *J. Polym. Sci. Part B* **4**, 1029 (1966).

35. A. Zambelli, *in* "NMR Basic Principles and Progress" (P. Diehl, E. Fluck, and R. Kosfeld, eds.), Vol. 4. Springer-Verlag, Berlin and New York, 1971.
36. F. A. Bovey and G. V. D. Tiers, *J. Polym. Sci.* **44**, 173 (1960).
37. D. W. McCall and W. P. Slichter, *in* "Newer Methods of Polymer Characterization" (B. Ke, ed.), pp. 321–346. Wiley (Interscience), New York, 1962.
38. J. C. Woodbrey, *J. Polym. Sci. Part B* **2**, 315 (1964).
39. G. Natta, M. Peraldo, M. Farina, and G. Bressan, *Makromol. Chem.* **55**, 139 (1962).
40. G. Natta, M. Farina, and M. Peraldo, *Atti Accad. Naz. Lincei Rend. Cl. Sci. Fis. Mat. Natur.* [8]**25**, 424 (1958).
41. M. Peraldo and M. Farina, *Chim. Ind.* (*Milan*) **42**, 1349 (1960).
42. T. Miyazawa and Y. Ideguchi, *J. Polym. Sci. Part B* **1**, 389 (1963).
43. A. Zambelli, *Chem. Commun.* **1967**, 1252 (1967).
44. A. Zambelli, M. G. Giongo, and G. Natta, *Makromol. Chem.* **112**, 183 (1968).
45. F. A. Bovey, *Pure Appl. Chem.* **15**, 349 (1967).
46. C. Schuerch, W. Fowells, A. Yamada, F. A. Bovey, and F. P. Hood, *J. Amer. Chem. Soc.* **86**, 4481 (1964).
47. T. Yoshino, M. Shinomiya, and J. Komiyama, *J. Amer. Chem. Soc.* **87**, 387 (1965).
48. T. Yoshino, J. Komiyama, and M. Shinomiya, *J. Amer. Chem. Soc.* **86**, 4482 (1964).
49. T. Yoshino and J. Komiyama, *J. Polym. Sci. Part B* **4**, 991 (1966).
50. T. Yoshino and K. Kuno, *J. Amer. Chem. Soc.* **87**, 4404 (1965).
51. T. Yoshino and J. Komiyama, *J. Polym. Sci. Part B* **3**, 311 (1965).
52. W. Fowells, C. Schuerch, F. A. Bovey, and F. P. Hood, *J. Amer. Chem. Soc.* **89**, 1396 (1967).
53. T. G. Fox, B. S. Garrett, W. E. Goode, S. Gratch, J. F. Kincaid, A. Spell, and J. D. Stroupe, *J. Amer. Chem. Soc.* **80**, 1768 (1958).
54. J. D. Stroupe and R. E. Hughes, *J. Amer. Chem. Soc.* **80**, 2341 (1958).
55. M. Szwarc, M. Levy, and R. Milkovich, *J. Amer. Chem. Soc.* **78**, 2656 (1956).
56. M. Szwarc, *Nature* **178**, 1168 (1956).
57. M. Szwarc, "Carbanions, Living Polymers, and Electron Transfer Processes." Wiley, New York, 1968.
58. T. E. Hogen-Esch and J. Smid, *J. Amer. Chem. Soc.* **88**, 307, 318 (1966).

Chapter IX

THE STUDY OF THE CONFORMATIONS OF VINYL POLYMER CHAINS AND MODEL COMPOUNDS BY NMR

1. Introduction

In Chapter III, the symmetry properties of 2,4-disubstituted pentanes and 2,4,6-trisubstituted heptanes were discussed in relation to polymer chain configuration, and in succeeding chapters the NMR spectra of α-monosubstituted vinyl polymers have been analyzed to yield chemical shifts and coupling constants. In this chapter, our task will be primarily to interpret the coupling constants observed for both small molecules and macromolecules in terms of chain conformation. We shall also consider other NMR data which cannot at present give quantitative information about the behavior of polymer chains, but which at least suggest certain qualitative conclusions. (Such information for polyethylene has already been considered in Chapter VII, Section 1.)

2. Model Compounds

A. The Averaging of Vicinal Coupling

It is well recognized (Chapter I, Section 15) that under conditions of rapid equilibration, i.e., when the characteristic time of bond rotation is substantially smaller than the reciprocal of the spectral line spacings (expressed in hertz), observed vicinal couplings are an average over the conformations present

$$J_{\text{obs}} = \sum_i X_i J_i \tag{IX-1}$$

where X_i and J_i represent the mole fraction and vicinal coupling of each conformer. This expression holds for ethanes of the type $X_2CH \cdot CHY_2$, but for ethanes of the type XCH_2CH_2Y, the spectrum is of the AA′BB′ type, which means that there are in general two observable vicinal couplings of the A and B protons; see Chapter III, Section 2. (Only in the fortuitous circumstances that the populations of the two mirror image *gauche* conformations and of the *trans* conformation are all equal will the system reduce to A_2B_2, i.e., only one vicinal coupling.) For ethanes with an asymmetric carbon atom $XCH_AH_B \cdot CH_CYZ$ (Chapter III, Section 2), Eq. (IX-1) applies to each of the distinguishable protons H_A and H_B.

For *meso*-2,4-disubstituted pentanes, there are nine possible conformers, including three pairs of equal-energy mirror image conformers [(4), (5), and (6) in Fig. IX-1] which are convertible into each other, i.e., (4a) into (4b),

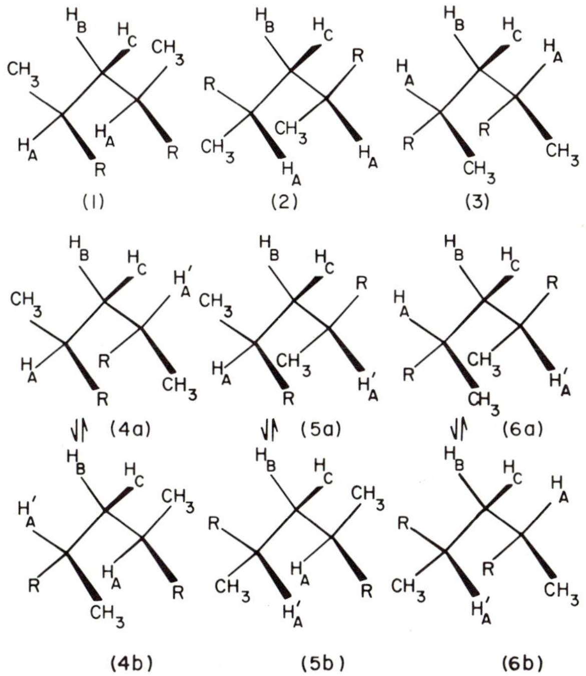

FIG. IX-1. Staggered conformers of *meso*-2,4-disubstituted pentanes.

(5a) into (5b), etc., by 120° rotations about the bonds between the α- and β-carbons. Conformers (1), (2), and (3) have a plane of symmetry; the "mirror-image" conformers (4), (5), and (6), have no plane or axis of symmetry.

For *racemic* 2,4-disubstituted pentanes (Fig. IX-2; only one of the two enantiomers is shown, since *d* and *l* isomers are ordinarily not distinguishable

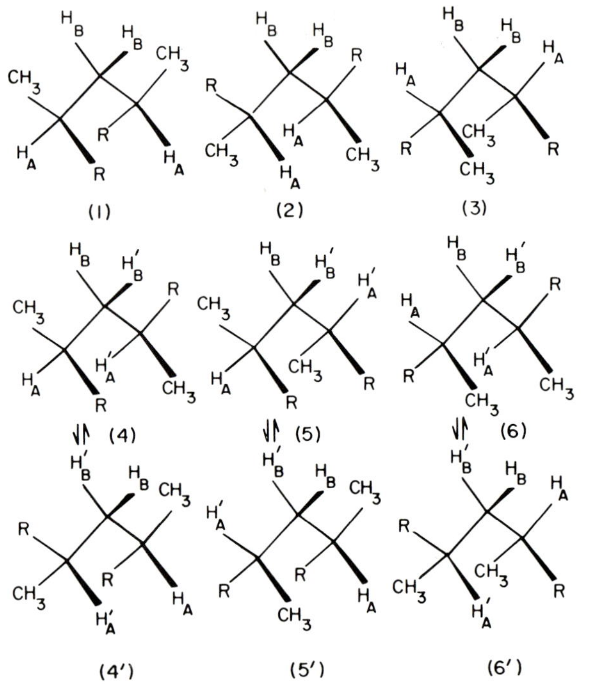

FIG. IX-2. Staggered conformers of *racemic* 2,4-disubstituted pentanes.

by NMR) conformers (1), (2), and (3) have twofold symmetry axes; the other three, (4), (5), and (6), are analogous to conformers (4), (5), and (6) of the *meso* isomer in having no plane or axis of symmetry, but differ in being converted into identical forms ("identomers") (4′), (5′), and (6′), rather than mirror images, by 120° rotations about the α–β bonds. This isomer can of course be converted to its mirror image only by inversion at both asymmetric carbon atoms.

In the *meso*-2,4,-disubstituted pentanes, the β-methylene protons H_B and H_c are in different environments in all conformations, and do not exchange environments upon conformational equilibration between mirror-image forms, as one can readily see from an inspection of models or from Fig. IX-1. (It may be noted, however, that H_A and H'_A are equivalent, either by symmetry or by conformational averaging.) Barring fortuitous equivalence, H_B and H_C

will thus always have different chemical shifts, as we have already seen in Chapter III (Section 2), and will in principle experience different couplings to the α-protons. In the asymmetrical conformers (4), (5), and (6), the couplings experienced by H_B and H_C to the α-protons will each be averaged by conformational equilibration between pairs of equal-energy mirror images. These considerations lead to the following expressions:

$$J_{AB} = X_1 J_{1\text{-}AB\text{-}g} + X_2 J_{2\text{-}AB\text{-}t} + X_3 J_{3\text{-}AB\text{-}g} + X_4 \bar{J}_{4\text{-}AB} + X_5 \bar{J}_{5\text{-}AB} + X_6 \bar{J}_{6\text{-}AB} \quad \text{(IX-2)}$$

$$J_{AC} = X_1 J_{1\text{-}AC\text{-}t} + X_2 J_{2\text{-}AC\text{-}g} + X_3 J_{3\text{-}AC\text{-}g} + X_4 \bar{J}_{4\text{-}AC} + X_5 \bar{J}_{5\text{-}AC} + X_6 \bar{J}_{6\text{-}AC} \quad \text{(IX-3)}$$

where the X_i of course include both mirror-image forms in conformers (4), (5), and (6), and the subscripts g and t denote *gauche* and *trans* couplings. (The designations of H_B and H_C here and in Fig. IX-1 are of course arbitrary and could be interchanged.) In the asymmetric conformers, the averaged couplings $\bar{J}_i$ are given by:

$$\bar{J}_{4\text{-}AB} = \tfrac{1}{2}(J_{4\text{-}AB\text{-}g} + J'_{4\text{-}AB\text{-}g}) \quad \text{(IX-4a)}$$

$$\bar{J}_{4\text{-}AC} = \tfrac{1}{2}(J_{4\text{-}AC\text{-}t} + J_{4\text{-}AC\text{-}g}) \quad \text{(IX-4b)}$$

$$\bar{J}_{5\text{-}AB} = \tfrac{1}{2}(J_{5\text{-}AB\text{-}g} + J_{5\text{-}AB\text{-}t}) \quad \text{(IX-5a)}$$

$$\bar{J}_{5\text{-}AC} = \tfrac{1}{2}(J_{5\text{-}AC\text{-}t} + J_{5\text{-}AC\text{-}g}) \quad \text{(IX-5b)}$$

$$\bar{J}_{6\text{-}AB} = \tfrac{1}{2}(J_{6\text{-}AB\text{-}g} + J_{6\text{-}AB\text{-}t}) \quad \text{(IX-6a)}$$

$$\bar{J}_{6\text{-}AC} = \tfrac{1}{2}(J_{6\text{-}AC\text{-}g} + J'_{6\text{-}AC\text{-}g}) \quad \text{(IX-6b)}$$

Primes are employed to distinguish *gauche* couplings within the same conformer which are in principle nonequivalent, although probably in fact nearly equal. Assuming equality of all *gauche* and of all *trans* couplings, we can then rewrite Eqs. (IX-2) and (IX-3) as

$$J_{AB} = X_1 J_g + X_2 J_t + X_3 J_g + X_4 J_g + X_5 (J_g + J_t)/2 + X_6 (J_g + J_t)/2 \quad \text{(IX-7a)}$$

$$J_{AC} = X_1 J_t + X_2 J_g + X_3 J_g + X_4 (J_t + J_g)/2 + X_5 (J_g + J_t)/2 + X_6 J_g \quad \text{(IX-7b)}$$

where J_g and J_t represent the *gauche* and *trans* couplings, respectively.

In the *racemic* conformers (1), (2), and (3), the β-methylene protons H_B are identical by symmetry, but experience two different observable couplings to the α-protons. In conformer (1) or (2), these would have the magnitudes expected of *trans* and *gauche* vicinal couplings if one conformer or the other were present alone. In the unsymmetrical conformers (4), (5), and (6), rapid equilibration between "identomers" averages the environments of H_B and H'_B

(and of H_A and H_A'), but two different couplings to the α-protons would still be observable (at least in principle) for each conformer, supposing again that it alone were present. The observed couplings will be given by

$$J_{AB} = X_1 J_{1\text{-}t} + X_2 J_{2\text{-}g} + X_3 J_{3\text{-}g} + X_4(J_{4\text{-}g} + J_{4\text{-}t})/2 + X_5(J_{5\text{-}g} + J_{5\text{-}t})/2 + X_6(J_{6\text{-}g} + J'_{6\text{-}g})/2 \tag{IX-8a}$$

$$J'_{AB} = X_1 J_{1\text{-}g} + X_2 J_{2\text{-}t} + X_3 J'_{3\text{-}g} + X_4(J_{4\text{-}g} + J_{4\text{-}t})/2 + X_5(J_{5\text{-}g} + J'_{5\text{-}g})/2 + X_6(J_{6\text{-}g} + J_{6\text{-}t})/2 \tag{IX-8b}$$

Assuming equality of all *gauche* and of all *trans* couplings, as before, we can simplify these equations.

$$J_{AB} = X_1 J_t + X_2 J_g + X_3 J_g + X_4(J_g + J_t)/2 + X_5(J_g + J_t)/2 + X_6 J_6 \tag{IX-9a}$$

$$J'_{AB} = X_1 J_g + X_2 J_t + X_3 J_g + X_4(J_g + J_t)/2 + X_5 J_g + X_6(J_g + J_t)/2 \tag{IX-9b}$$

which closely parallel Eqs. (IX-7a) and (IX-7b) for the *meso* isomer.

If there were no enthalpy differences among the conformers [or if the temperature were very high and there were no entropy differences other than those corresponding to the statistical weights of conformers (4), (5), and (6)], it follows from the above that the two couplings J_{AB} and J'_{AB} would be equal. This is also possible under less stringent conditions, as will be seen below. It cannot yet be stated whether J_{AB} or J'_{AB} will be the larger, since the assignment is arbitrary. Some independent indication of the conformer populations is required.

B. Conformer Populations

1. *Meso*-2,4-*Disubstituted Pentanes*

It is evident from Eqs. (IX-2) and (IX-3) or from (IX-7a) and (IX-7b) that in the *meso* spectrum J_{AB} and J_{AC} will differ substantially if either conformer (1) or conformer (2) is strongly preferred. If conformer (3) is preferred, the couplings will be nearly equal and will be of the order of 2–3 Hz. In Fig. III-1(b) is shown the 100 MHz spectra of *meso*-2,4-diphenylpentane at 35° in carbon tetrachloride.[1,2] The α-proton spectrum appears as a sextet at lower field, and the β-CH_2 spectrum as ten or eleven distinguishable peaks at higher field. Analysis of such spectra at varying temperatures as AA′BCDD′ spin systems (see Bovey *et al.*,[1] Bovey,[2] and Heatley and Bovey[3] for a detailed discussion) was carried out by comparison with computer simulated spectra. It was found that J_{AB} and J_{AC} are equal within experimental uncertainty and approximately 7.4 Hz near room temperature, decreasing slightly but remaining equal at elevated temperatures. (It should be noted that for other *meso*-2,4-disubstituted pentanes, J_{AC} is observed to exceed J_{BC} by as much as 1 Hz; we shall discuss this point later.) The symmetric conformers therefore cannot be present in significant proportion, a conclusion which is compatible with simple steric consideration, since the nonbonded repulsive interactions between phenyl and methyl groups will certainly tend to exclude these forms. Of the asymmetric conformers (4)–(6), steric considerations, supported by nonbonded interaction energy calculations,[4-11] indicate that in general conformer (5) will be strongly preferred. This is in agreement with the observed couplings. We then have from Eqs. (IX-2) and (IX-3)

$$J_{AB} - J_{AC} = \bar{J}_{5\text{-}AB} - \bar{J}_{5\text{-}AC}$$

which from Eqs. (IX-5a) and (IX-5b) becomes

$$J_{AB} - J_{AC} = (\Delta J_g + \Delta J_t)/2$$

where

$$\Delta J_g = J_{5\text{-}AB\text{-}g} - J_{5\text{-}AC\text{-}g} \qquad \text{and} \qquad \Delta J_t = J_{5\text{-}AB\text{-}t} - J_{5\text{-}AC\text{-}t}$$

Thus, if only conformer (5) is present, equal vicinal couplings will be observed if ΔJ_g and ΔJ_t are zero (or if their sum is zero). It seems very probable that both ΔJ_g and ΔJ_t will in fact be nearly zero in compounds of low polarity such as the diphenylpentanes, since there is little reason to expect the dihedral angles to differ greatly for protons H_B and H_C. For the dichloride, dibromide, and diol, McMahon and Tincher[4] have reported values of $|J_{AB} - J_{AC}|$ of the order of 0.5–1.0 Hz and attribute the difference to deviations from exact staggering. Doskocilova and Schneider[12] do not distinguish the *meso* vicinal couplings in the dichloride and diacetate, although in an earlier paper Doskocilova[13] reported two appreciably different couplings for the dichloride. More recently, Schneider *et al.*[14] have suggested that appreciable deviations from exact staggering, which could lead to marked deviations from expected coupling values, may occur. This will be discussed later.

2. *Racemic* 2,4-*Disubstituted Pentanes*

The spectra [see Fig. III-1(a)] of *racemic* 2,4-diphenylpentane have been determined at a number of temperatures.[1] At lower temperatures, the β-CH_2 resonance appears not as the triplet which might be expected but as a doublet of doublets. This part of the spectrum is very temperature dependent (as reported by McMahon and Tincher[4] for the dihalo pentanes), and becomes tripletlike at higher temperature. Analysis of the spectra shows that this isomer exhibits two widely different vicinal couplings, J_{AB} and J'_{AB} at lower temperatures, but that as the temperature is increased the difference between J_{AB} and

J'_{AB}, designated δJ, decreases. For example, at $-50°$ (CS_2), $\delta J = 8.0$ Hz; at $35°$ (carbon tetrachloride), $\delta J = 5.0$ Hz; at $100°$ (*o*-dichlorobenzene) it decreases to 3.3 Hz and finally to 1.0 Hz at $200°$. These observations find a ready explanation in terms of Eqs. (IX-8a) and (IX-8b). The large value of δJ at lower temperatures indicates that neither conformer (3) nor (4) can predominate, since this would lead to a small or zero value. A preference for conformer (5) or (6) might explain the data, but it would be difficult to account for a δJ as large as 8 Hz if J_t and J_g have the usual values; more significantly, these conformers appear to be strongly excluded from steric considerations. We may then rewrite Eqs. (IX-8a) and (IX-8b), eliminating all terms in conformers other than (1) and (2), to give

$$J_{AB} = X_1 J_t + X_2 J_g \qquad \text{(IX-10a)}$$

$$J'_{AB} = X_1 J_g + X_2 J_t \qquad \text{(IX-10b)}$$

from which it is evident that δJ will approach zero as X_1 and X_2 approach equality, since

$$|\delta J| = (X_1 - X_2)(J_t - J_g) \qquad \text{(IX-11)}$$

We also have

$$J_{AB} + J'_{AB} = J_t + J_g \qquad \text{(IX-12)}$$

and it is in fact found experimentally that the quantity $(J_{AB} + J'_{AB})$ is nearly independent of temperature, as would be expected. If it be assumed that all *gauche* couplings are equal and that all *trans* couplings are equal in both the *meso* and *racemic* isomers, and that the *meso* isomer is entirely as conformer (5), it further follows that

$$J_g + J_t = (J_{AB} + J'_{AB})_{rac} = (J_{AB} + J_{AC})_{meso} \qquad \text{(IX-13)}$$

This relationship is also obeyed within experimental error at all temperatures. These observations are unlikely to be merely fortuitous and provide support for the assumptions which must necessarily be made in order to draw any useful conclusions. Thus, $J_g + J_t$ is found to equal 14.8 ± 0.2 Hz. If we define $J_t - J_g$ as ΔJ and assume that δJ is positive, i.e., that conformer (1) predominates over the temperature range studies, it is readily shown that

$$(F_2 - F_1)_{rac} = \Delta F_{rac} = -RT \ln\left(\frac{\Delta J - \delta J}{\Delta J + \delta J}\right) \qquad \text{(IX-14)}$$

$$(H_2 - H_1)_{rac} = \Delta H_{rac} = -R\left[\partial \ln\left(\frac{\Delta J - \delta J}{\Delta J + \delta J}\right)\Big/\partial(1/T)\right]_P \qquad \text{(IX-15)}$$

and

$$(S_2 - S_1)_{rac} = \Delta S_{rac} = -(\partial \Delta F_{rac}/\partial T)_P \qquad \text{(IX-16)}$$

From the slope and intercept of an Arrhenius plot of the coupling data according to Eq. (IX-15), it is found that

$$\Delta H_{\mathrm{rac}} = 1.66 \pm 0.20 \quad \mathrm{kcal}$$

$$\Delta S_{\mathrm{rac}} = 3.2 \pm 0.5 \quad \mathrm{eu}$$

$$\Delta F_{\mathrm{rac}}(298^\circ) = 0.70 \pm 0.20 \quad \mathrm{kcal}$$

For this purpose, a value of ΔJ was chosen (10.0 Hz) such that a linear plot was obtained. Doskocilova and Schneider[15] and Pivcova *et al.*[16] reached similar conclusions.

It is seen that conformer (2) is favored by entropy, possibly because the phenyl groups are farther apart and can rotate more freely. (Such an effect would be absent for atomic substituents.) At temperatures above about 520°K conformer (2) will predominate.

In general, similar conclusions concerning conformational preferences have been reached for other 2,4-disubstituted pentanes. Exceptional behavior is shown by the 2,4-dihydroxypentanes[4, 17] in which the small size of the hydroxyl group and hydrogen bonding between the hydroxyl groups allow conformations which would otherwise tend to be strongly excluded.

Schneider *et al.*[14] have refined the analysis of the NMR and infrared spectra of the 2,4-dichloropentanes to take account of the fact that in these more polar compounds J_{AB} and J_{AC} are not quite equal for the *meso* compound and that Eqs. (IX-12) and (IX-13) do not hold, temperature dependence being appreciable. (Deviations from exact staggering cannot account for this observation, because they would leave J_{AB} and J_{AC} still equal.) They also observed small infrared peaks apparently corresponding to conformers previously excluded on energy grounds. They showed that if both torsional and nonbonded potentials are included, such forms as (4) for the *meso* isomer (see Fig. IX-1) and (5) for the *racemic* isomer (Fig. IX-2) may be allowed in small proportion, but with dihedral angles differing by several degrees from exact staggering.

3. 2,4,6-*Trisubstituted Heptanes*

The extension of model compound studies to the 2,4,6-trisubstituted heptanes has been discussed in general terms in Chapter III, and references to this work are given there. The most carefully studied compounds are the three trichloro stereoisomers (Chapter III, Section 2), the analogs of polyvinyl chloride.[17–19] The analysis of the NMR spectra follows the general lines described for the pentanes in the previous section, but with certain complications. For example, the protons H_A and H_A' in the syndiotactic isomer are nonequivalent, as we have already indicated. It is also found that the difference in chemical shift between the protons H_A and H_B in the isotactic isomer is markedly smaller (about 0.10 ppm) than in the *meso*-dichloropentane (about 0.25 ppm), illustrating a significant shielding influence of distant groups (see Chapter V, Section 1) which could be misleading if the pentane model were interpreted too literally as a guide to the polymer spectrum. Again, the heterotactic isomer offers some complications not present in the pentane models; for example, the "*meso*" portion of this isomer cannot be assumed to take

TABLE IX-1

CONFORMER PREFERENCES OF 2,4-DISUBSTITUTED PENTANES AND 2,4,6-TRISUBSTITUTED HEPTANES[a]

	Conformers	Substituent		
		Cl	C_6H_5	CO_2CH_3
Pentanes:				
meso	TG	1.00	1.00	1.00
racemic	TT	> .90	0.75[b]	<0.60
	GG	< .10	0.25	>0.40
Heptanes:				
isotactic	TGTG	0.80	0.78[c]	0.80
	GTTG	0.20	0.22	0.20
syndiotactic	TTTT	0.85	0.52[d]	0.38
	GGTT	0.15	0.48	0.62
heterotactic	TTTG	0.55	0.41[e]	0.32
	TTGT	0.45	0.44	0.43
	GGTG	<0.5	0.15	0.25

[a] Given as mole fractions as determined by NMR at about 300°K.[1,4,5,12,13,15–21]

[b] At 200° the TT/GG ratio is 0.54/0.46; for other measured pentanes, the ratio is less sensitive to temperature (or unknown).

[c] At 70°; decreases with increasing temperature.

[d] Decreases with increasing temperature.

[e] All conformers converge toward equality with increasing temperature.

on both "mirror-image" conformations with equal probability as does the *meso*-pentane because the molecular environment is unsymmetrical.

The conclusions as to conformational preferences of the trichloroheptanes, as deduced by Doskocilova *et al.*[19] from NMR and infrared studies, are summarized in Table IX-1 and Fig. IX-3. (The conclusions of Abe *et al.*[18] are generally in good agreement, but they do not attempt to give quantitative estimates of minor components.) The syndiotactic isomer strongly prefers the planar zigzag conformation. The isotactic isomer strongly prefers the TGTG $\rightleftarrows$ GTGT conformation, as might be anticipated from the behavior

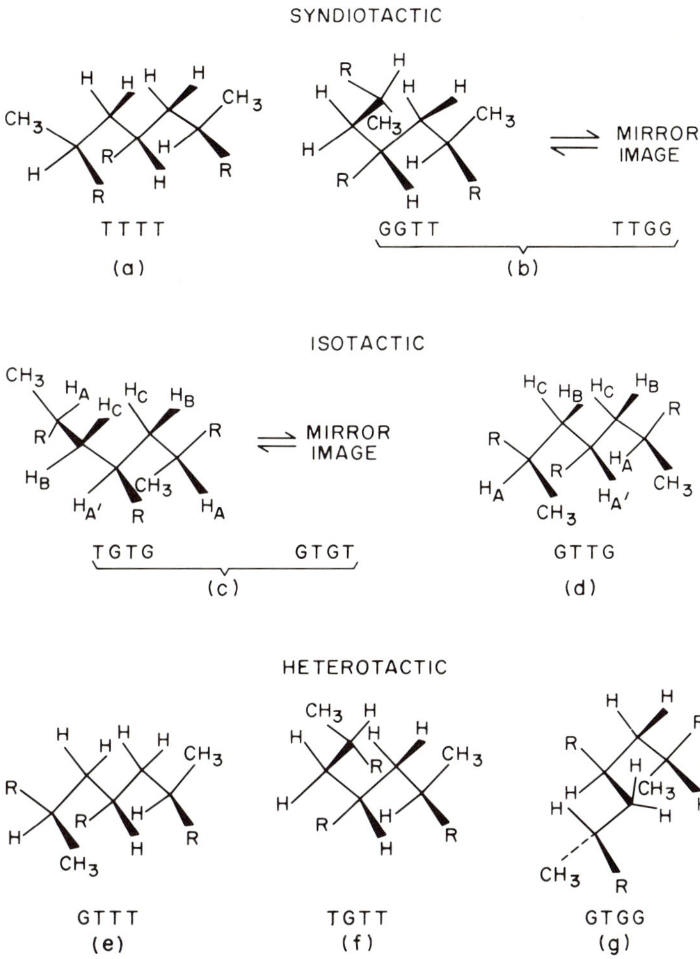

FIG. IX-3. Allowed conformations of 2,4,6-trisubstituted heptanes.

of the *meso*-pentane. As we shall see, both conformations are segments of 3_1 helices (i.e., helices that have three-fold symmetry and repeat in one turn). The heterotactic isomer is TTTG and TTGT with nearly equal probability.

Pivcova *et al.*[16] have analyzed the NMR spectra of the 2,4,6-triphenylheptanes described by Lim *et al.*[20] The conformational preferences at 300°K are shown in Table IX-1. They are similar to those of the chloro compounds, but less marked, the minor conformations being more prominent.

Doskocilova *et al.*[21] have reported the conformer preferences for the three-unit models of polymethyl acrylate, the carboxymethyl heptanes[22,23] (also the pentanes). These results are included in Table IX-1. The isotactic isomer behaves like those with Cl and C_6H_5 substituents, but the syndiotactic isomer actually prefers the folded GGTT form, and the GGTG form of the heterotactic isomer is much more strongly allowed than for the other compounds. These authors also reported NMR results for the hydroxy and acetoxy heptanes. The conformational preferences of the acetates are similar to those of the other compounds in Table IX-1, but the hydroxy compounds, like the pentanes, gave evidence of other conformers which are usually forbidden. These results are less quantitative and are not included in Table IX-1.

Semiquantitative interpretation of the conformer preferences of the tricyanoheptanes was given by Murano and Yamadera.[24]

Another approach to the study of conformational preferences of vinyl polymer chains and model compounds is that of achieving stereochemical equilibrium under conditions appropriate for the epimerization of asymmetric centers, and then measuring the resulting populations of diastereoisomers. The same intramolecular forces which determine conformational populations under conditions of fixed configuration also determine the diastereoisomeric populations which will be attained when the chemical environment allows inversion of asymmetric carbon atoms. The first experiments of this kind were carried out by Clark[25] on 2,4-dimethylglutaric acid, 2,4,6-tricarboxyheptane, and their methyl esters, and were extended to the 2,4-dichloropentanes by Billups and Kurtz[26] and Flory and Williams[11] and to polystyrene and its oligomers by Williams *et al.*[9] and Williams and Flory.[10] The experimental values of the conformer populations agree well with the predictions of a rotational isomeric state theory employing two adjustable parameters to express the steric interactions in a vinyl polymer chain[7,8,10] (see Section 3).

3. Polymer Chain Conformation

The classical methods of observing the conformations of polymer molecules in solution, such as light scattering and viscosity measurement, give information concerning the overall extension of the chains in solution, commonly expressed as the mean square end-to-end distance $\langle r^2 \rangle_0$. In contrast, spectroscopic observations, NMR in particular, provide information concerning local conformational preferences. If this information is sufficiently precise and is coupled with a measure of the molecular weight, one could in principle determine the extension in solution in this manner also. This is scarcely possible at present, however, as we shall see.

As we have seen, the analysis of the NMR spectra of model compounds involves the assumptions (a) that the conformer bonds are staggered in the sense of Figs. IX-1 and IX-2, or nearly so, and (b) that J_g and J_t are the same in all conformers. (These assumptions are closely related.) The first of these may be too stringent, for it appears that in both polar[14] and nonpolar[7, 8, 27, 28] chains departures from exact staggering may occur. We shall consider the second assumption further a little later.

Even if we accept these simplifying assumptions, we still face difficulties in the interpretation of polymer NMR spectra, mainly because of their relatively great line width. For example the spectra of *racemic* dyads do not appear as the well-resolved triplet shown in Fig. III-1(a) (still less as the multiline spectrum shown by this same compound when observed at 60 MHz rather than 100 MHz†; see Fig. 4d in Bovey *et al.*[1]), but as a much broader envelope, as for example that in Fig. V-4. We have seen in previous chapters that such spectra, despite their rather forbidding appearance, can in fact be solved by computer simulation on the basis of "dimer" spin models, or in the case of stereoirregular polymers, a summation of individually calculated tetrad and pentad subspectra. The margin for error in the determination of the magnitudes of the J couplings is naturally greater in such circumstances, but these spectra are still capable of yielding useful information. We shall now discuss the interpretation of the data thus obtained, which has already been given in earlier chapters on individual polymers.

A. Polystyrene

The chemical shifts and vicinal coupling constants for isotactic polystyrene in *o*-dichlorobenzene as a function of temperature were presented in Table VI-1. We note that the couplings J_{AC} and J_{BC}, which are equal for the 2,4-diphenylpentane at all temperatures, differ by 1.5 Hz at lower temperatures, decreasing to 1.0 Hz at the highest temperatures. The limited accuracy with which these couplings can be determined for the polymer make the significance of this apparent temperature trend rather doubtful.

In all other *meso*-2,4-disubstituted pentanes reported, J_{AB} and J_{AC} differ by about 1.0 Hz; in this respect, the *meso*-2,4-diphenylpentane is exceptional, as we have seen. It has been found[29] that in cyclic compounds and halogenated ethanes, *gauche* coupling constants are dependent on the substituents *trans* to the coupled protons to an extent sufficient to account for this difference. One may conclude that in the phenyl substituted model compounds and in polystyrene no such substituent effect occurs, and that the assumption of equality of all *gauche* and of all *trans* couplings is a fair approximation.

† Even in Fig. III-1(a), the triplet is deceptive, for the vicinal couplings actually differ substantially in magnitude, as we have seen in the previous section. It is a seeming anomaly that such spectra are often easier to comprehend and to match by computer simulation when obtained at *lower* observing frequencies, due to the greater number of observable lines, as just indicated.

We have seen that the coupling constants of the *meso*-pentane indicate that the GT $\rightleftarrows$ TG conformation [(5a) and (5b) in Fig. IX-1] is strongly preferred and that this conclusion is in harmony with energy calculations. It is reasonable to conclude that this conformation is the dominant one for the polymer chain as well. The GT conformation is a segment of the 3_1 helical conformation (Fig. IX-4) that the polymer is known to have in the crystalline state.[30–32]

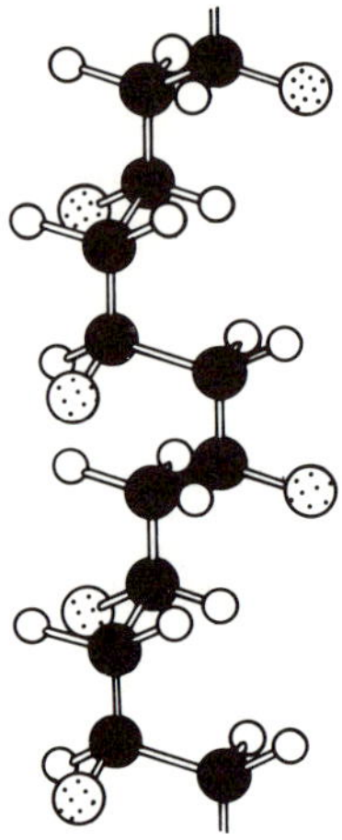

FIG. IX-4. The 3_1 helical conformation of an isotactic vinyl polymer chain. Darkest balls are backbone carbon atoms; lighter balls are substituent groups R; others are hydrogen atoms.

The corresponding heptane conformation TGTG $\rightleftarrows$ GTGT, is also strongly preferred, but the GTTG conformation is present in a concentration which cannot be ignored (Table IX-1). This conformation corresponds to the junction of two 3_1 helices of opposite sign:

$$\cdots\text{GTGTG}\underset{\rightarrow}{\text{TT}}\text{GTGTG}\cdots$$

To return to the helix conformation of the first sign (imagine arbitrarily that we are proceeding from left to right) it is necessary to attain a TGGT conformation:

$$\cdots\text{GTGT}\underset{\rightarrow}{\text{GG}}\text{TGTG}\cdots$$

These conformations are illustrated in Fig. IX-5. It will be noted that the second of these involves a severe overlap of substituents R (phenyl groups in the present case) if exact staggering is maintained, and that therefore the numbers of these junctions will be small. The energy of such junctions is a matter of dispute. It seems probable that the steric overlaps are actually relieved by deviations from exact staggering, as suggested, among others, by Corradini and Allegra.[27] We shall consider this question further in the discussion of polypropylene (Section 3C of this chapter), for which more precise information is available. The difference between vicinal couplings J_{AB} and J_{AC}

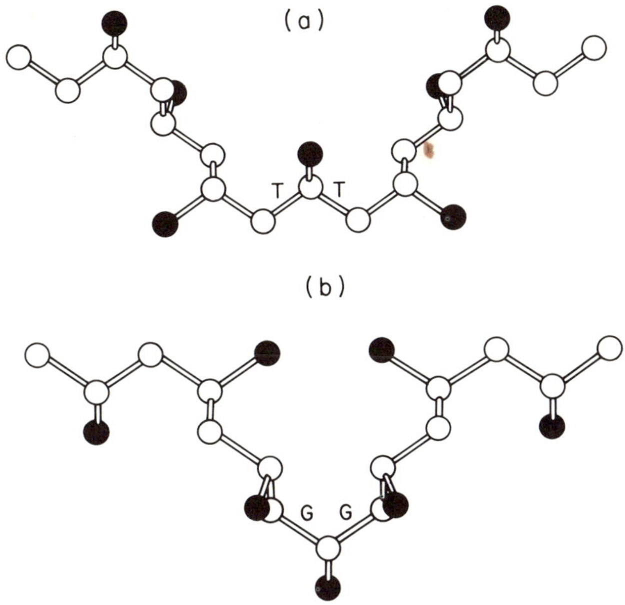

FIG. IX-5. Conformations of isotactic chains in (a) ···GTGTGTTGTGTG··· and (b) ···GTGTGGTGTG··· conformations. Hydrogen atoms are omitted for clarity (adapted from Corradini and Allegra[27]).

in isotactic polystyrene is probably due mainly to the presence of ···GTTG··· and ···TGGT··· conformations (or some modification of the latter).

NMR evidence for the predominance of the 3_1 helical conformation in isotactic polystyrene chains is consistent with ultraviolet, infrared, and other evidence for this conformation.[33–39] The ultraviolet evidence[35,36] rests on the observation of a distinct decrease in extinction in the near-ultraviolet aromatic band for the isotactic polymer compared to the atactic, corresponding to interaction of the phenyl groups in a regular layered arrangement such as occurs in the 3_1 helix. The infrared evidence is less direct.[33,34,37] Kobayashi *et al.*[37] observed that the infrared bands associated with helical conformations decreased markedly with increasing temperature, the effect having a ΔH of at least 1.5 kcal. These observations no doubt reflect a marked shortening of the average 3_1 helical domain at higher temperatures, although a full interpretation is not yet available.

Older measurements of solution dimensions[38,39] showed little difference between isotactic and atactic polystyrene. It would be worthwhile to repeat such measurements on polymers of more precisely known configuration, since even small proportions of GG *racemic* dyads will reduce the end-to-end distance very markedly (see Section 3C).

Liu and Ullman[40] observed that the line widths of the NMR spectra of polystyrene solutions decreased rather abruptly in the region of 50–60°, the effect being more marked for the isotactic polymer. This may reflect a partial loss of helical structure at this temperature, but since this transition is not reflected in the J couplings its interpretation is not entirely clear.

Fujiwara and Flory[41] have presented a careful and detailed calculation of the chemical shifts of the α-protons in atactic polystyrene, based on the rotational isomeric state model of the polymer chain.[7,8,10] The rotational state of each pentad sequence was calculated, i.e., the conformations of nearest and next-nearest neighboring dyads adjoining the α-CH were taken into account. The only chemical shift influence taken into account, however, was the shielding effect of the phenyl group when the α-proton was at the face of the benzene ring, i.e., in its shielding zone. The shielding anisotropy of neighboring C—C single bonds was ignored. This seemingly drastic simplification is probably justified, for in polypropylene (Chapter VII, Section 2), in which C—C single bond anisotropy must be the dominant influence, the α-CH chemical shifts are insensitive to the stereochemical configuration of the chain. This is evidently due to a fortuitous cancellation of effects which can be reasonably expected to hold in any nonpolar α-monosubstituted vinyl polymer.

The results of this calculation are summarized in Fig. IX-6. The configurational sequence was assumed to be Bernoullian with a P_m of 0.5. Figure IX-6

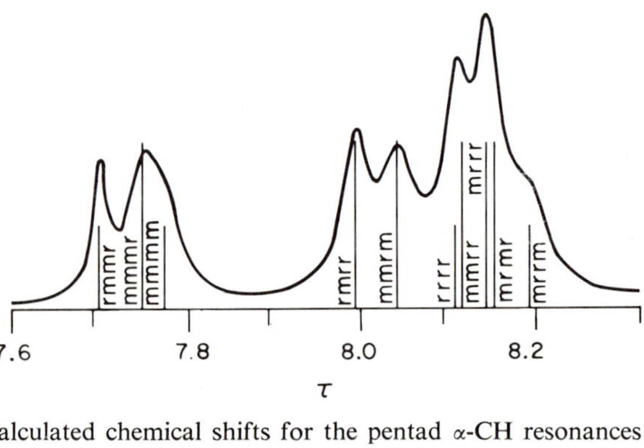

FIG. IX-6. Calculated chemical shifts for the pentad α-CH resonances of atactic polystyrene (adapted from Fujiwara and Flory[41]).

should be compared to Fig. VI-4. The agreement is probably as good as can be expected. The predicted spread of pentad chemical shifts is somewhat larger than observed and there are some discrepancies with respect to the assigned pentad positions shown in Fig. VI-4. These latter assignments are, however, tentative, as we have previously seen. The calculated relative triad chemical shifts are generally in agreement with those shown in Fig. VI-4, although with rather more overlap of *rr* and *mr* than shown there. Relative peak intensities differ because the experimental spectrum is that of a polymer characterized by a P_m of about 0.3.

B. Polyacrylates

Isotactic polyacrylate chains are known to have a 3_1 helical conformation in the crystalline state.[30–32, 42] Yoshino *et al.*[43] prepared isotactic polymers, using anionic initiators, from the monomers:

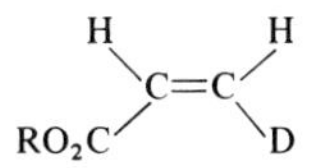

and

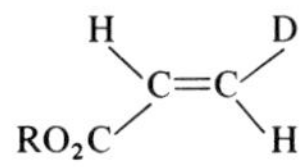

(a) (b)

The mole ratio of the *trans* (b) to *cis* (a) monomer was 3:2, R was —CH_3 and —$CH(CH_3)_2$. The disposition of *cis* and *trans* monomer units along the polymer chains was believed to be random, but whether this is the case or not or whether addition to the double bond is "*cis*" or "*trans*" (Chapter VIII, Section 4B) is actually immaterial. Replacement of one of the β-methylene protons by deuterium simplified the spectrum, and in principle should allow a less ambiguous estimate of the vicinal couplings to be made. The spectrum of the polymethyl-β-d_1-acrylate in

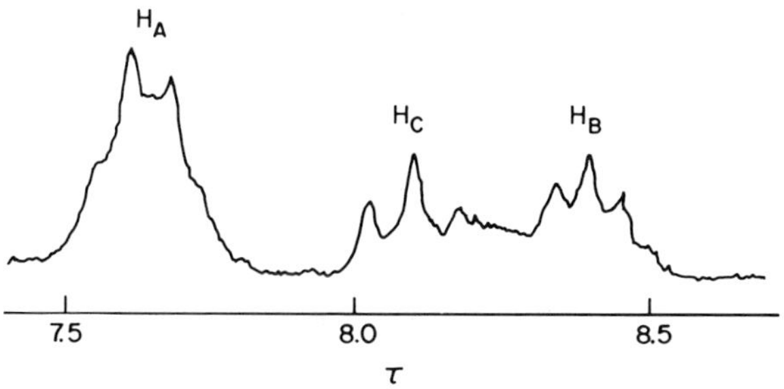

FIG. IX-7. 60 MHz spectrum of the α- and β-protons of isotactic polymethyl-β-d_1-acrylate, observed in methyl formate (Yoshino *et al.*[43]).

methyl formate solution is shown in Fig. IX-7. In addition to the α-proton resonance centered at about 7.6τ (and clearly not a simple triplet), triplet resonances of the *erythro* and *threo* β-protons appear at about 8.1 and 8.4τ, respectively. These triplets appeared to differ very markedly in spacing, the *erythro* protons being more strongly (9.1 Hz) coupled to the α-protons than the *threo* protons (5.6 Hz). Substantial populations of conformations normally considered to be very strongly excluded were suggested by Yoshino *et al.*[43] to explain this observation. Both their measurements and conclusions are in conflict with the results reported by Heatley and Bovey on isotactic polyisopropyl acrylate.[44] We have seen (Chapter IV, Section 2) that the spectrum of this polymer is consistent with values of J_{AC} and J_{BC} of 6.0 and 7.5 Hz, respectively. It is reasonable to interpret this as indicating a strong predominance of ···GTGT··· or 3_1-helical conformations, as in polystyrene. The results of the Yoshino work are at present unexplained. It should be noted that they analyzed their spectrum only by first-order measurements and not by simulation, but this alone does not appear to account for the discrepancy. It is of course possible that the bulkier side chain of polyisopropyl acrylate tends more strongly to exclude conformations other than GT, but this too seems an inadequate explanation, particularly in view of the measurements of its chain extension in solution, which we shall now consider.

Using the rotational isomeric state model, Flory *et al.*[7] have calculated the characteristic ratios $\langle r^2 \rangle_0 / nl^2$, i.e., the ratio of the actual mean square end-to-end distance to that expected for a random-flight chain of n links of length l, for vinyl polymer chains in which P_m varies from 0 to 1. The ratio is rather insensitive at lower values of P_m, remaining near 10 until P_m exceeds about 0.9, after which it increases rapidly toward a very high value characteristic of the ···GTGT··· helix (Fig. IX-8). According to the analysis of Heatley and

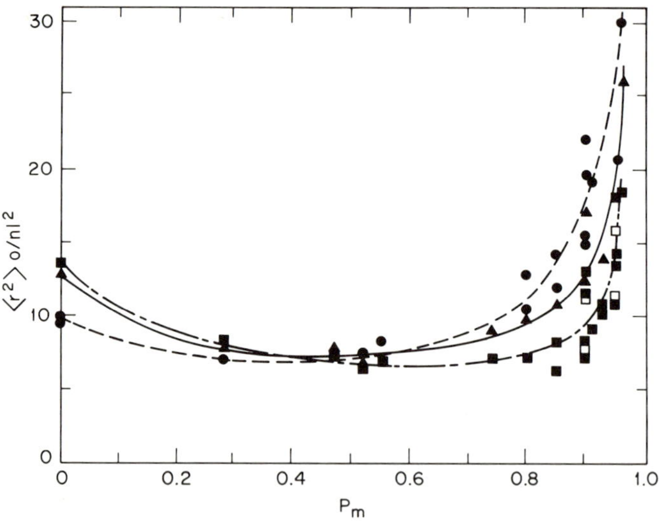

FIG. IX-8. Dependence of mean square radius of a vinyl polymer chain upon the fraction of m units, P_m. The allowed deviation in rotational angles is indicated by $\Delta\varphi$. Results for $\Delta\varphi = 20°$ calculated for a chain of 200 monomer units are indicated by *open* squares. The other calculated curves were obtained for chains of 100 monomer units (from Flory *et al.*[7]). Symbols used to indicate $\Delta\varphi$ values: ●, 0°; ▲, 10°; ■ and □, 20°.

Bovey,[44] P_m for the polyisopropyl acrylate they measured was 0.95. Mark and co-workers[45,46] reported a characteristic ratio of 9.7 for isotactic polyisopropyl acrylate. Although the NMR measurements available at the time were not as searching as those employed by Heatley and Bovey,[44] it is reasonable to conclude that the polymer is similar to that employed by Heatley and Bovey since the methods of preparation were closely similar. We may therefore conclude that the calculations of Flory *et al.* are in harmony with experiment in this case, particularly if the energy minima are allowed to be displaced by 10–20° from the 120° exact staggering values (see caption to Fig. IX-8).

Mark and co-workers[45,46] observed lower values of the characteristic ratio for "syndiotactic" and "atactic" polyisopropyl acrylate: 7.2 and 7.1, respectively. These polymers are probably both actually very similar in configuration with a P_m of about 0.3 (Chapter IV, Section 2). As can be seen from Fig. IX-8, the characteristic ratios are in good agreement with the energy calculations.

Muroga *et al.*[47] have observed that the vicinal coupling constants J_{AC} and J_{BC} of polyacrylic acid in aqueous solution are essentially the same as those for the polyalkyl acrylates in organic solvents. As the carboxyl groups are ionized by addition of base, the chemical shift difference between the β-methylene protons decreases very markedly but the coupling constants remain unchanged within experimental error, both being 7.0 Hz. Thus, despite the expansion of the chain and the consequent viscosity increase accompanying ionization, the *local* conformation remains predominantly that of a 3_1 helix. The expansion thus arises primarily from long-range Coulombic repulsion rather than local alterations of conformer populations. Of course, if one could measure the couplings with sufficiently high discrimination, some change would necessarily be noted, but it is evidently well within present experimental error.

Mark and Krimm[48] have suggested, on the basis of electrostatic energy calculations, that charged isotactic vinyl polymers, such as sodium polyacrylate, should have a conformation intermediate between a 3_1 helix and a five-fold helix.

C. Polypropylene

In Chapter VII, Section 2, the spectrum of a polypropylene was described[49] which was believed to be about 98% isotactic (although this claim has been disputed, as discussed there). The vicinal coupling constants (Table VII-1) are again, as for isotactic polystyrene and isotactic polyisopropyl acrylate, consistent with a predominantly 3_1 helical conformation. The characteristic ratio of this polymer has been reported[49] to be 4.7 in diphenyl ether at 140° (Θ conditions), whereas it should have a much higher value according to the calculations of Flory *et al.*[7] The predicted value depends strongly upon the value chosen for the parameter ω,[7, 50–52] which is a statistical weight reflecting all the second-order steric interactions in a vinyl chain—in an isotactic chain, principally that occurring in the conformation sequence TGGT (p. 194). As the value of $\omega \rightarrow 0$, the characteristic ratio increases rapidly to a value very much higher than the value of 5–10 usually found for supposedly highly isotactic polymers. If the estimate of 2% for the *r* content of the isotactic polypropylene discussed in Chapter VII, Section 2, is accepted, it follows[49] that ω for isotactic polypropylene is in the range 0.05–0.2, depending upon the deviation from exact staggering of the G and T conformations. This corresponds to an interaction energy of 1.2–2.4 kcal/mole, say 2.0 kcal as an approximate mean. This value is close to that calculated by Borisova and Birshtein, based on interatomic potential functions.[53] Flory[54] has maintained, however, that the weak triplet at 8.8τ in Fig. VII-7 should not be assigned to the *mrm* tetrad (see the discussion of "atactic" polypropylene below) but possibly to *rrr* and has further suggested that an *r* content substantially greater than 2% would not be discernible because of peak broadening by statistical variations in adjoining stereochemical sequences.† It may be that proton resonance measurements will not be able to resolve this disagreement but that careful ^{13}C measurements (Chapter VII, Section 2), preferably on enriched samples, may lead to a solution.

Syndiotactic dyads, whether in a syndiotactic chain or as a minor component of an isotactic chain, may assume either of the nonequivalent conformations TT or GG shown in Fig. IX-9, which are analogous to conformers

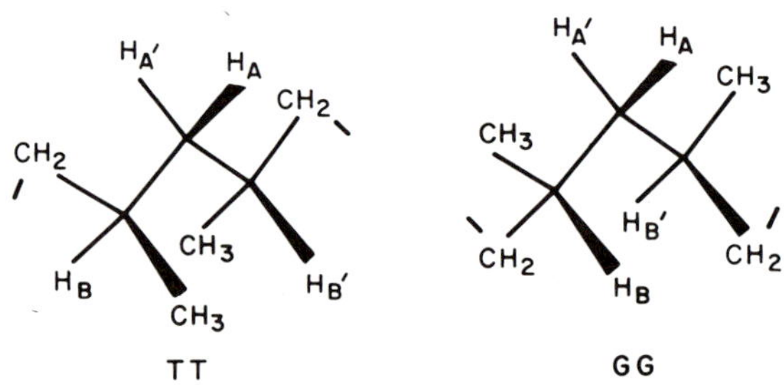

FIG. IX-9. Allowed conformations of *r* dyads in a vinyl polymer chain.

(1) and (2) (Fig. IX-2) of the *racemic* 2,4-disubstituted pentane. As we have seen, the conformational distribution may be determined experimentally from

† It may be noted that the suggestion of a substantial proportion of undetected *r* units in isotactic polypropylene and in isotactic vinyl polymers generally [Flory *et al.*[7]; also Flory and Baldeschwieler,[55] and Flory *et al.*[56]] has been strongly protested by Natta and Corradini,[57] who point out that x-ray diffraction results and energy calculations are inconsistent with such a view.

the vicinal couplings J_{AB} and $J_{AB'}$, the averaged values of which are given by

$$J_{AB} = J_{A'B'} = X_{TT}J_t + (1 - X_{TT})J_g \quad \text{(IX-17a)}$$

$$J_{AB'} = J_{A'B} = X_{TT}J_g + (1 - X_{TT})J_t \quad \text{(IX-17b)}$$

which are parallel to Eqs. (IX-10a) and (IX-10b); X_{TT} designates the mole fraction of the TT conformation. In Chapter VII, Section 2, the parameters for syndiotactic propylene (in *o*-dichlorobenzene at 110°) are given. The difference in vicinal coupling, δJ, is 3.5 Hz. Assuming reasonable values of 11 Hz and 2 Hz for J_t and J_g, respectively, we find from Eq. (IX-11) that $X_{TT} = 0.7$. For the observable residual *r* resonance in isotactic polypropylene we find (Table VII-1) that δJ is 3.0 Hz, giving $X_{tt} = 0.6$, the same as for the all-syndiotactic chain within experimental error. A substantial and equal proportion of GG is thus present, according to this evidence, whether the *r* dyad is part of an isotactic or a syndiotactic chain. This conclusion does not appear to be in harmony with the calculations of Flory and Fujiwara,[50,51] but of course depends upon the interpretation of the 8.8τ resonance in Fig. VII-7.

The assignment of tetrad chemical shifts in "atactic" polypropylene has been described in Chapter VII, Section 2. Flory and Fujiwara[51] and Flory[54] have calculated the expected chemical shifts for polypropylene β-methylene protons by a careful consideration of the probabilities of the accessible conformations of the neighboring (and more distant) dyads about the central dyad of each tetrad, and their effects on the central protons. The effects of such conformations depend upon the relative shielding anisotropies of the substituent R (=CH_3), the β-CH_2 group, and the solvent. The arguments are too lengthy to reproduce here, and the reader is referred to the original papers for a detailed discussion. The chemical shift predictions are strongly dependent on the value taken for ω, which is assumed on other grounds not to be greater than 0.10 and is actually taken as 0.04. In Fig. IX-10, the results of this calculation[54] are schematically compared to the experimental spectra and assignments of Heatley and Zambelli[58] and Zambelli and Segre[59] (see Fig. VII-6). These schematic spectra are presented as if the polymer were truly atactic, i.e., Bernoullian with a P_m of 0.5. We have seen that this is not the case for the deuteropolymers actually examined in detail and that in fact *m* placements predominate, and it is also possible that the configurational sequences may be block-like. Nevertheless, this representation is adequate, because we are principally concerned here with the positions of the peaks rather than their intensities. It will be noted that agreement of the experimental spectra and assignments with the calculation is very good with the principal exception of *mrm*, which the calculation places very close to the less shielded members of the two *mrr* and *rmr* pairs of resonances, rather than coincident with *rrr*.

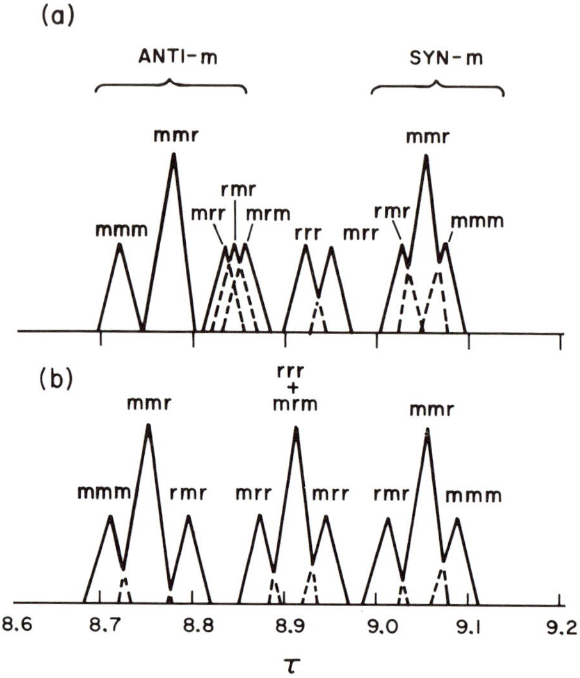

FIG. IX-10. Schematic representation of tetrad (β-methylene) chemical shifts in atactic polypropylene; (a) as calculated by Flory[54] (see also Flory and Fujiwara[51]); (b) as observed experimentally in *o*-dichlorobenzene at 145° and assigned by Heatley and Zambelli[58] (see also Zambelli and Segre[59]).

As we have already indicated, this would imply that the 8.8τ triplet in Fig. VII-7 may be *rrr*, seriously affecting the mechanistic conclusions discussed in Chapter VIII, Section 3E.

D. Polyvinyl Chloride

The spectral data shown in Table V-1 indicate that the *m* dyads in a stereo-irregular polyvinyl chloride chain have a strong preference for the TG $\rightleftharpoons$ GT conformation, in agreement with the data for the *meso* dimer; the vicinal couplings found for the polymer are the same as those reported for the dimer by a number of authors. The *r* dyads have a strong preference for the TT conformation, greater than that shown by the *racemic* dimer. In aromatic solvents, δJ is 9 Hz; analysis in the manner described for polypropylene (Section 3C) indicates the virtual exclusion of other conformations, despite the occurrence of a measurable proportion of GGTT in the syndiotactic trimer (Table IX-1). In pentachloroethane δJ is reduced from 9 Hz to 7

Hz, corresponding to a greater participation of GG and a more folded overall conformation. One might expect that δJ would provide a measure of solvent "goodness" and should correlate directly with the mean square radius. One would also expect that δJ should decrease with increasing temperature, as in the *racemic* dimer, corresponding to increasing contributions from GG and other energetically less favored conformations. As yet there appear to be no data with which to test these predictions. Since helix reversals of the type TGGT are likely to be of infrequent occurrence and since *r* sequences are almost as conformationally regular as *m* sequences, it would appear to follow that the majority of "kinks" in a polyvinyl chloride chain must occur at *mr* junctions of the form TGTT and GTTT and their mirror images.

REFERENCES

1. F. A. Bovey, F. P. Hood, E. W. Anderson, and L. C. Snyder, *J. Chem. Phys.* **42**, 3900 (1965).
2. F. A. Bovey, *Accounts Chem. Res.* **1**, 175 (1968).
3. F. Heatley and F. A. Bovey, *Macromolecules* **1**, 301 (1968).
4. P. E. McMahon and W. C. Tincher, *J. Mol. Spectrosc.* **15**, 180 (1965).
5. T. Shimanouchi and M. Tasumi, *Spectrochim. Acta* **17**, 775 (1961).
6. P. E. McMahon, *Trans. Faraday Soc.* **61**, 197 (1965).
7. P. J. Flory, J. E. Mark, and A. Abe, *J. Amer. Chem. Soc.* **88**, 639 (1966).
8. P. J. Flory, *J. Amer. Chem. Soc.* **89**, 1798 (1967).
9. A. D. Williams, J. I. Brauman, N. J. Nelson, and P. J. Flory, *J. Amer. Chem. Soc.* **89**, 4807 (1967).
10. A. D. Williams and P. J. Flory, *J. Amer. Chem. Soc.* **91**, 3111 (1969).
11. P. J. Flory and A. D. Williams, *J. Amer. Chem. Soc.* **91**, 3118 (1969).
12. D. Doskocilova and B. Schneider, *Coll. Czech. Chem. Commun.* **29**, 2290 (1964).
13. D. Doskocilova, *J. Polym. Sci. Part B* **2**, 421 (1964).
14. B. Schneider, J. Stokr, D. Doskocilova, S. Sykora, J. Jakes, and M. Kolinsky, *J. Polym. Sci. Part C* **22**, 1073 (1969).
15. D. Doskocilova and B. Schneider, *J. Polym. Sci. Part B* **3**, 213 (1965).
16. H. Pivcova, M. Kolinsky, D. Lim, and B. Schneider, *J. Polym. Sci. Part C* **22**, 1093 (1969).
17. T. Shimanouchi, M. Tasumi, and Y. Abe, *Makromol. Chem.* **86**, 43 (1965).
18. Y. Abe, M. Tasumi, T. Shimanouchi, S. Satoh, and R. Chujo, *J. Polym. Sci. Part A-1* **4**, 1413 (1966).
19. D. Doskocilova, J. Stokr, B. Schneider, H. Pivcova, M. Kolinsky, J. Petranek, and D. Lim, *J. Polym. Sci. Part C* **16**, 215 (1967).
20. D. Lim, M. Kolinsky, J. Petranek, and D. Doskocilova, *J. Polym. Sci. Part B* **4**, 645 (1966).
21. D. Doskocilova, S. Sykora, H. Pivcova, B. Obereigner, and D. Lim, *J. Polym. Sci. Part C* **23**, 365 (1968).
22. D. Lim, M. Kolinsky, E. Votavova, M. Ryska, and J. Lukas, *J. Polym. Sci. Part B* **4**, 573 (1966).
23. D. Lim, B. Obereigner, and D. Doskocilova, *J. Polym. Sci. Part B* **3**, 893 (1965).
24. M. Murano and R. Yamadera, *J. Polym. Sci. Part A-1* **6**, 843 (1968).

25. H. G. Clark, *J. Polym. Sci. Part C* **23**, 3455 (1968).
26. W. E. Billups and A. N. Kurtz, *J. Amer. Chem. Soc.* **90**, 1361 (1968).
27. P. Corradini and G. Allegra, *Atti Accad. Naz. Lincei Rend. Cl. Sci. Fis. Mat. Natur.* **30**, 516 (1961).
28. A. Zambelli, M. G. Giongo and G. Natta, *Makromol. Chem.* **112**, 183 (1968).
29. R. J. Abraham, L. Cavalli, and K. G. R. Pachler, *Mol. Phys.* **11**, 471 (1966).
30. G. Natta, *Makromol. Chem.* **35**, 94 (1960).
31. G. Natta, P. Corradini, and I. W. Bassi, *Nuovo Cimento* (*Suppl.*) **15**, 68 (1960).
32. R. L. Miller and L. E. Nielsen, *J. Polym. Sci.* **55**, 743 (1961).
33. M. Takeda, K. Iimura, A. Yamada, and Y. Imamura, *Bull. Chem. Soc. Jap.* **32**, 1151 (1959).
34. T. Onishi and S. Krimm, *J. Appl. Phys.* **32**, 2320 (1961).
35. M. T. Vala, Jr., and S. A. Rice, *J. Chem. Phys.* **39**, 2348 (1963).
36. J. W. Longworth, *Biopolymers* **4**, 1131 (1966).
37. M. Kobayashi, K. Tsumura, and H. Tadokoro, *J. Polym. Sci. Part A-2* **6**, 1493 (1968).
38. F. Danusso and G. Moraglio, *Makromol. Chem.* **38**, 250 (1958).
39. W. R. Krigbaum, D. K. Carpenter, and S. Newman, *J. Phys. Chem.* **62**, 1586 (1958).
40. K.-J. Liu and R. Ullman, *Polymer* **6**, 100 (1965).
41. Y. Fujiwara and P. J. Flory, *Macromolecules* **3**, 43 (1970).
42. T. Mamimoto, T. Tsuruta, and J. Furukawa, *Makromol. Chem.* **50**, 116 (1961).
43. T. Yoshino, Y. Kikuchi, and J. Komiyama, *J. Phys. Chem.* **70**, 1059 (1966).
44. F. Heatley and F. A. Bovey, *Macromolecules* **1**, 303 (1968).
45. J. E. Mark, R. A. Wessling, and R. E. Hughes, *J. Phys. Chem.* **70**, 1895 (1966).
46. R. A. Wessling, J. E. Mark, E. Hamori, and R. E. Hughes, *J. Phys. Chem.* **70**, 1910 (1966).
47. Y. Muroga, I. Noda, and M. Nagasawa, *J. Phys. Chem.* **73**, 667 (1969).
48. J. E. Mark and S. Krimm, *Macromolecules* **2**, 173 (1969).
49. F. Heatley, R. Salovey, and F. A. Bovey, *Macromolecules* **2**, 619 (1969).
50. P. J. Flory and Y. Fujiwara, *Macromolecules* **2**, 315 (1969).
51. P. J. Flory and Y. Fujiwara, *Macromolecules* **2**, 327 (1969).
52. P. J. Flory, "Statistical Mechanics of Chain Molecules," Chapters III and VI. Wiley (Interscience), New York, 1969.
53. N. P. Borisova and T. M. Birshtein, *Polymer Sci. USSR* **4**, 907 (1963).
54. P. J. Flory, *Macromolecules* **3**, 613 (1970).
55. P. J. Flory and J. D. Baldeschwieler, *J. Amer. Chem. Soc.* **88**, 2873 (1966).
56. P. J. Flory, J. E. Mark, and A. Abe, *J. Polym. Sci. Part B* **3**, 973 (1965).
57. G. Natta and P. Corradini, *Makromol. Chem.* **110**, 291 (1967).
58. F. Heatley and A. Zambelli, *Macromolecules* **2**, 618 (1969).
59. A. Zambelli and A. L. Segre, *J. Polym. Sci. Part B* **6**, 473 (1968).

Chapter X

VINYL COPOLYMERS

1. Introduction

High resolution NMR has proved particularly effective in the study of vinyl copolymers, revealing structural and sequence details which one could not expect to detect by any other means. Indeed, the information available is often so rich that it is difficult to sort out and interpret in detail. We shall therefore begin the discussion with relatively simple systems and then progress to those in which complications, particularly those arising from stereochemical configuration, must be dealt with. This general field has been ably reviewed,[1–5] and the present discussion will be concerned primarily with principles rather than attempting encyclopedic coverage.

The theory of copolymerization of vinyl and diene monomers was developed many years ago[6,7] and has been tested and supported by many subsequent investigations. This earlier work consisted almost entirely of analytical determinations of monomer ratios in copolymers as a function of monomer ratio in the feed, with little or no information concerning sequence lengths and none whatever concerning stereochemical configuration. The equations that successfully predict copolymer composition also predict sequence lengths. Nevertheless, direct measurement is highly desirable, for the subtlety of information available from NMR spectra greatly exceeds that available from analytical measurements. In addition, the occurrence of anomalous structures, such as head-to-head units, can often be detected.

It will be helpful to consider structural units which can be (at least in principle) distinguished by NMR. If we ignore stereochemistry, or employ comonomers of the type $CH_2{=}CX_2$ which do not give rise to asymmetric centers, we then have:

Dyads:	AA	AB(BA)	BB
Triads:	AAA		BBB
	BAA(AAB)		ABB(BBA)
	BAB		ABA
Tetrads:	AAAA	AABA(ABAA)	BBBB
	AAAB(BAAA)	AABB(BBAA)	BBBA(ABBB)
	BAAB	BABA(ABAB)	ABBA
		BABB(BBAB)	

Generalizing, we find the number $N(n)$ of distinguishable sequences of length n to be

n	2	3	4	5	6	7	8
$N(n)$	3	6	10	20	36	72	136

or

$$N(n) = 2^{n-1} + 2^m$$

where $m = (n-2)/2$ if n is even and $(n-1)/2$ if n is odd.

If stereochemical configuration is to be considered, we must consider the structures shown in Table X-1 for dyads and triads. The representation of longer sequences is omitted, for we find that the number of sequences $N'(n)$ increases very rapidly with n

n	2	3	4	5	6
$N'(n)$	6	20	72	272	1056

or

$$N'(n) = 2^{2(n-1)} + 2^{n-1}$$

There may thus be relatively little advantage in the observation of such copolymers at high field strengths unless one comonomer predominates strongly, for the sequence fine structure may otherwise either overwhelm analysis or be unresolvable. Indeed, it has been found in the author's laboratory[8] that methyl methacrylate–styrene copolymers (see below) give similar spectra at 40 MHz and 220 MHz, with little further detail in the latter.

If one of the monomers is of the type $CXY{=}CH_2$ (or $CHX{=}CH_2$) and the other of the type $CX_2{=}CH_2$ (X may be H, i.e., ethylene, and F may replace H), we have the more limited set of sequences shown in Table X-2.

The statistics of comonomer distribution are clearly closely related to the description of homopolymer configuration and may be expressed in terms of

TABLE X-1

CONFIGURATIONAL SEQUENCES IN COPOLYMERS

Dyads:	AA	AB (BA)	BB
m	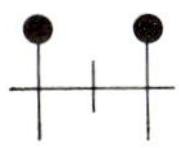	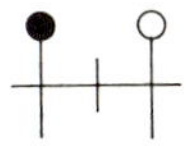	
r	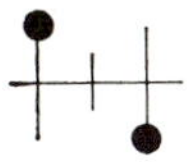	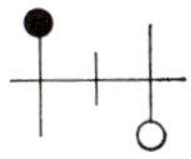	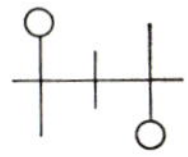
Triads:	**AAA**	**AAB (BAA)**	**BAB**
mm	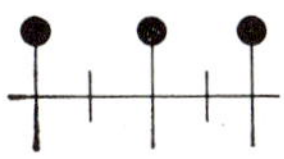	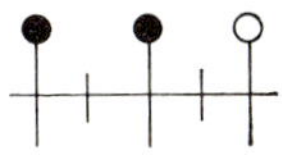	
mr	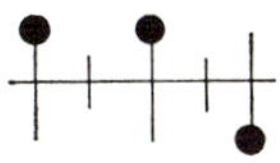	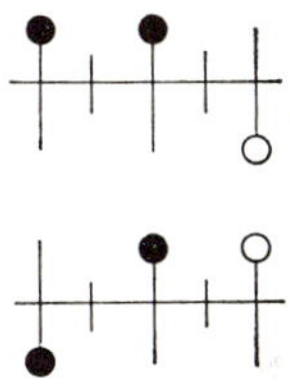	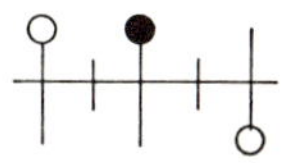
rr	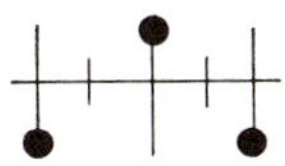	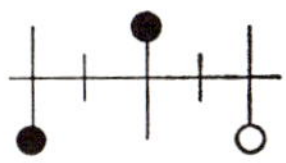	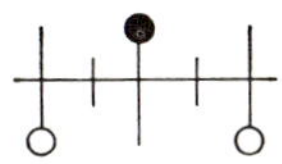

+ 10 others with ● and ○ reversed.

relative chain-end reactivities[1,2] or probabilities.[9–13] Without repeating the detailed descriptions given by the authors cited, we note the following:

1. If the reactivity of the growing chain is independent of the nature of the chain end or the past history of the chain, the comonomer ratio in the polymer is given by a single parameter, the comonomer feed ratio, and the chain statistics are Bernoullian. Such behavior is not common, and will be approximated only by comonomers which are chemically very similar. Deviations from Bernoullian propagation can be tested with dyad information.
2. If the reactivity of the growing chain is dependent only on the nature of the monomer unit at the growing end (most systems fit into this category), the sequence statistics can be described by two parameters, which may be

TABLE X-2

CONFIGURATIONAL SEQUENCES IN COPOLYMERS OF $CXY{=}CH_2$ (A) AND $CX_2{=}CH_2$ (B)[a]

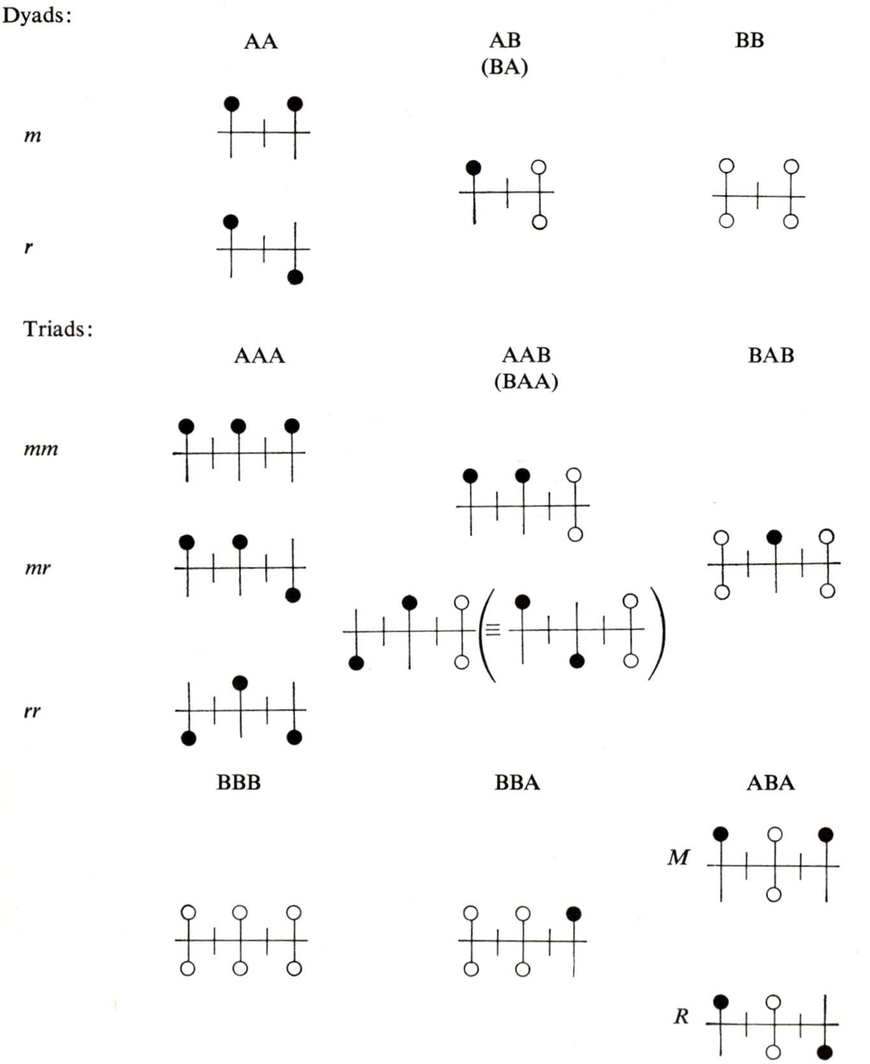

[a] See text. All triad sequences are explicitly shown.

taken as the reactivity ratios r_1 and r_2[6,7] or as the probabilities P_{12} and P_{21} (or P_{AB} and P_{BA}), i.e., as first-order Markov. These quantities can be deduced from dyad intensities. Triad or higher information is required to test for deviations.

Extension of such considerations to more complex statistics is self-evident. Since hexad sequences can be distinguished in favorable cases, quite searching tests of mechanistic proposals are possible if the spectrum can be adequately interpreted.

It is of course evident that NMR can be used as a means of analysis of copolymers in terms of overall comonomer composition. By this means, reactivity ratios can be determined. The method is less tedious and probably more accurate than more traditional analytical methods, and a number of studies of this kind have been published. These give none of the deeper insights which NMR is capable of providing, and are not reviewed here.

2. Vinylidene Chloride Copolymers

Copolymers of vinylidene chloride have been studied extensively[14–24] because this monomer introduces no asymmetric centers and causes marked deshielding of the protons of neighboring units. Its copolymers with isobutylene[16,17] are particularly amenable to analysis. A typical spectrum is presented in Fig. X-1, which shows the simple spectra of the homopolymers at the top and of a copolymer containing 70 mole % vinylidene chloride below. It is at once evident that tetrads are being resolved, for if dyads and triads represented the limit of discrimination there would be only three methylene peaks. In fact, seven can be clearly observed and are assigned[16,17] as shown, A being vinylidene chloride and B isobutylene. (The three remaining peaks correspond to BB-centered sequences and are in the same region as the methyl groups near 8.6τ.) Even at 60 MHz, the tetrad peaks are observably split,[17] so apparently hexad sequences can be resolved. At higher frequencies, there is little doubt that still longer sequences should be observed.

Analysis of peak intensities shows that while dyad data (i.e., lumping all AA-centered sequences together, all BB-centered sequences together, and all AB-centered sequences together) can be analyzed in terms of first-order Markov statistics to give values of r_1 and r_2 which are identical within experimental error to those obtained by conventional chemical analysis, the tetrad[16,17] and hexad[17] intensities show strong deviations from predictions based on these values. This result is somewhat unexpected, for the usual theories of "penultimate" effects do not predict deviations for this monomer pair.

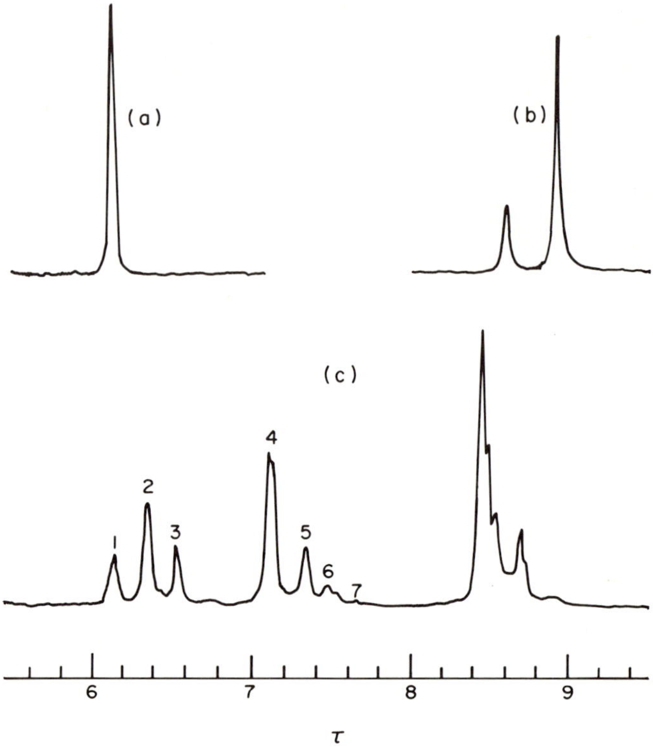

FIG. X-1. 60 MHz spectra of (a) polyvinylidene chloride, (b) polyisobutylene, and (c) vinylidene chloride (A)–isobutylene (B) copolymer (70 mole% vinylidene chloride). Peaks are identified with tetrad sequences as follows: (1) AAAA, (2) AAAB, (3) BAAB, (4) AABA, (5) BABA, (6) AABB, and (7) BABB (from Kinsinger *et al.*[17]).

Yamashita *et al.*[22] have measured the sequence distributions of a number of vinylidene chloride copolymers (including methyl and benzyl methacrylates, methyl acrylate, styrene, and vinyl acetate) and found no observable deviations from first-order Markov statistics, so the isobutylene system is evidently an exceptional one. This group has also succeeded in the NMR characterization of vinyldiene chloride copolymers as a function of conversion up to high conversions.[23,24]

A surprising structural complexity has been revealed in vinylidene chloride–vinyl chloride copolymers,[14,15] including evidence of "backward" (i.e., head-to-head) vinylidene chloride units not present in the homopolymer. It is possible that such structures may give rise, at least in part, to the apparent statistical deviations in the isobutylene copolymers.

3. Ethylene–Vinyl Ester Copolymers

Ethylene copolymers, like isobutylene copolymers, should be relatively simple to interpret in terms of sequence. Ethylene–vinyl formate and ethylene–vinyl acetate copolymers have been reported by Wu.[25–27] The triad sequences which are distinguishable in principle are shown in Table X-2, where A is the vinyl ester and B is ethylene (● is OCOR or OCHO; ○ is H). In the vinyl formate copolymer spectra the BBB, BBA, and ABA sequences are clearly distinguished in the 220 MHz methylene spectra; the AAA, AAB, and BAB sequences can be distinguished in the formyl spectrum and, upon decoupling from the neighboring methylenes, in the methine proton spectrum as well. There was no evident discrimination of configurational sequences. The observed vinyl formate and ethylene sequence lengths somewhat exceeded those calculated from the copolymer composition, indicating a deviation from first-order Markov statistics in the direction of a "blockier" structure.

The spacing of peaks in the formyl proton spectrum was dependent on the choice of solvent. This solvent dependence was observed also for the acetate methyl resonances of the vinyl acetate–ethylene copolymers and was explored in considerable detail for both systems.[26,27] Aromatic solvents—benzenes and alkylbenzenes—were believed to associate to form π-complexes with the methyl groups. From the temperature dependence of the acetate chemical shift, the enthalpy and entropy of complex formation were calculated: ΔH was found to be 2.65 ± 0.05 kcal and ΔS was -10.3 ± 0.1 eu for all the aromatic solvents employed. A different type of solvent dependence was observed for the BBB peak of the ethylene proton spectrum: this peak becomes doubled in aromatic solvents in a manner strongly reminiscent of that observed by Liu[28,29] for polyethylene in 1-chloronaphthalene solution (see Chapter IV, Section 1). A similar explanation probably applies, i.e., that the longer methylene chains adopt a folded form.

4. Styrene Copolymers and Related Copolymers

A number of studies of copolymers of styrene and related aromatic monomers have shown that sequence information can be obtained from the aromatic resonances and also from the effects of shielding of comonomer protons by the aromatic groups. Bovey *et al.*[30] observed that aromatic resonances in styrene–butadiene copolymers altered in form as the styrene sequences became shorter, the *ortho* protons becoming less shielded as in styrene oligomers (Chapter VI, Section 1). Similar effects have been observed in styrene–propylene copolymers[31] and a number of others, and have been employed by Mochel[32] to measure block styrene sequences in emulsion styrene–butadiene copolymers (see also Chapter XI, Section 4).

Of particular interest is the styrene–methyl methacrylate system, which has been reported by a number of authors.[1,10,11,23,33–39] The most significant area of the spectrum is that of the methoxyl protons, which exhibits at least three and as many as five peaks (Fig. X-2). We recall from the previous discussion and Table X-1 that if we are observing only compositional triads we should expect only three peaks, whereas if configurational information is being conveyed as well, as many as ten triad peaks may be observed. The fact that three peaks can still be distinctly seen when the proportion of BAB (A being methyl methacrylate) is over 90%[10] [Fig. X-2(b)], makes it clear that configurational effects are marked. The assignment in Table X-3 rationalizes the methoxyl peak intensities as a function of monomer feed ratio, observed

TABLE X-3

CHEMICAL SHIFT ASSIGNMENTS IN METHYL METHACRYLATE (A)–STYRENE (B) COPOLYMERS

	AAA	AAB	BAB
mm	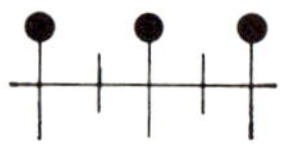	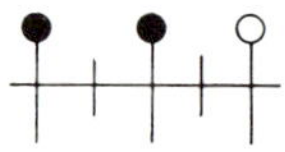	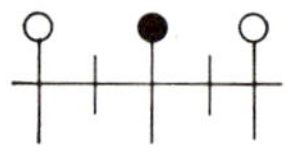
		7.0–7.2τ	7.6–7.9τ
mr	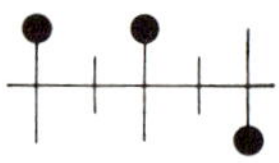	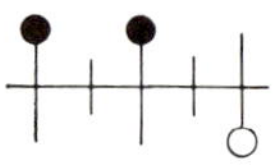 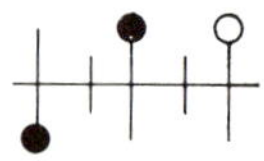	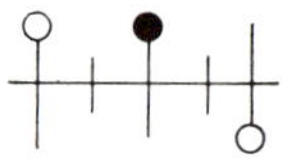
		6.4–6.8τ 7.0–7.2τ	7.0–7.2τ
rr	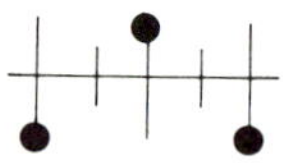	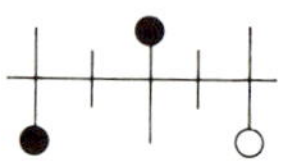	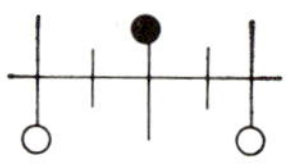
	all at about 6.6τ	6.4–6.8τ	6.4–6.8τ

in CCl_4 or $CDCl_3$ solution; the carbomethoxy group is represented by ● and the phenyl group by ○. This scheme may be summarized, at least in an approximate manner, by the statement that the shielding of a methoxyl group depends primarily on the number of neighboring styrene units that are "*co-meso*" to it, *mm*-BAB being the most shielded. Styrene units that are "*co-racemic*" have little effect, while the configuration of neighboring A units is immaterial. These conclusions have been supported by Ito *et al.*[38] and by Harwood and Ritchey,[11] but the latter investigators found that the spectra in aromatic solvents showed further complications which could not be so readily explained and may reflect pentad sequences.

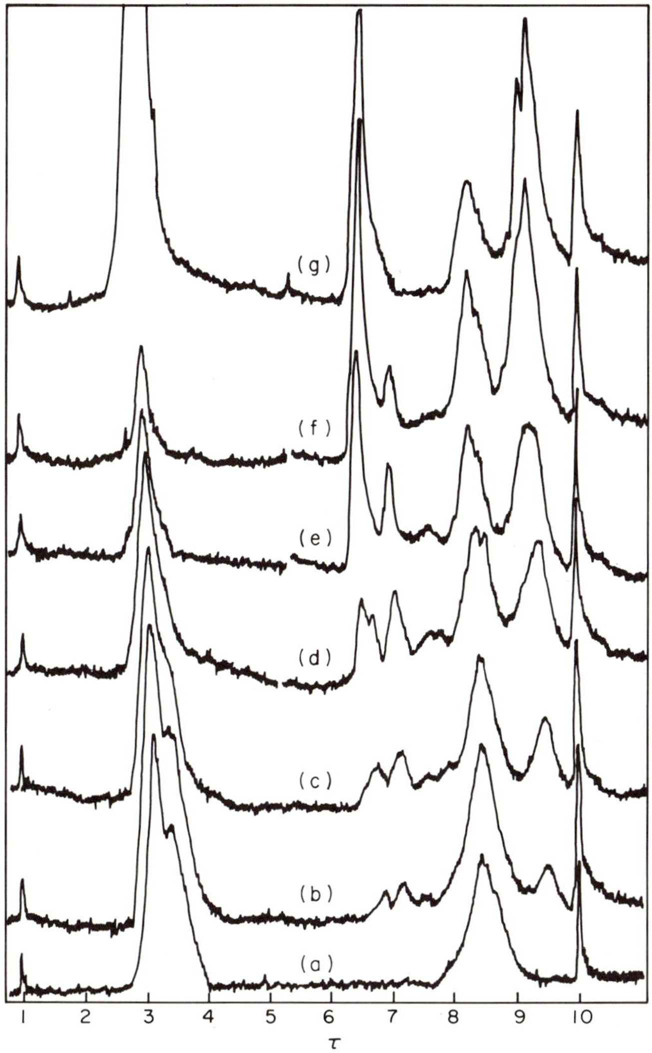

FIG. X-2. 40 MHz spectra of styrene–methyl methacrylate copolymers and homopolymers (about 20% w/v in CCl_4; observed at 90°): (a) polystyrene, (b) 10:90 (methyl methacrylate–styrene mole ratio in monomer feed), (c) 25:75, (d) 50:50, (e) 75:25, (f) 90:10, and (g) polymethyl methacrylate (in $CDCl_3$) (Bovey[10]).

It is evident that copolymers having long sequences of styrene and methyl methacrylate units will give spectra which approximate to summations of the homopolymer spectra, with little or no complexity in the methoxyl region. This is indeed found for anionic copolymers,[35,36] and thus NMR readily provides a most important mechanistic clue which would be awkward if not impossible to obtain otherwise.

Ito and Yamashita[40,41] have reported that styrene–methyl acrylate copolymers show analogous complexity in the methoxyl region. They have also studied a simplified system—copolymers of methyl acrylate and 1,1-diphenylethylene—in which the three methoxyl resonances reflect only the compositional sequences AAA, AAB, and BAB.

Bovey[10] reported that methyl methacrylate–α-methyl styrene show spectral properties similar to those of methyl methacrylate–styrene copolymers.

5. Acrylate and Methacrylate Copolymers

A number of studies of acrylate and methacrylate copolymers other than those discussed in the previous section have been reported. (References will be found in previous reviews.[4,5]) For the most part, these provide only compositional rather than sequence information, and are primarily of analytical significance. Less routine is a study of Johnsen and Kolbe[42] in which they demonstrated from a study of methyl methacrylate–acrylonitrile copolymers that both copolymerization parameters and constants of depolymerization can be obtained from NMR (and infrared) sequence information if the copolymers are prepared at temperatures high enough for appreciable depolymerization to occur. Klesper[43,44] and Klesper and Cantow[45] found six α-methyl peaks in the spectra of partially saponified syndiotactic polymethyl methacrylate, and were able to assign them to the six triad compositional sequences (p. 206). They concluded that saponification was essentially random.

6. Vinylidene Fluoride–Hexafluoropropylene Copolymers

A pioneering study of copolymer structure by NMR is that of Ferguson[46] on vinylidene fluoride–hexafluoropropylene copolymers.† The great structural sensitivity of ^{19}F chemical shifts (Chapter I, Section 16) permitted the resolution of many different arrangements of (a) and (b) units.

$$-CH_2-CF_2- \qquad\qquad -\underset{}{\overset{CF_3}{\overset{|}{C}}}F-CF_2-$$

(a) (b)

The comonomer sequences shown can be distinguished clearly in the CF_3

$$-CH_2CF_2\overset{CF_3}{\overset{|}{C}}FCF_2CH_2CF_2- \;; \qquad -CH_2CF_2CF_2\overset{CF_3}{\overset{|}{C}}FCH_2CF_2-$$

† This study was carried out at 40 MHz; further structural details were later observable at 94.1 MHz (see Ferguson and Phillips[47]).

region of the spectrum (least shielded). These two sequences also give rise to two CF resonances (most shielded). At intermediate positions in the spectrum, the CF_2 resonances show the most complex series of peaks, the strongest of which corresponds to head-to-tail vinylidene fluoride sequences:

$$—CH_2CF_2CH_2CF_2CH_2CF_2—$$

A number of other arrangements could also be recognized. There was apparently little or no branching.

7. Chlorotrifluoroethylene–Isobutylene Copolymers

Ishigure *et al.*[48] carried out an investigation of the chlorotrifluoroethylene–isobutylene system which revealed a number of points of interest. Both ^{19}F and proton spectra were observed at 94 MHz (100 MHz) and 56.4 MHz. Because isobutylene does not undergo homopolymerization under γ radiation, copolymers prepared at high isobutylene feed content approached an alternating 1:1 molar composition. The 56.4 and 94 MHz spectra of such a copolymer are shown in Fig. X-3. There are two possible structures, depending

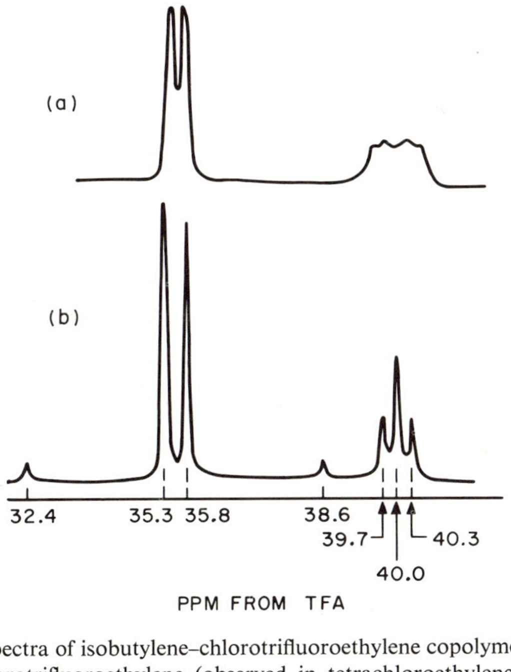

FIG. X-3. ^{19}F spectra of isobutylene–chlorotrifluoroethylene copolymer containing 0.478 mole fraction chlorotrifluoroethylene (observed in tetrachloroethylene solvent at 100°): (a) is the spectrum at 56.4 MHz and (b) the spectrum at 94 MHz (Ishigure *et al.*[48]).

upon whether the growing polymer radical attacks the chlorotrifluoroethylene monomer at the CF_2 (structure I) or CFCl (structure II) carbon atoms. In

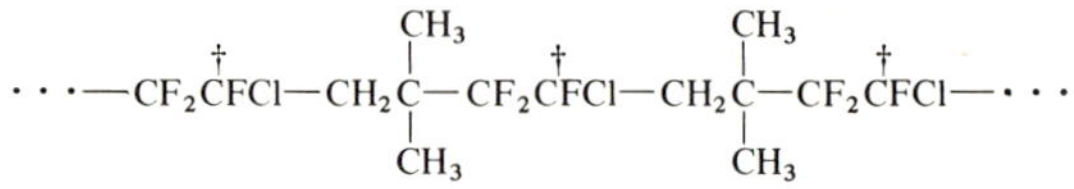

(I)

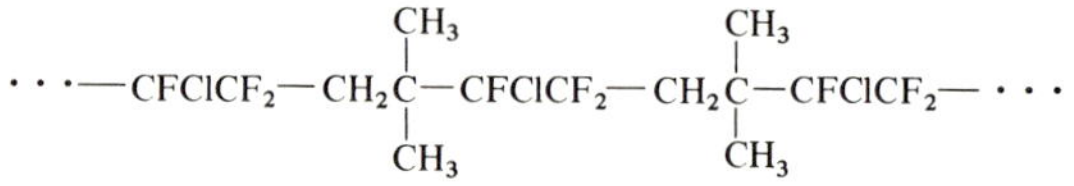

(II)

Fig. X-3, the CF_2 resonance appears as an AB quartet (outer peaks not resolved in the 56.4 MHz spectrum) the spacing of which (263 Hz) is appropriate for geminal ^{19}F–^{19}F coupling. No other couplings to CF_2 are resolved. The CFCl resonance, which seems to be the X part of an ABX spectrum, appears as two doublets in (a) and as a triplet in (b). These facts are consistent only with structure I. The presence of the asymmetric center ($C^{\dagger}$) differentiates the two CF_2 fluorines. Vicinal ^{19}F–^{19}F couplings are known to be very small (2–3 Hz), so no further structure is expected in the CF_2 resonance. The spacing of the outer peaks of the CFCl triplet (about 60 Hz) is appropriate for vicinal ^{19}F–^{1}H coupling, but the nature of the X resonance is such (see p. 105 *et seq.* of Bovey in the suggested further reading list at the end of Chapter I) that this spacing gives only $(J_{AX} + J_{BX})$. The couplings of the $CFCl^{19}F$ to the two CH_2 protons are in fact of rather different magnitudes, as indicated by the four line CFCl spectrum observed at 56.4 MHz (a).

The 100 MHz proton spectrum of this copolymer shows two methyl singlets, the $C(CH_3)_2$ groups being heterosteric due to the asymmetric centers marked $C^{\dagger}$. The appearance of only two methyl peaks suggests that these centers must have a consistent configuration, i.e., in terms of Table X-2, they must be either one of the two ABA sequences marked *m* or *r* (*meso* and *racemic* in an extended sense) but not both. This is very surprising in a free radical polymer, the more so since polychlorotrifluoroethylene is itself nearly random in configuration (Chapter V, Section 5).

8. Olefin–Sulfur Dioxide Copolymers

Fawcett *et al.*[49] have found that while the NMR spectrum of polyisoprene sulfone is fairly simple, corresponding to exclusive 1,4 addition,

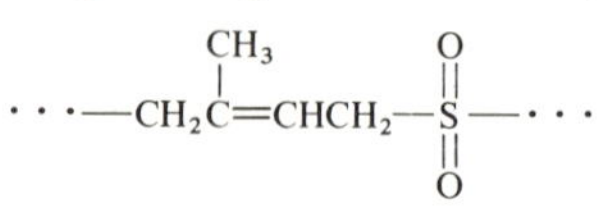

that of polypropene sulfone is surprisingly complex and could not be fully explained. Of particular interest was the conclusion, based on the employment of *cis*-1,2-d_2-propylene, that addition to the double bond occurs in a *trans* manner at low temperature, apparently because the free radical adds SO_2 before it has time to invert.

Bacskai *et al.*[50] have reported the NMR analysis of copolymers of higher olefins with sulfur dioxide.

REFERENCES

1. H. J. Harwood, *Angew. Chem. Int. Ed. Engl.* **4**, 1051 (1965).
2. U. Johnsen, *Ber. Bunsenges. Phys. Chem.* **70**, 320 (1966).
3. K. C. Ramey and W. S. Brey, Jr., *J. Macromol. Sci. Rev. Macromol. Chem.* **1**, 263 (1967).
4. H. A. Willis and M. E. A. Cudby, *Appl. Spectrosc. Rev.* **1**, 237 (1968).
5. J. C. Woodbrey, *in* "The Stereochemistry of Macromolecules" (A. D. Ketley, ed.), Vol. 3. Dekker, New York, 1968.
6. T. Alfrey Jr., J. J. Bohrer, and H. F. Mark, "Copolymerization." Wiley (Interscience), New York, 1952.
7. G. E. Ham, "Copolymerization." Wiley (Interscience), New York, 1964.
8. F. Heatley and F. A. Bovey, unpublished observations.
9. F. P. Price, *J. Chem. Phys.* **36**, 209 (1962).
10. F. A. Bovey, *J. Polym. Sci.* **62**, 197 (1962).
11. H. J. Harwood and W. M. Ritchey, *J. Polym. Sci. Part B* **3**, 419 (1965).
12. H. J. Harwood, *Angew. Chem.* **77**, 405, 1124 (1965).
13. J. Schaefer, *J. Phys. Chem.* **70**, 1975 (1966).
14. R. Chujo, S. Satoh, and E. Nagai, *J. Polym. Sci. Part A* **2**, 895 (1965).
15. J. L. McClanahan and S. A. Previtera, *J. Polym. Sci. Part A* **3**, 3919 (1963).
16. K. H. Hellwege, U. Johnsen, and K. Kolbe, *Kolloid-Z.* **214**, 45 (1966).
17. J. B. Kinsinger, T. Fischer, and C. W. Wilson III, *J. Polym. Sci. Part B* **4**, 379 (1966); **5**, 285 (1967).
18. U. Johnsen and K. Kolbe, *Kolloid-Z.* **220**, 145 (1967).
19. U. Johnsen, *Kolloid-Z.* **210**, 1 (1966).
20. S. Enomoto and S. Satoh, *Kolloid-Z.* **219**, 12 (1967).
21. K. Ito, S. Iwase, and Y. Yamashita, *Makromol. Chem.* **110**, 233 (1967).
22. Y. Yamashita, K. Ito, S. Ikuma, and H. Kada, *J. Polym. Sci. Part B* **6**, 219 (1968).
23. K. Ito and Y. Yamashita, *J. Polym. Sci. Part B* **6**, 227 (1968).
24. Y. Yamashita, K. Ito, H. Ishii, S. Hoshino, and M. Kai, *Macromolecules* **1**, 529 (1968).
25. T. K. Wu, *J. Phys. Chem.* **73**, 1801 (1969).
26. T. K. Wu, *Macromolecules* **2**, 520 (1969).
27. T. K. Wu, *Macromolecules* **3**, 610 (1970).
28. K.-J. Liu, *J. Polym. Sci. Part A-2* **4**, 155 (1966).
29. K.-J. Liu, *J. Polym. Sci. Part A-1* **3**, 1199 (1967).
30. F. A. Bovey, G. V. D. Tiers, and G. Filipovich, *J. Polym. Sci.* **38**, 73 (1959).
31. S. Kobayashi, Y. Kato, H. Watanabe, and A. Nishioka, *J. Polym. Sci. Part A-1* **4**, 245 (1966).
32. V. D. Mochel, *Macromolecules* **2**, 537 (1969).
33. A. Nishioka, Y. Kato, and N. Ashikari, *J. Polym. Sci.* **62**, S10 (1962).
34. Y. Kato, N. Ashikari, and A. Nishioka, *Bull. Chem. Soc. Jap.* **37**, 1630 (1964).

35. C. G. Overberger and N. Yamamoto, *J. Polym. Sci. Part B* **3**, 569 (1965).
36. K. Ito and Y. Yamashita, *J. Polym. Sci. Part B* **3**, 625, 631 (1965).
37. R. G. Bauer, H. J. Harwood, and W. M. Ritchey, *Polymer Preprints* **7**(2), 973 (1966).
38. K. Ito, S. Iwase, K. Umehara, and Y. Yamashita, *J. Macromol. Sci. Part A* **1**, 891 (1967).
39. S. Yabumoto, K. Ishii, and K. Arita, *J. Polym. Sci. Part A-1* **8**, 295 (1970).
40. K. Ito and Y. Yamashita, *J. Polym. Sci. Part B* **3**, 637 (1965).
41. K. Ito and Y. Yamashita, *J. Polym. Sci. Part A-1* **4**, 631 (1966).
42. U. Johnsen and K. Kolbe, *Makromol. Chem.* **116**, 173 (1968).
43. E. Klesper, *J. Polym. Sci. Part B* **6**, 313 (1968).
44. E. Klesper, *J. Polym. Sci. Part B* **6**, 663 (1968).
45. E. Klesper and H.-J. Cantow, *IUPAC Symp. Macromol. Chem., Toronto, 1968*, preprints, Sec. **A4**, no. 5 (1968).
46. R. C. Ferguson, *J. Amer. Chem. Soc.* **82**, 2415 (1960).
47. R. C. Ferguson and W. D. Phillips, *Science* **157**, 257 (1967).
48. K. Ishigure, Y. Tabata, and K. Oshima, *Macromolecules* **3**, 27 (1970).
49. A. H. Fawcett, K. J. Ivin, M. Navratil and N. A. Walker, *IUPAC Symp. Macromol. Chem., Toronto, 1968*, preprints, Sec. **A4**, no. 4 (1968).
50. R. Bacskai, L. P. Lindeman, D. L. Ransley, and W. A. Sweeney, *J. Polym. Sci. Part A-1* **7**, 247 (1969).

Chapter XI

DIENE POLYMERS AND COPOLYMERS

1. Introduction

High resolution NMR has proved useful in the determination of geometrical isomerism in diene polymer and copolymer chains (see Chapter II, Section 3). It is now at least as effective for such measurements as infrared spectroscopy, and in fact, when coupled with spectrum accumulation, is superior for the detection and identification of minor structures.

2. Isoprene

It is observed[1,2] that in carbon tetrachloride or carbon disulfide solution the methyl groups of *trans*-1,4-isoprene units (a) are about 0.07 ppm more

(a) (b)

shielded than those of *cis*-1,4 units (b). In benzene, the difference is slightly greater[1,3] [Fig. XI-1(a) and (b)], the *cis* methyl appearing at 8.37τ and the *trans* at 8.47τ.† The methylene resonances near 7.9τ show a nearly equal shielding difference (7.90τ for *cis* and 7.97τ for *trans*); there is no apparent

† The samples used were kindly supplied by Dr. D. Kates of Uniroyal, Wayne, New Jersey.

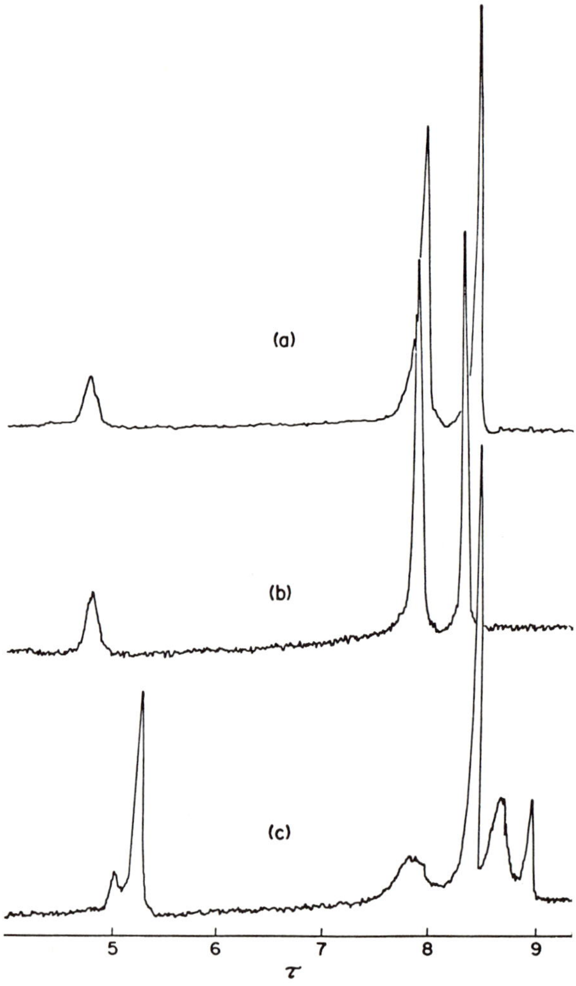

FIG. XI-1. 100 MHz spectra of polyisoprenes observed using 5% solutions in C_6D_6 at 74°: (a) poly-*trans*-isoprene, (b) poly-*cis*-isoprene, (c) polyisoprene containing about 80% of 3,4 units and about 20% of 1,2 units (Bovey and Brewster[3]).

difference in chemical shift between the two different methylene groups in each of the geometric isomers, nor is the coupling to the vinylene proton measurable, although there is some indication of fine structure in the *cis* spectrum. The position of the vinylene proton resonance is insensitive to the geometrical isomerism about the double bond, appearing at 4.80τ in both *cis* and *trans* structures. Its width suggests that the coupling to the neighboring methylene protons does not exceed about 3–4 Hz.

Spectrum (c) in Fig. XI-1 is that of a polyisoprene containing about 80% 3,4 units (c) and about 20% of 1,2 units (d). The 3,4-vinyl protons appear at

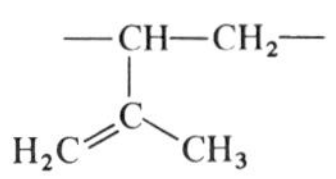

(c)

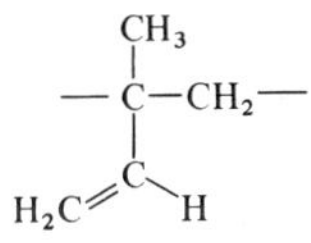

(d)

5.25τ. The 1,2-vinyl protons exhibit a peak at 5.05τ; intensity considerations indicate that a part of their resonance lies under the 3,4-vinyl proton peak (this is confirmed by the appearance of the vinyl resonance in poly-1,2-butadiene; see Section 4). The most useful resonances for analytical purposes are the methyl peaks appearing at 8.41τ for the pendant isopropenyl group of 3,4 units and at 8.95τ for the 1,2 units, in which the methyl groups are not adjacent to double bonds.

Golub and co-workers[4–6] have employed NMR to study isomerization in polyisoprene produced by chemical means or radiation. With the aid of deuterium labelling it was shown that no double bond migration takes place in the sulfur dioxide-catalyzed *cis–trans* isomerization of *cis*-1,4-polyisoprene. The carbonium ion-mediated cyclization of 3,4-polyisoprene was also elucidated.

The question of the presence of 1,2 or 3,4 structures in the predominantly 1,4 structure of natural rubber has been resolved by Chen,[7] who showed by time averaging that no 1,2 structures are present. Chen and Field[8] demonstrated by the same technique that the isoprene units present in extremely small proportion in commercial butyl rubber are in 1,4 configuration, as previously deduced by chemical means.[9]

Binder[10] has employed NMR to demonstrate the occurrence of interchain cyclization in poly-3,4-isoprene, under appropriate chemical treatment, yielding ladder polymers. The principal spectral change, aside from a general broadening (probably arising from a variety of isomeric structures) was a marked increase in intensity near 9τ, attributed to methylene groups of fused cyclohexane rings.

Chen[1] and Mochel[11] have analyzed the composition and isomeric structures of isoprene–butadiene copolymers.

Schué *et al.*[12–14] have employed NMR to study the structure of "living" polyisopropenyllithium oligomers. It is well known that the polymerization of isoprene by lithium alkyls gives stereoregular polymers. In hydrocarbon solvents, the polymer is *cis*-1,4 while in ethers it is principally 3,4. NMR offers the possibility of determining whether these preferences are inherent in the structure of the active end unit of the "living" chain, or come about during the addition of the next monomer unit. This information cannot be determined for the normal high-polymer growing chains because the resonances of the "dead" units mask those of the living end units. If *sec*- or *tert*-butyllithium is employed, however, it is possible to prepare a nearly 1:1 addition compound in which the addition of butyllithium has occurred almost exclusively in the direction shown[15] (R is butyl).

$$\begin{array}{c} \quad\;\; CH_3 \\ \quad\;\; | \\ RCH_2CHCCH_2(Li) \\ \delta \quad \gamma \;\; \beta\alpha \end{array}$$

Figure XI-2 shows the 100 MHz spectrum of this adduct at 25° in benzene.[14] The assignments indicated reflect the electropositive nature of the lithium

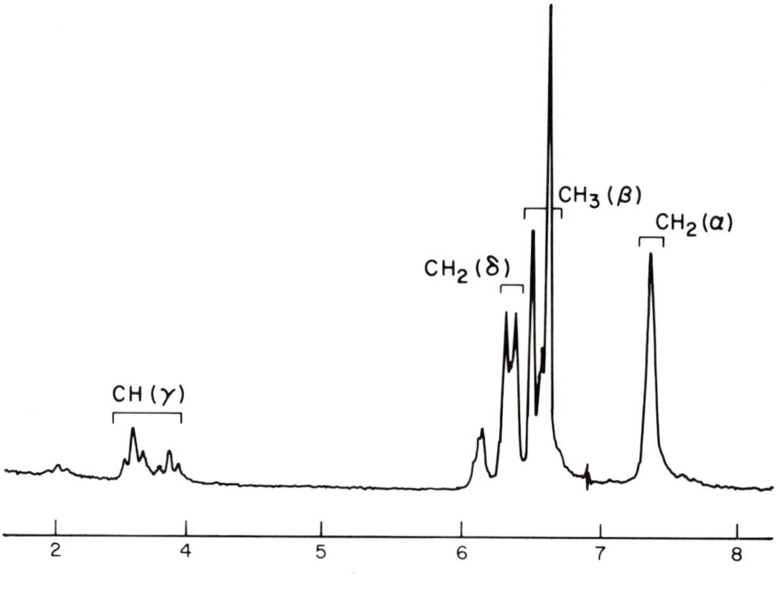

FIG. XI-2. 100 MHz spectrum of "living" lithium–isoprene adduct ($\overline{DP} \simeq 1.2$). The small signal at about 2 ppm comes from linking CH_2CH_2 groups in two isoprene chain units. The scale is in ppm upfield from internal benzene (Schué *et al.*[14]).

atom, which increases the shielding of the protons nearest to it, particularly that of the α-CH_2 group. It is observed that, with the exception of the latter, all the resonances are doubled. In other solvents, a doubling of the α-CH_2

resonance could be observed as well. This arises from the existence of *cis–trans* isomerism. Such geometrical isomerism was also observed for the end

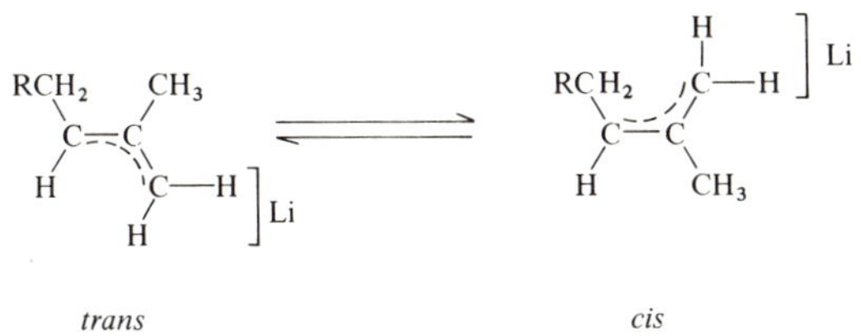

units of oligomers averaging about 2.7 and about 1.65 isoprene units per lithium atom. For these structures (but not for the 1:1 adduct), the *cis–trans* ratio varied with temperature, becoming nearly unity at higher temperatures. It was not possible to assign the peaks unambiguously to *cis* and *trans*, but the fact that such systems give a nearly all *cis* polymer (as we have seen) suggests that the larger peaks (about 2:1 ratio near 25°) correspond to *cis*. It is evident, however, that the structure of the end unit is not the all-determining factor in the stereochemistry of the finished chain.

The state of aggregation of the lithium complexes may influence their *cis–trans* ratio in hydrocarbon solvents. In ethers such as tetrahydrofuran, such aggregation does not occur, and the spectra are rather different. At low temperatures there is slow rotation about the C_α—C_β bond, giving a more complex α-proton resonance. Although in such solvents, the finished polymer chains have a predominantly 3,4 structure, the terminal unit contains only about 25% of this structure. Again, the state of the terminal unit does not determine the final structure of the polymer chain, which depends also on the propagation step itself.

3. Polychloroprene

Ferguson[16] observed that the olefinic protons of *cis*-1,4 units in polychloroprene appear at about 4.50τ, measured in carbon disulfide. It was further observed that the methylene resonance was split in a way which could only be accounted for as due to substantial proportions of head-to-head: tail-to-tail sequences as well as the dominant head-to-tail structures. Aufdermarsh and Pariser[17] established by this means that the polychloroprene prepared by chlorination of 1,4-poly-2-(tri-*n*-butyltin)-1,3-butadiene was indeed all *cis* as presumed, but was irregular with respect to structural sequences, containing about 20–25% each of head-to-head and tail-to-tail isomers.

4. Butadiene

NMR is somewhat less effective in determining the structure of polybutadienes than of polyisoprenes. Even at 220 MHz, the resonances of the vinyl and methylene protons in *trans*-1,4-units (a) are barely distinguishable from those of *cis*-1,4 units (b). Figure XI-3 shows the 220 MHz spectrum of a

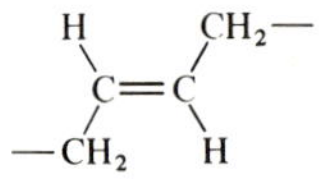

(a)

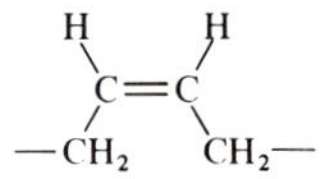

(b)

polybutadiene containing 52% of *trans*-1,4 and 48% of *cis*-1,4 units. Both the methylene resonance at 8.0τ and the vinylene resonance at about 4.7τ

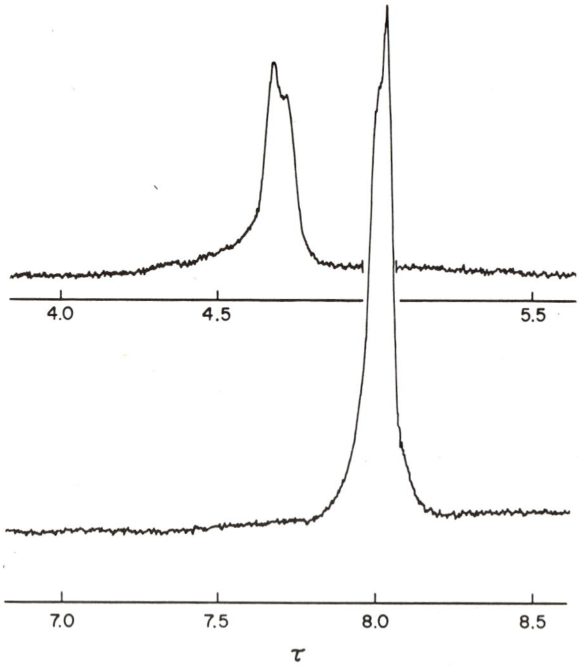

FIG. XI-3. 220 MHz spectrum (5% in CCl_4 at 25°) of a polybutadiene containing 52% of *trans*-1,4 and 48% *cis*-1,4 units (Bovey and Brewster[3]).

show distinct shoulders (arranged "back-to-back") which evidently reflect the *cis* units. At any lower observing frequency the two units would be indistinguishable. On the other hand, the presence of 1,2 units can be very clearly distinguished.[3] Figure XI-4 is the 100 MHz spectrum of a polybutadiene

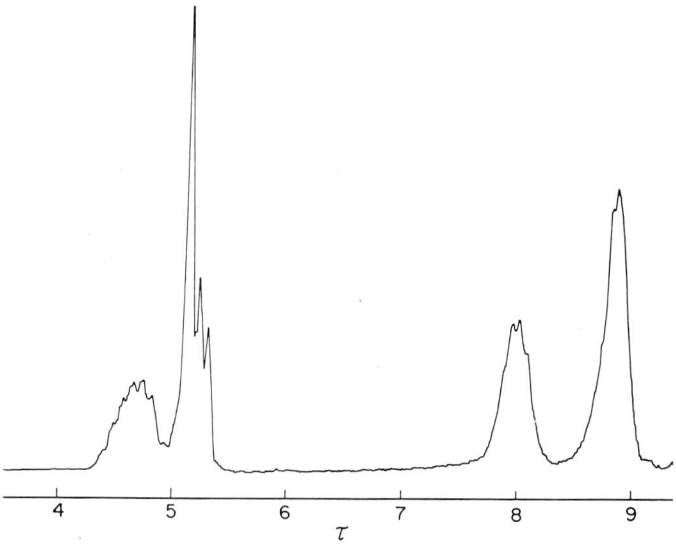

FIG. XI-4. 100 MHz spectrum (5% in CS_2 at 45°) of a polybutadiene containing about 98% of 1,2-linked monomer units (Bovey and Brewster[3]).

containing about 98% 1,2-linked monomer units. The vinyl resonance, appearing as a complex multiplet near 5τ, is analytically useful. The vinyl methylene protons are centered at about 5.2τ and the methine proton at 4.65τ. An analysis of this multiplet in terms of chemical shifts and couplings could yield information concerning the rotational state of the vinyl group with respect to the main chain. Configuration may also contribute.

Durbetaki and Miles[18] analyzed polybutadienes and epoxidized polybutadienes by both NMR and near infrared spectroscopy, and found a satisfactory correlation between them. This effort was directed primarily toward measurement of the degrees of cyclization and epoxidation rather than toward distinguishing isomeric diene units.

Bovey *et al.*[19] observed a number of butadiene–styrene copolymers of varying composition. At 40 MHz it was not possible to distinguish isomeric diene units clearly, but it was observed that the form of the styrene aromatic resonance varied markedly with its content. At high styrene contents, the *ortho* protons appeared as a distinguishable peak, as in polystyrene itself

(Chapter VI, Section 1); at low contents, the aromatic resonance approximated a singlet because the shielding of these protons was reduced in the absence of neighboring groups. Thus, sequence information of a sort could be obtained. The NMR data may also of course be used for simple compositional analysis.[20]

5. Other Dienes

Friedman *et al.*[21,22] have investigated the structure of poly-1,3-pentadiene (piperylene) by NMR and found it to be predominantly 1,4 with varying proportions of 1,2 units and possible cyclized groups. Schué and coworkers[23–25] have studied isomeric structure distribution in poly-2,3-dimethyl-1,3-butadiene. Sewall and Skidmore[26] have measured by use of the CAT very small concentrations of a variety of diene comonomers incorporated into ethylene–propylene copolymers to provide vulcanizability. The structure of polycyclopentadiene has been studied.[27–29]

6. ^{13}C Studies of Diene Polymers

It has been demonstrated that ^{13}C NMR can be very useful in the determination of the geometrical isomerism of diene polymers.[30–33] Duch and Grant[32] have shown that *cis*- and *trans*-1,4-polybutadiene and *cis*- and *trans*-1,4-polyisoprene give markedly different spectra in which the ^{13}C resonances (natural abundance) are widely spaced in chemical shift. Figure XI-5 shows the spectra of (a) *cis*- and (b) *trans*-1,4-polybutadiene, obtained under conditions described in the figure caption. Figure XI-6 shows corresponding spectra of polyisoprenes. (The protons are noise-decoupled in all spectra; see Chapter I, Section 18B.) In Table XI-1, the carbon chemical shifts, expressed relative to internal benzene [which in the case of Fig. XI-5(b) was also the solvent], are presented. The chemical shift assignments were made by comparison to small model molecules; the reader is referred to Duch and Grant[32] for details. It is evident that the vinylene carbons are considerably less sensitive to the chain isomerism than the saturated carbons. The latter should readily provide an accurate method for the analysis of such polymers, assuming of course that peak areas can be measured with some precision. It is of particular value for determining *cis* and *trans* units in polybutadienes, the proton spectra of which are barely distinguishable even at 220 MHz (Section 4, Fig. XI-3). At the present time, however, ^{13}C spectroscopy is probably not quite equal to proton spectroscopy for the determination of anomalous structures in minor or trace proportions.

A particularly significant and intriguing feature of the ^{13}C spectroscopy of polydienes (and one which no doubt will be observed for other rubbery

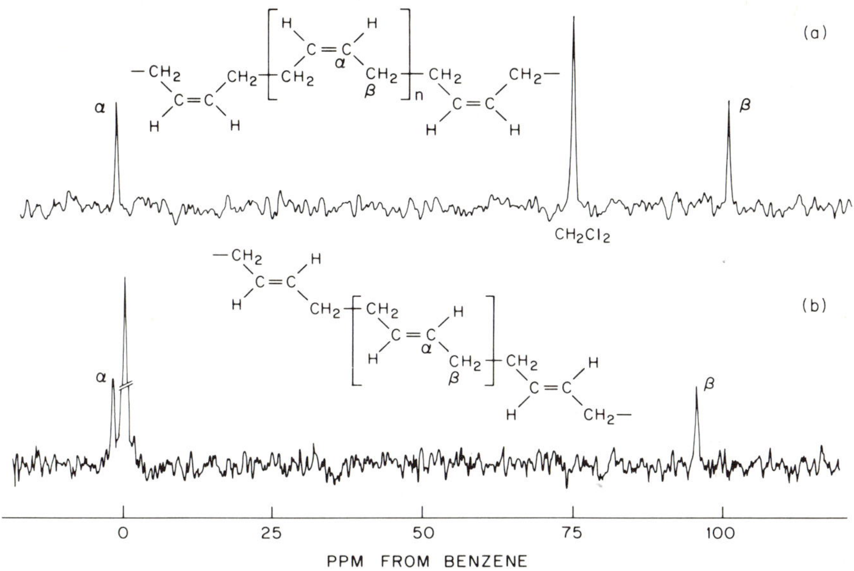

FIG. XI-5. Proton-decoupled 15.08 MHz ^{13}C spectra of (a) *cis*-1,4-polybutadiene and (b) *trans*-1,4-polybutadiene. Spectrum (a) was obtained using a 20% (w/v) solution in CH_2Cl_2 and 90 CAT scans; spectrum (b) was obtained using a benzene solution (gelatinous) and 224 CAT scans. All spectra were run at 45° and at a sweep rate of 21.56 Hz/sec (Duch and Grant[32]).

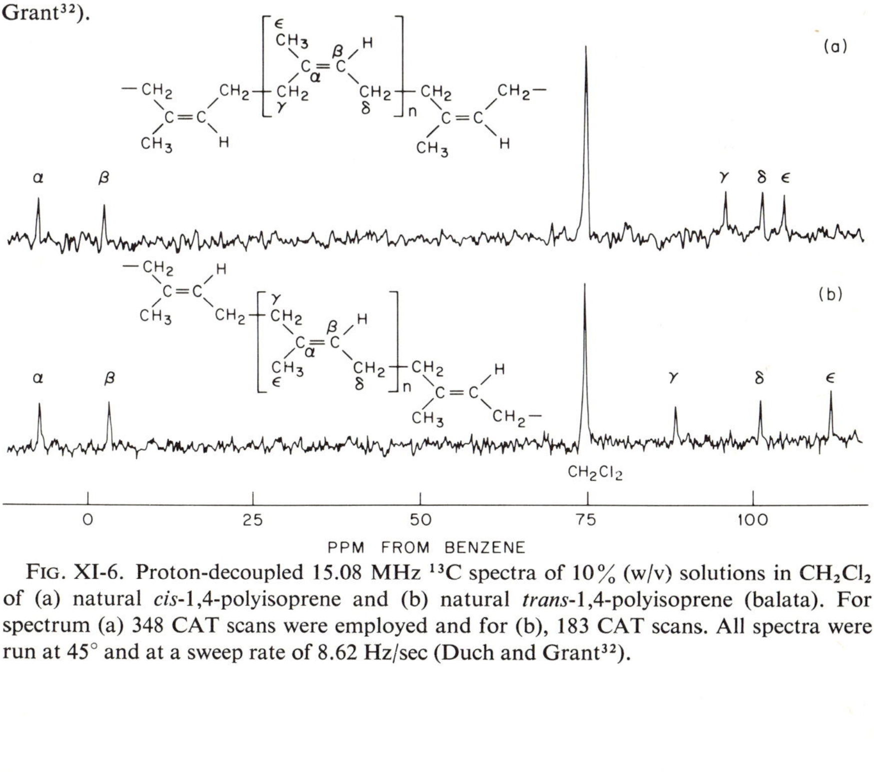

FIG. XI-6. Proton-decoupled 15.08 MHz ^{13}C spectra of 10% (w/v) solutions in CH_2Cl_2 of (a) natural *cis*-1,4-polyisoprene and (b) natural *trans*-1,4-polyisoprene (balata). For spectrum (a) 348 CAT scans were employed and for (b), 183 CAT scans. All spectra were run at 45° and at a sweep rate of 8.62 Hz/sec (Duch and Grant[32]).

TABLE XI-1

^{13}C CHEMICAL SHIFT DATA FOR 1,4-POLYDIENES[a, b]

	Carbon positions[c]				
	α	β	γ	δ	ε
Polybutadiene					
cis	−1.0	101.0	—	—	—
trans	−1.7	95.6	—	—	—
Polyisoprene[c]					
cis	−6.9	3.2	96.2	101.9	105.3
trans	−6.5	4.2	88.7	101.5	112.8

[a] Chemical shifts expressed in ppm from internal benzene.

[b] See Duch and Grant.[32]

[c] Refer to Figs. XI-5 and XI-6 for designations. The same chemical shifts within experimental error were found for natural and synthetic poly-1,4-isoprenes. Where a shift difference was reported, the values shown are those given for the synthetic polymer. A mixed *cis-trans* synthetic poly-1,4-isoprene showed slight shifts, mostly downfield, for some of the β- and γ-carbons compared to the sterically pure polymers.

polymers as well) is that quite well-resolved spectra can be observed for these materials even in the solid state.[30-33] For example, Fig. XI-7 shows the spectrum of a sample of natural rubber (poly-*cis*-1,4-isoprene) observed at 22.6 MHz in the form of a piece of vulcanized gum tubing. This spectrum may be compared to Fig. XI-6(a) (or to Fig. 3 of Duch and Grant[32]). The lines are somewhat broadened, but are still well resolved. Duch and Grant[32] observed that for solid balata (poly-*trans*-1,4-isoprene), which is partially crystalline, the spectrum was essentially the same as Fig. XI-6(b) (although with some line broadening) but required many more CAT scans to achieve a comparable signal-to-noise ratio. The temperature of their observations (45°) is well above the glass temperature of the amorphous portions of these polymers. These findings therefore suggest that the crystalline portions do not contribute to the spectrum, no doubt because the molecular motion in them is not sufficiently rapid to give peaks narrow enough to observe.

The fact that well-resolved ^{13}C spectra can be obtained for diene rubbers in the solid state, while the corresponding proton spectra are virtually featureless,[34] is attributable principally to the much larger chemical shift differences in ^{13}C spectra, but also in part to the smaller magnetic moment of the latter nucleus, which results in reduced dipolar broadening (Chapter I, Section 6). The irradiation of the protons cannot be expected to reduce the dipolar broadening appreciably, because it should have little or no effect on the direct

^1H–^{13}C dipole–dipole interactions. The use of deuterium-substituted polymers may, however, permit some increase in resolution of solid-state spectra.

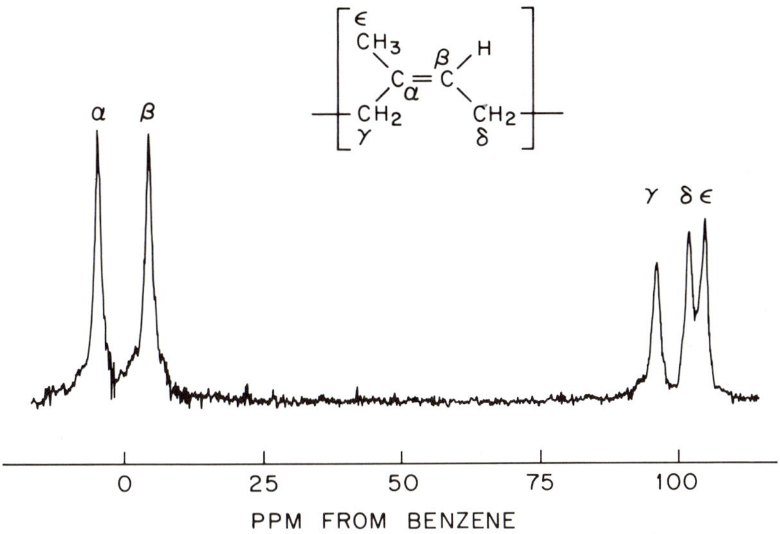

FIG. XI-7. 22.6 MHz ^{13}C spectrum of *cis*-1,4-polyisoprene (natural rubber). The sample is a piece of vulcanized gum vacuum tubing inserted into a 5 mm o.d. NMR tube. It was obtained through the kind cooperation of the Bruker Scientific Company, Elmsford, New York (Bovey[31,33]).

REFERENCES

1. H. Y. Chen, *Anal. Chem.* **34**, 1134, 1793 (1962).
2. M. A. Golub, S. A. Fuqua, and N. S. Bhacca, *J. Amer. Chem. Soc.* **84**, 4981 (1962).
3. F. A. Bovey and A. I. Brewster, unpublished observations.
4. M. A. Golub, *J. Polym. Sci. Part B* **4**, 227 (1966).
5. M. A. Golub and J. Heller, *J. Polym. Sci. Part B* **4**, 469 (1966).
6. M. A. Golub and C. L. Stephens, *J. Polym. Sci. Part A-1* **6**, 763 (1968).
7. H. Y. Chen, *J. Polym. Sci. Part B* **4**, 891 (1966).
8. H. Y. Chen and J. E. Field, *J. Polym. Sci. Part B* **5**, 501 (1967).
9. J. Rehner, Jr., *Ind. Eng. Chem.* **36**, 46 (1944).
10. J. L. Binder, *J. Polym. Sci. Part B* **4**, 19 (1966).
11. V. D. Mochel, *Rubber Chem. Technol.* **40**, 1200 (1967).
12. F. Schué, D. J. Worsfold, and S. Bywater, *J. Polym. Sci. Part B* 7, 821 (1969).
13. F. Schué and S. Bywater, *Macromolecules* **2**, 458 (1969).
14. F. Schué, D. J. Worsfold, and S. Bywater, *Macromolecules* **3**, 509 (1970).
15. F. Schué and S. Bywater, *Bull. Soc. Chim. Fr.* **1970**, 271 (1970).
16. R. C. Ferguson, *J. Polym. Sci. Part A* **2**, 4735 (1964).
17. C. A. Aufdermarsh and R. Pariser, *J. Polym. Sci. Part A* **2**, 4727 (1964).
18. A. J. Durbetaki and C. M. Miles, *Anal. Chem.* **37**, 1231 (1965).

19. F. A. Bovey, G. V. D. Tiers, and G. Filipovich, *J. Polym. Sci.* **38**, 73 (1959).
20. W. L. Senn, *Anal. Chim. Acta.* **29**, 505 (1963).
21. G. Friedman, F. Schué, M. Brini, A. Deluzarche, and A. Maillard, *Bull. Soc. Chim. Fr.* **1965**, 1343 (1965).
22. G. Friedman, *Bull. Soc. Chim. Fr.* **1967**, 698 (1967).
23. F. Schué and J. P. Dole-Robbe, *Bull. Soc. Chim. Fr.* **1963**, 975 (1963).
24. F. Assioma, J. Marchal, and F. Schué, *Compt. Rend.* **261**, 1315 (1965).
25. F. Schué, *Bull. Soc. Chim. Fr.* **1965**, 980 (1965).
26. P. R. Sewall and D. W. Skidmore, *J. Polym. Sci. Part A-1* **6**, 2425 (1968).
27. S.-P. S. Yen, Preprints of Polymer Div., p. 82. Amer. Chem. Soc. Meeting, New York, 1963.
28. M. A. Bonin, W. R. Busler, and F. Williams, *J. Amer. Chem. Soc.* **87**, 199 (1965).
29. C. Aso, T. Kinitake, K. Ito, and Y. Ishimoto, *J. Polym. Sci. Part B* **4**, 701 (1966).
30. L. F. Johnson, paper presented at a C-13 NMR Conference, Anaheim, Calif., October 1969.
31. F. A. Bovey, *Sciences* **66–67** (May, August), 4 (1970).
32. M. W. Duch and D. M. Grant, *Macromolecules* **3**, 165 (1970).
33. F. A. Bovey, *in* "NMR, Basic Principles and Progress" (P. Diehl, E. Fluck, and R. Kosfeld, eds.), Vol. 4, Springer-Verlag, Berlin, 1971.
34. W. P. Slichter, *Rubber Chem. Tech.* **5**, 1574 (1961).

Chapter XII

RING-OPENING POLYMERS, POLYAMIDES, POLYESTERS AND MISCELLANEOUS POLYMERS

1. Ring-Opening Polymers

A. Polyethylene Oxide (Polyethylene Glycol; Polyoxyethylene)

Ethylene oxide oligomers and polymers have received considerable study by NMR, mainly in an effort to determine their conformation in solution.[1-7] In such chains, only positional isomerism is possible. Studies at 60 MHz of

$$\underbrace{\mathrm{HOCH_2}}_{\mathrm{A}}\underbrace{\mathrm{CH_2O}}_{\mathrm{E_t}}\underbrace{\mathrm{(CH_2CH_2O)}_{n-2}}_{\mathrm{E_i}}\underbrace{\mathrm{CH_2}}_{\mathrm{E_t}}\underbrace{\mathrm{CH_2OH}}_{\mathrm{A}}$$

the oligomers from trimer to octamer[3] showed that in the pure liquids, the terminal "alcoholic" CH_2 protons A can be distinguished from the terminal E_t and internal E_i etherlike CH_2 protons only in the trimer and tetramer. Dilution with benzene or α-chloronaphthalene produced more complex spectra, since not only A and E_t protons are distinguished from E_i, but E_i protons also exhibit observable positional differences in chemical shift for the heptamer and hexamer but not the pentamer. No such resolution is observed in chloroform or aqueous solutions. Comparison with the spectra of the higher polymers led to the conclusion that, at least in some solvents, there is a change at $n \simeq 6$ to a conformation resembling the 7_2 helical structure of the polymer indicated by infrared and x-ray results[8-11] for the solid state. This structure requires a preference for the *gauche* over the *trans* conformer despite the higher Coulombic repulsion of the former.[12] Evidence was found[7] for a

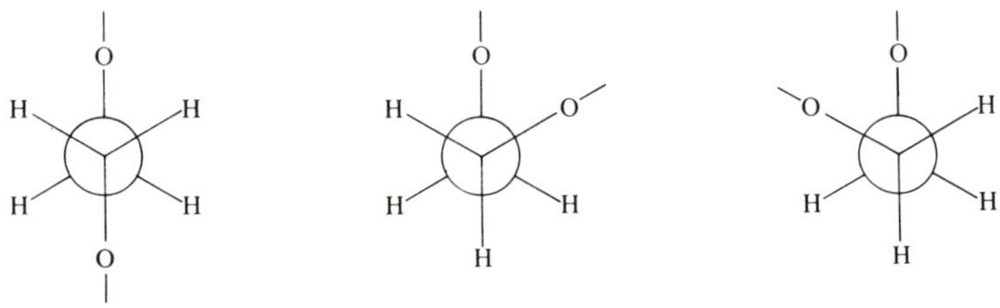

trans *gauche*

more ordered structure in aqueous solutions than in benzene or in the melt, possibly because of hydrate formation.

Connor and McLauchlan[1] measured the vicinal couplings in polyethylene oxide in aqueous and chloroform solutions and in the melt by observation of the ^{13}C satellites of the proton spectrum. These appear as a doublet straddling the main spectrum. Each member of the doublet is complex and may be treated as half of the AA′BB′ spectrum of the monomer unit:

$$\left[^{13}CH_2CH_2O \right]_n$$

Making the usual assumptions of constancy of J_g and J_t (Chapter IX), it can be shown that

$$J + J' = \tfrac{1}{2}[J_t + 3J_g + X_t(J_t - J_g)] \qquad \text{(XII-1)}$$

where J and J' are the two vicinal coupling constants and X_t is the mole fraction of the *trans* conformer. It was found that $J + J'$ increases with temperature, indicating that the *gauche* conformers must be of lower energy than the *trans*. However, the dependence of $J + J'$ upon temperature was found to be greater than could be accounted for by any single reasonable value for the energy difference, so that this approach requires further study and interpretation.

The interaction of potassium iodide with polyethylene oxide and oligomers in D_2O solution has been studied by both high resolution observations[4] and measurements of polymer proton T_1.[6] The results suggest a binding of the potassium ion to oligomers having $n \geqslant 7$, producing in effect a polyelectrolyte structure.

Schaefer and co-workers[13,14] have measured monomer sequence distributions in copolymers of ethylene oxide and maleic anhydride by natural-abundance ^{13}C and proton spectroscopy, combined with gas chromatographic analysis of the glycol ethers obtained by hydrolysis of the chains. These copolymers are polyether-esters generated by use of Lewis acid catalysts such as stannic chloride. Their limiting form is a 1:1 copolymer, because no sequences of maleic anhydride units are possible. The ethylene oxide region of the ^{13}C spectrum exhibited four peaks, assigned to

$$—CH_2CH_2O^{13}CH_2CH_2OCH_2CH_2—$$

$$(\text{or } —CH_2CH_2OCH_2{}^{13}CH_2OCH_2CH_2—)$$

$$—CH_2CH_2O^{13}CH_2CH_2O_2CCH{=}CHCO_2—$$

$$—CH_2CH_2OCH_2{}^{13}CH_2O_2CCH{=}CHCO_2—$$

and

$$—O_2CCH{=}CHCO_2{}^{13}CH_2CH_2O_2CCH{=}CHCO_2—$$

$$(\text{or } —O_2CCH{=}CHCO_2CH_2{}^{13}CH_2O_2CCH{=}CHCO_2—)$$

in order of increasing shielding. Effects of sequences longer than triads could be observed in the vinyl carbon lines, but the lines were only broadened by the dispersion of chemical shifts and were therefore not analytically useful. The carbonyl carbons, however, gave well-resolved lines corresponding to all six of the possible pentad sequences. The four ethylene oxide-centered triads and three of the six maleic anhydride-centered pentad sequences could also be discriminated in the 100 MHz proton spectra. By combining this information with that from chromatography of the hydrolyzates, which supplies a measure of the lengths of ethylene oxide sequences, it was concluded that the polymerization statistics are complex, in that there is an observable long-range nonrandomness of the sequences which is non-Markovian. Apparently, the catalyst can marshall groups of two to four monomers together prior to their addition to the chain.

B. Polypropylene Oxide

Propylene oxide monomer and polymer contain true asymmetric centers and the monomer units in the chain have a sense of direction (Chapter II, Section 5), so that, for example *ddl* is not equivalent to *ldd*. There are therefore four species of triad sequences, represented only in terms of relative configuration.† A further complication is provided by the possibility of both

† Note that in the planar zigzag projection used in this book the methyl groups appear on the *same* side of the zigzag plane in the syndiotactic polymer and on opposite sides in the isotactic polymer (see Chapter II, Section 5).

isotactic

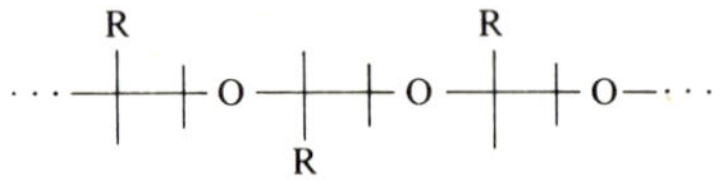

(*ddd* or *lll*)

heterotactic 1

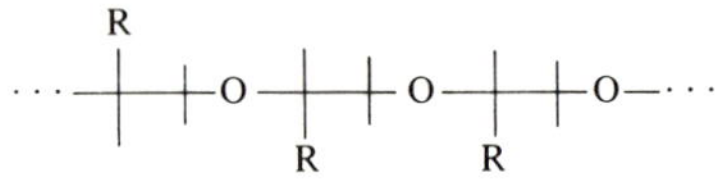

(*ddl* or *lld*)

heterotactic 2

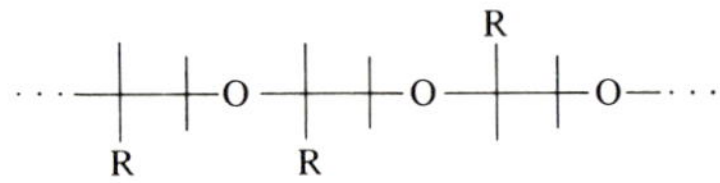

(*ldd* or *dll*)

syndiotactic

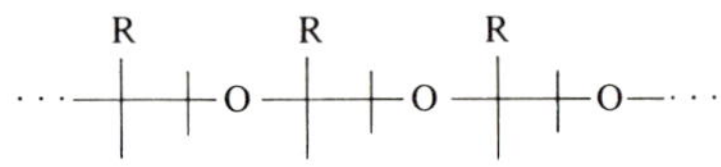

(*dld* or *ldl*)

head-to-tail and head-to-head:tail-to-tail addition. The proton spectra of stereoirregular polymers may thus be quite complex, even though there is no appreciable J coupling through the oxygen atoms. They consist of overlapping $ABCX_3$ subspectra. One might expect the principal differentiation in chemical shift of the methylene protons (for head-to-tail chains) to be that corresponding to the *meso* and *racemic* dyads

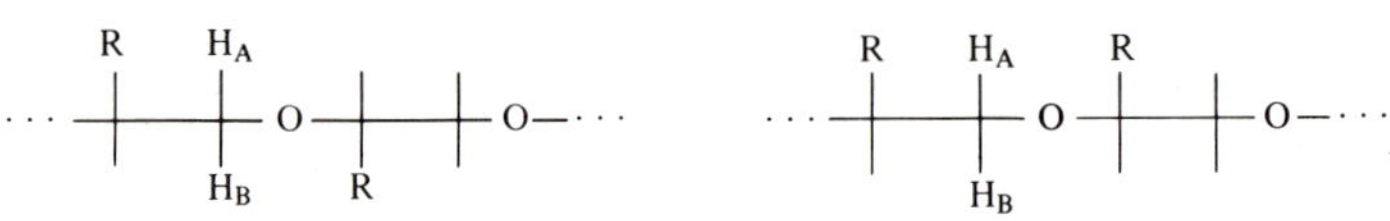

with possible triad fine structure, i.e., corresponding to the configuration of the next unit to the left of each of the sequences shown.

Ramey and Field[15] simplified the backbone proton spectrum to overlapping ABC spectra by irradiating the methyl groups. By comparison to the spectrum of a sample believed to be isotactic, they concluded that an "atactic" sample, a viscous liquid, was probably predominantly isotactic. The methyl resonances are not sensitive to configuration. Tani *et al.*[16] observed quartet methylene resonances in the spectra of poly-α-d_1-propylene oxide in chloroform solution and were able to distinguish quartets which they attributed to head-to-tail and head-to-head : tail-to-tail sequences, the latter in a minor proportion. In benzene solution the head-to-tail resonance was further resolved into what was believed (on the basis of solubility behavior of the samples compared) to be isotactic and syndiotactic dyad sequences.

Schaefer *et al.*[17] carried out an analysis of the 220 MHz spectra of propylene oxide dimers and trimers derived by hydrolysis of the maleic anhydride copolymers, which we shall discuss a little later. The ^{13}C spectrum of the trimer[18] provides a good example of the possible structural complexities in such oligomers, in which positional, directional, and configurational isomerism all combine. Of the 90 possible α- and β-^{13}C lines, about 50 can actually be resolved (Fig. XII-1). This spectrum was used to arrive at an approximate

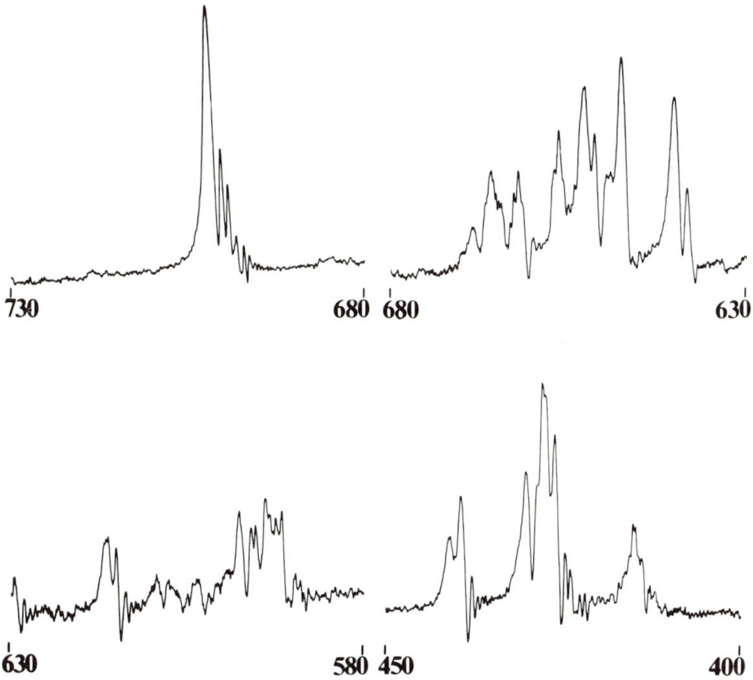

FIG. XII-1. White noise spin-decoupled ^{13}C NMR spectrum of the main-chain carbons of tripropylene glycol. The spectrum was taken in four parts and each part is the accumulation of 2500 scans, each scan taken at a sweep rate of 2 Hz/sec. The numbers give the separation in hertz downfield from the spin-decoupled line of methanol at 40°. The recorder gain for the parts of the spectrum between 730 and 680 Hz and between 450 and 400 Hz has been reduced by a factor of 2. The magnetic field increases from left to right. Expanded scale versions of the lowest field part of the spectrum indicate that the lines following the very intense, low-field line are real (Schaefer[18]).

statistical analysis of the ^{13}C spectra of the high polymers, shown in Fig. XII-2, which represent (a) the thermally precipitated crystalline fraction, (b) the noncrystalline fraction, and (c) the acetone-insoluble fraction of a polypropylene oxide prepared with stannic chloride catalyst. Arguments based largely

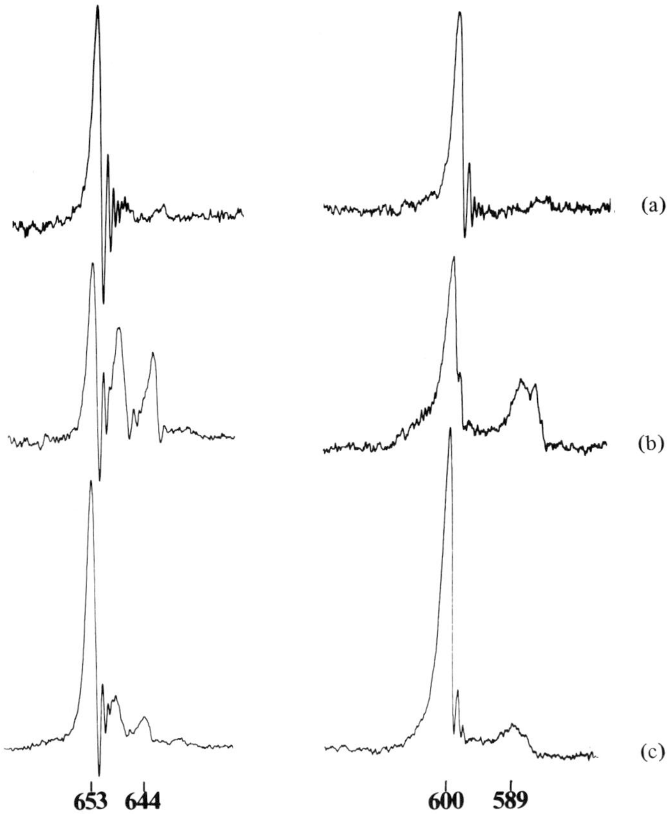

FIG. XII-2. White noise spin-decoupled ^{13}C NMR spectra of chloroform solutions of (a) the thermally precipitated crystalline fraction, (b) the noncrystalline fraction, and (c) the room-temperature, acetone-extracted, insoluble fraction of polypropylene oxide. Each spectrum is the accumulation of 2500 scans, each scan taken at a sweep rate of 2 Hz/sec. The concentrations of the solutions were not all identical. The numbers give the separation in hertz downfield from the spin-decoupled line of methanol at 40°. The magnetic field increases from left to right (Schaefer[18]).

on relative peak intensities and on chemical shifts were used to show that polymer (a) is 89% head-to-tail-isotactic. The peaks at lowest field correspond to the methine carbons and the more shielded peaks correspond to methylene carbons. Structural differences and steric differences contribute about equally to the methine carbon chemical shifts, and such irregularities account for the smaller peaks in this region in spectra (b) and (c). In the methylene carbon region, structural differences are the principal contributors to the smaller peaks of the spectra of the less regular polymers, steric differences having a smaller but observable effect. In these polymers, the sequence statistics do not appear to be Markovian of any finite order.

The structure and stereochemistry of polypropylene oxide and oligomers have also been studied by observation of the ^{19}F resonance of terminal trifluoroacetyl groups.[19] It was concluded that stereochemical irregularity at the chain end was the principal cause of the observed doubling of the fluorine resonances.

Schaefer *et al.* have investigated the proton spectra of copolymers of propylene oxide with maleic anhydride,[17,20] citraconic anhydride,[17] and sulfur dioxide.[21,22] In propylene oxide–maleic anhydride and propylene oxide–sulfur dioxide chains, the propylene oxide units have a sense of direction, as we have already seen. In propylene oxide–citraconic anhydride chains, both the propylene oxide units and the citraconic anhydride units

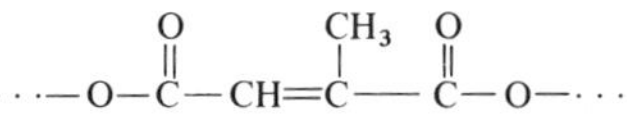

have a sense of direction and all four possible ways of combining them are observed. Triad compositional sequences are also discriminated. The methyl region of a propylene oxide–citraconic anhydride copolymer exhibited ten peaks at 220 MHz. By aid of such measurements, combined with gas chromatographic determination of propylene oxide sequence lengths in the manner indicated above for ethylene oxide–maleic anhydride copolymers, it was established that the structural isomer distribution of propylene oxide units around maleic anhydride and citraconic anhydride units and in closed runs of three propylene oxide units is random but that it is definitely nonrandom in closed runs of two propylene oxide units. Such a finding requires at least a third-order Markov mechanism for its interpretation. Here also, as in the ethylene oxide–maleic anhydride copolymers, the Lewis acid catalyst may be able to perform in several states of association with the monomer molecules.

The propylene oxide–sulfur dioxide copolymers, which are poly(sulfite esters), are also prepared with Lewis acid catalysts. It was found[21,22] that

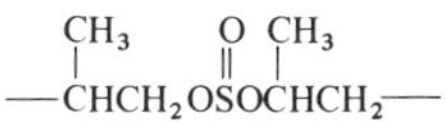

their chain structural statistics are still more complex and can only be explained by non-Markovian schemes.

C. Higher Epoxides and Copolymers

Price and Fukutani[23] have reported the 60 and 100 MHz spectra of poly-*tert*-butylethylene oxide prepared with basic catalysts. The ring proton spectrum is relatively simple, being approximately AMX at 100 MHz. Two different crystalline materials were obtained, which were believed to be isotactic and syndiotactic, respectively, as well as an amorphous product. *J* couplings were approximated by first-order interpretation of the rather poorly resolved spectra. The vicinal couplings in the isotactic polymer differ by at least 6 Hz, which, in the absence of firmly established peak assignments, is

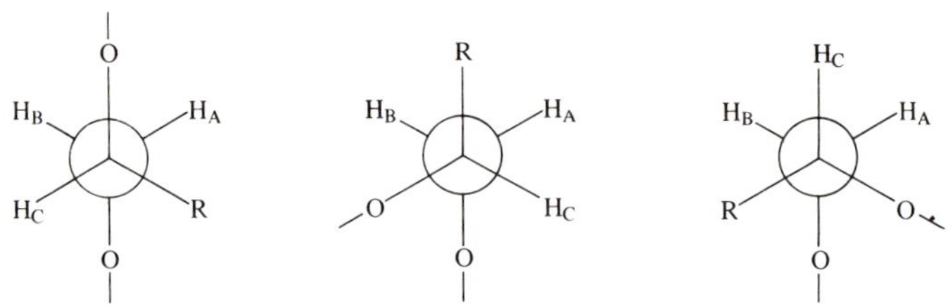

consistent with either a *trans* or *gauche* 1 conformation. In the spectrum of the "syndiotactic" polymer there were extra peaks which were attributed to the presence of more than one conformation. This appears to imply that, presumably because of steric hindrance, the conformations are long-lived on the NMR time scale, and that their spectra are not averaged. On the basis of molecular models, such an assumption appears unwarranted. It is more probable that this polymer was not in fact as stereoregular as assumed. There is also the possibility of head-to-head: tail-to-tail linkages, but the *tert*-butyl groups may discourage this.

Tani and Oguni[24] have reported the spectra of polymers of α-d_1-*tert*-butylethylene oxide. It was found that the assignments of Price and Fukutani[23] are in error, the α-proton being actually the most rather than the least shielded of the ring protons. This reduces the ambiguities of these spectra, but unfortunately the assignments of H_A and H_B are not yet secure. It would seem that a stereospecific β-d_1-monomer will be required to do this, and thereby to deduce the conformation of this polymer.

Yamashita *et al.*[25] have analyzed copolymers of 1,3,5-trioxane and 1,3-dioxolane. The 60 MHz spectra consist of singlets corresponding to —OCH_2CH_2O—, which can arise only from the dioxolane, and three species of formal unit: —$OCH_2O\mathbf{CH_2}OCH_2$— (from trioxane only); —$OCH_2O\mathbf{CH_2}$ OCH_2CH_2— (a junction of trioxane and dioxolane units); and —OCH_2CH_2

$OCH_2OCH_2CH_2$— (a junction of two dioxolane units). The comonomer sequences seemed to be initially formed in a random fashion but the statistics are complicated by the reversibility of the reaction, coupled with the fact that the dioxolane enters the chain more rapidly than the trioxane but forms structures of lower thermodynamic stability.

D. Polypropylene Sulfide

Ivin and Navratil[26] have reported the 100 MHz spectra of polymers of α-d_1-propylene sulfide, which were produced using (a) activated $CdCO_3$ and

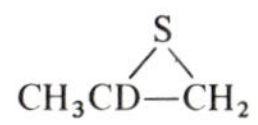

(b) activated $ZnCO_3$. The spectra consisted of two AB quartets of which the upfield doublets coincided. The low-field doublets appeared sensitive to stereochemical configuration. Polymer (a) was hard and was assumed to be predominantly isotactic; on this basis the more prominent (and least shielded) of the two downfield doublets was assigned to isotactic dyads. Polymer (b) was rubbery and showed equal downfield doublets. There was no evidence of heterotactic units; this is presumably because the methylene protons are sensitive only to dyad configuration (see p. 234). It is therefore not possible to tell from these spectra alone whether the configurations are distributed statistically or in runs. This question was answered by oxidation to the polysulfone (Chapter X, Section 8), in which triad sequences can be discriminated. The spectrum of the oxidized polymers consisted of four quartets, those from polymer (b) being nearly but not quite equal in intensity. These polymers are thus statistical rather than block in configurational character.

There appeared to be no indication of head-to-head: tail-to-tail sequences.

E. Poly-*N*-Acetylaziridine

Miron and Morawetz[27] have observed the methyl resonance of poly-*N*-acetylaziridine in an effort to determine whether the rate of rotation (Chapter

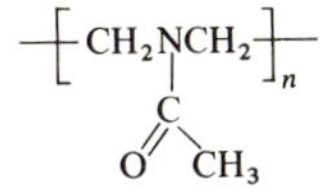

IV, Section 3; Chapter XIII, Section 3) about the C—N bond is affected by attaching the group to a polymer chain. Considered with respect to an isolated monomer unit, there is no difference in the environments of the methyl group in the two rotational states about this bond. Nevertheless, its 60 MHz spectrum is a doublet. This was explained in terms of the relative conformations of the group with respect to those of the nearest neighboring *N*-acetyl groups on each side in the polymer chain. Of the four possible relative conformations, only two are actually distinguished. The situation is reminiscent of that in polysarcosine (Chapter XIII, Section 6A).

2. Polyacetaldehyde

Polyacetaldehyde has the structure shown and is thus analogous to vinyl polymers in its stereochemistry. The absence of β-protons prevents absolute

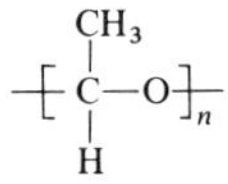

measurement of its configuration by NMR,[28–30] and therefore recourse must be had to analogies with model compounds. The spectrum of stereoregular polyacetaldehyde would consist only of an α-proton quartet near 5τ and a methyl doublet at higher field. The 100 MHz spectrum shown in Fig. XII-3 makes it immediately evident that the polymer is irregular. The methyl resonance appears as an apparent triplet with a shoulder on the low-field side, while the α-proton is a broad, poorly resolved overlap of multiplets. The decomposition of the "triplet" into *mm*, *mr*, and *rr* doublets, proposed by Goodman and Brandrup,[30] is shown in Fig. XII-3. It is supported by model compound results, particularly by the C–CH_3 chemical shifts of the *meso* and *racemic* "dimers" (a) and the central C–CH_3 chemical shifts of the *mm*, *mr*, and *rr* "trimer" models[31] (b). Polymers prepared with various initiators

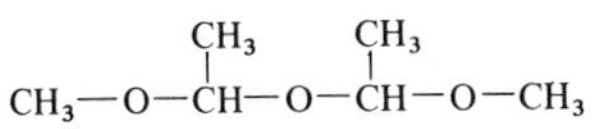

(a)

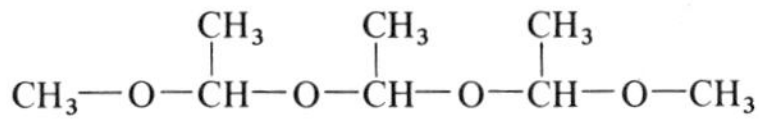

(b)

such as H_3PO_4 (–80°), $Zn(C_2H_5)_2$ (–80°), γ radiation at –196° (may be ionically initiated), or Al/$Zn(C_2H_5)_2$,[32] vary observably but not markedly in stereochemistry,[30] the *mm* sequence always tending to be excluded. Goodman and co-workers[29,30] have suggested an explanation for this in terms of association of monomer acetaldehyde into dimers prior to polymerization.

3. Polydimethylsiloxane

Liu and Ullman[33] have observed the NMR of polydimethylsiloxane and oligomers in benzene, d_6-benzene, chlorobenzene, and carbon tetrachloride

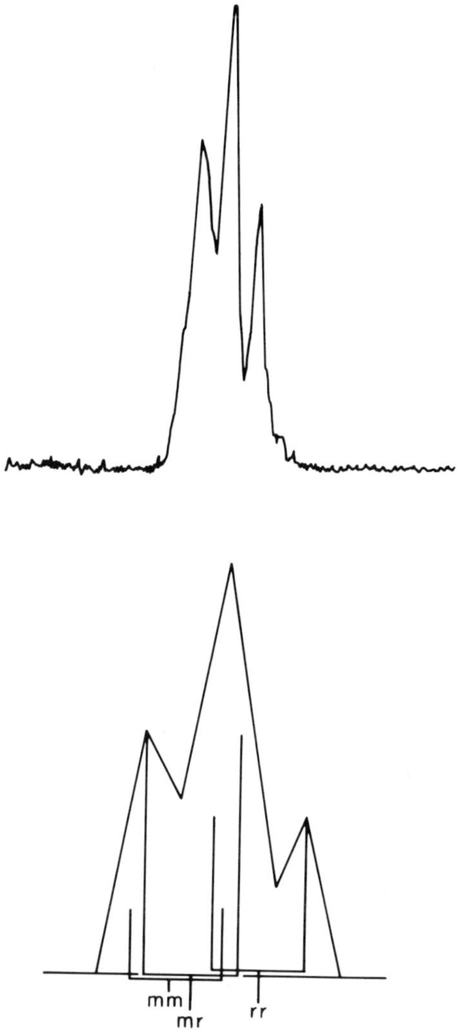

FIG. XII-3. 100 MHz spectrum of methyl groups of polyacetaldehyde (observed in aniline solution at 150°). The decomposition of the spectrum into *mm*, *mr*, and *rr* doublets is indicated. See Brandrup and Goodman[29] and Goodman and Brandrup.[30]

solutions. The spectra are very simple, the only possible complication arising from positional isomerism in the oligomers. Measurements of the spin–lattice relaxation time T_1 and the spin–spin relaxation time T_2 (Chapter I, Sections 5, 6, and 9) showed that the former is substantially independent of concentration, whereas T_2 decreases with increasing concentration. At zero concentration T_1 and T_2 approach nearly the same limiting value.

4. Polyamides

It is evident that NMR can be used analytically for the determination of the structure of polyamides and copolyamides. Such polymers are usually of relatively simple and known constitution and can be effectively studied by other means. For this reason, the NMR literature in this area is very sparse. (Polypeptides are a special case and are discussed in detail in Chapter XIII.)

We discuss here two investigations[34,35] which employ the specific power of NMR to determine conformational preferences in polyamide chains as well as the rates of conformational change about the —C(O)—N bonds (see Chapter XIII, Section 3, and Section 1E of this chapter). Miron *et al.*[34] observed the N—CH_2 resonances in the 60 MHz spectra of the piperazine polyamides (a), where $m = 2$ (succinyl), 4 (adipyl), and 8 (sebacyl); in polyterephthaloylpiperazine (b) and the isophthaloyl isomer; and in polysebacyl-

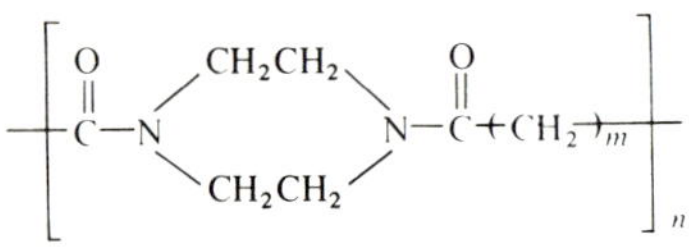

(a)

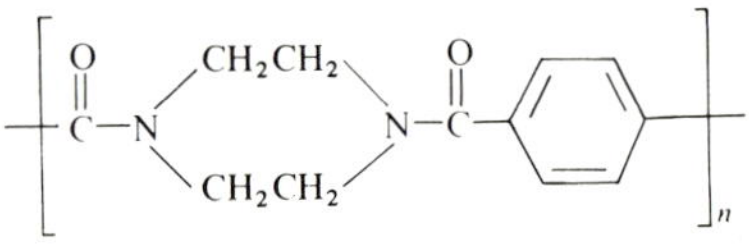

(b)

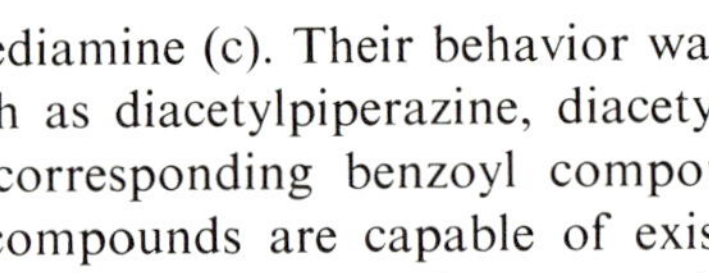

(c)

N,N′- dimethylethylenediamine (c). Their behavior was compared to that of model compounds such as diacetylpiperazine, diacetyl-*N,N′*-dimethylethylenediamine, and the corresponding benzoyl compounds. The piperazine polymers and model compounds are capable of existing in *cis* and *trans* forms. (Inversion at the nitrogen atoms also occurs, but is rapid compared to C—N bond rotation.) Peaks of nearly equal intensity can be seen for each

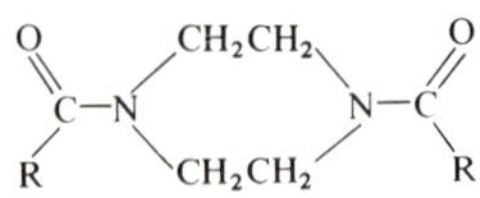

cis

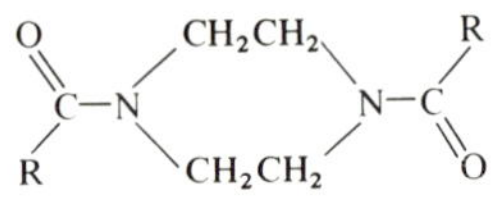

trans

form. The *N,N'*-dimethylethylenediamine compounds can exist in three conformations: *cis–cis*, *trans–trans*, and *cis–trans*. Peaks for each can be seen in the model compound spectrum but are not fully resolved in the polymer spectrum.

Measurement of the coalescence of the N–CH_2 peaks as a function of temperature showed that nearly all the compounds, polymers as well as models, exhibited free energies of activation of 18.0 ± 0.5 kcal for rotation about the C—N bond. The activation enthalpies were more scattered but were not consistently higher for the polymers. It is as though the C—N bonds acted as independent units. Clearly, however, there must be a degree of cooperativeness between them, even though this is not reflected in the present measurements, for otherwise isomerization about one bond would require a substantial portion of the polymer chain to be swept through the solution. Rotations about the N—CH_2 and CO—CH_2 bonds, which do not require high activation energies, can facilitate the necessary conformational adjustments.

Overberger *et al.*[35] have studied the polyamide

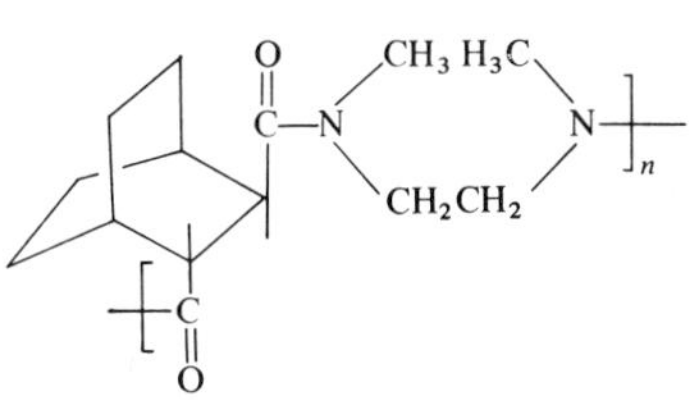

together with the model compounds I and II. All rotational isomers of I are

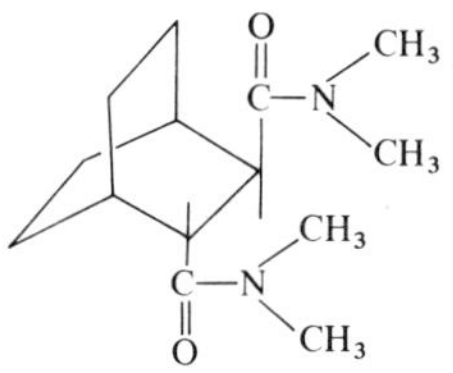

I

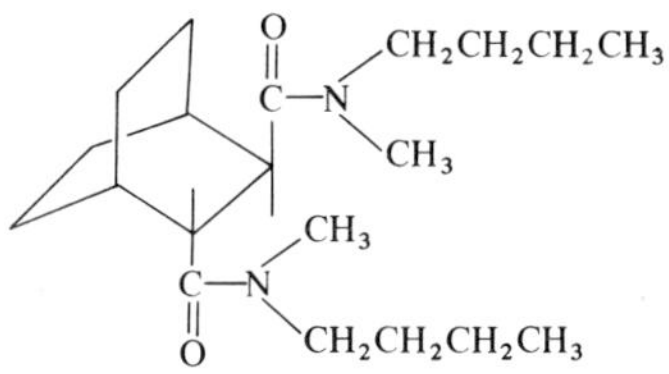

II

equivalent. Compound II may exist in three rotational states; that in which the larger alkyl group is *trans* to the bicyclooctane ring is believed to be somewhat preferred. It can be readily shown that there are four different methyl group environments in the three rotational states of II. The NMR spectrum shows only two, indicating that only the immediate local environment is significant. Upon increasing the observing temperature, there is some indication that the smaller upfield (*cis*) methyl peak decreases in intensity prior to coalescence, indicating an entropic preference for the *trans* conformation. The spectrum of the polymer showed this behavior in a more pronounced degree. An entropic preference for the *trans–trans* conformation of the polymer is understandable, since the *cis–cis* conformation would be substantially more folded. As with poly-L-proline (Chapter XIII, Section 6C), the *cis–trans* ratio should be strongly solvent dependent, as that of II was indeed observed to be.

5. Polyesters

The same remarks made in introducing the previous section apply to polyesters as well, indeed in a sense with greater force since the rotational barrier in the ester group does not appear to be sufficiently high to permit the observation of discrete conformer spectra. Nevertheless, NMR has been of some analytical utility in the determination of geometrical isomerism in polyesters and of sequence lengths in copolyesters. For example, Chien and Walker[36] obtained NMR evidence that the polyester of ethylene glycol and 1,4-cyclohexanedicarboxylic acid contained the same proportion of *cis* and *trans* units of the latter whether one begins with the *cis* or *trans* monomer. Unfortunately, their analysis was marred by serious interpretive errors. Yamadera and Murano[37] were able to detect hetero linkages in ethylene glycol copolyesters of terephthalic, isophthalic, and orthophthalic acid, among others, and thereby to measure the degree of randomness of their chains. Percival and Stevens[38] published useful data for the analysis of phthalate polyesters.

6. Other Polymers

Sumi *et al.*[39] have presented extensive data and spectra of polyurethanes and prepolymers. Suzuki *et al.*[40] have analyzed sequence distributions in polyester–urethane block copolymers.

Woodbrey *et al.*[41] and Hirst *et al.*[42] have shown that the structures of soluble phenol–formaldehyde resins can be quite completely analyzed by NMR. The characteristic chemical shifts of methylene protons in *p,p'*-dihydro-oxydiphenylmethane bridges allow the determination of the fraction of such structures.[42] Woodbrey *et al.*[41] studied both nonacetylated and acetylated resins and presented extensive data on small-molecule intermediates and model compounds. From such information they were able to give spectral assignments for the following functional groups in phenol–formaldehyde resins: benzyl-type hemiformals, hydroxymethyl groups, dibenzyl ether-type bridges, and diphenylmethane-type methylene bridges. Using this information, the spectra of soluble resins could be interpreted to provide measurements of the *o*-hydroxymethyl and *p*-hydroxymethyl groups and the ratio of *p,p'*- to *o,o'*- plus *o,p'*-methylene bridges. It was shown that (in the absence of ether bridges) quantitative measures could be provided for the average number of each functional group per aromatic ring, the average number of aromatic rings per molecule, and number average molecular weight. Similar structural information could be given for acetylated resins. Conclusions concerning the pathways of the condensation reaction were drawn which would probably be impossible to reach by any means other than NMR.

REFERENCES

1. T. M. Connor and K. A. McLauchlan, *J. Phys. Chem.* **69**, 1888 (1956).
2. K.-J. Liu, *Int. Symp. Macromol. Chem., Tokyo, Japan, 1966 Polym. Prepr. Amer. Chem. Soc. Div. Polym. Chem.* **7**, 48 (1966).
3. K.-J. Liu, *Macromolecules* **1**, 213 (1968).
4. K.-J. Liu, *Macromolecules* **1**, 308 (1968).
5. K.-J. Liu, *Makromol. Chem.* **116**, 146 (1968).
6. K.-J. Liu and J. E. Anderson, *Macromolecules* **2**, 235 (1969).
7. K.-J. Liu and J. L. Parsons, *Macromolecules* **2**, 529 (1969).
8. Y. Kuroda and M. Kubo, *J. Polym. Sci.* **36**, 453 (1959).
9. W. H. Barnes and S. Ross, *J. Amer. Chem. Soc.* **58**, 1129 (1936).
10. H. Tadokoro, Y. Chatani, T. Yoshihara, S. Tahara, and S. Murahashi, *Makromol. Chem.* **74**, 109 (1964).
11. H. Tadokoro, *Macromol. Rev.* **1**, 119 (1967).
12. P. J. Flory, "Statistical Mechanics of Chain Molecules," pp. 165–172. Wiley (Interscience), New York, 1969.
13. J. Schaefer, *Macromolecules* **2**, 210 (1969).
14. J. Schaefer, D. A. Bude, and R. J. Katnik, *Macromolecules* **2**, 289 (1969).
15. K. C. Ramey and N. D. Field, *J. Polym. Sci. Part B* **2**, 461 (1964).
16. H. Tani, N. Oguni, and S. Watanabe, *J. Polym. Sci. Part B* **6**, 577 (1968).
17. J. Schaefer, R. J. Katnik, and R. J. Kern, *Macromolecules* **1**, 101 (1968).
18. J. Schaefer, *Macromolecules* **2**, 533 (1969).
19. J. D. Ingham, D. D. Lawson, S. L. Manatt, N. S. Rapp, and J. P. Hardy, *J. Macromol. Chem.* **1**, 75 (1966).
20. J. Schaefer, R. J. Katnik, and R. J. Kern, *J. Amer. Chem. Soc.* **90**, 2476 (1968).
21. J. Schaefer, R. J. Kern, and R. J. Katnik, *Marcomolecules* **1**, 107 (1968).
22. J. Schaefer, *Macromolecules* **1**, 111 (1968).
23. C. C. Price and H. Fukutani, *J. Polym. Sci. Part A-1* **6**, 2653 (1968).
24. H. Tani and N. Oguni, *J. Polym. Sci. Part B* **7**, 803 (1969).
25. Y. Yamashita, T. Asakura, M. Okada, and K. Ito, *Makromol. Chem.* **129**, 1 (1969).
26. K. J. Ivin and M. Navratil, *J. Polym. Sci. Part B* **8**, 51 (1970).
27. Y. Miron and H. Morawetz, *Macromolecules* **2**, 162 (1969).
28. E. G. Brame, Jr., R. S. Sudol, and O. Vogl, *J. Polym. Sci. Part A* **2**, 5337 (1964).
29. J. Brandrup and M. Goodman, *J. Polym. Sci. Part B* **2**, 123 (1964).
30. M. Goodman and J. Brandrup, *J. Polym. Sci. Part A* **3**, 327 (1965).
31. M. Goodman and G. C.-C. Niu, *Macromolecules* **1**, 223 (1968).

32. F. Hiroyasu, J. Furukawa, T. Saegusa, and A. Kawasaki, *Makromol. Chem.* **40**, 226 (1960).
33. K.-J. Liu and R. Ullman, *Macromolecules* **2**, 525 (1969).
34. Y. Miron, B. R. McGarvey, and H. Morawetz, *Macromolecules* **2**, 154 (1969).
35. C. G. Overberger, R. A. Veneski, and G. Montando, *J. Polym. Sci. Part B* **7**, 877 (1969).
36. J. C. W. Chien and J. T. Walker, *J. Polym. Sci.* **45**, 239 (1960).
37. R. Yamadera and M. Murano, *J. Polym. Sci. Part A-1* **5**, 2259 (1967).
38. D. F. Percival and M. P. Stevens, *Anal. Chem.* **36**, 1574 (1964).
39. M. Sumi, Y. Chokki, Y. Nakai, M. Nakabayashi, and T. Kanzawa, *Makromol. Chem.* **78**, 146 (1964).
40. H. Suzuki, H. Ono, and T. Hongo, *Makromol. Chem.* **132**, 309 (1970).
41. J. C. Woodbrey, H. P. Higginbottom, and H. M. Culbertson, *J. Polym. Sci. Part A* **3**, 1079 (1965).
42. R. C. Hirst, D. M. Grant, R. E. Hoff, and W. J. Burke, *J. Polym. Sci. Part A* **3**, 2091 (1965).

Chapter XIII

POLYPEPTIDES

1. Introduction

A logical introduction to the study of proteins is the study of synthetic polypeptides with a single type of repeating unit or of copolypeptides with two or at most three types of units. Amino acids, simple amides, and short polypeptides (oligopeptides) may, in turn, be considered as models for polypeptides. In fact, the NMR spectra of oligopeptides may be more complex than those of long-chain polypeptides, and so we shall consider these toward the end of this chapter.

2. Amino Acids

Amino acids have frequently been studied as model compounds for proteins, but actually are more nearly models for end groups (although their *pK*'s differ markedly from those of the C and N termini of polypeptides) and side chains.

A. Effect of *p*H

The earliest studies[1,2] of the effect of *p*H on amino acid chemical shifts showed that all the nonexchanging protons, i.e., those bonded to carbon, become more shielded as the *p*H is increased. Figure XIII-1 shows the 100 MHz spectra of valine in strongly acid (a), nearly neutral (b), and basic (c) D_2O solution.[3] In Fig. XIII-2, the chemical shifts of the α-, β-, and methyl

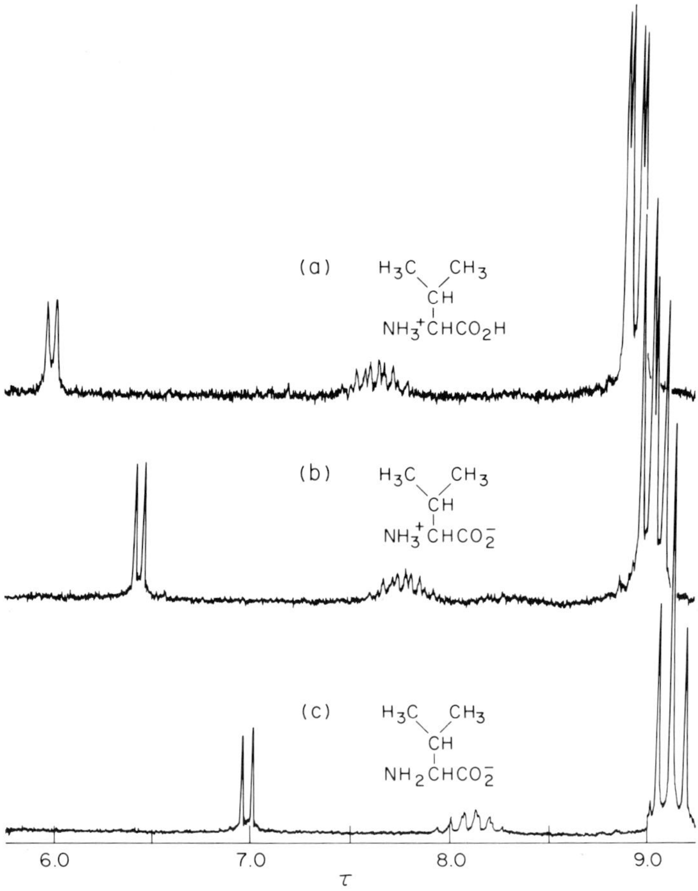

FIG. XIII-1. 100 MHz spectra of valine in D_2O as (a) cation (pD ~ 1), (b) zwitterion (pD ~ 6), and (c) anion (pD ~ 12) (Bovey and Brewster[3]).

protons are plotted versus pD. (The two methyl groups appear at different chemical shifts for reasons discussed in Chapter III and also later in this chapter.) The ammonium protons appear as a separate peak only in very acidic H_2O media; ordinarily they exchange rapidly with the solvent and appear in the H_2O or residual HDO peak near 5τ. The carboxyl protons exchange rapidly under all conditions. (Proton exchange of amino acids and its NMR observation is discussed in detail in Section 3G.) The HDO peak may be quite large if the solute has a substantial number of exchangeable protons. It will be broadened and shifted to lower field by the exchange processes. By repeated exchange with high-purity D_2O, it can be made small enough not to interfere seriously.

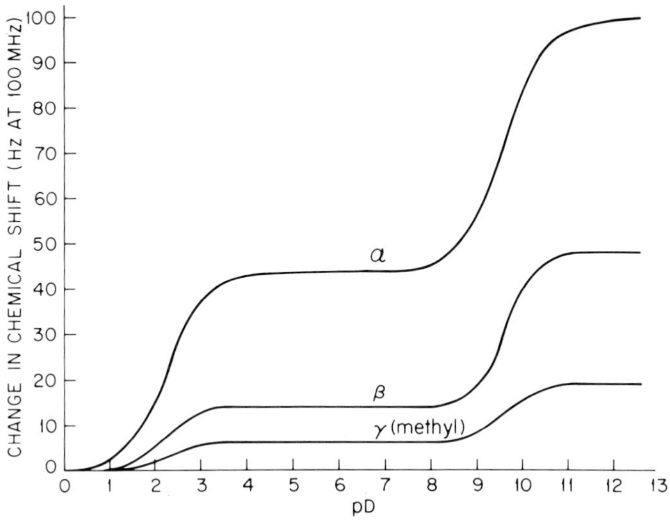

FIG. XIII-2. Chemical shifts (expressed in hertz at 100 MHz upfield from positions at lowest pH) of α-, β-, and γ- (methyl) protons of valine versus pD (Bovey and Brewster[3]).

The steps in the chemical shift plots in Fig. XIII-2 correspond to the processes shown (from left to right).

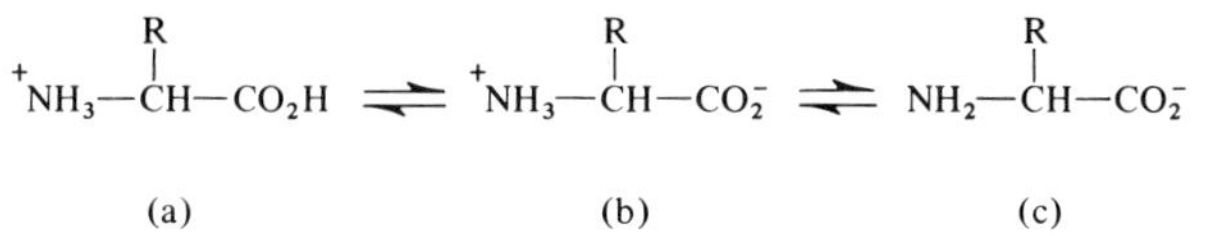

Through the step regions, two species are in rapid exchange (the third is also present in principle, but in vanishingly small concentration), and the chemical shifts are averages of those of each species.

A number of later studies [4–18] have confirmed and extended these findings. The chemical shift changes which take place upon proton gain or loss are in general greatest for the α-protons. As might be anticipated, the effect on the chemical shifts of side-chain protons (Fig. XIII-2) diminishes fairly rapidly with the number of intervening bonds. Exceptions to this behavior may of course be expected for side chains bearing functional groups, such as glutamic and aspartic acids, lysine, and arginine. In Fig. XIII-3 are shown the 220 MHz spectra of lysine in acid, neutral, and basic D_2O solution.[3] The γ-protons are least influenced by pD, and α- and ε-protons most affected.

The chemical shift effects can be readily rationalized, at least in an approximate manner. As the pH is raised, dissociation of the carboxyl group produces a negative charge density which increases the shielding of all protons; α-protons shift upfield by about 0.4τ. In addition, alterations occur in the intramolecular reaction field and in the anisotropic shielding influence of the carboxyl group. A somewhat larger shielding increase occurs upon removal of protons from amino groups, since the NH_3^+ group is quite potent in reducing local electron density. In proteins and polypeptides, the ϵ- and δ-protons of lysine

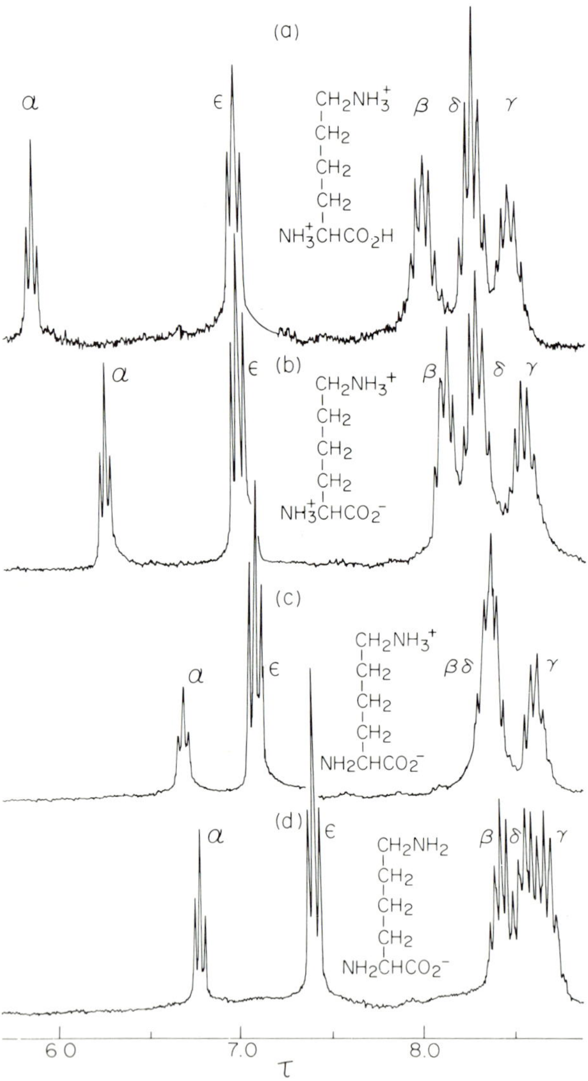

FIG. XIII-3. 220 MHz spectra of lysine in D_2O as (a) dication (pD ~1), (b) zwitterion (pD ~5), (c) zwitterion (pD ~9), (d) anion (pD ~12) (Bovey and Brewster[3]).

and ornithine, respectively, can be expected to shift by 0.5–0.6τ in the pH range 8–11. Amino acids with three functional groups show three steps in the pH (or pD) versus τ curve.

The imidazole ring protons of histidine residues are often particularly conspicuous in protein spectra (Chapter XIV), and in favorable instances may be employed to follow the titration of these groups. Figure XIII-4 shows

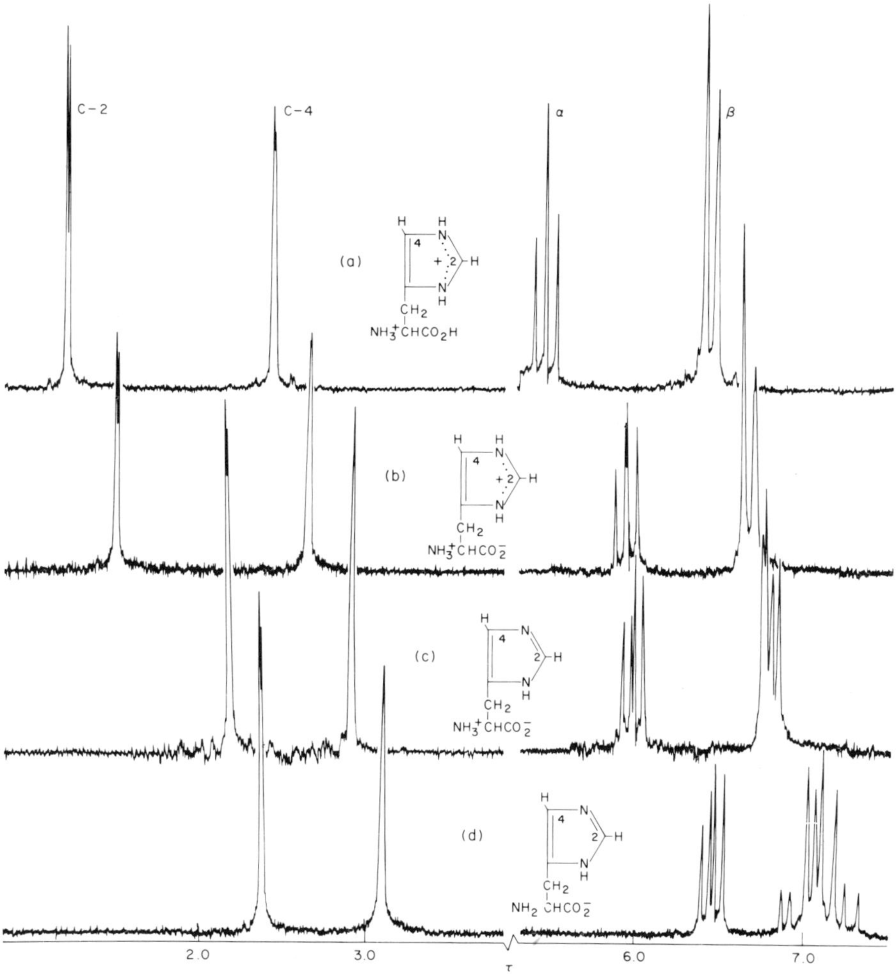

FIG. XIII-4. 100 MHz spectra of histidine in D_2O as (a) dication (pD ~ 1), (b) zwitterion with imidazole ring charged (pD ~ 5), (c) zwitterion with imidazole ring uncharged (pD ~ 8) (d) anion (pD ~ 12) (Bovey and Brewster[3]).

the 100 MHz spectra of histidine in acid, neutral, and basic D_2O solution. The chemical shifts are plotted versus *p*D in Fig. XIII-5. The least shielded proton and that most affected by increasing the *p*D is the one on C-2. It clearly reflects the state of ionization of the ring but not that of the amino and carboxyl groups.

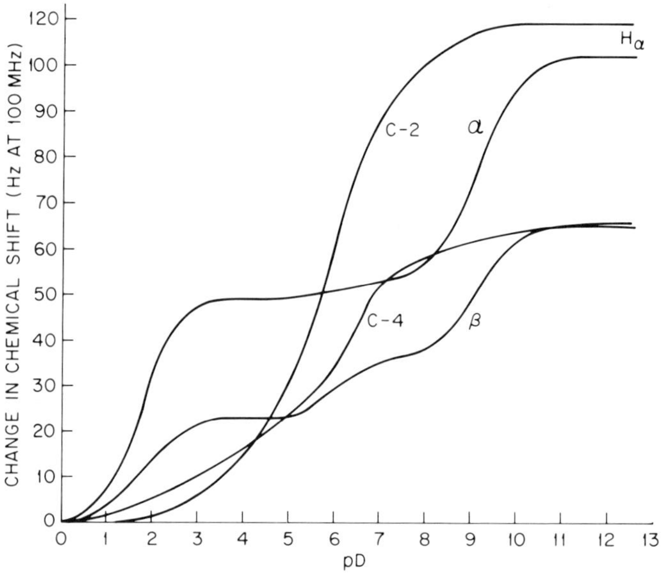

FIG. XIII-5. Chemical shifts (expressed in hertz at 100 MHz upfield from positions at lowest *p*H) of C-2, C-4, α-, and β-protons of histidine versus *p*D (Bovey and Brewster[3]).

The α-proton is not sensitive to the state of ionization of the ring. The β- and C-4 proton plots reflect all four states of ionization shown and exhibit three steps.[13, 18]

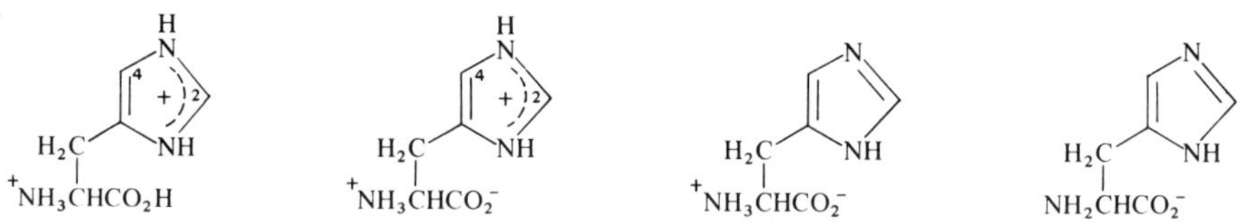

*p*H ≤ 1 *p*H = 3-5 *p*H = 7-8 *p*H ≥ 9

The chemical shifts of peptides and amino acids have been summarized and systematized by Nakamura[9] and Nakamura and Jardetzky.[16] McDonald and Phillips[19] have provided very useful data on amino acid (and certain peptide) chemical shifts in neutral D_2O solution, obtained at 220 MHz.

These data, somewhat rearranged and expressed on the τ scale, are shown in Table XIII-1.†

B. Spectra in Trifluoroacetic Acid

The 40 MHz spectra of a number of amino acids and peptides in trifluoroacetic acid were reported and discussed by Bovey and Tiers.[20] Trifluoroacetic acid is an excellent solvent for all amino acids and peptides, including cystine and tyrosine, which are nearly insoluble in water. A disadvantageous feature is that in this strong acid, the ammonium group protons exchange very slowly and their coupling complicates the spectrum considerably. At 40 MHz, resolution was sufficient to distinguish the general features of the spectra. Much more detailed chemical shift information was obtained in a parallel investigation by Bak *et al.*[21] at 220 MHz; this was used to aid in the interpretation of the spectrum of insulin, which (like all proteins) is a random coil in this solvent (Chapter XIV).

C. ^{13}C Spectra

The natural-abundance ^{13}C (Chapter I, Section 16C) spectra of glycine, glycylglycine, glycylglycylglycine, L-alanine, and L-alanylglycine have been observed by Horsley and Sternlicht.[22] The ^{13}C chemical shifts were measured by an internuclear double resonance (INDOR) technique which is practical only for relatively simple spectra.[23-25] This method employs the proton spectrum. One must locate one of the two ^{13}C satellites to the resonance of a proton directly coupled to the carbon atom of interest, and "sit" on this small peak, while simultaneously irradiating the system in the region of the ^{13}C resonance. As one passes through the precise frequencies of the carbon resonance lines, the satellite will appear to decrease in intensity due to splitting into multiple peaks. The ^{13}C spectrum can thus be traced out, and will of course be a multiplet for all carbons bearing hydrogens. In a later study, Horsley *et al.*[26] observed also the direct, natural-abundance ^{13}C spectra. This is of necessity the method which must be employed for more complex compounds.

The dependence of methylene ^{13}C and proton chemical shifts on *p*D is shown for glycine in Fig. XIII-6. Surprisingly, they move in opposite directions, although normally they show parallel effects in changing chemical environments. Protonation of the NH_2 and carboxylate groups causes increased shielding of the methylene ^{13}C nucleus and deshielding of the protons. The ^{13}C chemical shifts reflect primarily the local charge density, which evidently

† Table XIII-1 will be found at the end of this chapter.

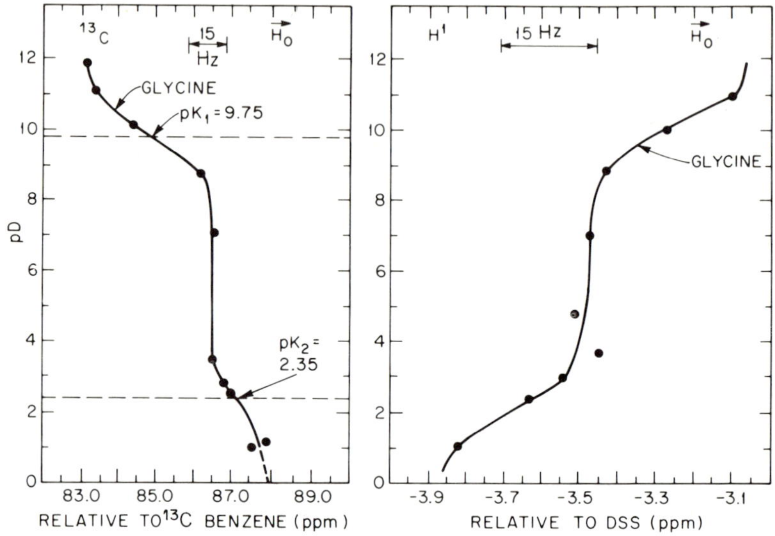

FIG. XIII-6. The ^{13}C (left-hand plot) and proton (right-hand plot) chemical shifts of the methylene carbon of glycine as a function of pD (from Horsley and Sternlicht[22]).

actually increases somewhat on protonation. Horsley and Sternlicht[22] have offered a rationalization of this behavior, but at present it appears both significant and puzzling. Their observations should have a healthy effect in inhibiting glib interpretations of more complex spectra.

D. Conformation Preferences

The presence of the asymmetric α-carbon atom means that α-amino acids have three possible staggered rotational isomers about the C_α—C_β bond. (α-imino acids, i.e., proline and hydroxyproline, are restricted by the ring to a much narrower range of conformations, not necessarily staggered.) Except

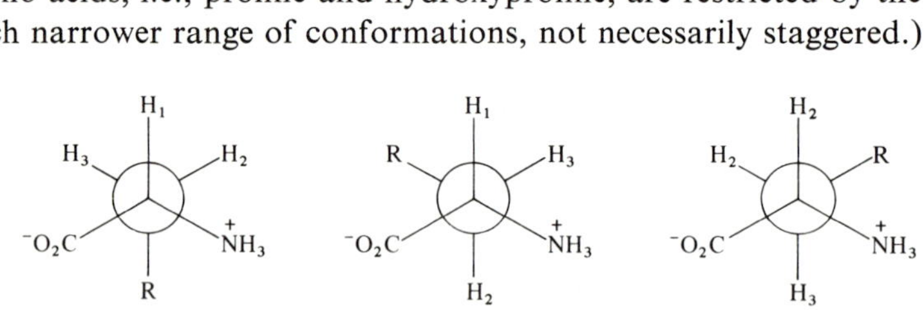

(I) (II) (III)

in alanine (where R = H), these conformers are nonequivalent and their energy levels will in general be unequally populated. The observed chemical shifts and coupling constants are averaged values weighted according to the fraction of each conformer. The $H–C_\alpha–C_\beta–H$ couplings are highly sensitive to the H–H dihedral angle (Section 3F) and are the principal basis for deductions concerning conformational preferences. The dependence of the couplings upon *p*H, and also in some cases upon solvent and temperature, has been measured for a number of amino acids, including aspartic acid,[4,5,8] cysteine,[4,5,6,8] histidine,[4,5,6,8] serine,[4,5,8,15] phenylalanine,[4,5,8,17] tyrosine,[4,5] and tryptophan,[4,5] the α,β-proton spectra of which can be treated as ABC or ABX. In Fig. XIII-7 is shown a representative spectrum, that of

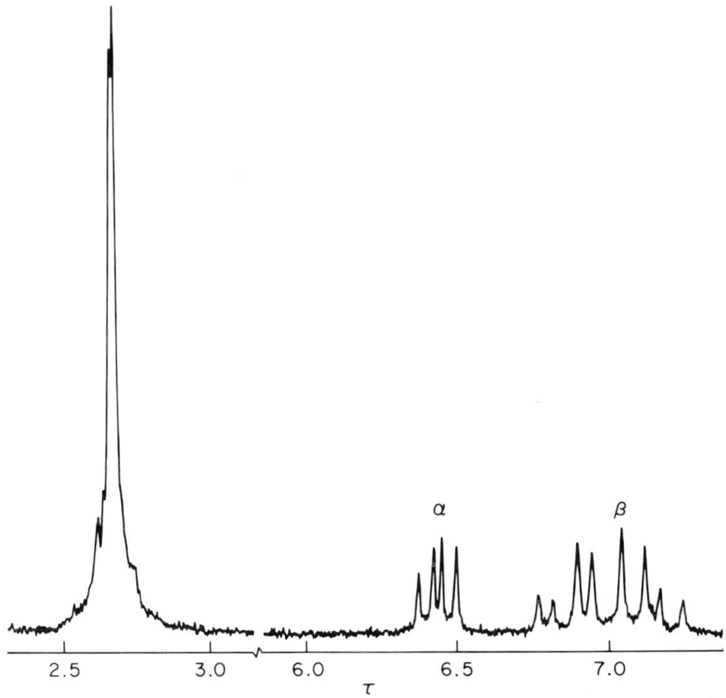

FIG. XIII-7. Phenylalanine anion in D_2O (observed at 100 MHz); the α-proton appears at 6.0τ and the β-protons appear near 7.0τ as the AB part of an approximately ABX spectrum; see Table XIII-2 (Bovey and Brewster[3]).

phenylalanine anion. Threonine[6] (or allothreonine[6]) and valine[4,5,6] have only one β-proton and can be regarded as AMX_3 and AMX_3Y_3, respectively; for these, $J_{\alpha\beta}$ can be obtained simply by measurement of the α-proton doublet spacing (see Fig. XIII-1). Typical data are shown in Table XIII-2 for phenylalanine and valine anion, zwitterion, and cation. In order to interpret these

TABLE XIII-2

J (Hz)	Phenylalanine[a] (22°)		
	C_6H_5–CH_2–$^+NH_3CHCO_2H$	C_6H_5–CH_2–$^+NH_3CHCO_2^-$	C_6H_5–CH_2–$NH_2CHCO_2^-$
J_{23}	−14.7	−14.5	−13.4
J_{13} (or J_{12})	5.6	5.1	5.4
J_{12} (or J_{13})	7.7	8.0	7.5
J(Hz)	Valine[b] (~20°)		
	$(H_3C)(CH_3)CH$–$^+NH_3CHCO_2H$	$(H_3C)(CH_3)CH$–$^+NH_3CHCO_2^-$	$(H_3C)(CH_3)CH$–$NH_2CHCO_2^-$
($J_{\alpha\beta}$)	4.3	4.4	5.1

[a] See Pachler[8] and Cavanaugh.[17]
[b] See Taddei and Pratt.[6]

data in at least semiquantitative terms, it is necessary to assume that the couplings between protons in *trans* conformations J_t, and those between protons in *gauche* conformations J_g, are constant for all conformers of a given amino acid. Pachler[8] has suggested that this assumption may be extended further, and that the following values of J_t and J_g can satisfactorily interpret the data for a number of amino acids:†

$$J_t = 13.5_6 \text{ Hz}, \qquad J_g = 2.6_0 \text{ Hz}$$

With this assumption, it is readily shown that the mole fraction of conformer I above is given by

$$X_{\mathrm{I}} = \frac{(J_t + J_g) - (J_{12} + J_{13})}{J_t - J_g} \qquad \text{(XIII-1)}$$

Thus in the ABX approximation, which is often adequate (see Fig. XIII-7), X_{I} is obtained directly from the outer spacing of the α-proton quartet, since this is equal to $J_{12} + J_{13}$. For phenylalanine, it is found that conformer I,

† We may note that if these values are accepted, $J_{\alpha\beta}$ for alanine should be

$$J_{\alpha\beta} = \tfrac{1}{3}(J_t + 2J_g) = 6.2_5 \text{ Hz}$$

The actual values are 7.2_8 Hz for the cation and zwitterion and 6.9_6 for the anion; thus alanine deviates somewhat from the norm.

which would appear to be the least favored sterically, actually is present in substantial proportion: X_I is 0.26 for the cation, 0.28 for the zwitterion, and 0.30 for the anion, not far short of the statistical value of 0.33. Unfortunately, proportions of conformers I and II cannot be unambiguously determined unless H_2 and H_3 can be specifically identified. (One way to do this would be by the stereospecific replacement of one of these protons by a deuteron, but this has not been done.) It is clear that I and II are not present in equal proportions, for then J_{12} and J_{13} would be equal or nearly so. It is usually assumed[8,13a,17] that conformer III, with the bulkiest substituents C_6H_5— and —CO_2H in the *trans* positions, is sterically favored. We then find the following proportions:

	X_I	X_{II}	X_{III}
Cation	0.26	0.19	0.55
Zwitterion	0.28	0.26	0.46
Anion	0.30	0.17	0.53

Assumptions concerning steric interactions must be treated with considerable reserve. We note, for example, that whatever assumption is made, the crowded conformer I is favored over one of the others, both of which seem sterically more likely. The case of valine is also relevant. Here, the staggered conformers are those shown:

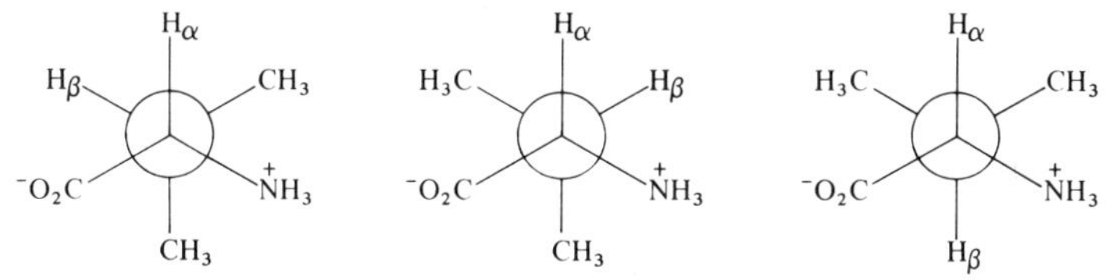

(I) (II) (III)

We find that:

$$X_{III} = \frac{J_{\alpha\beta} - J_g}{J_t - J_g} \qquad \text{(XIII-2)}$$

Using the same assumptions as for phenylalanine, we find that the proportions of III (which one would expect to be sterically most favored) are: cation, 0.15; zwitterion, 0.16; anion, 0.23. In the absence of any energy differences among the conformers, its proportion would of course be one-third. In this instance, one cannot determine the relative proportions of I and II from such data, no matter what assumption is made.

If the free energy differences between conformers are mainly enthalpy differences (an assumption customarily made in all studies of rotamer preferences), then as the temperature is increased one would expect J_{12} and J_{13} in amino acids of the type of phenylalanine (and $J_{\alpha\beta}$ in amino acids of the type of valine) to approach a common value of about 6.2 Hz (see footnote on p. 256). Actually, Cavanaugh[17] finds that in phenylalanine anion they diverge with increasing temperature. It must be concluded that, at least in aqueous systems, relative conformer free energies change with temperature (and in some cases also with concentration) and therefore the expected exponential dependence of conformer populations upon temperature cannot apply. It appears likely that the dominating contribution to conformer energies is solvation energy. This may explain why the usual considerations of steric interactions between neighboring groups seem to be quite inadequate to explain the behavior of these molecules.

E. Metal Complexes of Amino Acids and Simple Peptides; Relaxation and Contact Chemical Shifts

We have seen (Chapter I, Section 5) that paramagnetic ions can cause very marked decreases in both T_1 and T_2 because of the very large fluctuating magnetic fields generated by unpaired electron spins, of the order of 10^3 times the largest nuclear moments. The general theory of Bloembergen *et al.*[27] applied to the spin–lattice relaxation of water protons in solutions of paramagnetic ions gives[28]

$$\frac{1}{T_1} = \frac{4\pi^2 \gamma^2 \eta N_p \mu_{\text{eff}}^2}{kT} \tag{XIII-3}$$

where γ is the magnetogyric ratio of the proton, η is the solution viscosity, N_p is the number of paramagnetic ions per milliliter, and μ_{eff} is the "effective" magnetic moment of the paramagnetic species. This will be equal to the magnetic moment as measured by static susceptibility measurements provided that, as assumed in the derivation, each electronic moment maintains its orientation during the course of the diffusion together of ion and water molecule, i.e., that the spin–lattice relaxation time of the electron T_{1e}, is longer than the diffusional correlation time t. The latter is assumed to be given by the usual Stokes–Einstein approximation (Chapter I, Section 5). When $T_{1e} < t$, T_{1e} governs the relaxation process, and μ_{eff} may be smaller than the true μ. Some ions give values of μ_{eff}, calculated from Eq. (XIII-3), which are close to the static susceptibility measurements, while others give substantially smaller values (see Table 9-2 of Pople *et al.*[28]). Thus Cu^{2+}, Fe^{3+}, Mn^{2+}, and Cr^{3+}, having relatively long spin–lattice relaxation times, have values of μ_{eff} approximately equal to the static μ, while Co^{2+}, Fe^{2+}, Ni^{2+}, and most of the rare earths are much less effective in reducing T_1.

The contribution of paramagnetic species to $1/T_2$ parallels that to $1/T_1$ and so in most such systems $T_1 \simeq T_2$. Exceptions are found among those ions having unusually long spin–lattice relaxation times such as Mn^{2+} and Fe^{3+}, for which T_{1e} is of the order of 10^{-9} sec instead of the more commonly observed 10^{-11} to 10^{-12} sec. In such cases T_1/T_2 may considerably exceed unity. This is particularly likely to be the case where the paramagnetic ion forms a stable chelate or complex with the proton-containing molecule (see p. 261). Under these circumstances, the unpaired electron may give rise to an appreciable spin density actually at the nucleus, i.e., a "contact" interaction[29] will occur between nucleus and electron in addition to the dipole–dipole interaction. This same interaction is responsible for the nuclear hyperfine structure in electron spin resonance spectra and for chemical shifts to be discussed later. Bloembergen has shown[30] that under these circumstances the appropriate correlation time is T_{1e}; when its fluctuations occur with a frequency substantially less than the observing frequency, it will contribute more to $1/T_2$ than to $1/T_1$. The effect should be more marked the higher the frequency of the NMR instrument.

The relaxation of water protons by complexed paramagnetic ions, particularly heme proteins, has been employed to obtain information concerning the ligand structure; this is discussed in Chapter XIV.

Although Eq. (XIII-3) was derived specifically for aqueous solutions, its general form may be expected to hold for the interaction of paramagnetic ions with other molecules. In amino acid and peptide spectra, specific ligand formation may cause certain groups of protons to be relaxed much more effectively than others. This is well illustrated by the spectra in Fig. XIII-8,[31] which show the methylene protons of glycylglycine anion in D_2O solution containing 10^{-6}, 10^{-5}, and 10^{-4} *M* Cu^{2+} and 0.3 *M* NaOD. The more shielded protons are those of the N-terminal methylene group. The Cu^{2+} ions associate specifically with the amino group and broaden the nearest methylene resonance to the point of virtual disappearance with very little effect on the other. The ligand of course is in great excess and the paramagnetic ion exchanges rapidly among the amino groups. Such studies have been extended to triglycine and tetraglycine, glycylproline, and several amino acids.[31]

When the paramagnetic ion and the ligand are present in comparable molar concentrations and are capable of forming stable complexes, a quite different and very striking effect is observed. We have seen that when the spin–lattice relaxation time of the unpaired electrons T_{1e} is very short, $T_1 = T_2$ for the ligand protons. Since T_1 is of the order of 10^{-1} to 10^{-3} sec, nuclear resonances will be only moderately broadened under such circumstances and can be readily observed. Co^{2+} and Ni^{2+} have very short electronic relaxation times and their complexes are therefore favorable for observation. It is found† that the protons undergo highly variable and extremely large *contact shifts* both upfield and downfield from their normal positions, arising from

† For a comprehensive review of the NMR of paramagnetic metal complexes, see Eaton and Phillips.[32]

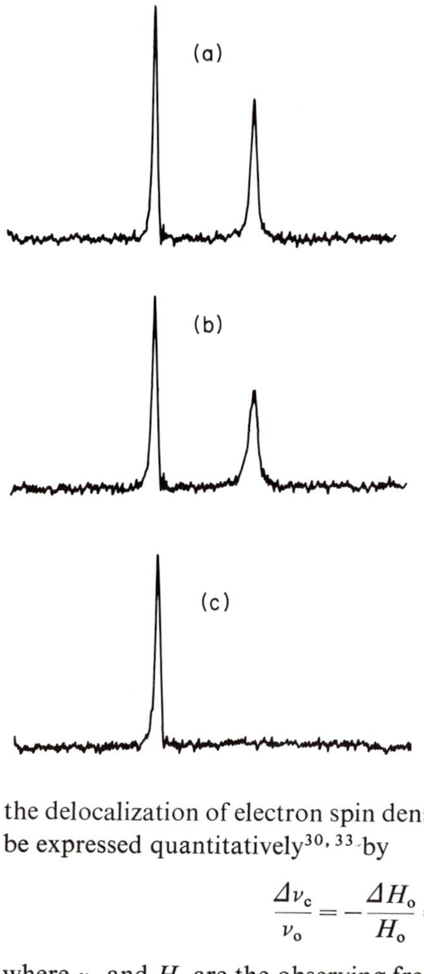

FIG. XIII-8. Proton spectrum of glycylglycinate anion $NH_2CH_2CONHCH_2CO_2^-$ (0.5 *M* in D_2O) as a function of Cu^{2+} concentration: (a) 10^{-6} *M*, (b) 10^{-5} *M*, (c) 10^{-4} *M* (from Li *et al.*[31]).

the delocalization of electron spin density as indicated previously.† These may be expressed quantitatively[30, 33] by

$$\frac{\Delta\nu_c}{\nu_0} = -\frac{\Delta H_o}{H_o} = A\frac{\gamma_e}{\gamma_p}\frac{S(S+1)}{3kT} \qquad \text{(XIII-4)}$$

where ν_0 and H_o are the observing frequency and field, A is the contact interaction constant, γ_e and γ_p are the magnetogyric ratios of the electron and proton, respectively, and S is the total electronic spin of the paramagnetic species. The value of A is typically 10^5–10^7 Hz, and since it is a condition for the observation of contact shifts that

$$\frac{1}{T_{1e}} \gg A \qquad \text{(XIII-5)}$$

we require that $T_{1e} \ll 10^{-8}$ sec.

† Under appropriate circumstances, similar chemical shift effects may be caused by "pseudocontact" interactions, arising from the magnetic dipole of the unpaired electron. This is discussed further in Chapter XIV, Section 2D.

Complexes of paramagnetic metals with histidine[18, 34, 35] are of significance as models for a number of proteins, particularly heme proteins (Chapter XIV) in which histidine residues serve as metal ligands, although the parallel is not exact since in the protein the histidine amino and carboxyl groups will in general not be available for complexing. McDonald and Phillips[18] carried out an extensive study of Co^{2+} complexes with histidine and marshalled evidence for four distinct species (Fig. XIII-9). Because of their ability to

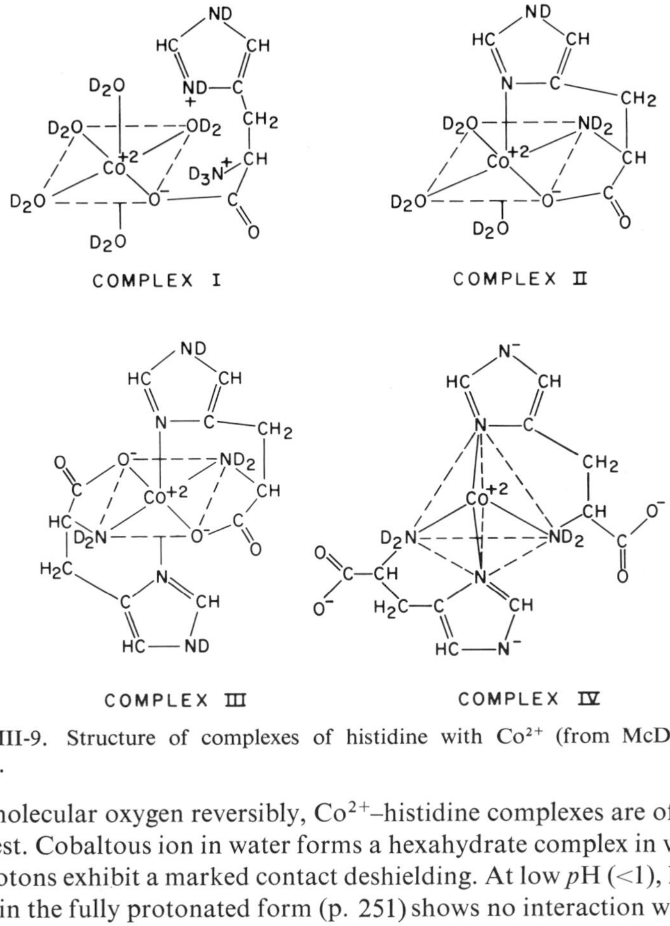

FIG. XIII-9. Structure of complexes of histidine with Co^{2+} (from McDonald and Phillips[18]).

absorb molecular oxygen reversibly, Co^{2+}–histidine complexes are of particular interest. Cobaltous ion in water forms a hexahydrate complex in which the water protons exhibit a marked contact deshielding. At low pH (<1), histidine, which is in the fully protonated form (p. 251) shows no interaction with Co^{2+}. As the pH is raised, complex I begins to be formed, being stable in the pH range 1.0–3.5. The α- and β-protons are shifted downfield but those of the ring are little affected and presumably not involved. This complex involves only the carboxyl group, which ionizes in this range, and there is rapid ligand exchange. As the pH is increased, the histidine peaks broaden and begin to exhibit very large shifts downfield; at *p*H 4.5, the spectrum of complex II shows the α-protons nearly 12,000 Hz downfield (at 60 MHz) and the β- and C-2 protons (see Table XIII-1) are nearly 4000 Hz downfield from the tetramethylammonium ion. At *p*H 5 a similar but different spectrum is observed and is attributed to complex III, in which water is entirely displaced from the cobaltous ion's coordination sphere. Finally at very high *p*H (>11.5), the ring becomes deprotonated, the α-proton resonance moves upfield, the solution becomes blue, and the static magnetic susceptibility decreases to a value characteristic of tetrahedral Co^{2+} (complex IV).

Complexes of diamagnetic ions with amino acids and peptides have also been studied by NMR and are of biological interest because many enzymes require such ions for activity. Jardetzky and Wertz[36] studied weak complexes of sodium ion with amino acids by observation of broadening of the ^{23}Na resonance; only histidine showed a marked effect. Complexes of cadmium,[31,37] zinc,[37,38,39] and magnesium [37,38] with glycine, alanine, and histidine generally show deshielding of the protons near the points of ligand attachment, probably due to both inductive and electric field effects (Chapter I, Section 11). An exception is the zinc complex of histidine at high *p*H, in which the imidazole protons appear at higher field than in free histidine.[38]

3. Amides and Dipeptides

A. General Structural Characteristics

Although formamide, the simplest of amides, has been shown by microwave studies to have a slightly pyramidal conformation about the nitrogen atom,[40] it is permissible to regard the bond system shown in the diagram as

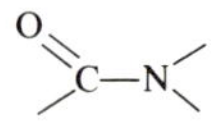

planar for most amides. This is particularly the case with respect to NMR observations, since on the NMR time scale, the pyramidal conformation will invert very rapidly.

Rotation about the C—N bond is slow, the barrier being of the order of 20 kcal mole^{-1}. This is attributed[41] to the partial double bond character derived from the structure

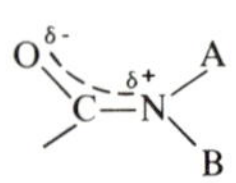

As a result, the environments of groups A and B are not averaged, and they can usually be observed separately by NMR even at temperatures well above

room temperature (Section 3C). These features of amide spectra are particularly significant in relation to proline units in polypeptide chains and also in relation to polymers of other N-substituted amino acids such as sarcosine and *N*-methylalanine (see Section 6).

B. *J* COUPLINGS

In most nitrogen-containing molecules, flipping of the ^{14}N quadrupole by the tumbling of the molecular framework, with resulting spin–lattice relaxation, is quite rapid. As a result, ^{14}N in effect decouples itself in varying degrees from the associated protons. The 1:1:1 triplet with about 50 Hz spacing, characteristic of protons directly bonded to ^{14}N in a symmetrical environment, is thus collapsed in varying degrees. This collapse is particularly marked in certain asymmetric molecules such as peptides, although it is not complete in all peptides.[42] Figure XIII-10 shows the 100 MHz spectra of five molecules

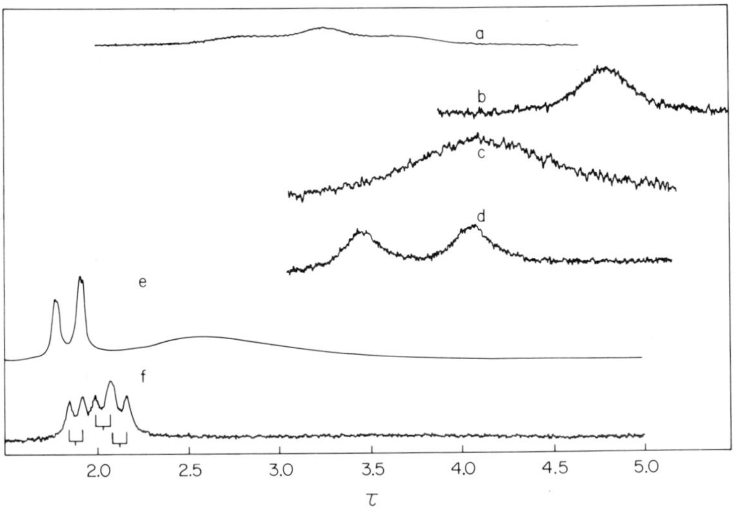

FIG. XIII-10. 100 MHz spectra illustrating NH resonances in various structures: (a) $^{+}NH_3CH_2CH_2CH_3$ (in trifluoroacetic acid); (b) *N*-methylacetamide (in $CDCl_3$, 25°); (c) acetamide (in d_6-DMSO, 81°); (d) acetamide (in d_6-DMSO, 25°); (e) formamide (neat, 25°); and (f) *N*-acetyl-L-alanyl-L-alanyl-L-alanyl methyl ester (in d_6-DMSO, 25°).

exhibiting varying degrees of collapse of CONH and NH_3^+ multiplets. The NH_2 resonance of formamide is the broad peak. In *N*-acetyl-L-alanyl-L-alanyl-L-alanine methyl ester, the CONH resonances indicate essentially complete decoupling from ^{14}N. They are doublets because of coupling to the α-protons.

Spin–lattice relaxation arising from the interaction of nuclear quadrupoles with fluctuating molecular electric field gradients may be expected to show a temperature dependence similar to that of spin–lattice relaxation arising from fluctuating magnetic dipole fields (Chapter I, Sections 5 and 10). For small molecules in mobile solvents, one would expect the collapsed NH multiplets to broaden with increasing temperature as the contribution of quadrupole relaxation to $1/T_1$ for ^{14}N becomes smaller. This is indeed observed. Thus, Roberts[43] found that the broad singlet resonance of the NH_2 group of formamide begins to assume a triplet character above 50°, while for acetamide and *N*-methylacetamide, whose lines are much narrower at room temperature (Fig. XIII-10), the transition is in the neighborhood of 200°. For *N*-methylformamide, the peak remained a singlet up to 250°, the highest temperature employed. Conversely, it is found that on the addition of glycerol to increase the viscosity of an aqueous formamide solution,[44] the broad NH_2 peak narrows and becomes a doublet. Likewise, the protons of the —$CONH_2$ group of polyacrylamide appear as a doublet.[44] Peptide NH resonances in proteins and polypeptides are thus actually narrowed rather than broadened by the restricted motion characteristic of large molecules. [In fact, as we shall see (Section 4), they have about the same line width as the α-proton resonances, both being determined primarily by dipolar broadening.]

Since ^{14}N couplings are difficult to observe directly, use has been made of ^{15}N (Table I-1), which has a spin of $\frac{1}{2}$ and thus no quadruple moment. The couplings have a constant proportionality to ^{14}N couplings, equal to the ratio of magnetogyric ratios, $\gamma_{^{15}N}/\gamma_{^{14}N}$, or about 1.41. Because the magnetic moment of ^{15}N is negative, ^{15}N couplings are opposite in sign to the corresponding ^{14}N couplings (see Chapter I, Section 16D).

In Table XIII-3,† ^{1}H–^{1}H and ^{1}H–^{15}N couplings are given for a number of simple amides and peptides. The *trans* ^{1}H–^{1}H coupling (J_{13} in entry 1) is comparable in magnitude to that of an olefin. The *cis* coupling (J_{12} in entries 1 and 3b) is considerably smaller than in an olefin. (These couplings have no direct relevance to polypeptide chains.) Both are positive. The geminal couplings (J_{23} in entries 1 and 2) are weak and positive. The four-bond couplings to methyl groups (J_{13} in entry 3; J_{12} and J_{13} in entry 4) are weak and negative, as in olefins. Thus, there is some similarity between amides and olefins as far as NMR parameters are concerned.

Nitrogen couplings to protons tend to fall off rapidly with the number of intervening bonds. The direct couplings are always negative to ^{15}N (positive to ^{14}N); there are indications that they increase with the fractional *s* character of the nitrogen bonding orbitals in a manner reminiscent of the well established

† Table XIII-3 will be found at the end of this chapter.

dependence of ^{13}C–H couplings. Two-bond nitrogen couplings are weaker to protons on tetrahedral carbon than to protons on trigonal carbon and are opposite in sign (compare $J_{^{15}N-H(1)}$ and $J_{^{15}N-H(3)}$ in entry 3b). It does not seem likely that either one- or two-bond nitrogen couplings will be significant in polypeptide studies. Three-bond NH couplings, however, might be capable of providing information concerning the conformation about the carbon–carbon bond, i.e., in the fragment shown. At present, there is little or no

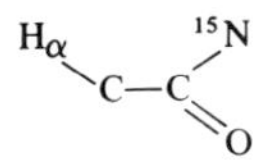

information on these couplings.

Measurements of ^{1}H–^{1}H couplings in peptides are relatively scarce. (There are no reported measurements of ^{15}N–^{1}H couplings.) Some values for *N*-acetyl amino acid esters are given in Table XIII-3, together with a few values for di- and tripeptides. The *N*-acetyl amino acid esters are useful models for amino acid residues in polypeptide and protein chains. The data given for cyclic dipeptides are a small selection from the extensive and valuable studies of Kopple and co-workers.[45,46]

C. Conformational Preferences at the C–N Bond in Amides; Chemical Shifts and Rotational Barriers

In N-monosubstituted amides, the equilibrium

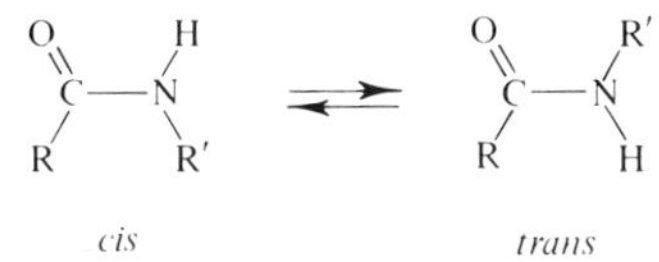

lies strongly to the right. Only in formamides (R = H) is there a fraction of *cis* conformer large enough to detect by NMR,[47] and here the preference for *trans* is still strong. Even a bulky *tert*-butyl group prefers, by a factor of 5 to 1, to be on the side of the carbonyl oxygen rather than on that of the sterically much less demanding formyl proton. For smaller groups, the preference is of course even greater. In peptides and polypeptides of α-amino acids it is customary and justifiable to consider only the *trans* conformation to be present.

Observed conformational preferences in unsymmetrically substituted *N*,*N*′-dialkylamides can be more rationally explained in terms of steric competition. In formamides, the bulkier substituent tends to be *cis* to the formyl proton.[47,48] In acetamides, the bulkier group tends to be *cis* to the carbonyl oxygen atom, but the preference is generally small.[47]

As we have seen (Section 3A), substituent groups on the amide nitrogen atom can be distinguished because amides are planar or nearly so and rotation about the C—N bond is slow. In formamide, the more shielded of the two NH protons is taken as that which is *cis* to the carbonyl oxygen atom. This can only be seen in the double resonance spectrum, i.e., with ^{14}N irradiation, not shown in Fig. XIII-10(e). The spectrum of acetamide at 25° [Fig. XIII-10(d)] is fundamentally similar except that here quadrupolar relaxation is more rapid and the NH_2 spectrum is a doublet without irradiation of the nitrogen. In *N,N'*-dialkylamides, the *N*-alkyl groups usually give distinguishable resonances, and it is from their relative magnitudes that conformational preferences are deduced. It is generally assumed that here also the more shielded resonance corresponds to the group which is *cis* to the carbonyl oxygen atom. For *N,N'*-dimethylformamide (Fig. XIII-11), this assignment has been demonstrated by nuclear Overhauser experiments, which show that the least shielded methyl group is spatially the closer to the formyl proton, and is therefore *cis* to it.[49] It is very probable that the differing shieldings of such groups arise primarily from the magnetic anisotropy of the C=O double bond, with a contribution from electric field effects arising from the strong electric dipole moment of the carbonyl group.[50–53] The relative contributions of these effects, and indeed the form of the shielding anisotropy, are at present in some doubt, for the most plausible form of the latter, according to the magnetic anisotropy calculations of Pople,[54] predicts relative shiedings opposite to those observed for *N,N'*-dimethylformamide.

Relative shieldings in sarcosine,[55] proline,[56,57] and *N*-methylalanine[58] derivatives have usually been assumed to follow the same pattern as *N,N'*-dimethylformamide. These questions are discussed in Section 6 of this chapter.

Another feature contributing to uncertainty in chemical shift assignments of *N*-substituents is the strong dependence on solvent observed for simple amides. Thus, for example, it is found[59] that in aromatic solvents such as benzene and anisole the relative chemical shifts of the N-methyl protons in *N,N'*-dimethylformamide and *N,N'*-dimethylacetamide are reversed from the assignment discussed above. This probably arises from a geometrically specific association of the aromatic ring with the amide group such that their planes are approximately parallel and both methyls are in the shielding region (Chapter I, Section 11) of the ring, but the one *trans* to the carbonyl oxygen atom is nearer the center of the ring. A similar and unexpected effect of chloroform on α-proton peaks in polyproline and proline oligomers is discussed in Section 6.

The rotational barrier about the C—N bond in amides has been the subject of numerous NMR measurements. The basis of these has been described in Chapter I, Section 15. Figure XIII-11 shows a series of spectra illustrating

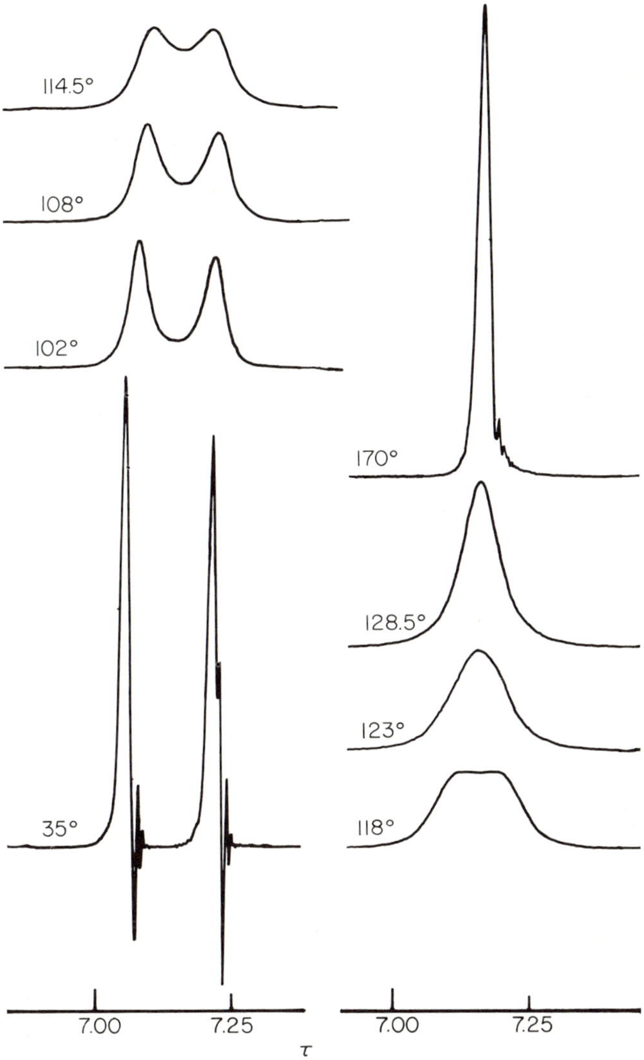

FIG. XIII-11. The 60 MHz methyl spectrum of *N,N'*-dimethylformamide (neat) as a function of temperature.[48a]

the collapse of the methyl doublet of *N,N'*-dimethylformamide as the temperature is raised and the rate of exchange of the methyl groups between *cis* and *trans* sites increases. Detailed analysis of these line shapes indicates a $\Delta H^{\neq}$ of about 20 kcal for this and most other *N,N'*-disubstituted amides.† This

† For a review of this and other aspects of amides see Robin *et al.*[60]

quantity is particularly significant in relation to the conformational isomerization of polyproline, polysarcosine, and other related polypeptides in which passage over the C—N rotational barrier is the rate-determining step. It also reminds us that in all polypeptide chains deviations from planarity of the peptide group are energetically very costly and can usually be excluded in realistic assessments of chain conformational states.

D. Chemical Shifts in Peptides

The α-proton in peptides is strongly deshielded by the CONH group on the N-terminal side; the CONH group on the C-terminal side has a weaker deshielding effect. Thus, when comparison is made between (a) and (b),

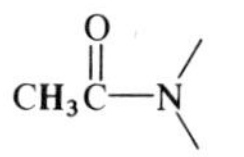

(a)

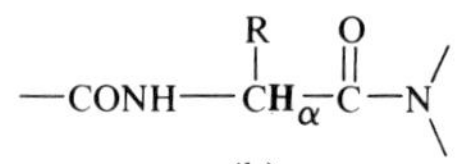

(b)

a deshielding of about 2 ppm is observed. The deshielding effect of the CONH is greater even than that of an ammonium group, so that replacement of (c) by (d) results in a deshielding of about 0.4 ± 0.1 ppm. The replacement of

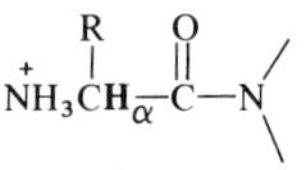

(c)

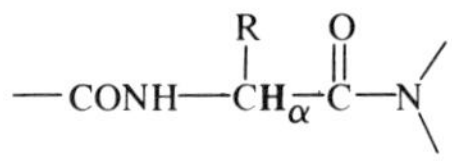

(d)

CO_2^- by CONH has about the same deshielding effect as protonation of the carboxylate group (Section 2), i.e., the influence of —CO_2H and —CONH are about equivalent. These effects are similar in both aqueous[16] and trifluoroacetic acid[20] solutions and probably in most other solvents as well. It seems probable that the CONH group deshields through both electric field and inductive effects.

The influence of dissociation of the carboxyl group in dipeptides and tripeptides has been carefully measured by Van Gorkom[61]; it can be detected through as many as eleven intervening bonds between the observed proton and the carboxyl proton, but is very small through five or more bonds.

As Nakamura and Jardetzky have pointed out,[16] the fairly high degree of regularity in the effects of alterations in state of ionization and backbone structure on peptide chemical shifts implies that sequence information (aside from end units) is not likely to be readily forthcoming from NMR data. It is possible, however, that by using high frequency instruments under conditions of rigid standardization of solvent, concentration, and temperature, sequences can be determined in polypeptides of perhaps four or five different residues. This possibility is encouraged by observations such as the following:

a. Diastereoisomeric peptides can be readily distinguished in many cases (in all cases in principle). Thus Bovey and Tiers[20] found that all of the chemical shifts differ for D-leucyl-L-tyrosine and L-leucyl-L-tyrosine, observed in trifluoroacetic acid solution; the difference was particularly obvious for the leucyl methyl groups, which were 0.43 ppm more shielded in the D-L diastereoisomer. It was suggested that the latter has a more compact structure in which the methyl groups are closer to the face of the aromatic ring of the tyrosine residue. Analogous observations have been made on alanyltyrosine diastereoisomers.[62] In nonaromatic peptides, the differences are generally smaller. Thus, Van Gorkom[61] could detect no spectral differences between diastereoisomeric alanyl dipeptides and tripeptides in aqueous solution. Halpern *et al.*,[63] however, were able to detect differences in the methyl chemical shifts of these and a number of diastereoisomeric alanyltyrosyl and alanylphenylalanyl dipeptides. Bystrov *et al.*,[64] in a study to be discussed more fully in Section 3F, observed differences in the chemical shifts of nearly all the protons in diastereoisomeric *N*-acyl alanylalanyl dipeptide methyl esters, and more significantly, differences in coupling constants as well.

b. Glycine α-proton resonances can differ markedly depending upon the position and nature of the neighboring residue.[14,61] The glycine methylene group is in principle heterosteric (or "diastereotopic") in any polypeptide containing an asymmetric amino acid residue, i.e., any residue other than glycine or sarcosine. N-terminal glycine units exhibit observable nonequivalence of the α-protons in a number of L-aminoacylglycines.[14] The nonequivalence is particularly marked when the asymmetric residue is aromatic and the dipeptide is in the zwitterion state. It tends to diminish toward zero in the cation and anion state, although still observable in the anion of L-phenylalanylglycine. C-terminal glycine residues show much smaller nonequivalence, usually undetectable at lower instrumental frequencies. These effects cannot yet be adequately explained, but are undoubtedly due principally to readjustments of conformational populations. One might expect the zwitterion to exist in a somewhat more folded and compact state in which shielding differences are accentuated. Unfortunately, the J coupling information necessary for such conclusions is not at present at hand, most of the reported data being concerned with chemical shifts.

c. Sheinblatt[65,66] has demonstrated alterations in chemical shift as a function of pH, many of which we have already discussed, by which the sequence in dipeptides and tripeptides can be determined. Sequence can also be determined[66] by measurement of rate of exchange of peptide hydrogen; this is discussed further in Section 3G.

E. ^{13}C Chemical Shifts in Peptides

Horsley and Sternlicht[22] have observed that in glycylglycine zwitterion, the N-terminal α-carbon is about 3 ppm more shielded than the C-terminal α-carbon, as might be expected. Upon protonation of the carboxyl group, the C-terminal α-carbon becomes more shielded; upon loss of a proton from the NH_3^+ group, the N-terminal α-carbon is deshielded (see Fig. XIII-12).

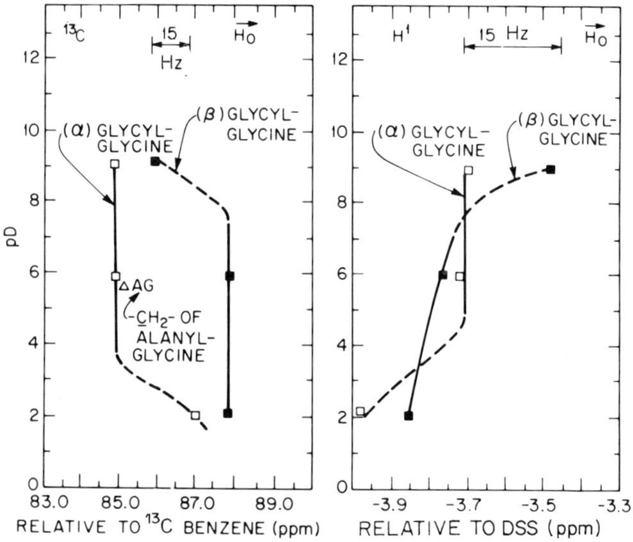

Fig. XIII-12. The ^{13}C (left-hand plot) and proton (right-hand plot) chemical shifts of the α- and β-methylene carbons (i.e., C-terminal and N-terminal units) of glycylglycine as a function of *p*D. Also included is a single ^{13}C point for L-alanylglycine (from Horsley and Sternlicht[22]).

As in the amino acids (Section 2C), these α-carbon shielding effects are opposite to those observed for the α-protons; both undoubtedly reflect the same influences, which at present are obscure.

F. *J* Couplings and Conformation Preferences at the N—C$_\alpha$ Bond

One expects (Chapter I, Section 14) that the magnitude of the vicinal coupling in H—N—C$_\alpha$—H, designated $J_{N\alpha}$, should provide a measure of the conformational preference about the N—C$_\alpha$ bond. By convention,[67] the main chain bonds in peptides are designated as shown. The chain is represented in the

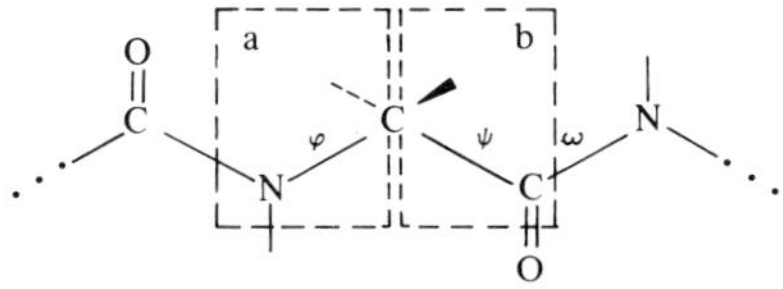

planar zigzag form, in which state the three dihedral angles φ, ψ, and ω are defined as zero. It is convenient to picture the two planar peptide fragments as rotating about the α-carbon as a common swivel point. Since the peptide group is planar or nearly so (p. 262), ω is either 0° or 180°; in the *trans* conformation shown, it is defined as zero, as just stated. If we imagine ourselves to be standing on the α-carbon atom and looking along the N—C_α bond, φ will increase from zero as we rotate the N-terminal peptide plane "a" clockwise about this bond. If we then look along the C_α—C bond from this same point, ψ will increase as we rotate the plane "b" about the C_α—C bond. (There is at present no known NMR measure or any other direct measure of ψ.)

It is customary to define the dihedral angle in any three-bond fragment, also commonly called φ, as zero when the protons are eclipsed. For the present discussion, we shall adopt this convention, and shall call this angle φ'. It is evident that the relationship between φ and φ' will be as shown by the upper and lower abscissae of Fig. XIII-13. Because there is obviously no single

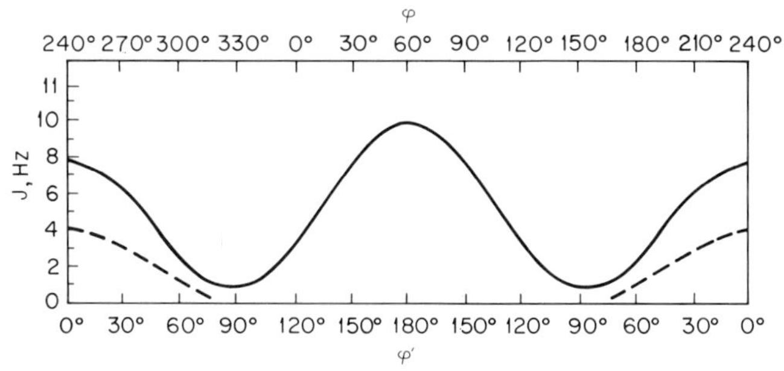

FIG. XIII-13. The dependence of $J_{N\alpha}$ on the H–N–C_α–H dihedral angle expressed as a function of φ (upper scale) and φ' (lower scale; see text) according to the relationship proposed by Bystrov *et al.*[64] The dotted curves are the relationships which appear to be necessary to describe observed values of $J_{N\alpha}$ in diketopiperazines (Kopple and co-workers[45,46]).

value of φ about which the chain conformations are symmetrical, φ is defined through a full 360°. Despite this however, φ' is defined through only 180° and it is assumed that any function of φ, in particular $J_{N\alpha}$, will be symmetrical about $\varphi' = 180°$. (It is unfortunate that still a third angle convention has now been promulgated[67a] in which the planar zigzag conformation corresponds to $\varphi = -180°$ and $\psi = -180°$. The energy map is unchanged, but the 0° values of φ and ψ now occur in the middle of the abscissa and ordinate, respectively. This convention has only the rather artificial advantage of defining similar helices of opposite chirality by equal angles of opposite sign, but still departs from the true measure of the dihedral angle given by φ'.)

The dependence of $J_{N\alpha}$ upon φ is not well established because the data upon which such a correlation might be based are very scanty and not entirely self-consistent. In order to make use of even this limited information, it is helpful first to decide on the most probable form of the potential energy function for rotation about the N—C_α bond. The most plausible assumption seems to be that it is analogous to that of the conformers of propene[68] and acetaldehyde,[69] as deduced from microwave studies. In these molecules, the preferred conformer is that in which one of the bonds to the tetrahedral carbon atom eclipses the vinyl or carbonyl group. For the fragment —CONH—$C_\alpha HR_1R_2$, the analogous conformations of minimum potential energy are as shown, assuming a threefold potential barrier. Reference to

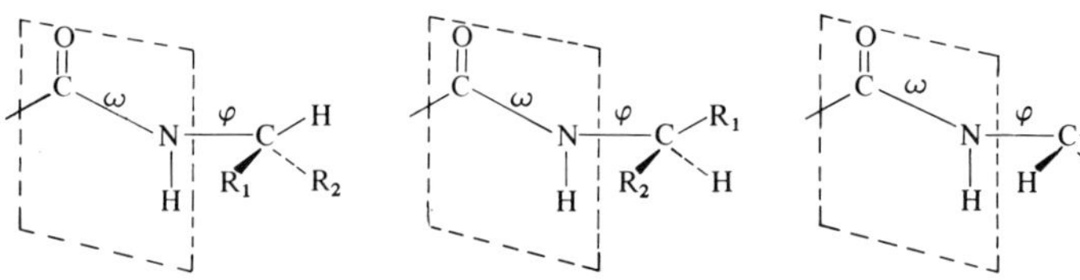

(a) (*trans*) $\left\{\begin{matrix}\varphi' = 180^\circ \\ \varphi = 60^\circ\end{matrix}\right\}$ (b) (*gauche* 1) $\left\{\begin{matrix}\varphi' = 60^\circ \\ \varphi = 300^\circ\end{matrix}\right\}$ (c) (*gauche* 2) $\left\{\begin{matrix}\varphi' = 60^\circ \\ \varphi = 180^\circ\end{matrix}\right\}$

propene and other analogous compounds suggests that the barrier heights are probably of the order of 1–2 kcal.[70, 71] There may also be a contribution of sixfold periodicity, as in nitromethane,[72] but this can be neglected because such potential barriers are usually very small.†

The dependence of $J_{N\alpha}$ upon φ' proposed by Bystrov *et al.*[64] is shown in Fig. XIII-13. This is expressed as

$$J_{N\alpha} = 8.9\cos^2\varphi' - 0.9\cos\varphi' + 0.9\sin^2\varphi' \tag{XIII-6}$$

The vinyl–allyl proton coupling in fragment (a), as deduced by Garbisch,[73] is closely analogous. There is likewise evidence‡ that a similar dependence

† It may be noted that Bystrov *et al.*,[64] in a carefully thought-out discussion based in part on considerations of the conformations of internally hydrogen-bonded dipeptides (to be discussed), have concluded that the most probable threefold minima are those in which the bonds to the α-carbon eclipse the N—H bond. Their arguments are plausible but not conclusive.

‡ See, for example, Garbisch[74] and Hogeveen *et al.*[74a]

occurs in oxygen-containing structures (b). All are similar to that proposed

(a) (b)

by Karplus for protons on neighboring tetrahedral carbon atoms[75] and amply confirmed, at least in its general form, by a large number of observations. Bystrov *et al.* have supported Eq. (XIII-6) by careful arguments concerning the conformation of *N*-acyl alanine dipeptide esters such as entries 17 and 19 in Table XIII-3. Their conformation is shown. Hydrogen bonding between the

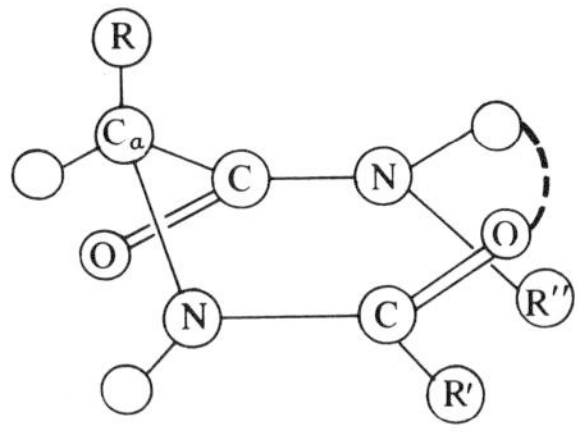

C-terminal peptide NH and the N-terminal acyl carbonyl group appeared to be indicated by infrared measurements of the NH stretching frequency[76]† in chloroform and carbon tetrachloride solution. Models indicate such a hydrogen bond to be highly nonlinear. The conformation shown corresponds to $\varphi \simeq 240°$ ($\varphi' \simeq 0°$) and $\psi \simeq 120°$, a highly unfavorable position on the φ–ψ potential energy plot, but consistent with the observed value of $J_{N\alpha}$ ($H_{(\alpha_2)}$–$NH_{(2)}$) in entries 17 and 19 of Table XIII-3—indeed the principal basis for the value chosen for $J_{N\alpha}$ at $\varphi' = 0$ in Fig. XIII-13. In H_2O, where the hydrogen bond might be expected to be broken, this $J_{N\alpha}$ decreases markedly (entry 19) to a value not inconsistent with that expected for an α-helix-like conformation ($\varphi \simeq \varphi' \simeq 120°$; $\psi \simeq 130°$; see Section 4), corresponding to an opening up of the hydrogen-bonded ring. For other di- and tripeptides however, (entries 30–31) values of $J_{N\alpha}$ similar to those observed by Bystrov *et al.*[64] in the presumably internally hydrogen-bonded structure are observed in both chloroform and trifluoroacetic acid, an effective hydrogen bond breaker. There is no marked conformational change at the φ bonds as judged by this criterion. There remains the possibility that in Bystrov's compound, φ' is actually on the other side of the $J_{N\alpha}$ minimum, i.e., at 150–180°, equally consistent with the observed $J_{N\alpha}$. This conformation is energetically more reasonable, but is not consistent with the proposed hydrogen bond.

† Many measurements of such intramolecular —CO···NH— hydrogen bonds have been reported. See Mizushima,[76] Mizushima *et al.*,[76a] Portnova *et al.*,[76b] and extensive further references quoted in Bystrov *et al.*[64]

Despite these seeming inconsistencies, support for the general form of Fig. XIII-13 is furnished by the data of LaPlanche and Rogers[47] on N-substituted formamides. In entries 3, 7, and 8 in Table XIII-3 we observe a marked increase in J_{23}, i.e., in $J_{N\alpha}$, as the bulkiness of the alkyl group increases. Closely analogous behavior has been observed by Bothner-By *et al.*[77] for a series of alkylethylenes. The most likely explanation is in terms of something similar to Bystrov's suggested $J_{N\alpha}$–φ' relationship and Garbisch's[73] function for vinyl–allyl couplings, i.e., that in *N*-isopropylformamide the *trans* conformation shown, in which φ' is 180°, tends to be strongly preferred. The

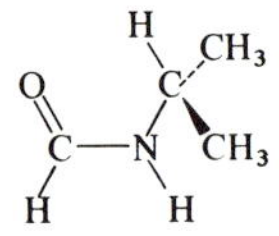

lower value for the *N*-ethyl compound corresponds to increasing populations of the *gauche* conformers (Section 2D) in which $J_{N\alpha} \simeq 2$–3 Hz. For *N*-methylformamide we have

$$J_{N\alpha} = \tfrac{1}{3}(2J_{N\alpha\text{-gauche}} + J_{N\alpha\text{-trans}}) = 5.0 \text{ Hz}$$

Bystrov's curve was in part based on $J_{N\alpha}$ for *N*-methylformamide, and so is consistent with it.

It might be reasonably hoped that cyclic dipeptides, i.e., diketopiperazines, would serve as useful model compounds for polypeptide chains, and indeed in some ways they do. The number of conformations attainable by the six-membered ring must be very limited, and so arguments relating conformation to NMR and other spectral properties are somewhat less indefinite than those involving open-chain peptides. Kopple and co-workers[45,46] have made careful NMR measurements of a large number of diketopiperazines, including diketopiperazine itself. Some typical data are shown in Table XIII-3, entries 32–35. It can be seen that $J_{N\alpha}$ is always much smaller than in open-chain peptides, indeed too small to be consistent with Bystrov's curve in Fig. XIII-13. In diketopiperazine (entry 32), we may suppose that the ring is either planar, as it is known to be in the crystal[78,79] or rapidly inverting between boat conformers. If planar, $\varphi' = 60°$ for each α-proton, and $J_{N\alpha}$ should be about

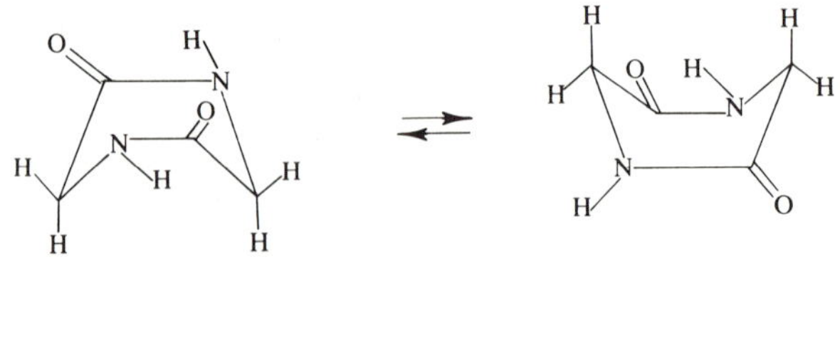

2.5 Hz; for inverting boat forms, each $J_{N\alpha}$ will be averaged over values of φ' of about 0–20° and about 110–90°, and $J_{N\alpha}$ should be 3.5–4.0 Hz. The unobservably small value in trifluoroacetic acid is not consistent with either of these alternatives. Similar difficulties are found with the substituted rings. In the L-tyrosylglycyldiketopiperazine (entry 33), a boat conformation with axial substituents is strongly indicated by marked differential shielding of the glycyl protons[46, 80] due to the phenyl group's being folded over the ring, and by different values of $J_{N\alpha}$ for each glycyl proton. The 3.0 Hz coupling of the

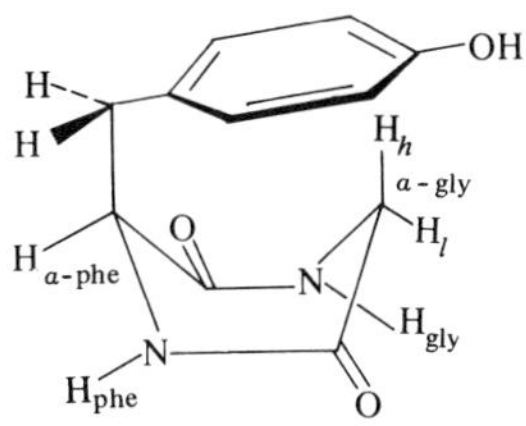

low-field glycine proton (H_l) to NH_{gly} is too small for $\varphi' \simeq$ 0–20°; the approximately zero coupling of H_h is reasonable. But in trifluoroacetic acid, *both* couplings are about zero, which again is inexplicable in terms of Bystrov's curve. Kopple and Ohnishi[46] have suggested a twist boat conformation which relieves the inconsistencies somewhat but requires a degree of nonplanarity of the peptide group which seems somewhat improbable. At present, the most likely explanation seems to be that the dependence of $J_{N\alpha}$ on φ (or φ') is affected also by the value of ω. In diketopiperazines, of course, ω is necessarily not far from 180° rather than 0° as in open-chain peptides. A $J_{N\alpha}$–φ' relationship for such *cis* peptide residues is suggested in Fig. XIII-13. The coefficient in Eq. (XIII-6) is reduced to about half for the region $0 < \varphi < 90°$; in diketopiperazines φ' cannot attain values much beyond about 90°. It thus appears that cyclic dipeptides are probably not valid models for polypeptide chains at least so far as proton–proton couplings are concerned.

G. Protolysis and Protonation of Amino Acids, Peptides, and Related Compounds

An important characteristic of amino acids and peptides, which profoundly affects the appearance of their NMR spectra and can provide valuable structural information, is the rate at which their functional groups exchange protons with the solvent and possibly also with other functional groups. We shall discuss here the behavior of small molecules in this respect as a background for later discussions of proton exchange in polypeptides (Section 4) and proteins (Chapter XIV).

Proton exchange may be measured by NMR in three ways:

(a) By use of D_2O or other solvents containing labile deuterons, exchange with substrate protons may be measured analytically by observing the disappearance of the substrate proton peaks. This method does not employ the rate-dependent characteristics inherent in NMR spectra (Chapter I, Section 15), but is very useful for slower exchange rates such as those of peptide protons.

(b) By observation of the collapse of spin multiplets.

(c) By observation of the broadening and possible coalescence of substrate and solvent peaks.

Sheinblatt has measured the rate of exchange of amino protons in aqueous solutions of sarcosine as cation,[81] zwitterion,[82] and methyl ester,[82] and of glycine and glycine methyl ester.[83] The procedure is similar for all of these compounds and is based on the study of ammonium ion protolysis by Grunwald *et al.*[84] and Loewenstein and Meiboom.[85] The lifetimes of protons in the ammonium group, τ, and in water, τ_{H_2O}, are estimated by methods (b) and (c), respectively. In method (b), the α-proton resonance, which is a quartet from coupling to the NH_3^+ protons (Fig. XIII-14), is observed. The form of this

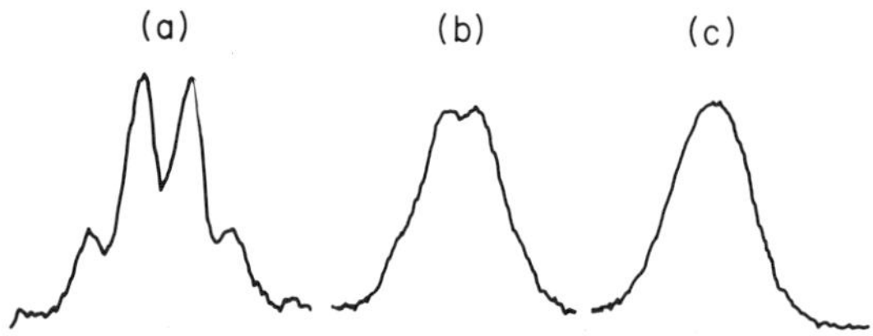

FIG. XIII-14. 60 MHz spectra of the methylene spectrum of glycine in H_2O at varying $[H^+]$ (HCl): (a) $[H^+] = 4.3$ *M*, (b) $[H^+] = 3.0$ *M*, and (c) $[H^+] = 2.5$ *M* (Sheinblatt and Gutowsky[83]).

multiplet, recorded as a function of the solution variables, principally the *p*H, is compared to theoretical line shapes (see Grunwald *et al.*,[84] Fig. 2) calculated by an adaptation of the equations derived by Gutowsky *et al.*[86] under the assumption that the ammonium protons are exchanged one at a time.

The principal result is that the proton lifetimes in the NH_3^+ group become longer as the acidity of the solution is increased (Fig. XIII-14). In strongly acid solutions it may be of the order of minutes; the rate of exchange is inversely proportional to $[H^+]$. At lower acidities it tends to become independent of $[H^+]$, particularly at low concentrations of the amino acid. In aqueous solutions the process is thus evidently not catalyzed. The dominant reactions appear to be as follows:

$$R^+NH_2H^* \cdots OH_2 \underset{k_{-1}}{\overset{k_1}{\rightleftarrows}} RNH_2 \cdots H^*OH + H^+ \qquad \text{(XIII-7)}$$

$$RNH_2 \cdots H^*OH \xrightarrow{k_2} RNH_2 + H^*OH \qquad \text{(XIII-8)}$$

Equation (XIII-7) represents the reversible migration of the ammonium proton (the asterisk denoting the particular one whose fate is being followed) to a hydrogen-bonded water molecule in the hydration sphere of the cation, accompanied by dissociation. Exchange with the solvent cannot occur by this process as such, since the same proton is returned to NH_2 by the reverse step. Exchange occurs by reaction (XIII-8), in which the water molecule containing H* is lost from the hydration sphere and diffuses away. Under the assumption of a steady state in $RNH_2 \cdots H_2O$, it is found that

$$\frac{1}{\tau} = \frac{k_1 k_2}{k_2 + k_{-1}[H^+]} \qquad \text{(XIII-9a)}$$

At very high acidities, $k_2 \ll k_{-1}[H^+]$ and so

$$\frac{1}{\tau} = \frac{k_1 k_2}{k_{-1}[H^+]} = K_{a_1} \frac{k_2}{[H^+]} \qquad \text{(XIII-9b)}$$

where K_{a_1} is the dissociation constant of the ammonium group, about 10^{-8}. The results are summarized in Table XIII-4. It is clear that the slow step in

TABLE XIII-4

PROTOLYSIS RATE CONSTANTS FOR AMINO GROUPS OF GLYCINE, SARCOSINE, AND DERIVATIVES[a]

Compound	k_1, sec^{-1}	k_{-1}, sec^{-1}	k_2, sec^{-1}
Glycine cation	370	2.0×10^{10}	0.85×10^{10}
Glycine zwitterion	~5	—	—
Glycine methyl ester	470	2.5×10^{10}	1.52×10^{10}
Sarcosine cation	110	4.4×10^{10}	1.15×10^{10}
Sarcosine zwitterion	4.5	—	—
Sarcosine methyl ester	120	4.8×10^{10}	2.9×10^{10}

[a] See Sheinblatt[81, 82] and Scheinblatt and Gutowsky.[83]

exchange is the transfer of the proton from NH_3^+ to the hydration shell. The subsequent ionization is assumed to be very fast. High acidity retards exchange by forcing the equilibrium in Eq. (XIII-7) to the left.

Equation (XIII-9a) implies that at higher pH, the exchange rate would become independent of $[H^+]$, and that $1/\tau$ would become equal to k_1. The mechanism as a whole further implies that there is no exchange of amino group protons in which water does not participate, and that therefore τ_{H_2O} can be calculated directly from τ and the $[H_2O]/[NH_3^+]$ ratio. The second of these predictions is in fact true at high acidities. But at lower acidities, both predictions break down to some extent because of increasing participation by bimolecular reactions, the most rapid of which is

$$RNH_2\overset{+}{H}^* + RNH_2 \rightleftarrows RNH_2 + RNH_2\overset{+}{H}^* \qquad \text{(XIII-10)}$$

Such reactions, excluded at very low pH because of the negligible concentration of RNH_2, have rate constants of the order 10^8 sec^{-1} and can thus compete effectively as dissociation of the ammonium group becomes appreciable and as the substrate concentration is increased. In native proteins, it seems possible that intramolecular exchange of lysine and ornithine ammonium protons might rival exchange with the solvent.

The rate constants for exchange of carboxyl protons with aqueous solvents have not been determined for amino acids but the results of Luz and Meiboom[87] for aqueous acetic acid can furnish a good estimate. The carboxyl proton would be expected to exhibit at most a very weak coupling to the α-protons. Furthermore, the exchange evidently does not proceed via a slow step analogous to Eq. (XIII-7), and is always so fast, regardless of pH, that the water and carboxyl protons appear as a single peak not detectably broadened over that of water alone. Thus, neither method (b) nor (c) (p. 276) is directly applicable. Luz and Meiboom[87] were able to make NMR measurements of the exchange rate by using H_2O enriched in ^{17}O by about 10–20 fold over the normal abundance of 0.037%, and by working near pH 7 where the rate of direct proton exchange between water molecules, not mediated by carboxyl groups, is minimal. The ^{17}O provides the spin–spin splitting of the proton resonance necessary for exchange broadening of the resonance line, which was measured by the spin-echo technique.[88] (The original references should be consulted for details.) The pseudo-first-order rate constant for proton exchange between acetic acid and water was found to be about $1/N \times 10^8$ sec^{-1}, where N is the number of water molecules associated with the event. This cannot be identified with the rate of dissociation of the carboxyl group

$$RCO_2H + H_2O \rightarrow RCO_2^- + H_3O^+ \qquad \text{(XIII-11)}$$

which is known to be two orders of magnitude slower.[88] It most probably corresponds to the *second* step in the following scheme, i.e., to cyclic proton transfer within a hydrogen-bonded complex (here, $N = 2$). No ionized species is involved.

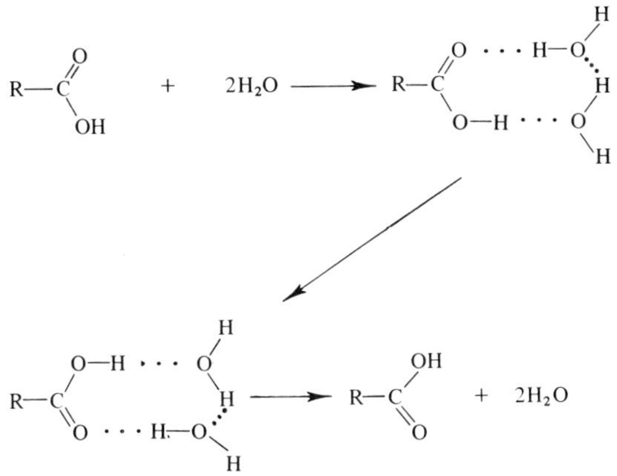

Other rapidly exchanging protons in aqueous media are those of the hydroxyl groups of tyrosine, serine, and threonine. The rate of exchange of the hydroxyl protons of phenol, a model for tyrosine, was measured by Luz and Meiboom[89] by the same procedure used for acetic acid.[87] The principal exchange process is rapid and is second-order in phenol; it involves proton interchange between phenol and phenolate ion via hydrogen-bonded water molecules as intermediates. (There may also be exchange processes between phenol molecules that do not involve water molecules, but these would not be detected by the method.) Exchange can be expected to be much retarded in acidic solutions.

There are no measurements of the rate of exchange of alcoholic hydroxyl protons with water in predominantly aqueous systems. It is undoubtedly quite rapid but probably slower than that of phenolic protons. It is very likely that it proceeds within hydrogen-bonded alcohol–H_2O species, as with carboxylic acids. It is too fast for straightforward multiplet collapse measurement, yet too slow for the Luz–Meiboom ^{17}O technique.

There is little information on the rates of exchange of sulfhydryl protons (as in cysteine) with protonic solvents. Both acid and base catalysis of sulfhydryl proton interchange in neat benzyl mercaptan have been reported,[90] but in trifluoroacetic acid the sulfhydryl group of cysteine appears as a well resolved triplet ($J = 9$ Hz) with no indication of broadening.[20] Exchange with aqueous solvents is reported to be rapid.[2]

The guanidinium protons of arginine appear to undergo an exchange in aqueous media which is faster than that of peptide protons but slower than ammonium protons.[91] In proteins, it is believed that their exchange may be very slow (see Chapter XIV, Section 2).[91]

By far the most intensively studied exchange process in proteins is that of the peptide protons, the exchange of which is generally much slower than the processes we have so far discussed.† It is also very highly dependent on structure, conformation, and solution variables. The fundamental mechanism of this process has been elucidated by NMR studies of *N*-methylacetamide.[93,94] From the collapse of the *N*-methyl doublet it is found that CONH exchange with the aqueous solvent is catalyzed by both acid (HCl) and base, in contrast to NH^+ proton exchange. The dominant reactions in basic solution are

$$CH_3CONHCH_3 + OH^- \xrightarrow{k_1} CH_3CON^-CH_3 + H_2O \qquad (XIII\text{-}12a)$$

$$CH_3CON^-CH_3 + H_2O \xrightarrow{k_{-1}} CH_3CONHCH_3 + OH^- \qquad (XIII\text{-}12b)$$

In acid solution:

$$CH_3CONHCH_3 + H_3O^+ \xrightarrow{k_2} CH_3CON^+H_2CH_3 + H_2O \qquad (XIII\text{-}13a)$$

$$CH_3CON^+H_2CH_3 + H_2O \xrightarrow{k_{-2}} CH_3CONHCH_3 + H_3O^+ \qquad (XIII\text{-}13b)$$

In each case, the first reaction is relatively slow and is not diffusion controlled. The reverse reaction, i.e., the attack of water on the ionized amide, is very rapid, corresponding to the fact that the amide is very weak both as an acid and as a base. Exchange results from the low probability that the same water molecule will participate in both steps. Direct exchange between amide molecules and not involving water does not occur to an appreciable extent, for the rates determined from NH doublet collapse and from the water signal broadening are the same within experimental error.[93] There is also no indication of an uncatalyzed exchange.

$$CH_3CONH^*CH_3 + HOH \underset{k_{-3}}{\overset{k_3}{\rightleftharpoons}} CH_3CONHCH_3 + H^*OH \qquad (XIII\text{-}14)$$

The rate constants are summarized in Table XIII-5. Because the base catalyzed process is faster than the acid catalyzed, the exchange rate has a minimum on the acid side of neutrality, at about *p*H 5.

The protolysis processes in glycylglycine[95] and glycylglycylglycine[96] are similar in mechanism to that of *N*-methylacetamide but interestingly different in their rates. In glycylglycine zwitterion (Table XIII-5, entry 2), k_2 is two orders of magnitude greater than in *N*-methylacetamide (entry 1), but in glycylglycine methyl ester (entry 3) it is even greater. It is evident that while the carboxylate group retards alkaline protolysis, the charged ammonium group, by increasing the acid strength of the peptide group, accelerates the process by an even greater factor. These effects are further illustrated by the greater rate of protolysis of the N-terminal CONH than of the C-terminal

† Reviews of protein hydrogen exchange, with some reference also to small model molecules, have been given by Hvidt and Nielsen,[92] and Englander.[92a]

TABLE XIII-5

PROTOLYSIS RATE CONSTANTS FOR CONH GROUPS OF *N*-METHYLACETAMIDE,[a] GLYCYLGLYCINE[b] AND GLYCYLGLYCYLGLYCINE[c]

	Compound	k_1(liter-mole^{-1}-sec^{-1})	k_2(liter-mole^{-1}-sec^{-1})	k_3(sec^{-1})
1	$CH_3COHNCH_3$	5.2×10^6	3.8×10^2	0.0
2	$NH_3^+CH_2CONHCH_2CO_2^-$	7.8×10^8	0.0	3.5
3	$NH_3^+CH_2CONHCH_2CO_2CH_3$	1.2×10^9	0.0	—
4	$NH_3^+CH_2CONHCH_2CONHCH_2CO_2^-$	8.4×10^7	0.0	~2
5	$NH_2CH_2CONHCH_2CONHCH_2CO_2^-$	0.8×10^7	0.0	~45
6	$NH_3^+CH_2CONHCH_2CONHCH_2CO_2^-$ / $NH_2CH_2CONHCH_2CONHCH_2CO_2^-$	6.7×10^9	0.0	5

[a] See Berger *et al.*[93]
[b] See Sheinblatt.[95]
[c] See Sheinblatt.[96] Here, the NH concerned is printed boldface.

CONH in glycylglycylglycine zwitterion (compare entries 4 and 6). Comparison of 4 and 5 with 6 reveals the rather puzzling fact that charging of the amino group markedly accelerates the exchange process in the distant CONH but does not affect the nearer one. The identification of groups seems clear;[65] not so the explanation.

There are two other marked differences between *N*-methyl amides and the glycyl peptides. The first is that, being a weaker base, the peptide exhibits no measurable acid-catalyzed exchange (column 3, Table XIII-5). The second is the observation (last column) of readily measurable uncatalyzed protolysis.

Let us consider what these rates may mean in terms of our ability to observe the CONH resonance itself—a most important feature of polypeptide and protein spectra. In aqueous solutions, the CONH and H_2O peaks are usually separated by about 3τ, a $\Delta\nu$ of 300 Hz at 100 MHz. The peaks will merge if $1/\tau \geqslant \pi\,\Delta\nu$, i.e., if $1/\tau \geqslant$ about 1000 sec^{-1} (Chapter I, Section 13), supposing them to be equal in intensity. (The NH peak will usually be much weaker than the H_2O peak and so will broaden and disappear correspondingly sooner.) If $k_1 = 10^8$ liter-mole^{-1}-sec^{-1}, then according to Eq. (XIII-12a), in neutral solution ($[OH^-] = 10^{-7}$ mole-liter^{-1}), we have

$$\frac{1}{\tau} \simeq 10^8 \times 10^{-7} \simeq 10$$

There may be a perceptible broadening under these conditions. If k_1 is in fact greater, as Table XIII-5 indicates it may well be, or if the pH much exceeds neutrality, then we can expect that exposed CONH groups will not give distinguishable resonances. Since acid catalysis is negligible, it may be advantageous to employ a *p*H less than neutral, provided this is not incompatible with the integrity and function of the molecule for some other reason. We also see from these rough estimates that if D_2O is employed, free CONH proton signals can be expected to disappear within 0.1 sec or less. As we shall see (Section 5 of this chapter and Chapter XIV), hydrogen bonding and other circumstances may lengthen this interval by many orders of magnitude.

Protolysis and protonation of amino acids, peptides, and amides in solvents other than water has received much less study. Exchange processes in methanol are usually quite rapid and probably generally comparable to those in aqueous systems. Peptide protons exchange sufficiently slowly in such solvents as methanol, trifluoroethanol, hexafluoroisopropanol $(CF_3)_2CHOH$, and hexafluoroacetone sesquihydrate $(CF_3)_2CO \cdot 1.5H_2O$, that peptide resonances are generally well resolved, but fast enough so that if the deutero-oxy alcohol ROD is employed, their peaks disappear within a few seconds unless specially bonded. (These questions will be discussed further in Sections 4 and 5 of this chapter.) Dimethyl sulfoxide has no labile protons itself and tends to retard the intermolecular (and intramolecular) exchange of solute protons by forming strong hydrogen bonds with them. In trifluoroacetic acid (TFA), ammonium group, peptide, and sulfhydryl (p. 279) protons exchange slowly and give well-resolved resonances, while hydroxyl and carboxyl protons exchange rapidly and appear in the trifluoroacetic acid carboxyl resonance.[20,21,97] However, it has been found that in TFA–chloroform solutions, the *N*-methyl doublet of *N*-methylacetamide collapses to a singlet when the mole ratio of TFA and amide is approximately 1:1. It has been suggested[97] that NH proton exchange with the solvent is particularly rapid in this range because of the formation of a hydrogen-bonded complex in which the oxygen atom of the amide is protonated.

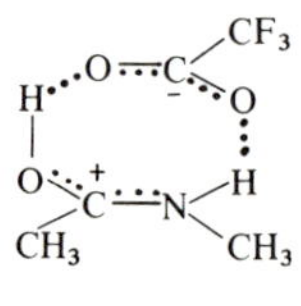

The extent to which amides and peptides are protonated by strong acids has been a much vexed question. The position of the N-protonation equilibrium represented by Eqs. (XIII-13a) and (XIII-13b) is undoubtedly far to the left. However, it is well established that under some conditions O-protonation can occur, as represented above. This suggestion was first put forward by Hantzsch,[98] and has since been strongly supported by other studies, principally NMR.[93,99–107] (For a review, see Jones and Katritzky.[99]) Cryoscopic[101] and ultraviolet spectroscopic[98,105,107] evidence show nearly complete protonation of a variety of amides in 100% sulfuric acid. NMR evidence proves that this cannot occur on nitrogen (or at most only to a very small extent). For example, the spectrum of *N*-methylacetamide in sulfuric acid exhibits the normal methyl group doublet. If the nitrogen atom were extensively protonated, as in Eq. (XIII-15a), this resonance would be either a triplet or, if exchange with the solvent were rapid, a singlet. A methyl doublet is also observed for *N*-

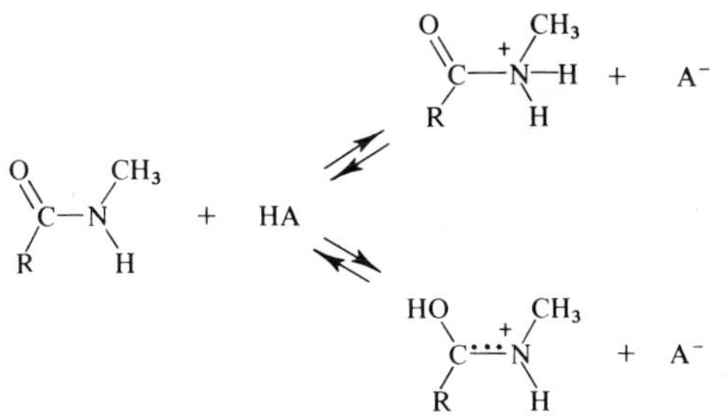

(XIII-15a)

(XIII-15b)

methylacetamide in 72% $HClO_4$ solutions in water–dioxane. In none of these NMR studies however, was the supposed OH proton actually observed, this of course being assumed to be the result of rapid exchange with the solvent. Direct evidence for the O-protonated form has been provided by Gillespie and Birchall,[106] who studied a number of amides in fluorosulfonic acid FSO_3H, which is even stronger than 100% sulfuric acid and has the additional advantage of being liquid to $-89°$. Figure XIII-15 shows the spectrum of acetamide in fluorosulfonic acid at 25, -80, and $-92°$. At room temperature the NH_2 group appears as a broad peak near 2τ, and the solvent as a large peak near -2τ. At $-80°$, the NH_2 resonance is resolved into a doublet because of slower molecular tumbling and more rapid quadrupolar relaxation of ^{14}N (p. 263); there is also observable at this temperature a broad peak near to but *upfield* from the solvent peak. At $-92°$, this peak, corresponding to one proton, is clearly resolved. This can only be the proton on the carbonyl oxygen. At higher temperatures, protonation at this position causes the solvent resonance to move *upfield* from its normal position, due to the exchange process.

In this connection, the behavior of the solvent proton in solutions of *N*-methylacetamide in TFA–$CDCl_3$, discussed above, is significant. As shown in Fig. XIII-16, it exhibits a sharp minimum at a concentration of TFA corresponding to the 1:1 mole ratio of TFA:*N*-methylacetamide at which proton exchange is a maximum. Since the solvent proton shift is in the opposite direction to that which must necessarily correspond to the O-protonated form observed by Gillespie and Birchall,[106] Stewart *et al.*[97] proposed that it corresponded to the O-protonated ion pair shown on p. 282, while at higher acid concentration only the simple O-protonated form shown in Eq. (XIII-15b) was present. However, there seems to remain the possibility, particularly

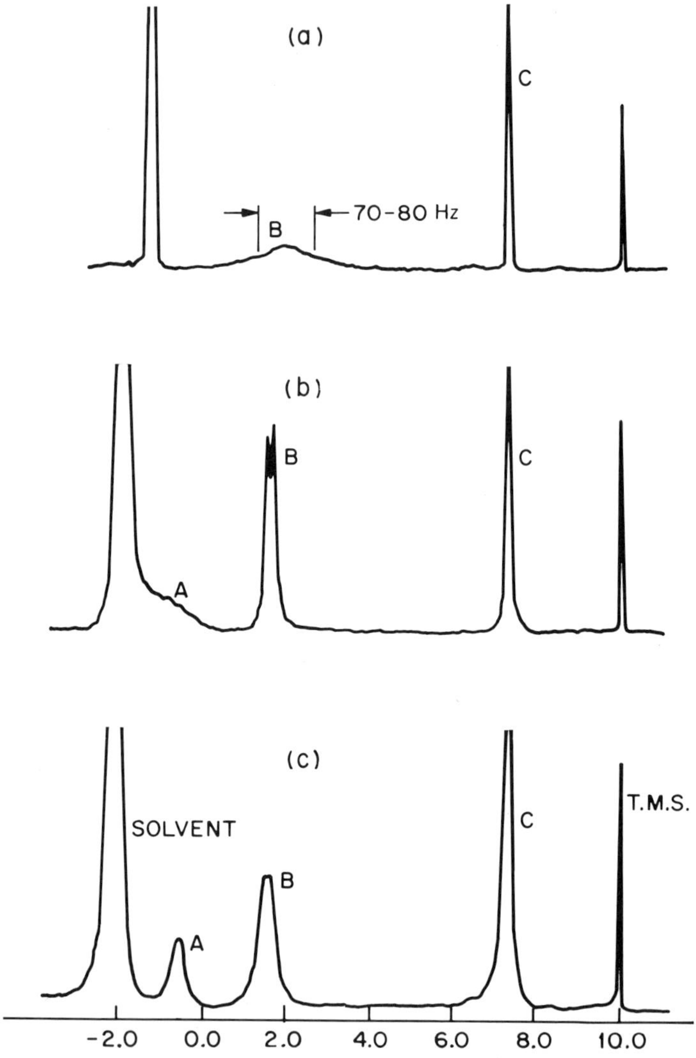

FIG. XIII-15. 60 MHz spectra of acetamide in fluorosulfonic acid at (a) 25°, (b) −80°, and (c) −92°. Peak A is that of the proton on the carbonyl oxygen; B is the NH_2 resonance, and C, the methyl resonance (from Gillespie and Birchall[106]).

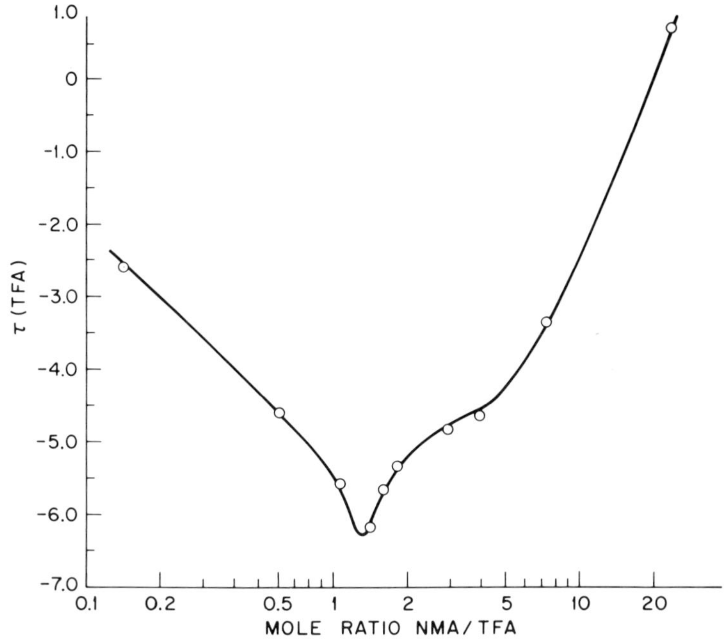

FIG. XIII-16. The dependence of the chemical shift of the TFA proton on the *N*-methylacetamide/TFA mole ratio (from Stewart *et al.*[97]).

in view of other evidence that TFA does not protonate amides extensively (see below), that what is observed is the hydrogen-bonded species.

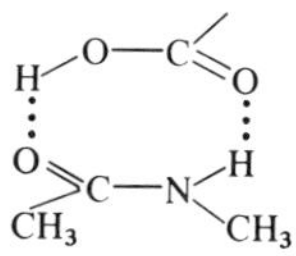

It is somewhat puzzling that the N-protonated form, which at moderate acidities mediates protolysis and may also mediate rotation about the C—N bond[93,97] (although this latter point is somewhat moot at present[44]), should play no part in highly acidic environments. There are features of these equilibria which still remain to be explained.

Hanlon *et al.*,[108] Klotz *et al.*,[109] and Hanlon[110] have presented a variety of evidence—particularly from near infrared spectroscopy—that model amides such as *N*-methylacetamide can be protonated by trifluoroacetic acid and presumably by other acids such as dichloroacetic acid which share the ability to cause the helix-coil transition in polypeptide chains (Section 4). It is possible that some protonation may occur under these conditions, but peptide groups, being weaker bases (as witnessed by their failure to undergo observable acid-catalyzed protolysis) will be protonated to a substantially lesser degree. We shall consider the question of the protonation of polypeptide chains further in Section 4 of this chapter. We may note here (see Bovey[107] for details) that in the model amide D-oxolupanine the n–π^* transition in the circular dichroism

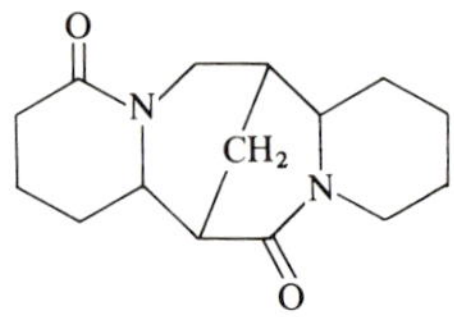

D-oxolupanine

spectrum remains unaffected in TFA solution. When protonation does occur, as in 100% sulfuric acid, the n–π^* disappears as expected, the nonbonding electrons of the carbonyl group being now bonded; it reappears on dilution with water.

4. Poly-α-Amino Acids

The first NMR observations of a polypeptide were those of Bovey *et al.*[111] on poly-γ-benzyl-L-glutamate. After a considerable interval, other authors reported on this and related systems[112,113] and more recently the flow of published studies has swelled to near torrential proportions.[107,114–131] Principal attention has centered on the conformations of the chains, in particular on the α-helix–random-coil transition. This phenomenon has of course been intensively studied both experimentally and theoretically (for reviews see Bovey,[107] Schellman and Schellman,[132] Urnes and Doty,[133] and Yang[134]), the most powerful means of investigation heretofore being optical rotation and circular dichroism measurements. NMR has now been found capable of giving qualitative and quantitative information which in general supports that from optical studies, and also goes beyond this.

In Fig. XIII-17 is shown the now familiar structure of the α-helix.[135] The structure shown is that of the right-handed α-helical conformation of poly-L-alanine. As it occurs in nature and in many synthetic polypeptides, the α-helix has approximately 18-fold symmetry. There are 3.6 residues per turn and a residue translation of 1.49 Å, i.e., as we pass from a point in one residue to the corresponding point in the next, we move 1.49 Å along the helical axis. In Fig. XIII-18, the energy contours of poly-L-alanine are shown as functions of φ and ψ (p. 271), as calculated by Brant.[136] (For further information on such calculations, see the papers in Ramachandran.[137]) These calculations include, in addition to torsional and van der Waals potentials and the interactions of peptide dipoles, the stabilizing effect of the >C═O···H—N< hydrogen bonds (not shown in Fig. XIII-17) between each residue and the one four units removed along the chain. The minimum at $\varphi = 126°$ and

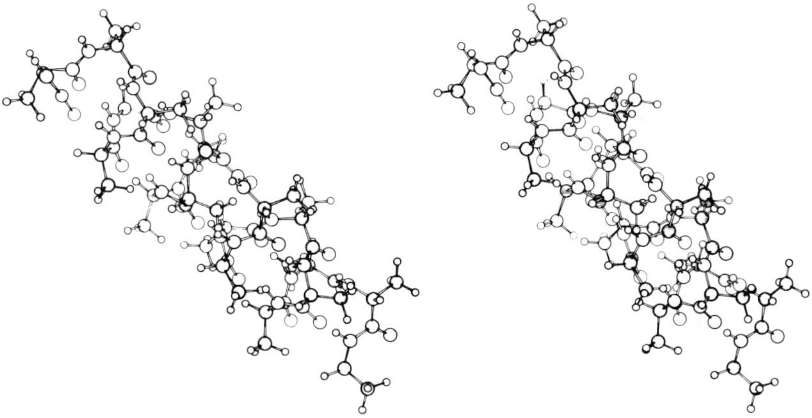

FIG. XIII-17. Stereo views of the α-helix.

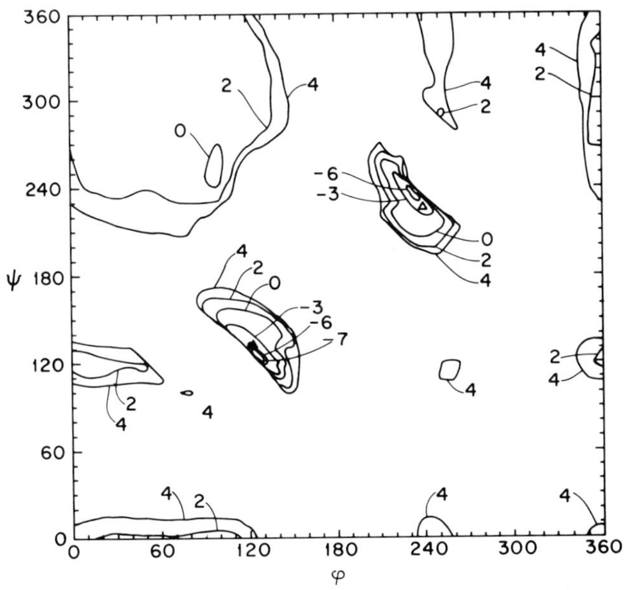

FIG. XIII-18. Energy contours for poly-L-alanine as a function of the dihedral angles φ and ψ (see text); absolute energies are shown in kcal-mole^{-1}. ▲ designates the right-handed α-helix and △ the left-handed α-helix (from Brant[136]).

$\psi = 126°$ corresponds to the right-handed α-helix; the nearly equally deep minimum approximately at the complements of these angles ($\varphi = 232°$; $\psi = 236°$) represents the left-handed helix.

The random-coil conformation obviously cannot be described by a single point on Fig. XIII-18, i.e., by a single conformation. It must be regarded as being represented by the entire φ–ψ plane, each point being exponentially weighted according to its energy. In practice, only certain ranges of angles, corresponding to energies of not more than 5 kcal, need to be considered; the permitted conformations cluster in the upper left-hand portion of the energy field. In the absence of the intramolecular hydrogen bonds, which must be presumed broken in the random coil, the energy minimum in this region is deeper in relation to that of the α-helix than in Fig. XIII-18.[138] Confidence in this view of the random coil is provided by the fact that calculations based upon it correctly predict the signs of its optically active transitions[139] and, as we shall see (Sections 4C, 4D), the magnitude of $J_{N\alpha}$ within experimental error.[140]

A. Polyglutamate Esters

In inert or weakly hydrogen-bonding, helix-supporting solvents such as chloroform or trichloroethylene[111] poly-γ-benzyl-L-glutamate (PBLG)[107,113,115,116,119,129] exhibits behavior typical of α-helix polypeptides under these conditions. As can be seen in Fig. XIII-19(a), the peaks are very

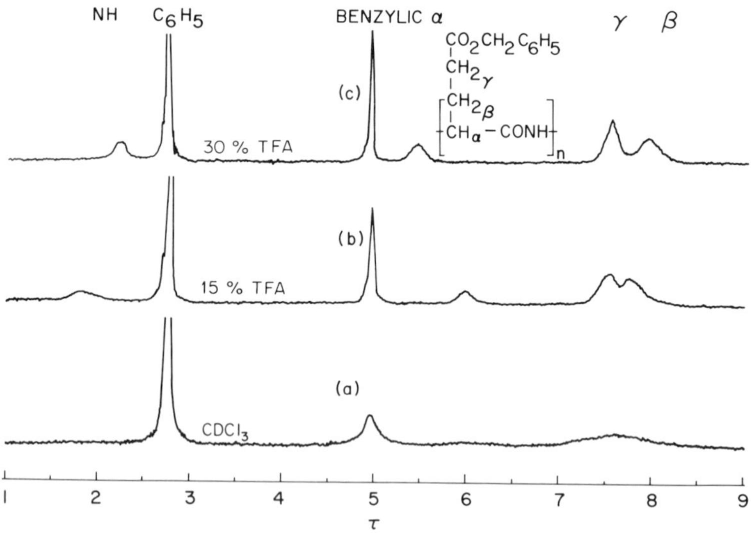

Fig. XIII-19. 100 MHz spectra of PBLG (10% w/v) of $\overline{DP}$ 55 in $CDCl_3$ alone and in 15% and 30% TFA in $CDCl_3$ at 50°. Assignments are given in the text (from Bovey[107]).

broad, only the phenyl (2.76τ) and benzylic (4.96τ) protons being clearly visible. The peaks of the NH (about 2τ) and α-CH (about 6τ) protons are so broadened that they seem virtually to have disappeared, as noted by Goodman and Masuda[112] for polyethyl-L-glutamate. However, their resonance intensity is, of course, actually unchanged and they can be readily detected by increasing the spectrometer gain and particularly by operating in the integral mode (Chapter I, Section 17C). When the $\overline{DP}$ of the polymer is increased by some 20-fold [Fig. XIII-20(a)], the broadening becomes so extreme that no

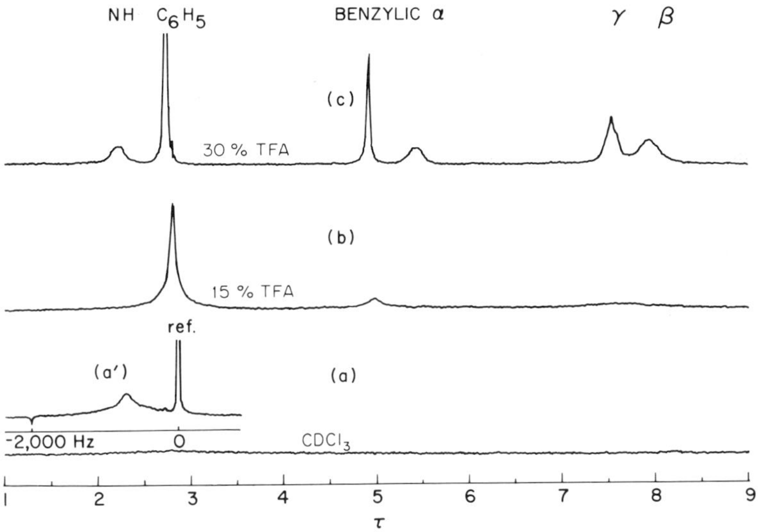

FIG. XIII-20. 100 MHz spectra of PBLG (10% w/v) of $\overline{DP}$ 1000 in $CDCl_3$ alone and in 15% and 30% TFA in $CDCl_3$ at 50° (from Bovey[107]).

spectrum at all is evident. The small inset spectrum, recorded over an approximately 3000 Hz sweep, shows that the spectrum now approaches a wideline spectrum. Measurable intensity is spread out over at least 3000 Hz and the only visible peak, the phenyl resonance, is now nearly 250 Hz in width. Upon addition of 15% TFA, the peaks all narrow dramatically [Fig. XIII-19(b) and Fig. XIII-20(b)] but remain strongly dependent on molecular weight. In the polymer of $\overline{DP}$ 55, the NH, α-CH, β-CH_2 (about 7.7τ) and γ-CH_2 (about 7.5τ) peaks all become clearly visible. Their positions remain about the same, however, and optical measurements show that the chains are still α-helical. At 30% TFA [Fig. XIII-19(c) and Fig. XIII-20(c)] the peaks narrow again but relatively much less, and optical measurements show that the polymer is now random coil.[107,141] In this conformation, the line width is entirely independent of molecular weight.

The extreme broadening in chloroform must be due to aggregation.[107, 111, 118] The existence of PBLG in a liquid crystalline state under these conditions has been established.[142] In more dilute polymer solutions (say, under 5%), a mesophase may not actually be present, but association of helices persists, for it is observed that line broadening is not sensitive to concentration. The dipolar broadening exhibited (Chapter I, Section 5) is closely analogous to that shown by native proteins (Chapter XIV), but somewhat more extreme. The molecular weight of the liquid-crystal domains or swarms appears to be of the order of 10^6 for the material of $\overline{DP}$ 55 and perhaps approaches 10^8 for the $\overline{DP}$ 1000 polymer; the size of the swarms is thus closely related to the size of the helical subunits. Upon adding TFA, these aggregates are broken up, but the helices remain intact. Finally, at 25–30% TFA, the chains attain segmental freedom and peak widths become independent of molecular weight. Under all conditions, broadening is greatest for the main-chain protons and decreases progressively for the side-chain protons as we move out toward the end.[124] This is intuitively reasonable, because one would expect their correlation times to decrease in this order. The details of the theory of multiple correlation times in such systems have not been worked out, although the framework within which this can be done is well established[143] (see also Chapter I, Section 5, Bloembergen *et al.*,[6] Kubo and Tomita,[7] and Solomon[8]).

In addition to changes in line width, there are also changes in chemical shift on passing through the helix–coil transition. These are largest for the α-CH and NH resonances, but are observable for all the others as well, again decreasing outward along the side chain.[107, 116] Markley *et al.*[116] reported that at 60 MHz the α-CH and NH peaks appear to move downfield and upfield, respectively, by about 0.5 ppm when the helix is transformed to the disordered state; Bradbury *et al.*[120] and Ferretti and Ninham[129] were able to discriminate separate peaks in the transition region, apparently corresponding to the α-helical and random-coil conformations, as first reported by Ferretti[117] for poly-β-methyl-L-asparatate and poly-L-leucine (see below). These peaks are more easily resolved for polymers of lower molecular weight, and are not resolved at 60 MHz for polymers of $\overline{DP}$ 340[116] or $\overline{DP}$ 640.[119] At 100 MHz and 220 MHz they can be readily seen even for materials of relatively high molecular weight. In Fig. XIII-21 are shown α-proton spectra of PBLG of $\overline{DP}$ 40, 160 (both at 100 MHz), and 300 (at 220 MHz).[129]† In the spectra of the $\overline{DP}$ 40 polymer, the central peak is probably that of the N-terminal residues and does not change in position or intensity. The transition is very broad and is partially accomplished even at 4% TFA; at higher TFA concentrations, the random-coil peak shows a partial resolution into multiplet

† For reasons not at present entirely clear, these workers were able to resolve spectral features at 100 MHz not evident in the similar system reported in Bovey[107] (Fig. XIII-19).

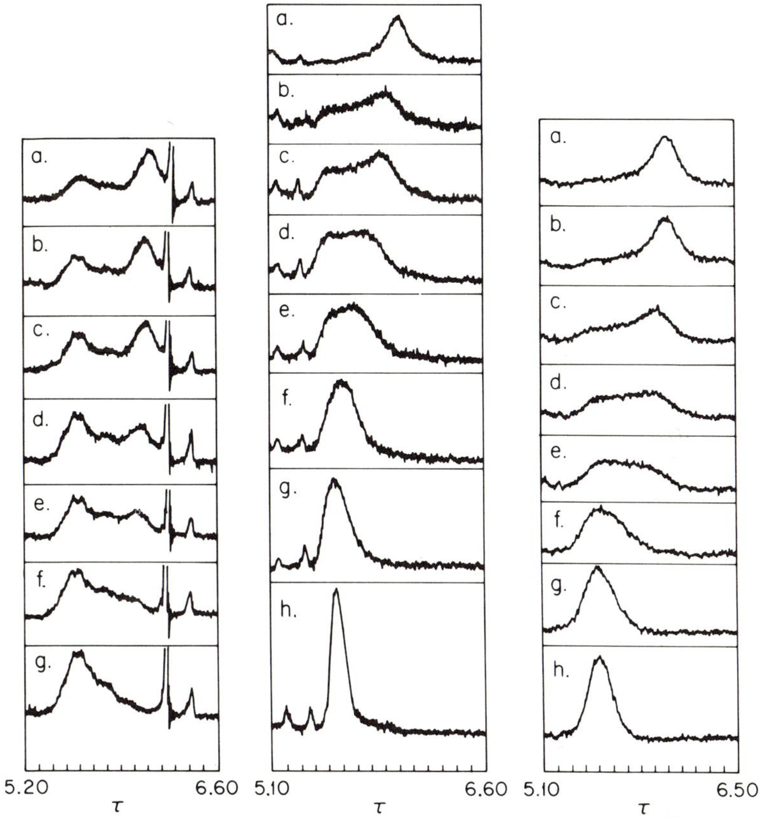

FIG. XIII-21. The α-proton spectra of PBLG of varying molecular weight in $CDCl_3$ with added TFA. For polymer of $\overline{DP}_w$ 40 observed at 100 MHz (left): (a) 4% TFA, (b) 8% TFA, (c) 10% TFA, (d) 12% TFA, (e) 14% TFA, (f) 16% TFA, (g) 18% TFA. (The peak at 6.20τ is an impurity.) For polymer of $\overline{DP}_w$ 300 observed at 220 MHz (middle): (a) 15.0% TFA, (b) 16.0% TFA, (c) 16.5% TFA, (d) 17.0% TFA, (e) 17.5% TFA, (f) 18.0% TFA, (g) 18.5% TFA, (h) 19.0% TFA. For polymer of $\overline{DP}_w$ 160 observed at 100 MHz (right); (a) 11% TFA, (b) 13% TFA, (c) 15% TFA, (d) 16% TFA, (e) 17% TFA, (f) 18% TFA, (g) 19% TFA, (h) 20% TFA (from Ferretti and Ninham[129]).

components, approximately a binomial quartet. The polymer of $\overline{DP}$ 160 also clearly shows two peaks through the transition, but exhibits the additional phenomenon of a downfield movement of the helical peak as the equilibrium is shifted toward the random coil; an indication of this behavior has also been provided by Bradbury *et al.*[119] The $\overline{DP}$ 300 polymer shows this effect in an accentuated degree, the helix peaking merging with the coil peak at lower TFA concentrations.

The explanation of these findings clearly cannot be expected to be a simple one. We shall consider possible hypotheses at the end of this section after we have reviewed parallel phenomena in other polypeptides.

NMR has shed some light on the stability of structures formed by poly-glutamate chains which are not configurationally regular. Tsuboi *et al.*[144] have reported that fibers of "random" copoly-γ-benzyl-D,L-glutamate, prepared by polymerization of a mixture of the D- and L-*N*-carboxy anhydrides, show x-ray and infrared properties very similar to those of poly-γ-benzyl-L-glutamate; they concluded, nevertheless, that the chains contain both α-helical and random-coil portions. Chloroform solutions of such polymers exhibit only single NH and α-CH peaks in the normal α-helical position,[121,131] and the broadening characteristic of associated helices.[131] Upon addition of TFA, these peaks exhibit all the phenomena we have just seen to be characteristic of the helix-to-random-coil transition, except that they occur at about 3% rather than 20% acid concentration. Very similar behavior was observed[131] for a regularly alternating poly-γ-benzyl-D,L-glutamate containing a 53:47 D:L ratio prepared from the DL dipeptide *p*-pentachlorophenyl ester, and by Marlborough and Rydon[128] for an alternating poly-γ-*tert*-butyl-D,L-glutamate having a 45:55 D:L ratio. It thus appears that these configurationally irregular chains are capable of assuming conformations at least very closely similar to that of an α-helix. It is known[145] that in the polymerization of mixed D- and L-*N*-carboxy anhydrides, the growing chain end prefers to add monomer of its own handedness. The NMR evidence permits an independent judgment of the strength of this preference. If substantial blocks of right-handed α-helical L units and of left-handed α-helical D units were formed, they would be expected to undergo the helix–coil transition at about the same acid concentration as chains of single chirality. The fact that the stability of such chains is much less indicates that such blocks cannot on the average exceed about 8–10 units, and may well be shorter.

B. Polyaspartate Esters

Poly-L-asparate chains appear to be unique among polypeptides of L-amino acids in being capable of assuming a left-handed α-helical conformation in helix-supporting solvents. This was first deduced from b_0 measurements[146] and has been confirmed by the observation of positive n–π^* CD bands for poly-β-methyl-L-aspartate and poly-β-benzyl-L-aspartate.[107] Sternlicht and Wilson[147] have performed calculations suggesting that electric field influences (Chapter I, Section 11) arising from the peptide group would lead to different shielding effects upon the α-CH in the helix–coil transition of right-handed and left-handed helices. In fact, however, the effects in poly-β-methyl-L-aspartate[117] and poly-β-benzyl-L-aspartate[107,116,121] (PBLA) are rather similar to those in PBLG. The increase in the shielding of the NH proton in PBLA upon disruption of the helix is markedly greater than in PBLG. Separate peaks for helix and random-coil α-CH and NH protons are observed through the transition region. The most striking difference from right-handed helices is that the concentration of TFA necessary to achieve the disordered state is much less; for example, the half-way point of the transition in $CDCl_3$:TFA solutions of PBLA is at about 3% TFA.[107] The stability of these helices is thus comparable to that of poly-γ-benzyl glutamate helices of mixed residue chirality (Section 4A).

C. Polyalanine

Polyalanine is the simplest optically active poly-α-amino acid. Poly-L-alanine has been shown by x-ray analysis to have a right-handed α-helical conformation in the crystalline state.[148,149] Optical measurements in organic solvents or on block copolymers in aqueous media are consistent with this conformation.[107,150,151,152] Energy minimization calculations also indicate that the right-handed α-helix is preferred over the left-handed.† Although the (negative) n–π^* CD band is somewhat weaker than in most other α-helical polypeptide spectra[152] (and the value of b_0 correspondingly somewhat less strongly negative), the conclusion from the present evidence is that poly-L-alanine is fully α-helical in the solvent systems in which it dissolves to a sufficient extent to be observed.

Optical[114,120,138] and NMR[114,116,120,126] studies indicate that the α-helical conformation of polyalanine is unusually stable, but the two methods do not give concordant estimates of the helical content. Although difficult to see at 60 MHz, doubled NH and α-CH peaks can be clearly seen at 100 MHz[120] and 220 MHz[120,126] in the spectra of polymers of lower molecular weight. Figure XIII-22(a) shows a poly-L-alanine of $\overline{DP}_n$ equal to 35 in 80:20 TFA:$CDCl_3$. Helix and coil NH peaks appear at about 2.2 and 2.3τ (doublet), respectively. At 2.7τ is the resonance of the terminal NH_3^+ protons and at 2.73τ the sharp spike of residual $CHCl_3$. The α-CH protons appear at about 5.3τ (coil) and 5.5τ (helix), the latter being (rather surprisingly) the more clearly resolved into multiplet components from $J_{\alpha\beta}$ and $J_{N\alpha}$. The methyl resonance at about 8.4τ shows a clear splitting into two doublets. For the helix, $J_{N\alpha}$ is about 4–5 Hz; for the coil, the spacing of the NH doublet gives $J_{N\alpha}$ about 6.5 Hz. This is in agreement with the value calculated for poly-L-alanine by an averaging over possible random-coil conformations and the assumption of a Karplus-like dihedral angle function.[140] For polymer of $\overline{DP} \geqslant 300$ [Fig. XIII-22(b)], the NH_3^+ peak can no longer been seen (it serves as a useful measure

† See Brant *et al.*[138] and references therein.

of the $\overline{DP}_n$ of shorter chains), and doubled NH and α-CH peaks can no longer be resolved at 100 MHz, although the NH resonance is somewhat asymmetrical. They can be seen at 220 MHz. From their relative areas, the per cent helicity can be determined as a function of TFA concentration in TFA:$CDCl_3$ mixtures for $\overline{DP}_n = 35$ and $\overline{DP}_n \geqslant 300$ polymer. Judged by this criterion, the latter polymer is almost completely helical even in TFA and even the polymer of low molecular weight is over 50% helical. No clear indication

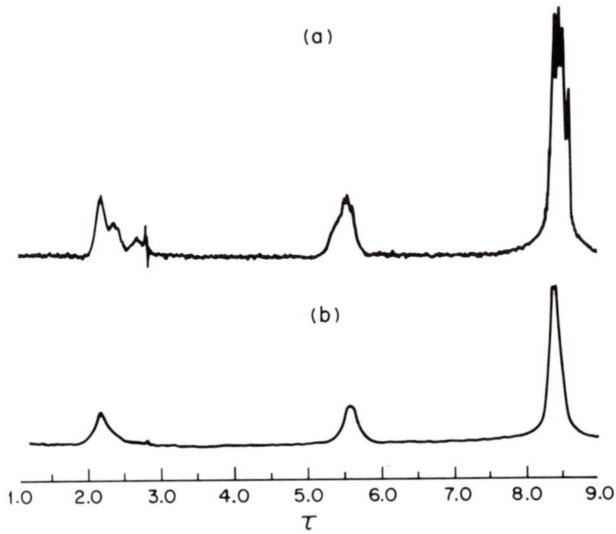

FIG. XIII-22. 100 MHz spectra of poly-L-alanine of (a) low molecular weight ($\overline{DP}_n = 35$) and (b) high molecular weight ($\overline{DP}_n \geqslant 300$), in 80:20 (v/v) TFA:$CDCl_3$ (from Ferretti and Paolillo[126]).

of a transition can be seen. In contrast, the b_0 for poly-L-alanine of high molecular weight is only about –250 to –300 in 60% TFA,[133] indicating a helix content of not more than 50%, while in 80:20 TFA:DCA (dichloroacetic acid), the b_0 of poly-D-alanine is scarcely more than +100.[120] It appears at present that the NMR estimate is the more meaningful and that at higher acid concentrations the n–π^* transition may be greatly reduced in intensity without a corresponding disruption of structure. No comparable disparity has yet been reported for other polypeptides.

The partial resolution of the α-CH resonance of the low molecular weight polymer in helix form [Fig. XIII-22(a)] into spin components appears to imply that the lifetime of the helix in equilibrium with the coil is equal to or greater than $\sqrt{2}(2\pi J_{N\alpha})^{-1}$, or of the order of at least 0.1 sec. This observation, as well as the solvent carboxyl proton chemical shifts in the polyalanine system, are discussed in Sections 4H and 4I.

D. Poly-L-Methionine

Markley *et al.*[116] have reported the 60 MHz spectra of poly-L-methionine ($\overline{DP}$ 280) in $CDCl_3$:TFA mixtures. Its behavior is typical. At about 45 vol% TFA, a sharp transition occurs accompanied by a 0.4 ppm downfield shift of the α-CH resonance. A broadening of this resonance in the transition region was attributed to special relaxation mechanisms, but in the light of the more recent findings discussed above is undoubtedly due to α-helical and random-coil peaks which are incompletely resolved at this frequency.

In the random-coil state, poly-L-methionine gives an exceptionally well-resolved spectrum. The S–CH_3 peak is very narrow (and in fact is only slightly broader in the α-helical state) and the NH and α-CH resonances show coupling fine structure. Markley *et al.*[116] reported a $J_{N\alpha}$ of 5.3 Hz from the spacing of the NH doublet. Bovey and Hood[153] observed the somewhat larger value of 6.3 ± 0.3 Hz, which is in agreement with that observed for poly-L-alanine and a value of 6.5 Hz calculated from an averaging over possible random-coil conformations.[140] Poly-L-methionine would repay careful study at higher frequencies.

E. Poly-L-Leucine, Poly-D-Norleucine, and Poly-D-α-Aminobutyric Acid

Bradbury *et al.*[118] have observed the 60 MHz spectra of these polypeptides in $CDCl_3$:TFA mixtures. All show a downfield shift of the α-CH resonance in the random-coil state. Poly-D-α-aminobutyric acid and poly-D-norleucine show a clear doubling of the peak in the transition region. Poly-L-leucine shows doubling at 220 MHz.[117] There are differences in helical stability as judged by the concentration of TFA required to disrupt them. In order of decreasing stabilizing effect, the hydrocarbon side chains can be ranked as shown. The difference between the last two does not appear to be marked, as judged from the published spectra:

$$CH_3{-} > (H_3C)_2CHCH_2{-} > CH_3CH_2{-} > CH_3CH_2CH_2CH_2{-}$$

F. Poly-L-Phenylalanine

Conti and Liquori[125] have observed the 100 MHz spectra of poly-L-phenylalanine of unstated molecular weight in $CDCl_3$:TFA mixtures. Because of its insolubility in helix-supporting solvents, there are no published optical data on this polypeptide, but the optical rotatory dispersion (ORD) data on block copolymers[154,155] suggest a right-handed α-helical conformation. The NMR findings are consistent with this conclusion in that they are very similar to those for other polypeptides known to be α-helical in this system, i.e., there is a downfield shift of the α-CH resonance and an upfield shift of the NH resonance upon addition of a critical proportion of TFA. The transition to the random coil occurs at a low TFA content, about 4–6%, indicating a relatively low order of helical stability. The β-CH_2 and aromatic resonances also shift at the transition, but in opposite directions, the former moving *upfield.* This behavior of the β-protons probably reflects an alteration in the preferred conformation of the benzene ring upon disruption of the helix. All peaks except that of the NH reflect a solvent sensitivity as well, moving downfield with increased TFA content after the transition has been passed.

G. Polypeptides and Copolypeptides in Aqueous Media

The NMR observation of polypeptides in aqueous media encounters the twin disadvantages that if H_2O is employed, the α-CH resonance is more or less completely obscured by the large nearby water proton peak, whereas if D_2O is employed, the NH proton disappears because of exchange. Markley *et al.*[116] and Bradbury *et al.*[122] have reported the 60 MHz spectra of poly-L-glutamic acid in D_2O solution over the *p*D range from 4.4 to 5.9, during which ionization of the carboxyl groups is known from optical measurements to cause a transition to a disordered state, primarily as a result of Coulombic repulsion.[133,156] Upon increasing the *p*D, the peaks narrow in a similar manner to that observed in nonaqueous media and the α-CH resonance moves downfield. The change in its chemical shift is somewhat smaller (0.12 ppm) than that exhibited by most polypeptides in organic solvents, and is dependent upon polymer concentration for polymer of lower molecular weight, being smaller at higher concentration.[116] For polymer of higher molecular weight it is nearly independent of concentration. The β- and γ-peaks move upfield somewhat upon ionization, in a manner already described for amino acids (Section 2) and not primarily dependent on conformation. The helix–coil transition is highly temperature dependent, the helix being the stable form at lower temperatures, in contrast to what is usually observed in organic solvents.

Poly-L-lysine[122] behaves in much the same manner as poly-L-glutamic acid, although of course the *p*H (or *p*D) dependence is inverted, because the charging of the amino groups at *p*H about 10 and below causes the disruption of the helix.[133,157] There is the additional complication that at very high *p*H, β-structures may be formed with possible precipitation.[157,158] This was avoided by making measurements at 5°,[122] but with some loss of resolution. At *p*D 12.9 the α-CH appears as a broad singlet near 6τ, the ϵ-CH_2 at about 7.4τ, and the β-, γ-, and δ-protons as a series of poorly resolved peaks at 8–9τ. Upon protonation, the latter move together to form a single broad peak, the ϵ-CH_2 peak undergoes the expected deshielding, and the α-CH narrows (as do the others to a lesser degree) and moves downfield by about 0.1–0.2 ppm, the exact amount being difficult to determine because of limited resolution.

Bradbury *et al.*[122] reported the spectrum of block copoly(L-glutamic acid,$^{(42)}$ L-lysine,$^{(28)}$ L-alanine$^{(30)}$) in TFA and in D_2O. The spectra are a composite of the expected peaks of the three different amino acid residues. The b_0 measurements[122, 159, 160] show a decrease in helical content with increasing pD, occurring in two sharp transitions at pD 4 and pD 10. The side-chain protons adjacent to amino and carboxyl groups show the expected changes in chemical shift upon ionization, but at 33° there is little or no alteration in peak widths, although there is a broadening when helix content is increased by lowering the temperature; equally surprisingly, the α-protons, which are all roughly coincident, show no change of chemical shift with changes of helicity. The behavior finds no obvious explanation and merits further study, particularly at higher frequencies.

Bradbury and Scheraga[13] reported a portion of the titration curve of the C-2 and C-4 imidazole ring protons of poly-L-histidine in D_2O solution, but could not follow it beyond pD 6 because of limited solubility. The imidazole ring appeared to have a pK similar to that of histidine itself, i.e., about 6.7, and was thus not markedly perturbed by being attached to a polymer chain.

H. Kinetics of the Helix–Coil Interconversion

The kinetics of the helix–coil interconversion is of considerable interest not only in itself but because of its implications for rates of conformational change in proteins and their bearing on enzyme action and other biological processes. The observation of separate resonances for helix and coil in equilibrium (Section 4A), separated by chemical shift differences of the order of 10^2 Hz, implies lifetimes of each form of about 10^{-2} sec or greater; the resolution of J coupling fine structure in these peaks, as in the spectrum of poly-L-alanine, implies lifetimes of at least 10^{-1} sec, as we have seen. But on the other hand, temperature jump,[161] dielectric relaxation,[162–164] and ultrasonic absorption[165–167] measurements all point to lifetimes of 10^{-6} sec or less, and theoretical treatments of the process give estimates of this same order of magnitude.[163, 168–170] Here, then, is an apparent gross contradiction which clearly cannot be resolved merely by dismissing some part of the experimental observations.

Ullman[171] has suggested that the helix–coil equilibration rate is actually very fast and that the doublet peaks arise from the difference in helicity of residues near the ends of the chain in comparison with those near the middle, and from a polydispersity of molecular weight. The transition from helix to coil with change of solvent is known to begin sooner and to be broader for short chains, so that in a polydisperse system there will be a point where the shorter chains are in the random-coil state while the longer chains are largely helical. An explanation of the NMR observations for chains of short to moderate length ($\overline{DP} \leqslant 100$) can be provided in this way, but requires rather artificial assumptions of bimodal or trimodal molecular weight distributions. Ferretti and Milstien,[172] however, have observed typical peak doubling in PBLG samples having narrow, normal molecular weight distributions and even in fractions ($\overline{DP}_n \simeq 40$; $\overline{DP}_w/\overline{DP}_n \leqslant$ about 1.1) obtained by preparative gel permeation chromatography. An even more serious objection to this view is that it does not explain the qualitatively different results (doubled peaks with the helix peak moving toward the coil peak during the transition to coil; see p. 291) which are obtained with long chains.

Ferretti *et al.*[130] have proposed the following hypothesis to resolve the apparent rate discrepancies and to explain all aspects of the NMR observations. They suggest that the helix–coil system cannot be explained in strictly classical "bond-formed, bond-broken" terms,[173,174] and that the Eyring activated-complex formalism is likewise inappropriate. The helix in equilibrium with the coil (in contrast to the helix in a fully helix-supporting solvent) is regarded as existing in a set of low-lying excited torsional states which represent an average over the hydrogen-bonded and nonhydrogen-bonded states of the amino acid residues. The physical state of the molecules is characterized essentially by two processes having time scales which are widely different and also chain-length dependent. These two processes are: (a) a slow process which describes transitions from the random coil to any or all of the excited helix states; and (b) a fast, stepwise, random-walk process between states having one or more helical units. These states compose the average excited helical state. For *short* chains, the nature of the average excited helical state is close to that of a fully formed helix and remains essentially constant throughout the transition. In contrast, for *long* chains, where many more torsional modes are possible, the nature of the average excited helical state changes continuously through the transition region, becoming essentially equivalent to a random coil at the end of the transition. The NMR experiment provides a measure of process (a); the observation of separate peaks is interpreted as demonstrating that this process is indeed slow. The temperature jump and other nonequilibrium experiments measure process (b). (So far, such experiments have been confined to long-chain polymers.) Ferretti *et al.*[130] have solved the mathematical model underlying this physical picture and have shown that it predicts correctly and in detail the NMR results. Particularly satisfying is the prediction of the dependence on chain length. Their model is somewhat complex and requires some thought to grasp it in all its implications; but the experimental findings are themselves complex, and it does not seem likely at present that any simple model will suffice.†

I. Origin of the Helix–Coil Transition

In aqueous solutions of polypeptides having ionizable side chains, the cause of the disruption of the helix lies clearly in Coulombic repulsion of these groups. It is not so obvious what causes the helix–coil transition in organic systems, in which strong organic acids most commonly serve as the disrupting agent. As Stewart *et al.*[97,114] have pointed out, the proton of the acid carboxyl group behaves differently in solutions containing small model amides and in solutions of polypeptides. In the latter, no minimum in shielding is observed (p. 283 *et seq.*) but instead the carboxyl proton peak moves steadily to higher field as the polymer: acid mole ratio is increased.[107,114,118] This may indicate the formation of quite different hydrogen-bonded complexes in the two systems, since protonation of the amide group in either case seems to be unlikely unless very strong acids such fluorosulfonic acid are employed (Section 3G). At present, it does not appear possible to offer any alternative to the familiar hydrogen-bond competition, pushed to the right by increasing acid–polymer ratios, to explain the helix–coil transition.

$$\underset{\text{helix}}{\mathrm{HA}\cdots\mathrm{HA} + \text{—CO}\cdots\mathrm{NH}} \rightleftharpoons \underbrace{\text{—C=O}\cdots\mathrm{HA} + \mathrm{HA}\cdots\mathrm{NH}}_{\text{coil}}$$

5. α-Amino Acid Oligomers

Open-chain α-amino acid oligomer spectra are more complex than those of long-chain polypeptides because each residue gives a distinctive spectrum according to its position in the chain. As the discussion of di- and tripeptide spectra in Section 3 suggests, such positional effects become too small to resolve when the residues are more than 3 or 4 units from the N- or C-terminal and therefore are not usually evident when the chains exceed about 30–40 residues in length. Cyclic oligomers of course do not exhibit end effects, but their spectra can be complex if all the residues are not identical in type and configuration. Both open-chain and cyclic oligomers have been examined by NMR, but at present the number of reported studies is very limited.

† Note added in proof: Very recently Miller[174a] has proposed an Ising-model kinetic description which is also in good accord with experiment.

A. Open-Chain Oligomers

Goodman *et al.*[175] have reported the spectra of carbobenzoxy-γ-ethyl-L-glutamate oligomers, where $n = 2, 3, 4, 5, 6, 7, 8, 9$, and 12. In Fig. XIII-23

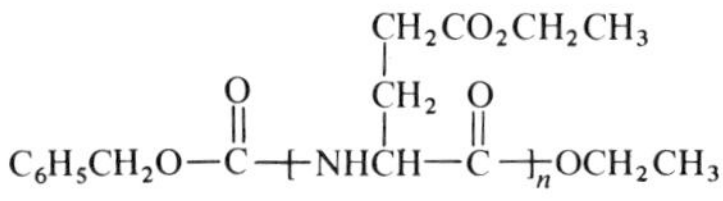

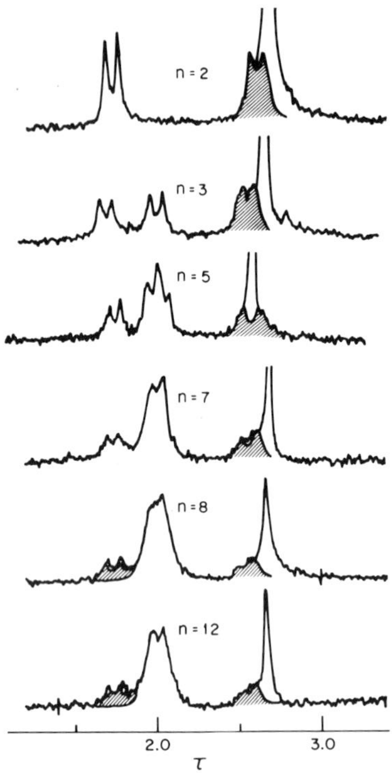

FIG. XIII-23. 100 MHz spectra of the NH protons of carbobenzoxy-γ-ethyl-L-glutamate oligomers of varying number of residues (n) in d_6-DMSO (from Goodman *et al.*[175] and unpublished measurements).

are shown the 100 MHz peptide NH spectra of the oligomers of $n = 2, 3, 5, 7, 8$, and 12 in d_6-DMSO, which optical measurements[175] show is not a helix-supporting solvent. The increase of spectral complexity with n is evident. Model studies indicate that the doublets at about 1.7 and 2.6τ correspond to C-terminal and N-terminal NH protons, respectively; the internal NH protons appear at about 2.0τ. Their positions are the same in all spectra. It is thus evident that in random-coil conformations, the peptide proton environments are not affected by chain length.

In trifluoroethanol, a helix-supporting solvent, the appearance of the spectra is quite different. In Fig. XIII-24, the 220 MHz NH spectra of the oligomers from $n = 2$ through 9 are shown. The aromatic protons of the N-terminal blocking group appear at 2.70τ and provide a convenient reference. All NH resonances are resolved for the dimer through the hexamer; for the heptamer and higher peptides, several of the internal NH peaks overlap. The relative assignment to C-terminal and N-terminal units is the same as in d_6-DMSO. The

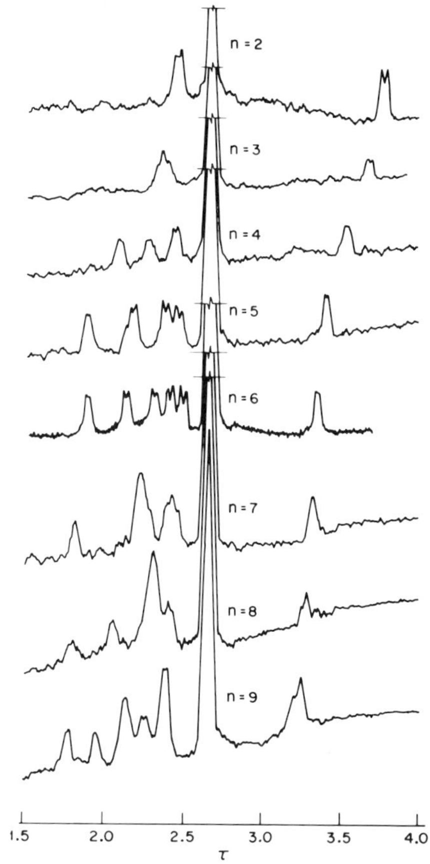

FIG. XIII-24. 220 MHz spectra of the NH protons of carbobenzoxy-γ-ethyl-L-glutamate oligomers of varying number of residues (n) in trifluoroethanol (from Goodman *et al.*[175]).

most striking observation is that these peaks are now very sensitive to chain length; they both move to lower field as n is increased through 7 and then become nearly constant in position for the longer chains (Fig. XIII-25). Circular dichroism measurements show a rapid increase in the intensity of the n–π^* Cotton band at 222 mμ for the hexamer through the decamer, doubtless due to the onset of the α-helical conformation. These structural changes are reflected in the shielding of both C- and N-terminal NH protons, although the latter cannot be involved in internal hydrogen bond formation.

The NMR measurements suggest that helix formation is nearly complete in the heptamer and higher members of the series, a conclusion not so clearly indicated by the CD measurements.

Open-chain oligomers of more complex structure have also been examined by NMR. Since these are more closely related to proteins, they will be considered in Chapter XIV.

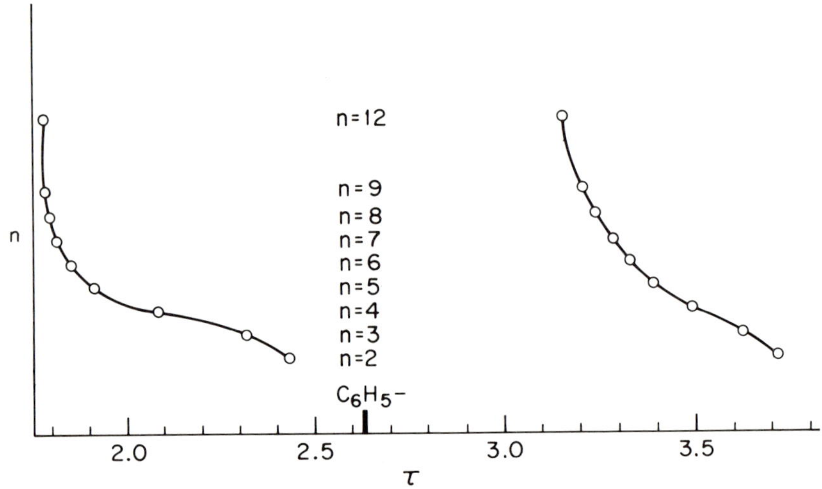

FIG. XIII-25. Chemical shifts of C-terminal NH (low field) and N-terminal NH (high field) resonances in trifluoroethanol (Fig. XII-24) as a function of the number of residues *n*. (from Goodman *et al.*[175]).

B. Cyclic Oligomers

Our interest here in oligomers rests primarily on their possible usefulness as models for the behavior of polypeptides and proteins. The pertinence of studies of cyclic olgomers in this regard is probably rather less than that of the open-chain compounds, but they deserve mention because they appear to have relatively limited range of accessible conformations (possibly only one in some cases) and in consequence their spectra can often be interpreted in a satisfying way. Structural correlations can thus be established which may have a broader applicability.

Kopple *et al.*[176] have made a thorough study of the cyclic hexapeptide cyclopentaglycyl-L-tyrosyl and a specifically C-deuterated derivative, cyclo-d_2-glycyl-d_2-glycyltyrosyltriglycyl. In these molecules there is no element of symmetry and each residue should give a separate spectrum. The six NH resonances were observed (in H_2O and d_6-DMSO solution) to be divided into a high-field group (about 2.3τ) of two protons and a low-field group (centered at about 1.8τ) of four protons. By homonuclear decoupling, the assignments of NH and α-protons were established, and it was shown that both the high-field NH protons are associated with glycine residues. Both groups of protons shifted upfield with temperature (behavior general for peptide NH peaks) but the temperature dependence was markedly greater for the low-field resonances. From these findings it was concluded that in H_2O and d_6-DMSO the conformation of the ring is stabilized by two internal hydrogen bonds, in the general manner suggested by Schwyzer[177, 178, 179] and co-workers for cyclic hexapeptides (Fig. XIII-26). Rather surprisingly, (see below) no marked difference in exchange rate with D_2O could be detected between the two groups of NH protons.

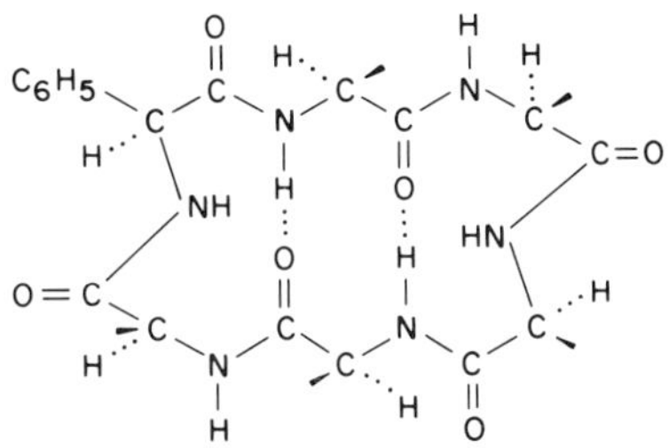

FIG. XIII-26. Conformation of cyclopentaglycyl-L-tyrosyl (according to Kopple *et al.*[176]).

The glycine methylene groups are in principle heterosteric (p. 269) and should appear as the AB portions of ABX spectra and as AB quartets on NH irradiation. In d_6-DMSO only the glycine on the carboxyl side of the tyrosine is observably split at 100 MHz, a result contrasting with that observed in open-chain peptides, in which the glycine on the N-terminal side is the more strongly split. In H_2O, three glycines had significant splittings and in TFA, two. These splittings, which are not discernible in the corresponding open-chain peptide, triglycyl-d_2-glycyl-d_2-glycyltyrosyl, are probably not due entirely to the anisotropy of the tyrosyl aromatic ring, for the α–β couplings in the tyrosine residue show that the ring is not in the folded conformation which is found for L-tyrosylglycyl diketopiperazine (p. 275) and which is favorable to the differentiation of the methylene proton chemical shifts. Presumably the anisotropies of the peptide groups also make a significant contribution.

Stern *et al.*[180] have employed NMR to establish in some detail the conformation of gramicidin S in methanol and DMSO solution. This molecule is a decapeptide. Because of the repeating antiparallel arrangement of the residues (see Fig. XIII-28) it has a twofold symmetry axis. The Phe residues are D, all others L. The members of each pair of like residues are equivalent and give identical spectra, thereby considerably lightening the task of sorting out their resonances. The assignments shown in Fig. XIII-27 were established from

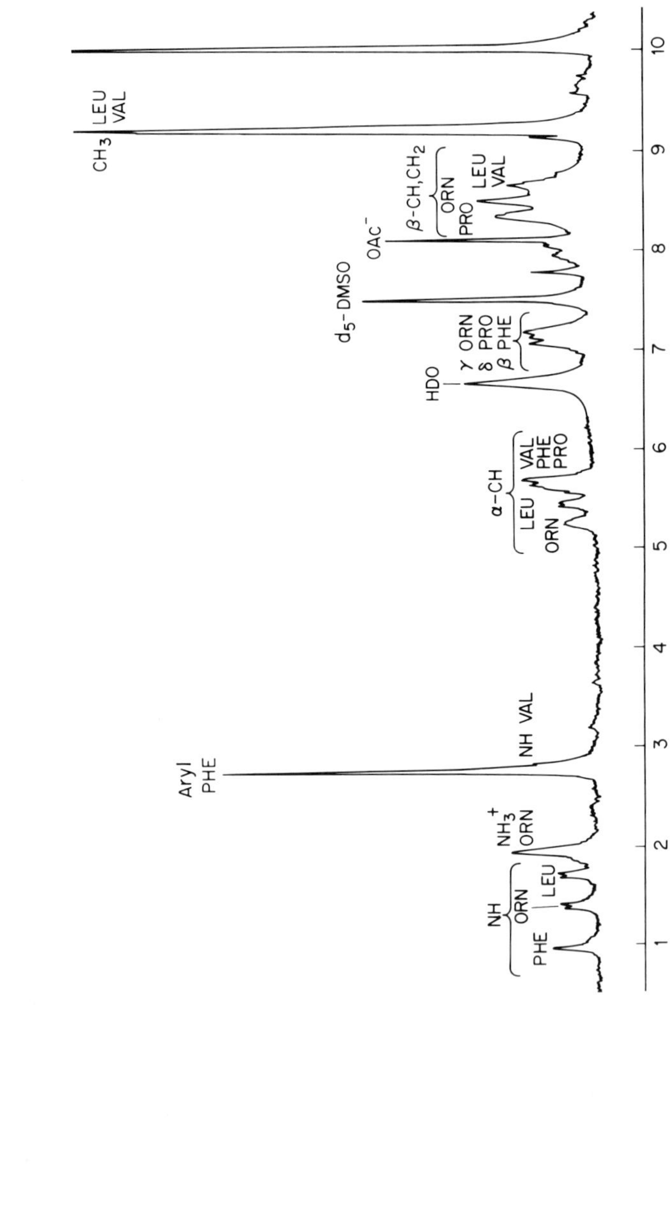

FIG. XIII-27. 220 MHz spectrum of gramicidin S in d_6-DMSO (from Stern *et al.*[180]).

known chemical shifts (Table XIII-1) and from α-β and α-NH proton decoupling experiments. The rates of NH exchange in CD_3OD and with D_2O in d_6-DMSO were found to be much lower for the Val and Leu residues than for the Phe and Orn residues. The Val NH protons are the most shielded, like those of the hydrogen-bonded glycines in Kopple's compound (above), but the Leu NH peaks are not. All the NH resonances appear to show about equal temperature coefficients of chemical shift. Thus generalizations cannot be easily made concerning these spectral features.

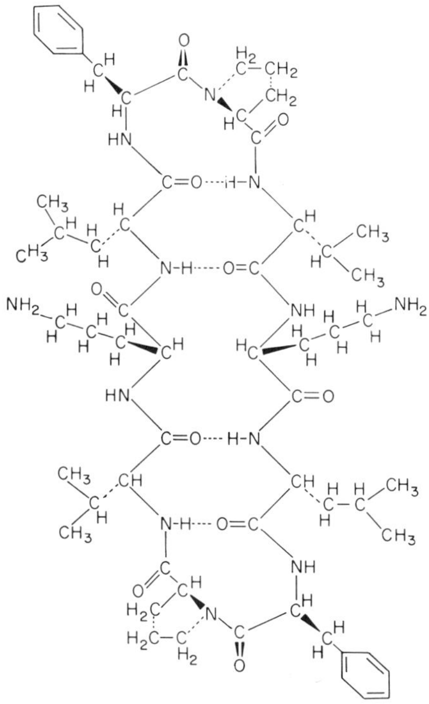

FIG. XIII-28. Conformation of gramicidin S according to Stern *et al.*[180] with ω angles all 0° except Leu ≃ 10° and Pro ≃ 180°; φ angles of Val, Orn, and Leu = 30°; Phe = 150°; ψ angles of Val, Orn, and Leu = 0°; Phe = 150°; and Pro = 130°. Dashed lines O---H indicate hydrogen bonds.

The spacing of the NH doublets was interpreted in terms of an approximate Karplus (or Bystrov[64]) function to give the approximate conformation of the ring by providing limiting values of φ for each residue (with the exception of the prolines, in which the ring ensures no great departure from 120°). For the Phe units, small values (about 2 Hz) for $J_{N\alpha}$ indicate φ near 150° (φ′ ≃ 90°); all the other units showed $J_{N\alpha}$ values of 8.5–9.0 Hz, corresponding to NH—C_αH bond conformation of 0–30° (φ′ = 180°–150°), i.e., not far from *trans*. The Phe–Pro peptide bonds are represented as *cis*. A structure consistent with these conclusions is shown in Fig. XIII-28. It is not in agreement with that deduced from NMR by Liquori and Conti,[181] whose conclusions, however, are based in part on serious interpretive errors.[182] Of the many other structures proposed (see Stern *et al.*[180] for the appropriate literature), it agrees most closely with those of Hodgkin and Oughton[183] and Schwyzer.[184] It may be regarded as having an overlapping "double-Schwyzer" structure, analogous to that in Fig. XIII-26. Schwyzer and Ludescher[185] have offered confirmatory evidence of this structure by demonstrating upfield shifts in the phthaloyl protons of gramicidin S in which phthaloyl groups have been substituted on the ornithine amino groups. These shifts are attributed to a close interaction of the Phe and phthaloyl benzene rings, consistent with the proposed conformation.

NMR has been used to elucidate the conformation of valinomycin, a cyclic depsipeptide (i.e., containing both ester and peptide linkages)[186] and of actinomycin D, which contains two depsipeptide rings attached to an actinocin ring.[187]

6. Poly-α-Imino Acids and Oligopeptides

We have seen that whereas the α-amino peptide group is invariably *trans* in open chains, α-imino residues may be *cis* or *trans* at the ω bond (Section 3C), the definition being as usual in terms appropriate to a polymer chain. When

the nitrogen atom is disubstituted there is of course no peptide hydrogen atom available for hydrogen bonding. If such chains are capable of maintaining fixed helical structures (and we shall see that most of them are) they must do so under stabilizing influences other than those provided by hydrogen bonds.

A. Polysarcosine

Polysarcosine is the simplest such chain and the one in which steric restrictions might be expected to be the least demanding. It has no asymmetric carbons and so shows no optical activity. The sarcosine unit is present in certain proteins and natural polypeptides, but its occurrence is rather rare. In Fig. XIII-29 are shown the 60 MHz spectra of (a) polysarcosine of $\overline{DP}$ 94 and of

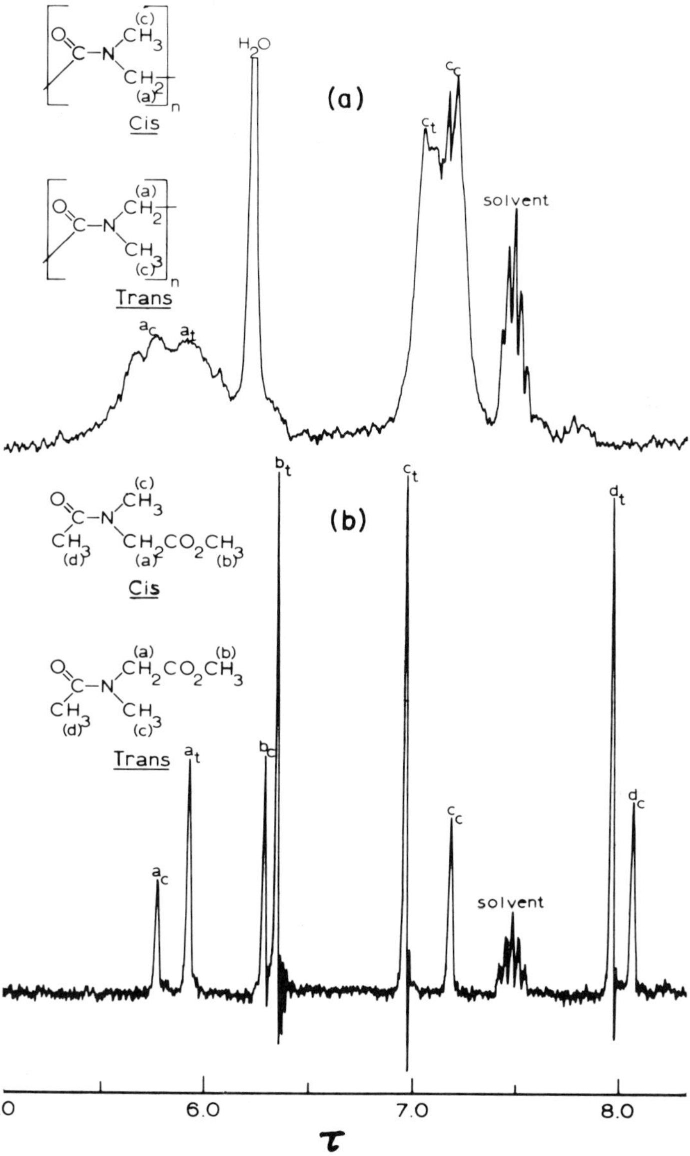

FIG. XIII-29. 60 MHz spectra of (a) polysarcosine ($\overline{DP}$ 94) and (b) *N*-acetylsarcosine methyl ester in d_6-DMSO (about 10% w/v) at 35°. The peak assignments to *cis* (c) and *trans* (t) are indicated by subscripts (from Bovey *et al.*[55]).

(b) the model compound *N*-acetylsarcosine methyl ester[55] employed to assist in making assignments in the polymer spectrum, both observed at about 35° in d_6-DMSO. In the spectrum of the latter, it is observed that all the peaks are doubled, due to the presence of both *cis* and *trans* forms. The peak assign-

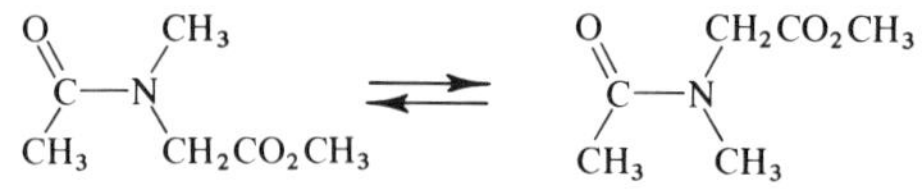

cis *trans*

ments are indicated in Fig. XIII-29 and are based on the relative chemical shifts established for the methyl groups of *N*,*N*′-dimethylformamide, as discussed in Section 3C. The N–CH_3 and N–CH_2 peaks exhibit the "back-to-back" intensity pattern to be expected if each experiences similar relative shielding effects. The *trans–cis* ratio varies somewhat in other solvents (d_4-methanol, $CDCl_3$, TFA, trifluoroethanol, pyridine) but the *trans* form is usually preferred.

The polymer spectrum (a) shows the same general features, except that of course there are no acetyl or methoxyl proton peaks. The methyl and methylene resonances show several peaks each. The methyl resonance is shown as observed at 220 MHz in Fig. XIII-30. A set of eight peak positions and intensities which match the observed spectrum are shown as "sticks" beneath the spectrum. It is believed that these peaks reflect the fact that N–CH_3 shielding is not only a function of the conformation of the residue to which it is attached but also of the conformations of the nearest neighboring residues as well. We can then recognize eight possible conformational triads: *trans-trans-trans*, *trans-trans-cis*, *cis-trans-trans*, *cis-trans-cis*, *cis-cis-cis*, *cis-cis-trans*, *trans-cis-cis*, and *trans-cis-trans*. There would be eight equal peaks if the chain were random with respect to this bond. This is not strictly the case, yet it is clear that there is no strong conformational preference. Assignments are speculative, but it seems reasonable that the more shielded peaks correspond to *cis*. The difference is nearly 1 kcal in some cases, suggesting that the ω bond conformation of a residue may influence that of its neighbors to some degree. This, although reasonable, cannot be clearly demonstrated until at least some specific peak assignments are made. Other solvents (methanol, chloroform, water, trifluoroethanol, TFA) show different patterns, but a multiplicity of peaks is observed in all, indicating that no regular structure can be present. This confirms indications from other physical measurements[188,189] that the chains are random coils.

Dale and Titlestad[190] have reported the NMR spectra of cyclic oligopeptides of sarcosine having 2, 3, 4, 5, 6, and 8 residues. The dimer (i.e., *N*,*N*′-dimethyldiketopiperazine) and trimer are sterically constrained to be all *cis*, and their

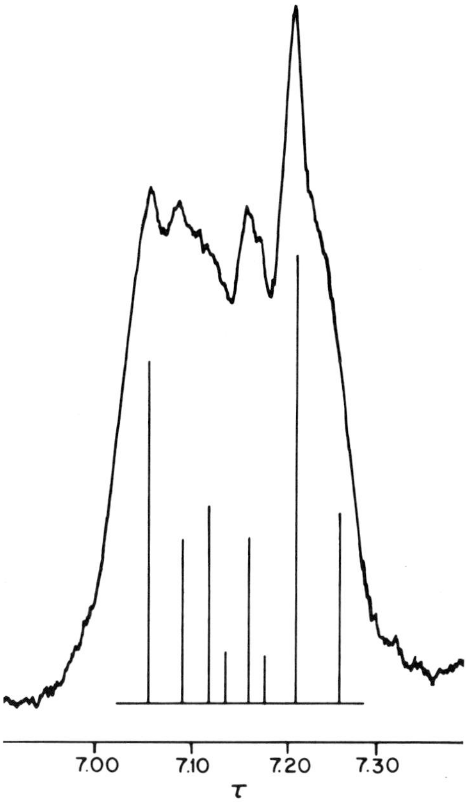

FIG. XIII-30. 220 MHz spectrum of the *N*-methyl resonance of polysarcosine in d_6-DMSO (from Bovey *et al.*[55]).

spectra are correspondingly simple, each showing a single methyl resonance. The tetramer and octamer show equal numers of *cis* and *trans* ω linkages; the pentamer and hexamer show more complex patterns. In all except the dimer, the methylene groups are heterosteric. This must be due to high resistance to ring inversion, no doubt arising from strong transannular interactions.

B. POLY-*N*-METHYL-L-ALANINE

Goodman and Fried[58] have prepared poly-*N*-methyl-L-alanine and com-

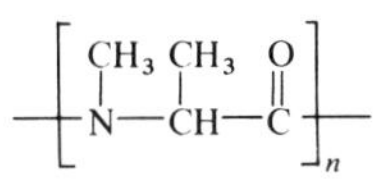

pared its NMR and CD spectra with those of the model compound, *N*-acetyl-

N-methyl-L-alanine methyl ester, which can assume *cis* and *trans* conformations entirely analogous to those of the sarcosine derivative discussed previously. In

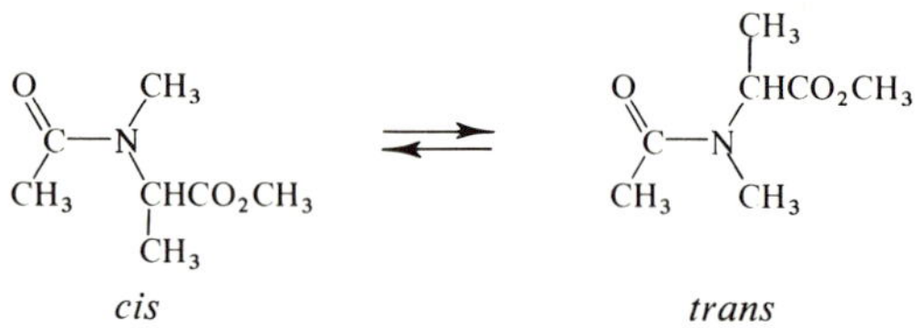

Fig. XIII-31 are shown the 60 MHz spectra of the polymer (a) and model compound (b). In the latter, the N-methyl resonance appears as a large peak

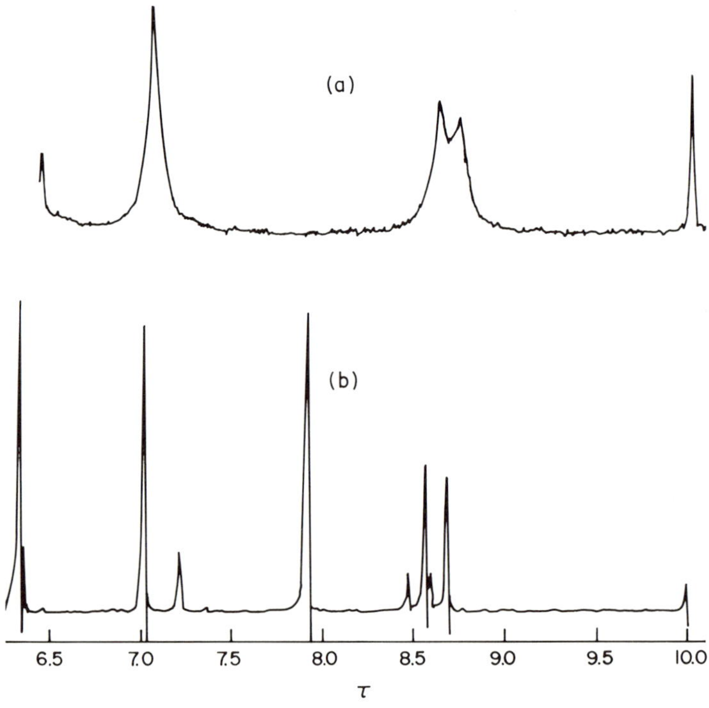

FIG. XIII-31. 60 MHz spectra of (a) poly-*N*-methyl-L-alanine and (b) *N*-acetyl-*N*-methyl-L-alanine methyl ester in trifluoroethanol solution (from Goodman and Fried[58]).

at 7.0τ and a smaller peak at 7.2τ, as in Fig. XIII-29. The C-methyl group appears as two doublets near 8.5τ. The two sets of peaks again have the expected "back-to-back" relationship, and are interpreted as indicating a preferred *trans* conformation. In the polymer spectrum (a), the N-methyl and

C-methyl resonances appear in the *trans* position, and there is no evidence of the other conformation—a marked contrast to the behavior of polysarcosine. The presence of the additional methyl group evidently restricts the chains to a single regular structure, presumably helical. This is further supported by a marked hypochromism in the absorption spectrum at 201 mμ and by a broad intense negative CD band centered at 223 mμ. Energy calculations[191–193] indicate that there is indeed a very striking restriction of the potential energy space in this structure; regions representing 5 kcal-mole^{-1} or less take up only 2.5% of the φ-ψ conformational space. The most probable structure is a right-handed, approximately threefold *trans* helix, shown in stereo view in Fig. XIII-32.

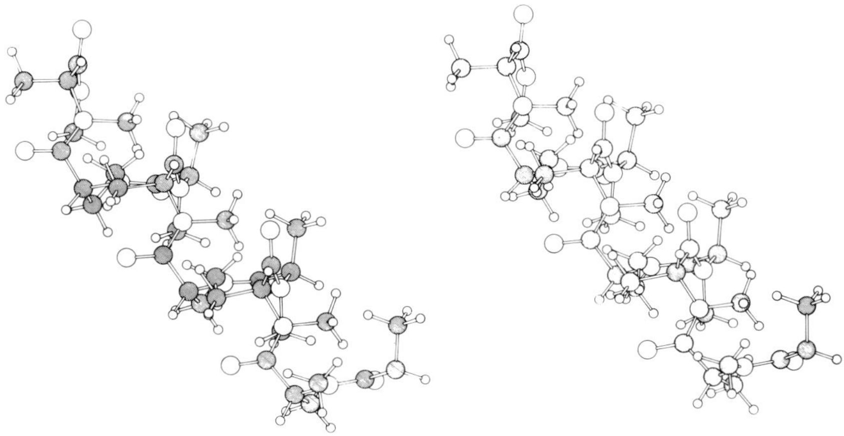

FIG. XIII-32. Stereo views of the most probable helical conformation of poly-*N*-methyl-L-alanine (Mark and Goodman[191, 192]).

In trifluoroacetic acid, poly-*N*-methyl-L-alanine gives a complex NMR spectrum not unlike that of polysarcosine.[194] Evidently, steric restrictions are somewhat relaxed in this solvent, permitting a "scrambling" of the peptide bond conformation.

C. POLY-L-PROLINE

Poly-L-proline is well known to be able to assume two distinct helical conformations.† In water, organic acids, and certain organic solvents such as

† For recent reviews, see Carver and Blout[195] and Mandelkern.[195a]

trifluoroethanol it exists as a left-handed, highly extended *trans* helix having threefold symmetry. This structure, determined by x-ray studies of the crystalline form, is designated form II and is shown in Fig. XIII-33. In

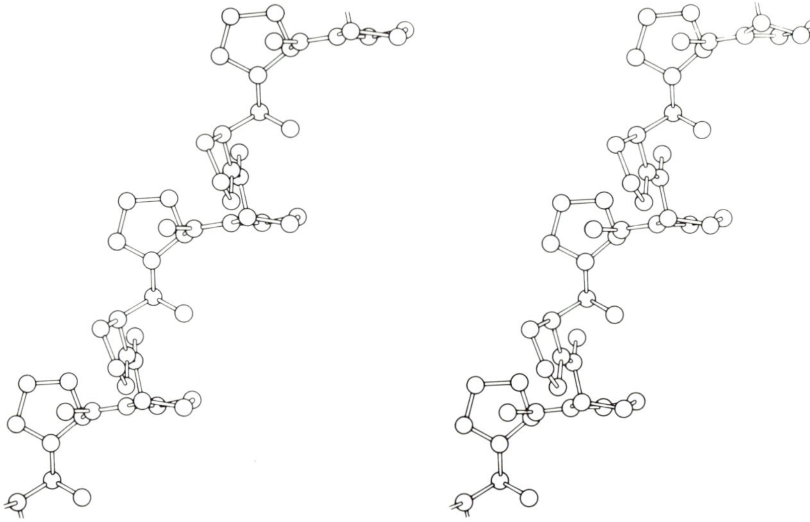

FIG. XIII-33. Stereo views of the helix of poly-L-proline II.

this conformation, $\varphi = 120°$, $\psi = 300°$, and $\omega = 0°$. As normally prepared by polymerization of the *N*-carboxy anhydride and precipitation from pyridine, it exists as a much more compact right-handed *cis* helix with approximately tenfold symmetry, designated form I. The x-ray structure is shown in Fig. XIII-34; in this conformation, φ and ψ are approximately the same as in form

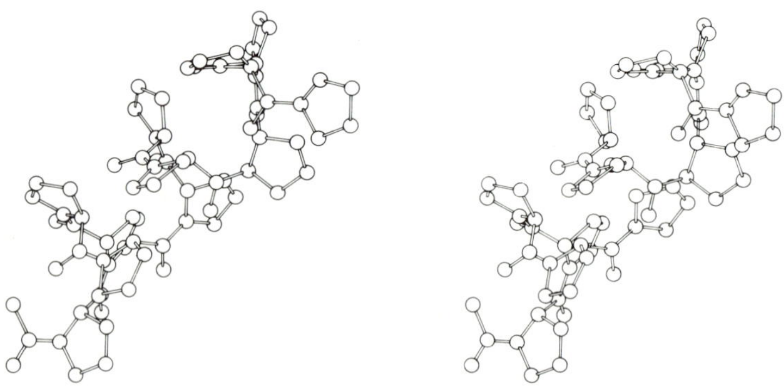

FIG. XIII-34. Stereo views of the helix of poly-L-proline I.

II, but $\omega = 180°$. It is a crowded structure with very close intramolecular contacts and is stable only in relatively poor solvents such as ethyl or benzyl alcohol in which polymer–polymer contacts are preferred to polymer–solvent contacts.[196] Form I is strongly dextrorotatory while form II is strongly levorotatory; when form I is dissolved in a solvent which favors form II, the conformational isomerization can be readily measured by observing the rate of mutarotation.† It can also be followed by NMR.[56, 57] When isomerization of I to II occurs, the α- and δ-protons of the proline ring exchange positions with respect to the carbonyl group. From the relative shieldings observed for

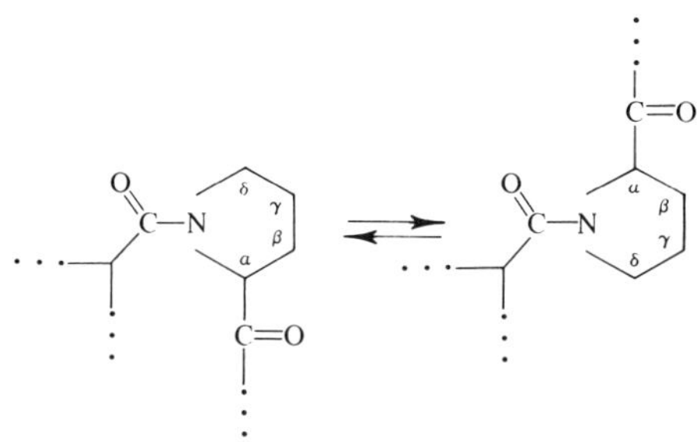

polysarcosine and poly-*N*-methyl-L-alanine (Sections 6A and 6B) one would expect the α-proton (which is the most easily distinguished resonance in the complex spectrum of this polymer) to be more shielded in form II than in form I. Actually, the reverse is the case in acetic acid, in which II is known to be the stable form. In Fig. XIII-35 are shown the spectra of a poly-L-proline of $\overline{DP} \simeq 25$, dissolved initially as form I in CD_3CO_2D, and observed after 10 min, 4 hr, and 100 hr.[57] The upfield α-proton peak, ~5.4τ, is eventually replaced by the downfield peak, ~5.2τ. Somewhat similar behavior is observed in $CDCl_3$,[57] except that the isomerization takes place very much more slowly and never proceeds to a completely *trans* state.

The reversal of the relative shieldings of poly-L-proline α-protons from that expected means that one must be very careful in making conformational assignments in such polymers and in the use of model compounds for this purpose. For example, the α-proton of the *cis* conformer of *N*-acetyl-L-proline methyl ester is less shielded than that of the *trans* in all the solvents tested (TFA, D_2O, CD_3CN, CD_3CO_2D, CD_3OD, d_6-DMSO, and trifluoroethanol) except $CDCl_3$, in which the peaks are reversed, as in the polymer. As discussed previously in Section 6A and 6B, the *cis* and *trans* assignments are based on *N*,*N*′-dimethylformamide, and it is found that the *trans* conformation is preferred by at least 3:1 in all solvents. (Measurements in mixed solvents

† For a review of the optical properties see Carver and Blout,[195] and also Bovey.[197]

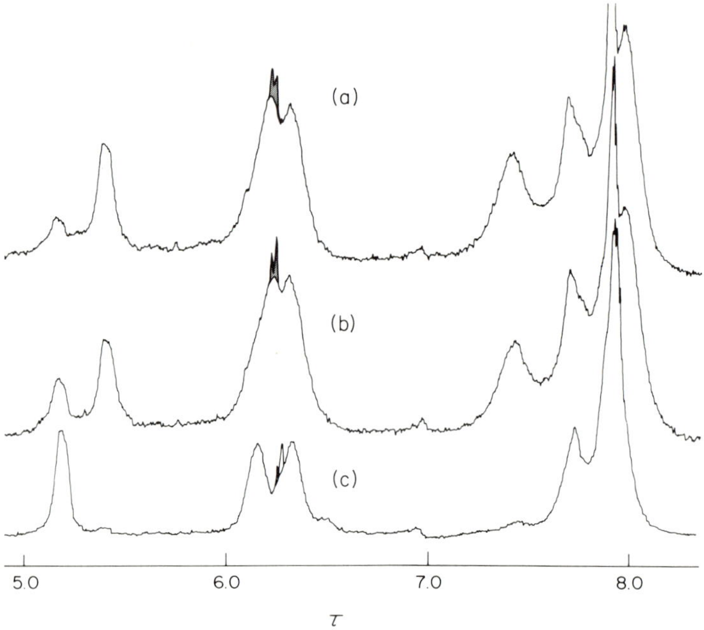

FIG. XIII-35. 220 MHz spectra of poly-L-proline ($\overline{DP}$ about 25) in CD_3CO_2D (about 30°); times after initial dissolution: (a) 10 min, (b) 4 hr, (c) 100 hr (from Deber *et al.*[57]).

show that the peak reversal in $CDCl_3$ is not an intensity reversal, i.e., is not due to an alteration in conformational preference.)

A series of L-proline oligomers studied by Deber *et al.*[57] proved particularly significant for the establishment of conformational preferences and the conditions for helix formation in non-hydrogen-bonded polypeptides. These oligomers had the structure shown with n equal to 2, 3, 4, 5, and 6. Partial

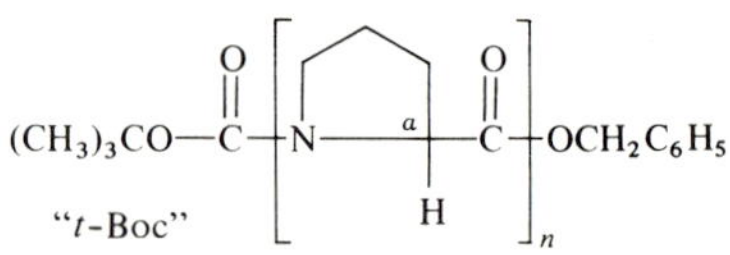

220 MHz spectra, taken in $CDCl_3$ at 25°, showing the α-protons, the methylene resonance of the benzyl group, and the "*t*-Boc" resonance, are shown in Fig. XIII-36. The remaining portions of the spectra, with the exception of the *tert*-butyl resonances at about 8.6τ, are too complex for interpretation. The benzylic resonance is centered at about 4.8τ and the α-protons appear in the

5.2–5.6τ region. The latter are sensitive to both the chain position and conformation of the proline residues. The benzylic protons signal the state of the C-terminal residue: when it is *cis* this group is fortuitously homosteric and gives a singlet residue; when it is *trans*, these protons appear as an AB quartet.

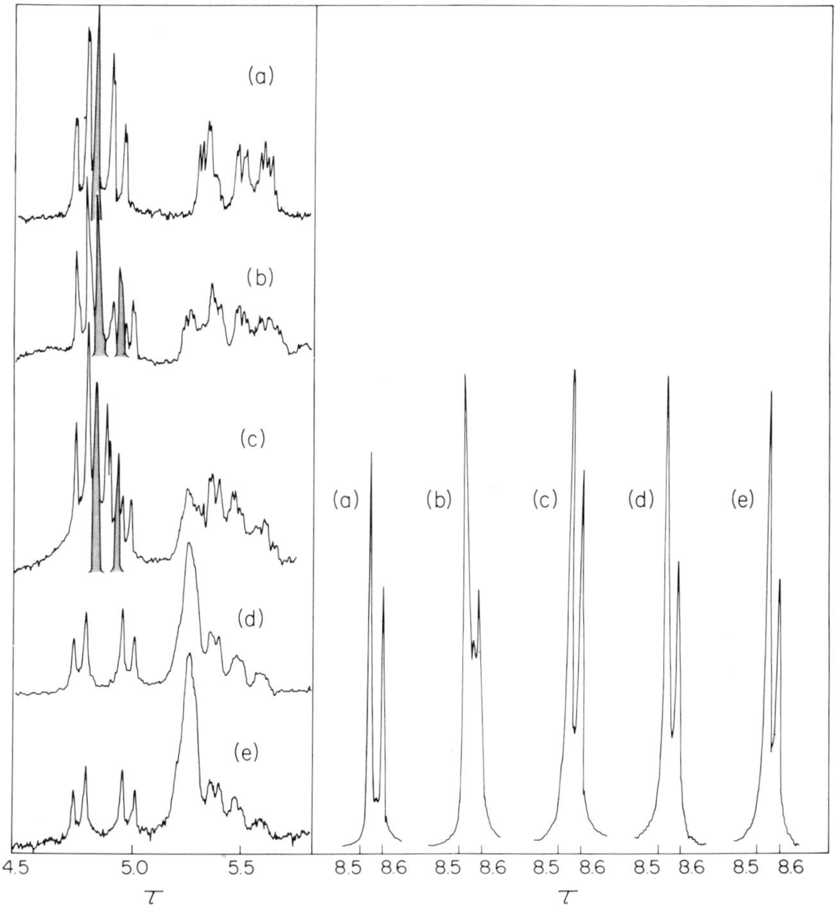

FIG. XIII-36. 220 MHz spectra of *t*-Boc-(Pro)$_n$-OBz oligomers in $CDCl_3$. At the left are the benzyl methylene and α-CH resonances; at the right, the *t*-Boc resonances; (a) $n = 2$, (b) $n = 3$, (c) $n = 4$, (d) $n = 5$, (e) $n = 6$.

The state of the residue next to the C-terminal residue is signalled by a marked splitting of these singlet and quartet resonances. Even the state of the N-terminal residue is indicated by very small splittings of the quartet peaks. The splitting of the "*t*-Boc" resonance also reflects the state of this residue. The

α-proton spectrum shows well separated resonances for *cis* and *trans* N-terminal residues, appearing in the most shielded region at 5.60 and 5.48τ, respectively. There is a slight preference for *trans* which is nearly insensitive to chain length. The C-terminal α-proton appears at 5.38τ, the *cis* and *trans* resonances being nearly overlapping. The "middle" α-protons (for $n \geqslant 3$) are the least shielded; the *trans* units appear at 5.25τ, as in form II poly-L-proline, and the *cis* units at slightly higher field. All the α-proton resonances are four-line spectra (not always clearly resolved), reflecting J couplings of differing magnitude to the two β-protons. Their similar form suggests that the proline ring has about the same shape regardless of the rotational state of the main chain.

As the chain increases in length, the spectra increase in complexity through $n = 4$. Of the 16 possible ω bond conformational isomers of the tetramer, 8 are unmistakably present. The ambiguity concerns only the residue next to the N-terminal residue; the conformational state of this unit is not clearly revealed in the spectrum. It seems highly probable that all possible conformers are present in appreciable proportions. The chain conformation is thus thoroughly "scrambled" although not strictly random, for the *trans* conformation is somewhat preferred. When one more unit is added to the chain, there is a dramatic simplification of the benzyl spectrum, which now appears as a single AB quartet, with no trace of any other conformations. All residues except the N-terminal now appear to be entirely *trans*, strongly suggesting that the chain has assumed a form II helical conformation, into which only the N-terminal unit is not firmly incorporated. This conclusion is greatly strengthened by a very marked increase in the intensity of the negative Cotton band characteristic of this helix on going from $n = 4$ to $n = 5$.

We find that the onset of the helical state in poly-L-proline chains takes place suddenly and completely at the pentamer (in chloroform), rather than gradually over the hexamer–decamer range as in the ethyl-L-glutamate oligomers (Section 5). It will clearly be of interest to extend these studies to other solvents, particular those which stabilize the form I helical state of the longer chains. It does not appear that the energy minimization studies so far reported for poly-L-proline chains[196, 198–203] have taken account of all significant contributions, since they do not suggest the conclusion that there is a critical minimum in the number of units capable of sustaining the form II helical conformation.

Deber *et al.*[57] have also reported the NMR spectra of the first member of a series of alternating Gly-Pro oligomers as well as the cyclic-L-proline and L-Pro-L-Pro-L-Hydroxyproline trimers.

TABLE XIII-1

Proton Resonance Positions of Amino Acids[a] and Peptides in Neutral D_2O Solution at 40°[b, c]

Amino acid	Proton type	Equiv. protons per residue	Compound	Resonance position, τ
Alanine				
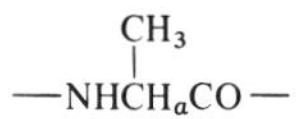	α-CH	1	L-Alanine	6.24
			Glycyl-L-alanine	5.84
	CH_3	3	L-Alanine	8.56
			Glycyl-L-alanine	8.66
			L-Tyrosyl-L-alanine	8.67
			Glycyl-L-phenylalanyl-L-alanine	8.69
			L-Tyrptophanyl-L-alanine	8.64
Arginine				
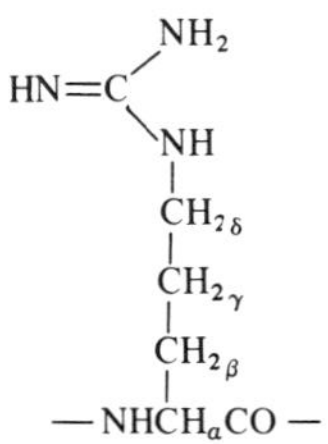	α-CH	1	L-Arginine	6.26
	β-CH_2	2		8.13
	γ-CH_2	2		8.33
	δ-CH_2	2		6.78

TABLE XIII-1 continued

Amino acid	Proton type	Equiv. protons per residue	Compound	Resonance position, τ
Asparagine $CONH_2$–$CH_{2\beta}$– —$NHCH_{\alpha}CO$—	α-CH("A")	1	L-Asparagine	6.08
			Glycyl-L-asparagine	5.48
	β-CH("B")	1	L-Asparagine	7.09
			Glycyl-L-asparagine	7.21
(α- and β-protons form an ABC spin system)	β-CH("C")	1	L-Asparagine	7.15
			Glycyl-L-asparagine	7.36
Aspartic Acid CO_2H–$CH_{2\beta}$– —$NHCH_{\alpha}CO$—	α-CH	1	L-Aspartic acid	6.15
	β-CH_2	2		7.30
Cysteine SH–$CH_{2\beta}$– —$NHCH_{\alpha}CO$—	α-CH	1	L-Cysteine	6.05
	β-CH_2	2		6.95
Cystine $S]_2$–$CH_{2\beta}$– —$NHCH_{\alpha}CO$—	α-CH	1	L-Cystine	(~6.0)
	β-CH_2	2		(~7.0)

Glutamic acid				
CO_2H – $CH_{2\gamma}$ – $CH_{2\beta}$ – —$NHCH_{\alpha}CO$—	α-CH	1	L-Glutamic acid	6.28
	β-CH_2	2		7.94
	γ-CH_2	2		7.67
Glutamine				
$CONH_2$ – $CH_{2\gamma}$ – $CH_{2\beta}$ – —$NHCH_{\alpha}CO$—	α-CH	1	L-Glutamine	6.27
	β-CH_2	2		7.90
	γ-CH_2	2		7.57
Glycine				
—$NHCH_{2\alpha}CO$—	α-CH_2	2	Glycine	6.47
			Glycylglycine	6.21, 6.20
			Glycylglycylglycine	6.24, 6.15, 5.99
			Glycyl-L-tyrosine	6.23
			Glycyl-L-leucine	6.21
			Glycyl-L-alanine	6.22
			Glycyl-L-asparagine	6.20
Histidine				
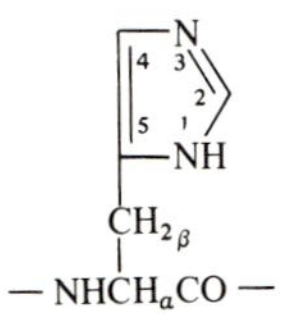	α-CH	1	L-Histidine	6.02
			L-Histidyl-L-leucine	~5.93
	β-CH_2	2	L-Histidine	6.82
			L-Histidyl-L-leucine	6.81
			Glycyl-L-histidylglycine	6.86
	C-2 imidazole	1	L-Histidine	2.16
			L-Histidyl-L-leucine	1.94
			Glycyl-L-histidylglycine	2.01
	C-4 imidazole	1	L-Histidine	2.92
			L-Histidyl-L-leucine	2.86
			Glycyl-L-histidylglycine	2.92

TABLE XIII-1 continued

Amino acid	Proton type	Equiv. protons per residue	Compound	Resonance position, τ
Isoleucine				
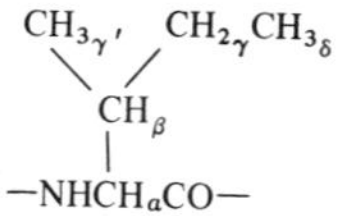	α-CH	1	L-Isoleucine	6.36
	β-CH	1		8.50
	γ-CH_2	2		8.55, 8.76
	δ-CH_3	3		9.08
(γ-protons are distinguishable in chemical shift)	γ'-CH_3	3		9.01
Leucine				
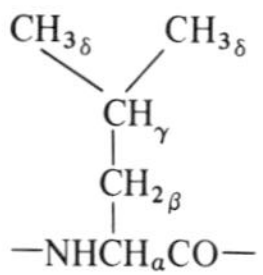	α-CH	1	L-Leucine	6.31
			Glycyl-L-leucine	5.82
			L-Histidyl-L-leucine	~5.87
	β-CH_2 + γ-CH	3	L-Leucine	8.30
			Glycyl-L-leucine	8.41
			L-Histidyl-L-leucine	8.43
	δ-CH_3	6	L-Leucine	9.05
			Glycyl-L-leucine	9.10
			L-Histidyl-L-leucine	9.12
Lysine				
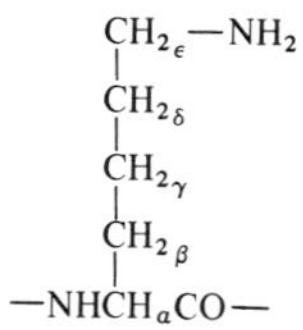	α-CH	1	L-Lysine	6.27
			Poly-L-Lysine	5.70
	β-CH_2	2	L-Lysine	8.13
			Poly-L-lysine	8.29
	γ-CH_2	2	L-Lysine	8.54
			Poly-L-lysine	8.57
	δ-CH_2	2	L-Lysine	8.30
			Poly-L-lysine	8.29

	ϵ-CH_2	2	L-Lysine	6.98
			Poly-L-lysine	7.00
Methionine				
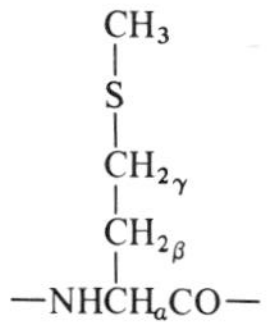	α-CH	1	L-Methionine	6.17
	β-CH_2	2		7.88
	γ-CH_2	2		7.44
	CH_3	3		7.88
Phenylalanine				
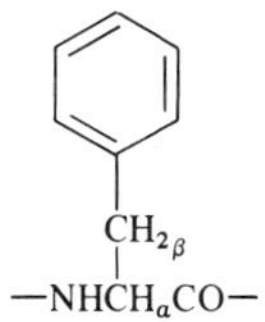	α-CH ("A")	1	L-Phenylalanine	6.02
	β-CH ("C")	1	L-Phenylalanine	6.89
			Glycyl-L-phenylalanine	~7.04
	β-CH ("B")	1	L-Phenylalanine	6.73
			Glycyl-L-phenylalanyl-L-alanine	~6.81
(α- and β-protons form an ABC spin system)	Aromatic:			
	C-3, C-4, C-5	3	L-Phenylalanine	2.61
	C-2, C-6	2		2.66
	Aromatic: all protons	5	Glycyl-L-phenylalanine	2.70
Proline				
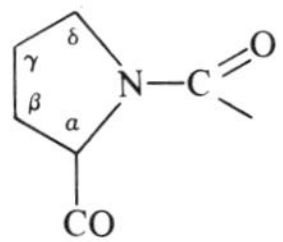	α-CH	1	*N*-Acetyl-L-proline	5.38
	β-CH_2	2		~7.9
	γ-CH_2	2		~7.9
	δ-CH_2	2		6.50
cis				
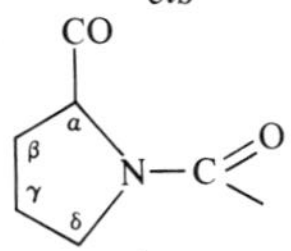	α-CH	1	*N*-Acetyl-L-proline	5.58
	β-CH_2	2		~7.9
	γ-CH_2	2		~7.9
	δ-CH_2	2		6.50
trans				

TABLE XIII-1 continued

Amino acid	Proton type	Equiv. protons per residue	Compound	Resonance position, τ
Serine				
$CH_{2\beta}OH$ / $-NHCH_{\alpha}CO-$	α-CH	1	L-Serine	6.18
	β-CH_2	2		6.07
Threonine				
$CH_{3\gamma}$ / $CH_{\beta}OH$ / $-NHCH_{\alpha}CO-$	α-CH	1	L-Threonine	6.49
	β-CH	1		5.78
	γ-CH_3	3		8.78
Tryptophan				
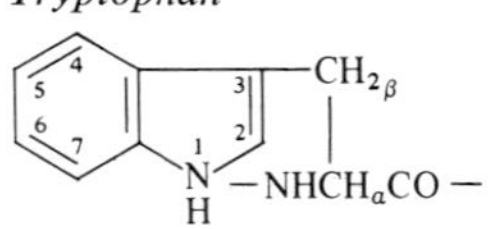	α-CH	1	L-Tryptophanyl-L-alanine	5.76
	β-CH_2	2		6.62
	C-2	1		2.68
	C-5, C-6	1, 1		2.72, 2.81
	C-4, C-7	1, 1		2.32, 2.46
Tyrosine				
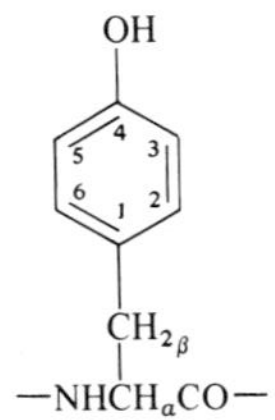	α-CH	1	Glycyl-L-tyrosine	5.58
			L-Tyrosyl-L-alanine	5.88
	β-CH_2	2	Glycyl-L-tyrosine	6.90
	C-2, C-6	2	L-Tyrosine (80°)	2.81
			Glycyl-L-tyrosine	2.86
			L-Tyrosyl-L-alanine	2.83
	C-3, C-5	2	L-Tyrosine (80°)	3.12
			Glycyl-L-tyrosine	3.16
			L-Tyrosyl-L-alanine	3.12

Valine 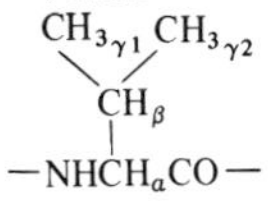 (γ_1 and γ_2 not assigned)	α-CH	1	L-Valine	6.41
	β-CH	1		7.76
	γ-CH_3	3		9.03
	γ-CH_3	3		8.98

[a] See Table XIII-1A.

[b] Data from McDonald and Phillips,[19] except as indicated.

[c] Resonance positions are expressed in τ with sodium 2,2′-dimethyl-2-silapentane-sulfonate (DSS) as internal reference (Tiers and Coon[19a]).

TABLE XIII-1A[a]

One-letter	Three-letter	Residue	One-letter	Three-letter	Residue
A	Ala	Alanine	N	Asn	Asparagine
B	Asn or Asx (not distinguished)	Asparagine (or unspecified Asp deriv)	P	Pro	Proline
			Q	Gln	Glutamine
C	Cys	Cysteine	R	Arg	Arginine
D	Asp	Aspartic acid	S	Ser	Serine
E	Glu	Glutamic acid	T	Thr	Threonine
F	Phe	Phenylalanine	V	Val	Valine
G	Gly	Glycine	W	Trp	Tryptophan
H	His	Histidine	X	—	Undetermined or atypical
I	Ilu	Isoleucine			
K	Lys	Lysine	Y	Tyr	Tyrosine
L	Leu	Leucine	Z	Gln or Glx (not distinguished)	—
M	Met	Methionine			

[a] These three-letter and one-letter designations for amino acid residues are currently accepted.[19b, 19c] The three-letter code will be used in this chapter and Chapter XIV. The one-letter code is used where space is limited, as in sequence listings and x-ray structures. In discussing protein structures, the L-configuration is assumed.

TABLE XIII-3

J COUPLINGS IN AMIDES, *N*-ACYLAMINO ACIDS, AND PEPTIDES (Hz)

Entry no.	Compound	Solvent	^1H–^1H		^1H–^{15}N		Ref.
1.	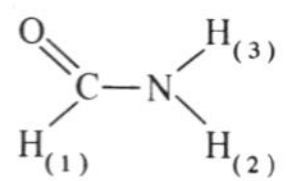	Neat	$H_{(1)}$–$H_{(2)}$:	2.1			*a*
			$H_{(1)}$–$H_{(3)}$:	13.3			
			$H_{(2)}$–$H_{(3)}$:	2.1			
		H_2O	$H_{(1)}$–$H_{(2)}$:	2.2	^{15}N–$H_{(1)}$:	−19.0	*b, c*
			$H_{(1)}$–$H_{(3)}$:	13.8	^{15}N–$H_{(2)}$:	−91.9	
			$H_{(2)}$–$H_{(3)}$:	2.2	^{15}N–$H_{(3)}$:	−95.4	
2.	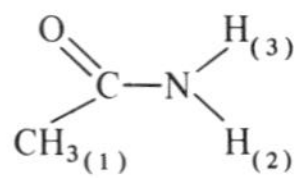	H_2O	$H_{(1)}$–$H_{(2)}$:	<0.2	^{15}N–$H_{(2)}$:	−89	*d, e*
			$H_{(1)}$–$H_{(3)}$:	<0.2	^{15}N–$H_{(3)}$:	−89	
			$H_{(2)}$–$H_{(3)}$:	2.4			
3.	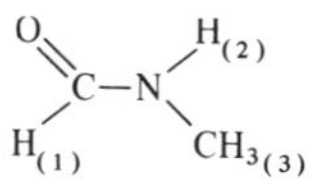	Neat	$H_{(2)}$–$H_{(3)}$:	4.5			*f*
		H_2SO_4	$H_{(1)}$–$H_{(3)}$:	(−) 0.9			*f*
	cis						
	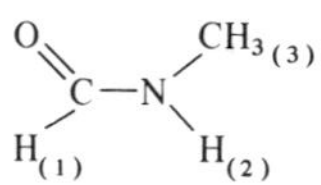	Neat	$H_{(1)}$–$H_{(2)}$:	1.8			*f, g*
			$H_{(1)}$–$H_{(3)}$:	−0.9			
			$H_{(2)}$–$H_{(3)}$:	5.0			
	trans	H_2O	$H_{(1)}$–$H_{(2)}$:	2.3	^{15}N–$H_{(1)}$:	−14.1	*c, e*
			$H_{(1)}$–$H_{(3)}$:	0.8	^{15}N–$H_{(2)}$:	−93.6	*f, g*
			$H_{(2)}$–$H_{(3)}$:	5.0	^{15}N–$H_{(3)}$:	1.4	
		H_2SO_4	$H_{(1)}$–$H_{(2)}$:	5.0			*f*
			$H_{(1)}$–$H_{(3)}$:	1.3			
			$H_{(2)}$–$H_{(3)}$:	5.6			

	Compound	Solvent	Coupling	J (Hz)	Coupling	J (Hz)	Ref.
4.	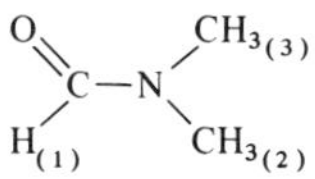	All solvents except H_2SO_4	$H_{(1)}$–$H_{(2)}$:	−0.4			g, h
			$H_{(1)}$–$H_{(3)}$:	−0.7			i, j
			$H_{(2)}$–$H_{(3)}$:	<0.1			
		H_2SO_4	$H_{(1)}$–$H_{(2)}$:	1.2			j
			$H_{(2)}$–$H_{(3)}$:	1.7			
		Neat			^{15}N–$H_{(1)}$:	−15.6	e
					^{15}N–$H_{(2)}$:	1.1	
					^{15}N–$H_{(3)}$:	1.2	
5.	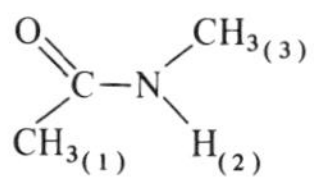	Neat	$H_{(2)}$–$H_{(3)}$:	4.7			d
		H_2O	$H_{(1)}$–$H_{(3)}$:	0.4			k
6.	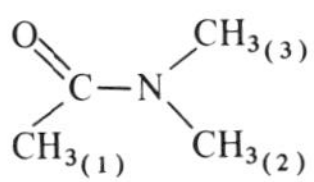	$HCONH_2$	$H_{(1)}$–$H_{(2)}$:	~0.0			i, l
			$H_{(1)}$–$H_{(3)}$:	0.4			
		CCl_4	$H_{(2)}$–$H_{(3)}$:	~0.0			
7.	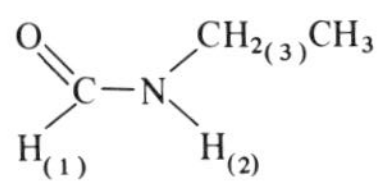	Neat	$H_{(1)}$–$H_{(3)}$:	0.7			f
			$H_{(2)}$–$H_{(3)}$:	6.0			
8.	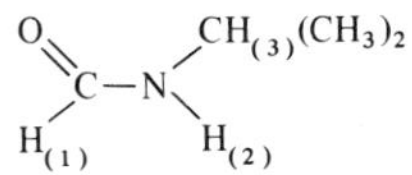	Neat	$H_{(1)}$–$H_{(3)}$:	1.0			f
			$H_{(2)}$–$H_{(3)}$:	7.7			
9.	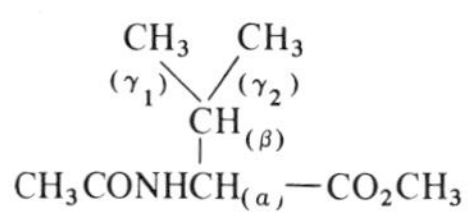	d_6-DMSO (30°)	NH–$H_{(\alpha)}$:	8.0			m
			$H_{(\alpha)}$–$H_{(\beta)}$:	6.5			
			$H_{(\beta)}$–$H_{(\gamma_1)}$:	7.0			
			$H_{(\beta)}$–$H_{(\gamma_2)}$:	7.0			

TABLE XIII-3 continued

Entry no.	Compound	Solvent	^1H–^1H		Ref.
10.	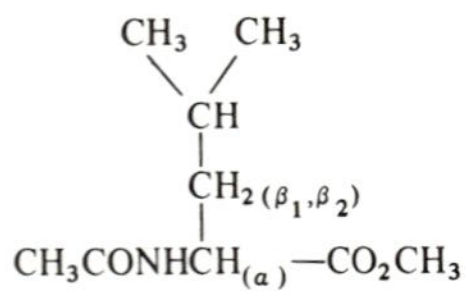	d_6-DMSO	NH–$H_{(\alpha)}$: $H_{(\alpha)}$–$H_{(\beta_1)}$: $H_{(\alpha)}$–$H_{(\beta_2)}$: $H_{(\beta_1)}$–$H_{(\gamma)}$: $H_{(\beta_2)}$–$H_{(\gamma)}$: $H_{(\beta_1)}$–$H_{(\beta_2)}$:	7.3 9.8 4.8 8.0 8.0 −14.0	*m*
11.	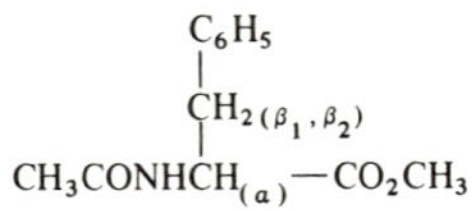	d_6-DMSO	NH–$H_{(\alpha)}$: $H_{(\alpha)}$–$H_{(\beta_1)}$: $H_{(\alpha)}$–$H_{(\beta_2)}$: $H_{(\beta_1)}$–$H_{(\beta_2)}$:	7.8 9.5 5.5 −13.5	*m*
12.	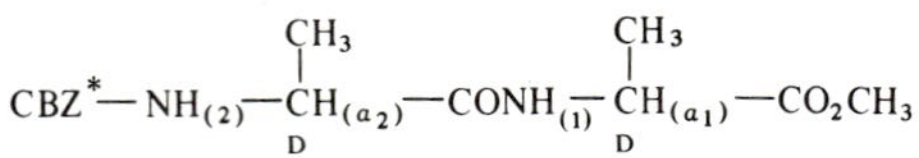	$CDCl_3$	$H_{(\alpha_1)}$–$NH_{(1)}$: $H_{(\alpha_2)}$–$NH_{(2)}$:	7.5 7.6	*n*
13.	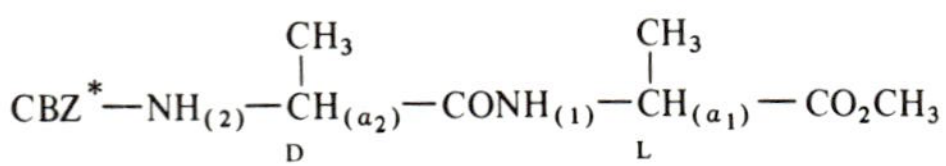	$CDCl_3$	$H_{(\alpha_1)}$–$NH_{(1)}$: $H_{(\alpha_2)}$–$NH_{(2)}$:	7.3 7.6	*n*
14.	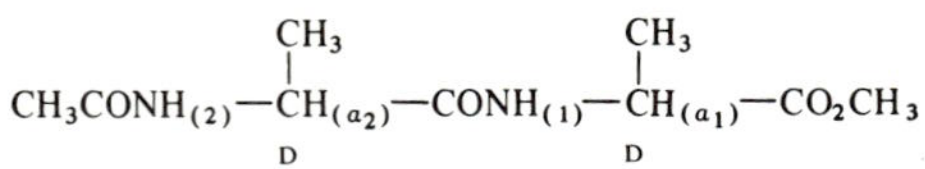	$CDCl_3$	$H_{(\alpha_1)}$–$NH_{(1)}$: $H_{(\alpha_2)}$–$NH_{(2)}$:	7.6 7.7	*n*
		DMSO	$H_{(\alpha_1)}$–$NH_{(1)}$: $H_{(\alpha_2)}$–$NH_{(2)}$:	7.2 7.6	
		H_2O	$H_{(\alpha_1)}$–$NH_{(1)}$: $H_{(\alpha_2)}$–$NH_{(2)}$:	6.4 5.1	

No.	Compound	Solvent	Coupling	J	
15.	$CH_3CONH_{(2)}—CH_{(\alpha_2)}(CH_3)$ [D] $—CONH_{(1)}—CH_{(\alpha_1)}(CH_3)$ [L] $—CO_2CH_3$	DMSO	$H_{(\alpha_1)}–NH_{(1)}$:	7.2	n
			$H_{(\alpha_2)}–NH_{(2)}$:	7.8	
		H_2O	$H_{(\alpha_1)}–NH_{(1)}$:	7.2	
			$H_{(\alpha_2)}–NH_{(2)}$:	5.8	
16.	$CBZ^*—NH_{(2)}—CH_{(\alpha_2)}(CH_3)$ [D] $—CON(CH_3)—CH_{(\alpha_1)}(CH_3)$ [D] $—CO_2CH_3$	$CDCl_3$	$H_{(\alpha_2)}–NH_{(2)}$:	7.7	n
		CCl_4	$H_{(\alpha_2)}–NH_{(2)}$:	8.0	
17.	$CBZ^*—NH_{(2)}—CH_{(\alpha_2)}(CH_3)$ [D] $—CON(CH_3)—CH_{(\alpha_1)}(CH_3)$ [L] $—CO_2CH_3$	$CDCl_3$	$H_{(\alpha_2)}–NH_{(2)}$:	7.6	n
		CCl_4	$H_{(\alpha_2)}–NH_{(2)}$:	7.6	
18.	$CH_3CON(CH_3)—CH_{(\alpha_2)}(CH_3)$ [D] $—CONH_{(1)}—CH_{(\alpha_1)}(CH_3)$ [D] $—CO_2CH_3$	$CDCl_3$	$H_{(\alpha_1)}–NH_{(1)}$:	7.6	n
		H_2O	$H_{(\alpha_1)}–NH_{(1)}$:	6.0	
19.	$CH_3CON(CH_3)—CH_{(\alpha_2)}(CH_3)$ [D] $—CONH_{(1)}—CH_{(\alpha_1)}(CH_3)$ [L] $—CO_2CH_3$	$CDCl_3$	$H_{(\alpha_1)}–NH_{(1)}$:	7.3	n
		H_2O	$H_{(\alpha_1)}–NH_{(1)}$:	6.5	
20.	$CH_3CONH_{(2)}—CH_{(\alpha_2)}(CH_3)$ [D] $—CON(CH_3)—CH_{(\alpha_1)}(CH_3)$ [D] $—CO_2CH_3$	$CDCl_3$	$H_{(\alpha_2)}–NH_{(2)}$:	7.1	n
		CCl_4	$H_{(\alpha_2)}–NH_{(2)}$:	7.6	
		H_2O	$H_{(\alpha_2)}–NH_{(2)}$:	5.8	
21.	$CH_3CONH_{(2)}—CH_{(\alpha_2)}(CH_3)$ [D] $—CON(CH_3)—CH_{(\alpha_1)}(CH_3)$ [L] $—CO_2CH_3$	$CDCl_3$	$H_{(\alpha_2)}–NH_{(2)}$:	7.0	n
		CCl_4	$H_{(\alpha_2)}–NH_{(2)}$:	7.4	
		H_2O	$H_{(\alpha_2)}–NH_{(2)}$:	6.2	

* Carbobenzoyloxy (CBZ) = $C_6H_5CH_2OCO—$

TABLE XIII-3 continued

Entry no.	Compound	Solvent	^1H–^1H	Ref.
22.	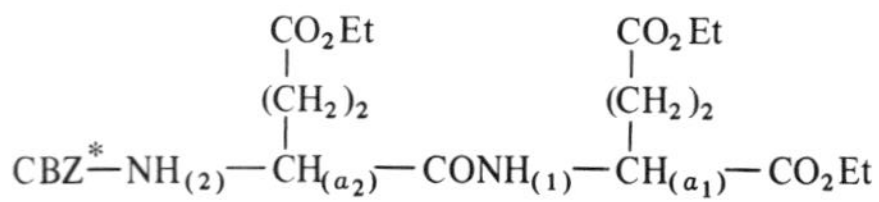	TFA	$H_{(\alpha_1)}$–$NH_{(1)}$: (broad) $H_{(\alpha_2)}$–$NH_{(2)}$: 8.2	*o*
		d_6-DMSO	$H_{(\alpha_1)}$–$NH_{(1)}$: 7.2 $H_{(\alpha_2)}$–$NH_{(2)}$: (masked)	*o*
		$CDCl_3$	$H_{(\alpha_1)}$–$NH_{(1)}$: 8.0 $H_{(\alpha_2)}$–$NH_{(2)}$: 8.0	
23.	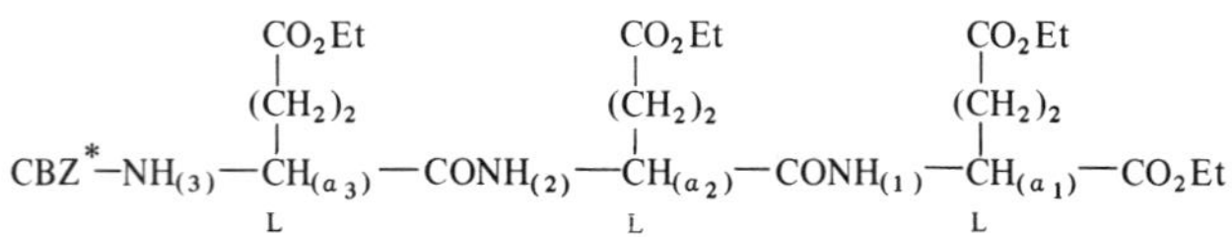	TFA	$H_{(\alpha_1)}$–$NH_{(1)}$: (broad) $H_{(\alpha_2)}$–$NH_{(2)}$: 8.0 $H_{(\alpha_3)}$–$NH_{(3)}$: 7.0	*o*
		d_6-DMSO	$H_{(\alpha_1)}$–$NH_{(1)}$: 7.6 $H_{(\alpha_2)}$–$NH_{(2)}$: 7.0 $H_{(\alpha_3)}$–$NH_{(3)}$: 7.6	*o*
		$CDCl_3$	$H_{(\alpha_1)}$–$NH_{(1)}$: 7.0 $H_{(\alpha_2)}$–$NH_{(2)}$: (masked) $H_{(\alpha_3)}$–$NH_{(3)}$: (masked)	*o*
24.	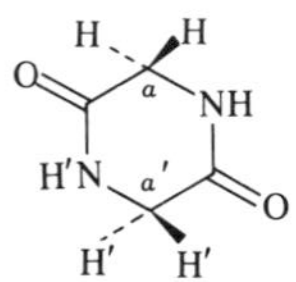	DMSO	$H_{(\alpha)}$–NH = $H_{(\alpha')}$–NH′: 2.2	*q, r, t*
		TFA	$H_{(\alpha)}$–NH = $H_{(\alpha')}$–NH′: <1	

No.	Compound	Solvent / residue	Protons	J (Hz)	Ref.
25.	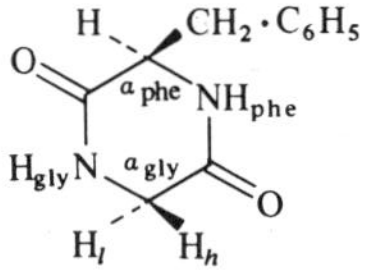	DMSO			
		gly†	$H_{(\alpha l)}$–$NH_{(gly)}$:	3.0	q, r, t
			$H_{(\alpha h)}$–$NH_{(gly)}$:	⩽0.5	
			$H_{(\alpha h)}$–$H_{(\alpha l)}$:	−17.3	
		phe†	$H_{(\alpha)}$–$NH_{(phe)}$:	—	
			$H_{(\alpha)}$–$H_{(\beta l)}$:	4.7	
			$H_{(\alpha)}$–$H_{(\beta h)}$:	5.3	
			$H_{(\beta l)}$–$H_{(\beta h)}$:	−13.5	
		TFA			
		gly†	$H_{(\alpha l)}$–$NH_{(gly)}$:	⩽0.5	q, r, t
			$H_{(\alpha h)}$–$NH_{(gly)}$:	⩽0.5	
			$H_{(\alpha h)}$–$H_{(\alpha l)}$:	−18.8	
		phe†	$H_{(\alpha)}$–$NH_{(phe)}$:	—	
			$H_{(\alpha)}$–$H_{(\beta l)}$:	4.5	
			$H_{(\alpha)}$–$H_{(\beta h)}$:	4.5	
			$H_{(\beta l)}$–$H_{(\beta h)}$:	−14.3	
26.	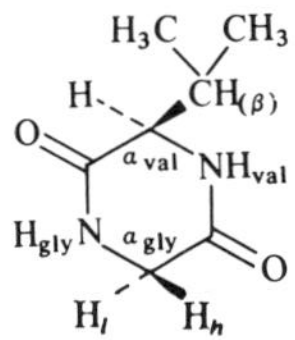	DMSO			
		gly†	$H_{(\alpha l)}$–$NH_{(gly)}$:	~1	q, t
			$H_{(\alpha h)}$–$NH_{(gly)}$:	3.0	
			$H_{(\alpha h)}$–$H_{(\alpha l)}$	−17.5	
		val	$H_{(\alpha)}$–$H_{(\beta)}$:	4.2	
		D_2O			
		gly†	$H_{(\alpha l)}$–$NH_{(gly)}$:	⩽0.5	q, t
			$H_{(\alpha h)}$–$NH_{(gly)}$:	⩽0.5	
			$H_{(\alpha h)}$–$H_{(\alpha l)}$:	−18.5	
		val	$H_{(\alpha)}$–$H_{(\beta)}$:	3.7	
		TFA			
		val	$H_{(\alpha)}$–$H_{(\beta)}$:	3.5	

* Carbobenzoyloxy (CBZ) = $C_6H_5CH_2OCO$—

† Subscripts *l* and *h* designate α- and β-protons which are, respectively, at lower and higher fields.

TABLE XIII-3 continued

Entry no.	Compound	Solvent	^1H–^1H	Ref.
27.	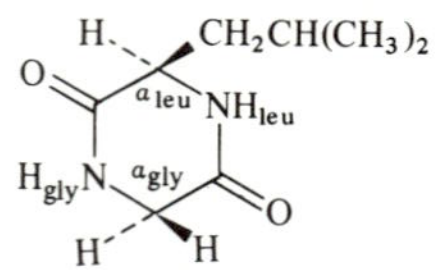	DMSO gly†	$H_{(\alpha_l)}$–$NH_{(gly)}$: 1.5 $H_{(\alpha_h)}$–$NH_{(gly)}$: 3.0 $H_{(\alpha_h)}$–$H_{(\alpha_h)}$: −17.5	*q*, *t*

† Subscripts *l* and *h* designate α-protons which are, respectively, at lower and higher fields.

References for Table XIII-3

a. J. B. Stothers and T. G. Hill, private communication; see also L. H. Piette, J. D. Ray, and R. A. Ogg, *J. Mol. Spectrosc.* **2**, 66 (1958).
b. H. Kamei, *Bull. Chem. Soc. Jap.* **41**, 2269 (1968).
c. B. Sunners, L. H. Piette, and W. G. Schneider, *Can. J. Chem.* **38**, 681 (1960).
d. H. Kamei, *Bull. Chem. Soc. Jap.* **38**, 1212 (1965).
e. G. Binsch, J. B. Lambert, B. W. Roberts, and J. D. Roberts, *J. Amer. Chem. Soc.* **86**, 5564 (1964); this paper also reports some longer-range ^{15}N–H couplings.
f. L. A. La Planche and M. T. Rogers, *J. Amer. Chem. Soc.* **86**, 337 (1964).
g. A. J. R. Bourn and E. W. Randall, *Mol. Phys.* **8**, 567 (1964); A. J. R. Bourn, D. G. Gillies, and E. W. Randall, *in* "Nuclear Magnetic Resonance in Chemistry" (B. Pesce, ed.), p. 277. Academic Press, New York, 1965.
h. J. V. Hatton and R. E. Richards, *Mol. Phys.* **5**, 139 (1962).
i. R. C. Neuman, Jr. and L. B. Young, *J. Phys. Chem.* **69**, 1777 (1965).
j. G. Fraenkel and C. Franconi, *J. Amer. Chem. Soc.* **82**, 4478 (1960).
k. A. Berger, A. Loewenstein, and S. Meiboom, *J. Amer. Chem. Soc.* **81**, 62 (1959).
l. D. G. De Kowalewski and V. J. Kowalewski, *Ark. Kemi* **16**, 373 (1969).
m. A. Stern, J. J. Ryan, R. L. Kornegay, and F. A. Bovey, unpublished data.
n. V. F. Bystrov, S. L. Portnova, V. I. Tsetlin, V. T. Ivanov, and Yu. A. Ovchinnikov, *Tetrahedron* **25**, 493 (1969).
o. M. Goodman, A. S. Verdini, C. Toniolo, W. D. Phillips, and F. A. Bovey, *Proc. Nat. Acad. Sci.* (*U.S.*) **64**, 444 (1969).
q. K. D. Kopple and D. H. Marr, *J. Amer. Chem. Soc.* **89**, 6193 (1967).
r. F. A. Bovey and F. P. Hood, unpublished data.
t. K. D. Kopple and M. Ohnishi, *J. Amer. Chem. Soc.* **91**, 962 (1969).

REFERENCES

1. M. Takeda and O. Jardetzky, *J. Chem. Phys.* **26**, 1346 (1957).
2. O. Jardetzky and C. D. Jardetzky, *J. Biol. Chem.* **233**, 383 (1958).
3. F. A. Bovey and A. I. Brewster, unpublished observations.
4. S. Fujiwara, Y. Arata, N. Hayakawa, and H. Monoi, *Bull. Chem. Soc. Jap.* **35**, 1658 (1962).
5. S. Fujiwara and Y. Arata, *Bull. Chem. Soc. Jap.* **36**, 578 (1963); **37**, 344 (1964).
6. F. Taddei and L. Pratt, *J. Chem. Soc.* **1964**, 1553 (1964).
7. R. J. Abraham and W. A. Thomas, *J. Chem. Soc.* **1964**, 3739 (1964).
8. K. G. R. Pachler, *Spectrochim. Acta* **19**, 2085 (1963); **20**, 581 (1964).
9. A. Nakamura, *J. Chem. Soc. Jap. Pure Chem. Sect.* **86**, 780 (1964).
10. M. Mandel, *J. Biol. Chem.* **240**, 1586 (1965).
11. R. B. Martin and R. Mathur, *J. Amer. Chem. Soc.* **87**, 1065 (1965).
12. R. B. Martin and V. J. Morlino, *Science* **150**, 493 (1965).
13. J. H. Bradbury and H. A. Scheraga, *J. Amer. Chem. Soc.* **88**, 4240 (1966).

13a. K. G. R. Pachler, *Z. Anal. Chem.* **224**(2), 211 (1967).

14. V. J. Morlino and R. B. Martin, *J. Amer. Chem. Soc.* **89**, 3107 (1967).
15. H. Ogura, Y. Arata, and S. Fujiwara, *J. Mol. Spectrosc.* **23**, 76 (1967).
16. A. Nakamura and O. Jardetzky, *Proc. Nat. Acad. Sci. U.S.* **58**, 2212 (1967).
17. J. R. Cavanaugh, *J. Amer. Chem. Soc.* **89**, 1558 (1967); **90**, 4533 (1968).
18. C. C. McDonald and W. D. Phillips, *J. Amer. Chem. Soc.* **85**, 3736 (1963).
19. C. C. McDonald and W. D. Phillips, *J. Amer. Chem. Soc.* **91**, 1513 (1969).

19a. G. V. D. Tiers and R. I. Coon, *J. Org. Chem.* **26**, 2097 (1961).

19b. M. O. Dayhoff, ed., "Atlas of Protein Sequence and Structure," Vol. 4. National Biomedical Research Foundation, Silver Spring, Maryland, 1969.

19c. R. E. Dickerson and I. Geis, "The Structure and Action of Proteins." Harper, New York, 1969.

20. F. A. Bovey and G. V. D. Tiers, *J. Amer. Chem. Soc.* **81**, 2870 (1959).
21. B. Bak, C. Dambmann, F. Nicolaisen, and E. J. Pedersen, *J. Mol. Spectrosc.* **26**, 78 (1968).
22. W. J. Horsley and H. Sternlicht, *J. Amer. Chem. Soc.* **90**, 3738 (1968).
23. W. A. Anderson and R. Freeman, *J. Chem. Phys.* **37**, 85 (1962).
24. R. Freeman, *J. Chem. Phys.* **43**, 3084 (1965).
25. J. K. Becconsall and P. Hampson, *Mol. Phys.* **10**, 21 (1965).
26. W. Horsley, H. Sternlicht, and J. S. Cohen, *J. Amer. Chem. Soc.* **92**, 680 (1970).
27. N. Bloembergen, E. M. Purcell, and R. V. Pound, *Phys. Rev.* **73**, 679 (1948).
28. J. A. Pople, W. G. Schneider, and H. J. Bernstein, "High Resolution Nuclear Magnetic Resonance," p. 207. McGraw-Hill, New York, 1959.
29. E. Fermi, *Z. Phys.* **60**, 320 (1930).
30. N. Bloembergen, *J. Chem. Phys.* **27**, 572, 595 (1957).
31. N. C. Li, R. L. Scruggs, and E. D. Becker, *J. Amer. Chem. Soc.* **84**, 4650 (1962).
32. D. R. Eaton and W. D. Phillips, *in* "Advances in Magnetic Resonance" (J. S. Waugh, ed.), Vol. 1, p. 103. Academic Press, New York, 1965.
33. H. M. McConnell and D. B. Chesnut, *J. Chem. Phys.* **28**, 107 (1958).
34. R. S. Milner and L. Pratt, *Disc. Faraday Soc.* **32**, 88 (1962).
35. N. C. Li, E. Doody, and J. M. White, *J. Amer. Chem. Soc.* **79**, 5859 (1957).
36. O. Jardetzky and J. E. Wertz, *J. Amer. Chem. Soc.* **82**, 318 (1960).

37. N. C. Li, L. F. Johnson, and J. Shoolery, *J. Phys. Chem.* **65**, 1902 (1961).
38. R. H. Carlson and T. L. Brown, *Inorg. Chem.* **5**, 268 (1966).
39. D. Leussing and C. K. Stanfield, *J. Amer. Chem. Soc.* **86**, 2805 (1964).
40. C. C. Costain and J. M. Dowling, *J. Chem. Phys.* **32**, 158 (1960).
41. L. Pauling, "The Nature of the Chemical Bond." Cornell Univ. Press, Ithaca, New York, 1948.
42. G. V. D. Tiers and F. A. Bovey, *J. Phys. Chem.* **63**, 302 (1959).
43. J. D. Roberts, *J. Amer. Chem. Soc.* **78**, 4495 (1956).
44. F. A. Bovey and G. V. D. Tiers, *J. Polym. Sci. Part A* **1**, 849 (1963).
45. K. D. Kopple and D. H. Marr, *J. Amer. Chem. Soc.* **89**, 6193 (1967).
46. K. D. Kopple and M. Ohnishi, *J. Amer. Chem. Soc.* **91**, 962 (1969).
47. L. A. La Planche and M. T. Rogers, *J. Amer. Chem. Soc.* **86**, 337 (1964).
48. C. Franconi, *Z. Elektrochem.* **65**, 645 (1961); *Sci. Tec. N.S.* **4**, 183 (1960).
48a. F. A. Bovey, *Chem. Eng. News* **43**, 98 (August 30, 1965).
49. F. A. L. Anet and A. J. R. Bourn, *J. Amer. Chem. Soc.* **87**, 5250 (1965).
50. P. T. Narasimhan and M. T. Rogers, *J. Phys. Chem.* **63**, 1388 (1959).
51. J. W. ApSimon, W. G. Craig, P. V. Demarco, D. W. Mathieson, A. K. G. Nasser, L. Saunders, and W. B. Whalley, *Chem. Commun.* **1966**, 754 (1966).
52. H. Sternlicht and D. Wilson, *Biochemistry* **6**, 288 (1967).
53. H. Paulson and K. Todt, *Angew. Chem. Int. Ed. Engl.* **5**, 899 (1966).
54. J. A. Pople, *J. Chem. Phys.*, **37**, 60 (1962).
55. F. A. Bovey, J. J. Ryan, and F. P. Hood, *Macromolecules* **1**, 305 (1968).
56. F. Conti, M. Piatelli, and P. Viglino, *Biopolymers* **7**, 155 (1969).
57. C. M. Deber, F. A. Bovey, J. P. Carver, and E. R. Blout, *J. Amer. Chem. Soc.* **92**, 6191 (1970).
58. M. Goodman and M. Fried, *J. Amer. Chem. Soc.* **89**, 1264 (1967).
59. J. V. Hatton and R. E. Richards, *Mol. Phys.* **5**, 139 (1962).
60. M. B. Robin, F. A. Bovey, and H. Basch, Molecular and Electronic Structure of the Amide Group *in* "The Chemistry of the Amides" (J. Zabicky, ed.). Wiley, New York, 1970.
61. M. van Gorkom, *Tetrahedron Lett.* **1966**, 5433 (1966).
62. T. Wieland and H. Bende, *Chem. Ber.* **98**, 504 (1965).
63. B. Halpern, D. E. Nitecki, and B. Weinstein, *Tetrahedron Lett.* **1967**, 3075 (1967).
64. V. F. Bystrov, S. L. Portnova, V. I. Tsetlin, V. T. Ivanov, and Yu. A. Ovchinnikov, *Tetrahedron* **25**, 493 (1969).
65. M. Sheinblatt, *J. Amer. Chem. Soc.* **88**, 2845 (1966).
66. M. Sheinblatt, *in* "Magnetic Resonance in Biological Systems" (A. Ehrenberg, B. G. Malmström, and T. Vänngard, eds.), p. 35. Pergamon Press, Oxford, 1967.
67. J. T. Edsall, P. J. Flory, J. C. Kendrew, A. M. Liquori, G. Ramachandran, and H. A. Scheraga, *Biopolymers* **4**, 121 (1966); *J. Biol. Chem.* **241**, 1004 (1966); *J. Mol. Biol.* **15**, 399 (1966).
67a. J. C. Kendrew, W. Klyne, S. Lifson, T. Miyazawa, G. Nemethy, D. C. Phillips, G. N. Ramachandran, and H. A. Scheraga, *Biochemistry* **9**, 3471 (1970); *J. Biol. Chem.* **245**, 6489 (1970); *J. Mol. Biol.* **52**, 1 (1970).
68. D. R. Herschbach and L. C. Krisher, *J. Chem. Phys.* **28**, 728 (1958).
69. R. W. Kilb, C. C. Lin, and E. B. Wilson, *J. Chem. Phys.* **26**, 1965 (1957).
70. D. A. Brant and P. J. Flory, *J. Amer. Chem. Soc.* **87**, 2788 (1965).
71. R. A. Scott and H. A. Scheraga, *J. Chem. Phys.* **42**, 2209 (1965).
72. L. Pauling and R. B. Corey, *Proc. Nat. Acad. Sci. U.S.* **37**, 729 (1951).
73. E. W. Garbisch, Jr., *J. Amer. Chem. Soc.* **86**, 5561 (1964).

74. E. W. Garbisch, Jr., *J. Amer. Chem. Soc.* **85**, 1696 (1963).
74a. H. Hogeveen, A. F. Bickel, C. W. Hilbers, E. L. Mackor, and C. MacLean, *Chem. Commun.* **1966**, 898 (1966).
75. M. Karplus, *J. Chem. Phys.* **30**, 11 (1959); **33**, 1842 (1960); *J. Amer. Chem. Soc.* **85**, 2870 (1963).
76. S. Mizushima, "Structure of Molecules and Internal Rotation." Academic Press, New York, 1954.
76a. S. Mizushima, T. Shimanouchi, M. Tsuboi, and T. Azakawa, *J. Amer. Chem. Soc.* **79**, 5357 (1957).
76b. S. L. Portnova, V. F. Bystrov, V. I. Tsetlin, V. T. Ivanov, and Yu. A. Ovchinnikov, *Abh. Obs. Chim.* **38**, 428 (1968).
77. A. A. Bothner-By, C. Naar-Colin, and H. Gunther, *J. Amer. Chem. Soc.* **84**, 2748 (1962).
78. R. Degeilh and R. E. Marsh, *Acta Crystallogr.* **12**, 1007 (1959).
79. R. B. Corey, *J. Amer. Chem. Soc.* **60**, 1598 (1938).
80. F. A. Bovey, "Nuclear Magnetic Resonance Spectroscopy," pp. 67–68. Academic Press, New York, 1969.
81. M. Sheinblatt, *J. Chem. Phys.* **36**, 3103 (1962).
82. M. Sheinblatt, *J. Chem. Phys.* **39**, 2005 (1963).
83. M. Sheinblatt and H. S. Gutowsky, *J. Amer. Chem. Soc.* **86**, 4814 (1964).
84. E. Grunwald, A. Loewenstein, and S. Meiboom, *J. Chem. Phys.* **27**, 630 (1957).
85. A. Loewenstein and S. Meiboom, *J. Chem. Phys.* **27**, 1067 (1957).
86. H. S. Gutowsky, D. W. McCall, and C. P. Slichter, *J. Chem. Phys.* **21**, 279 (1953).
87. Z. Luz and S. Meiboom, *J. Amer. Chem. Soc.* **85**, 3923 (1963).
88. Z. Luz and S. Meiboom, *J. Chem. Phys.* **39**, 366 (1963).
89. Z. Luz and S. Meiboom, *J. Amer. Chem. Soc.* **86**, 4766 (1964).
90. M. Sheinblatt and Z. Luz, *J. Phys. Chem.* **66**, 1535 (1962).
91. A. Wishnia and M. Saunders, *J. Amer. Chem. Soc.* **84**, 4235 (1962).
92. A. Hvidt and S. O. Nielsen, *Advan. Protein Chem.* **21**, 287 (1966).
92a. S. W. Englander *in* "Poly-α-Amino Acids" (G. D. Fasman, ed.), Vol. I, p. 339. Dekker, New York, 1967.
93. A. Berger, A. Loewenstein, and S. Meiboom, *J. Amer. Chem. Soc.* **81**, 62 (1959).
94. M. Takeda and E. O. Stejskal, *J. Amer. Chem. Soc.* **82**, 25 (1960).
95. M. Sheinblatt, *J. Amer. Chem. Soc.* **87**, 572 (1965).
96. M. Sheinblatt, *J. Amer. Chem. Soc.* **88**, 2123 (1966).
97. W. E. Stewart, L. Mandelkern, and R. E. Glick, *Biochemistry* **6**, 150 (1967).
98. A. Hantzsch, *Chem. Ber.* **64**, 661 (1931).
99. R. A. Y. Jones and A. R. Katritzky, *Chem. Ind. (London)* **1961**, 722 (1961).
100. J. L. O'Brien and C. Niemann, *J. Amer. Chem. Soc.* **79**, 1386 (1957).
101. G. Fraenkel and C. Niemann, *Proc. Nat. Acad. Sci. U.S.* **44**, 688 (1958).
102. G. Fraenkel and C. Franconi, *J. Amer. Chem. Soc.* **81**, 62 (1959).
103. G. Fraenkel and C. Franconi, *J. Amer. Chem. Soc.* **82**, 4478 (1960).
104. G. Fraenkel, A. Loewenstein, and S. Meiboom, *J. Phys. Chem.* **65**, 700 (1961).
105. J. T. Edward, H. S. Chang, K. Yates, and R. Stewart, *Can. J. Chem.* **38**, 1518 (1960).
106. R. J. Gillespie and T. Birchall, *Can. J. Chem.* **41**, 148 (1963).
107. F. A. Bovey, *Pure Appl. Chem.* **16**, 417 (1968).
108. S. Hanlon, S. F. Russo, and I. M. Klotz, *J. Amer. Chem. Soc.* **85**, 2024 (1963).
109. I. M. Klotz, S. F. Russo, S. Hanlon, and M. A. Stake, *J. Amer. Chem. Soc.* **86**, 4774 (1964).
110. S. Hanlon, *Biochemistry* **5**, 2049 (1966).

111. F. A. Bovey, G. V. D. Tiers, and G. Filipovich, *J. Polym. Sci.* **38**, 73 (1959).
112. M. Goodman and Y. Masuda, *Biopolymers* **2**, 107 (1964).
113. D. I. Marlborough, K. G. Orrell, and H. N. Rydon, *Chem. Commun.* **1965**, 518 (1965).
114. W. E. Stewart, L. Mandelkern, and R. E. Glick, *Biochemistry* **6**, 143 (1967).
115. K.-J. Liu, S. J. Lignowski, and R. Ullman, *Biopolymers* **5**, 375 (1967).
116. J. L. Markley, D. H. Meadows, and O. Jardetzky, *J. Mol. Biol.* **27**, 25 (1967).
117. J. A. Ferretti, *Chem. Commun.* **1967**, 1030 (1967).
118. E. M. Bradbury, C. Crane-Robinson, and H. W. E. Rattle, *Nature* **216**, 862 (1967).
119. E. M. Bradbury, C. Crane-Robinson, H. Goldman, and H. W. E. Rattle, *Nature* **217**, 812 (1968).
120. E. M. Bradbury and H. W. E. Rattle, *Polymer* **9**, 201 (1968).
121. E. M. Bradbury, B. G. Carpenter, C. Crane-Robinson, and H. W. E. Rattle, *Nature* **220**, 69 (1968).
122. E. M. Bradbury, C. Crane-Robinson, H. Goldman, and H. W. E. Rattle, *Biopolymers* **6**, 851 (1968).
123. E. M. Bradbury and C. Crane-Robinson, *Nature* **220**, 1079 (1968).
124. J. H. Bradbury and G. J. Stubbs, *Nature* **218**, 1049 (1968).
125. F. Conti and A. M. Liquori, *J. Mol. Biol.* **33**, 953 (1968).
126. J. A. Ferretti and L. Paolillo, *Biopolymers* **7**, 155 (1969).
127. J. H. Bradbury and M. D. Fenn, *Aust. J. Chem.* **22**, 357 (1969).
128. D. I. Marlborough and H. N. Rydon, "Some Newer Physical Methods in Structural Chemistry" (Bonnett and Davis, eds.). United Trade Press, London, 1967.
129. J. A. Ferretti and B. W. Ninham, *Macromolecules* **3**, 30 (1970).
130. J. A. Ferretti, B. W. Ninham, and V. A. Parsegian, *Macromolecules* **3**, 34 (1970).
131. F. A. Bovey, J. J. Ryan, G. Spach, and F. Heitz, *Macromolecules* **4**, 433 (1971).
132. J. A. Schellman and C. Schellman, *in* "The Proteins" (H. Neurath, ed.), Vol. II, pp. 1–128. Academic Press, New York, 1964.
133. P. J. Urnes and P. Doty, *Advan. Protein Chem.* **10**, 401 (1961).
134. J. T. Yang, *in* "Newer Methods of Polymer Characterization" (B. Ke, ed.), pp. 103–153. (Interscience), Wiley, New York, 1964.
135. L. Pauling, R. B. Corey, and H. R. Branson, *Proc. Nat. Acad. Sci. U.S.* **37**, 205 (1951).
136. D. A. Brant, *Macromolecules* **1**, 291 (1968).
137. G. N. Ramachandran, ed., "Conformation of Biopolymers," Vol. 1, parts A & D. Academic Press, New York, 1967.
138. D. A. Brant, W. G. Miller, and P. J. Flory, *J. Mol. Biol.* **23**, 47 (1967).
139. A. E. Tonelli, *Macromolecules* **2**, 635 (1969).
140. A. E. Tonelli and F. A. Bovey, *Macromolecules* **3**, 410 (1970).
141. F. Quadrifoglio and D. W. Urry, *J. Phys. Chem.* **71**, 2364 (1967).
142. C. Robinson, J. C. Ward, and R. B. Beevers, *Disc. Faraday Soc.* **25**, 29 (1958).
143. D. E. Woessner, *J. Chem. Phys.* **42**, 1855 (1965).
144. M. Tsuboi, Y. Mitsui, A. Wada, T. Miyazawa, and N. Nagashima, *Biopolymers* **1**, 297 (1963).
145. R. D. Lundberg and P. Doty, *J. Amer. Chem. Soc.* **79**, 3961 (1957).
146. R. H. Karlson, K. S. Norland, G. D. Fasman, and E. R. Blout, *J. Amer. Chem. Soc.* **82**, 2268 (1960).
147. H. Sternlicht and D. Wilson, *Biochemistry* **6**, 2881 (1967).
148. A. Elliot and B. R. Malcolm, *Proc. Roy. Soc.* (*London*) **A249**, 30 (1959).
149. S. Arnott and A. J. Wonacott, *J. Mol. Biol.* **21**, 371 (1966).
150. W. B. Gratzer and P. Doty, *J. Amer. Chem. Soc.* **85**, 1193 (1963).

151. N. Lotan, A. Berger, E. Katchalski, R. T. Ingwall, and H. A. Scheraga, *Biopolymers* **4**, 239 (1966).
152. F. Quadrifoglio and D. W. Urry, *J. Amer. Chem. Soc.* **90**, 2755 (1968).
153. F. A. Bovey and F. P. Hood, unpublished observations.
154. H. E. Auer and P. Doty, *Biochemistry* **5**, 1715 (1966).
155. H. J. Sage and G. D. Fasman, *Biochemistry* **5**, 286 (1966).
156. G. D. Fasman, *in* "Poly-α-Amino Acids" (G. D. Fasman, ed.), pp. 537–557, 587. Dekker, New York, 1967.
157. J. Applequist and P. Doty, *in* "Polyamino Acids, Polypeptides and Proteins" (M. Stahmann, ed.), p. 161. Univ. of Wisconsin Press, Madison, Wisconsin, 1962.
158. P. K. Sarker and P. Doty, *Proc. Nat. Acad. Sci. U.S.* **55**, 981 (1966).
159. E. Friedman, T. J. Gill, and P. Doty, *J. Amer. Chem. Soc.* **84**, 3485 (1962).
160. K. Morita, E. R. Simons, and E. R. Blout, *Biopolymers* **5**, 259 (1967).
161. R. Lumry, R. Legare, and W. G. Miller, *Biopolymers* **2**, 434 (1964).
162. G. Schwarz, *Ber. Bunsenges. Physik. Chem.* **68**, 843 (1964).
163. G. Schwarz, *J. Mol. Biol.* **11**, 64 (1965).
164. G. Schwarz and J. Seeling, *Biopolymers* **6**, 1263 (1968).
165. R. C. Parker, K. Applegate, and L. J. Slutsky, *J. Phys. Chem.* **70**, 3018 (1966).
166. K. Saksema, B. Michels and R. Zama, *J. Chim. Phys. Physicochim. Biol.* **65**, 597 (1968).
167. G. G. Hammes and P. B. Roberts, *J. Amer. Chem. Soc.* **91**, 1812 (1969).
168. G. Schwarz, *Biopolymers* **6**, 873 (1968).
169. D. Poland and H. A. Scheraga, *J. Chem. Phys.* **45**, 2071 (1966).
170. D. Poland and H. A. Scheraga, *in* "Poly-α-Amino Acids" (G. D. Fasman, ed.), p. 467. Dekker, New York, 1967.
171. R. Ullman, Paper presented to Div. of Polymer Chem. 158th Meeting of Amer. Chem. Soc., New York City, Sept. 1969. (Preprint booklet, p. 1442.)
172. J. A. Ferretti and J. Milstien, private communication.
173. B. H. Zimm and J. K. Bragg, *J. Chem. Phys.* **31**, 526 (1959).
174. S. Lifson and A. Roig, *J. Chem. Phys.* **34**, 1964 (1961).
174a. W. G. Miller, *Polym. Prepr. Amer. Chem. Soc. Div. Polymer Chem.* **12**, 558 (1971).
175. M. Goodman, A. S. Verdini, C. Toniolo, W. D. Phillips, and F. A. Bovey, *Proc. Nat. Acad. Sci. U.S.* **64**, 444 (1970).
176. K. D. Kopple, M. Ohnishi, and A. Go, *J. Amer. Chem. Soc.* **91**, 4264 (1969).
177. R. Schwyzer, P. Sieber, and B. Gorup, *Chimia* **12**, 90 (1958).
178. R. Schwyzer, *Rec. Chem. Progr.* **20**, 147 (1959).
179. R. Schwyzer, *Helv. Chim. Acta* **47**, 441 (1964).
180. A. Stern, W. A. Gibbons, and L. C. Craig, *Proc. Nat. Acad. Sci. U.S.* **61**, 734 (1968).
181. A. M. Liquori and F. Conti, *Nature* **217**, 635 (1968).
182. F. Conti, *Nature* **221**, 777 (1969).
183. D. C. Hodgkin and B. M. Oughton, *Biochem. J.* **65**, 752 (1957).
184. R. Schwyzer, *in* "Amino Acids and Peptides with Antimetabolic Activity" (G. E. W. Wolstenholme, ed.), p. 171. Churchill, London, 1958.
185. R. Schwyzer and U. Ludescher, *Biochemistry* **7**, 2514, 2519 (1968).
186. D. H. Haynes, A. Kowalsky, and B. C. Pressman, *J. Biol. Chem.* **244**, 502 (1969).
187. T. A. Victor, F. E. Hruska, K. Hikichi, S. S. Danyluk, and C. L. Bell, *Nature* **223**, 302 (1969).
188. J. H. Fessler and A. G. Ogston, *Trans. Faraday Soc.* **47**, 667 (1951).
189. A. N. Glazer and K. Rosenheck, *J. Biol. Chem.* **237**, 3674 (1962).
190. J. Dale and K. Titlestad, *Chem. Commun.* **1969**, 656 (1969).

191. J. E. Mark and M. Goodman, *J. Amer. Chem. Soc.* **89**, 1267, (1967).
192. J. E. Mark and M. Goodman, *Biopolymers* **5**, 809 (1967).
193. A. M. Liquori and P. De Santis, *Biopolymers* **5**, 815 (1967).
194. M. Goodman and F. Prince, Jr., private communication (1969).
195. J. P. Carver and E. R. Blout, *in* "Treatise on Collagen" (G. N. Ramachandran, ed.), Vol. I. Academic Press, New York, 1967.
195a. L. Mandelkern *in* "Poly-α-Amino Acids" (G. D. Fasman, ed.), p. 675. Dekker, New York, 1967.
196. P. R. Schimmel and P. J. Flory, *Proc. Nat. Acad. Sci. U.S.* **58**, 52 (1967).
197. F. A. Bovey, "Polymer Configuration and Conformation," Chapter V. Academic Press, New York, 1969.
198. P. De Santis, E. Giglio, A. M. Liquori, and A. Ripamonti, *Nature* **206**, 456 (1965).
199. S. J. Leach, G. Nemethy, and H. A. Scheraga, *Biopolymers* **4**, 369 (1966).
200. J. M. Rifkind and J. Applequist, *J. Amer. Chem. Soc.* **90**, 3651 (1968).
201. G. Holzwarth and K. Backman, *Biochemistry* **8**, 883 (1969).
202. G. Holzwarth and R. Chandrasekaran, *Macromolecules* **2**, 245 (1969).
203. A. E. Tonelli, *J. Amer. Chem. Soc.* **92**, 6187 (1970).

Chapter XIV

PROTEINS

1. Introduction

Without doubt, x-ray diffraction is the prime method for the determination of the structure of proteins in the native, crystalline state. Indeed, it is difficult at present to imagine any other technique capable of providing the thousands of parameters necessary to achieve this spectacular accomplishment. For the study of proteins in solution however, other methods must be used. In the past, a great variety of physical and chemical means have been employed to elucidate conformations and conformational changes, substrate and cofactor binding, enzymatic specificity, and many other aspects of protein structure and function. The sum of knowledge which has been obtained in this way is very great. Only very recently has high resolution NMR emerged as a method which can also provide a large number of structural parameters, some of which are peculiar to itself and give information not otherwise accessible. It cannot precisely be said to rival x-ray diffraction, but is proving capable of supplying very significant complementary knowledge.

NMR studies have been devoted to three principal aspects of protein structure: first, the direct observation of the protein molecule itself, with particular emphasis on effects produced by interaction of the chains in the native or folded state and by the process of unfolding or denaturation; second, the observation of the binding of small molecules—including substrates, inhibitors, cofactors, and the solvent itself—to the protein; and third, the study of paramagnetic active sites in enzymes and electron-transfer proteins through their effect on the chemical shifts of neighboring protons and in relaxing the nuclei of the solvent or other associated molecules. This last is one of the earliest biological uses of NMR, but will be dealt with only in passing here because our main interest is in polymer structure as such.

Very logically, the bulk of this work has dealt with proteins whose structures in the native state have been determined by x-ray diffraction: lysozyme, ribonuclease, myoglobin, hemoglobin, and insulin. Some attention has also been given to trypsin, chymotrypsin, and their zymogen precursors, and the cytochromes, for which partial structures or at least amino acid sequences are available. The principal nucleus observed has of course been the proton, but ^{19}F, ^{31}P, ^{35}Cl, and a few other nuclei have found limited application in specialized studies.

2. Direct Observations of Protein Structure

The first observation of the NMR spectrum of a protein was that of ribonuclease in D_2O by Saunders *et al.*[1] These measurements were made at 40 MHz and revealed four broad peaks (Fig. XIV-1) but no fine structure. The

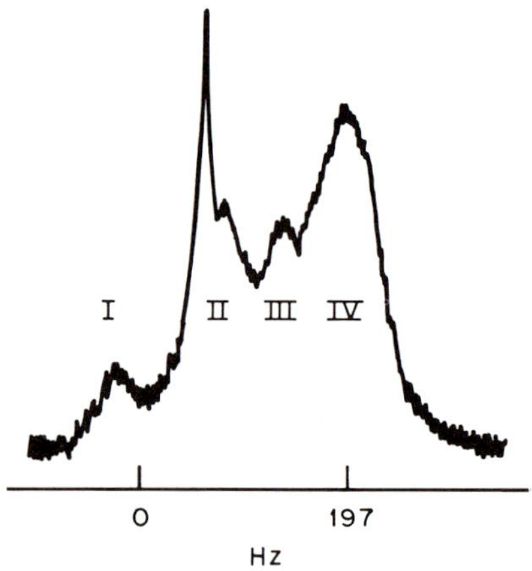

FIG. XIV-1. 40 MHz spectrum of ribonuclease in D_2O. The reference zero is the aromatic resonance of external toluene (Saunders *et al.*[1]).

peaks were assigned in a general way by Jardetzky and Jardetzky.[2] Bovey *et al.*[3] reported the 40 MHz spectra of a number of proteins in trifluoroacetic acid, a solvent in which they are completely unfolded. As many as nine peaks could be distinguished and assigned. The narrowing of the peaks of the spectrum of bovine serum albumin in D_2O upon unfolding in the presence of urea was also demonstrated. Kowalsky observed a number of proteins at 56.4 MHz[4] and made a particularly significant study of cytochrome C at 60 MHz.[5] In the latter, he observed and recognized contact-shifted peaks appearing at positions several ppm upfield from the DSS reference (see Section 2D). Bradbury and

Scheraga[6] were able to use the C-2 proton peaks of the imidazole group of the histidine residues[7] (Chapter XIII, Section 2) in the 60 MHz spectrum of ribonuclease to establish apparent *pK* values for these groups.

These pioneering studies at lower frequencies clearly indicated the possibilities of NMR for protein investigations, but peak overlap was so severe that, except for special resonances well removed from the main spectrum, it was not possible to obtain significant detailed structural information. Mandel[7, 8] showed that the gain in sensitivity and chemical shift discrimination possible at 100 MHz expanded the potential information available substantially, and a number of important 100 MHz studies will be cited in later discussion. At 220 MHz[9, 10] the spectral resolution becomes sufficient to offer the promise of structural information concerning a major proportion of the protein molecule, although so far only glimpses of these potentialities are available. Denaturation and small-molecule binding are particularly attractive subjects of investigation. At this frequency, the observation of anomalous peaks is of course also correspondingly improved.

A. General Peak Assignments

Protein spectra all bear a strong family resemblance. It is the differences in detail, particularly those produced by small-molecule binding, chain folding and unfolding, and other structural alterations, which are important. The general assignments of their proton resonances follow those of analogous small molecules, amino acids and peptides, described in Chapter XIII and given in Table XIII-1. In Table XIV-1 these are shown for the twenty commonly occurring amino acid residues. These data are in large part taken from the careful and comprehensive work of McDonald and Phillips[11] and include minor adjustments to take account of deviations in chemical shifts for amino acids in long polypeptide chains as compared to the free acids or short peptides. The values given are to be expected for protein chains in the completely unfolded, random-coil state and on the assumption (very nearly valid) of no interactions between neighboring residues. For groups significantly affected by the state of protonation, the range of chemical shifts to be expected over the *p*H range of about 1–13 are shown; this applies to the histidine ring protons, and to methylene groups next to side-chain amino and carboxyl groups. End units are ignored as are those of possible prosthetic groups such as heme units. Protons of NH_2, OH, CO_2H, and imidazole NH groups are not shown, since they normally appear in the solvent peak because of rapid exchange (Chapter XIII, Section 3G). Chemical shifts of specific residues (aside from titration ranges) are expressed as bars with an arbitrary width of 0.1 ppm for convenience in plotting. Alpha-protons (with the exception of those of glycine) and peptide NH protons are lumped together and the range normally exhibited is shown.

TABLE XIV-1

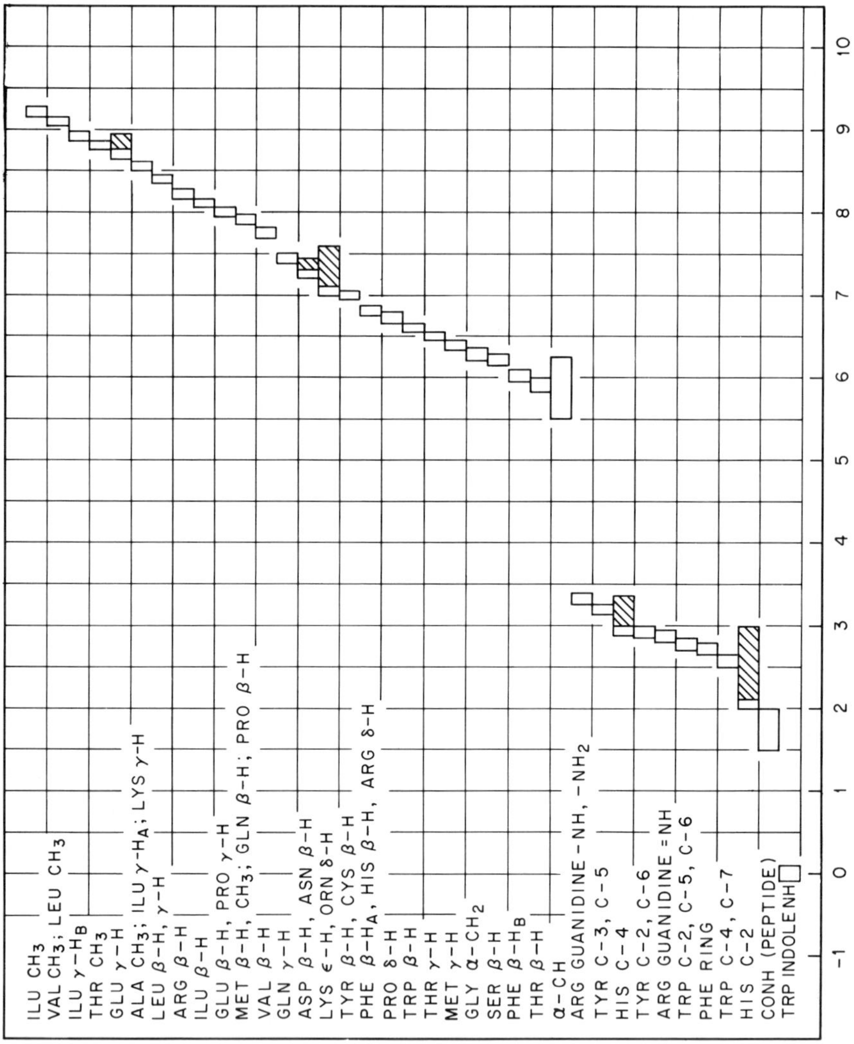
ILU CH3
VAL CH3; LEU CH3
ILU γ-HB
THR CH3
GLU γ-H
ALA CH3; ILU γ-HA; LYS γ-H
LEU β-H, γ-H
ARG β-H
ILU β-H
GLU β-H, PRO γ-H
MET β-H, CH3; GLN β-H; PRO β-H
VAL β-H
GLN γ-H
ASP β-H, ASN β-H
LYS ε-H, ORN δ-H
TYR β-H, CYS β-H
PHE β-HA, HIS β-H, ARG δ-H
PRO δ-H
TRP β-H
THR γ-H
MET γ-H
GLY α-CH2
SER β-H
PHE β-HB
THR β-H
α-CH
ARG GUANIDINE -NH, -NH2
TYR C-3, C-5
HIS C-4
TYR C-2, C-6
ARG GUANIDINE =NH
TRP C-2, C-5, C-6
PHE RING
TRP C-4, C-7
HIS C-2
CONH (PEPTIDE)
TRP INDOLE NH
-1
0
1
2
3
4
5
6
7
8
9
10
τ

The principal characteristics of the protein spectrum in the random-coil state are that (a) the side chains are in a solvent environment, and (b) the correlation times (Chapter I, Section 5) of both side-chain and backbone protons are at a minimum. The resolution of individual peaks is therefore optimal, as we have seen, and their positions are most readily predictable. Such spectra offer the best opportunity to identify all resonances, and serve as a norm to which the molecules in their folded and biologically functional states can be compared. McDonald and Phillips[11] have shown that spectra computed on the basis of known amino acid compositions (but without regard for sequence) and on the basis of model compound chemical shifts for neutral aqueous solutions match (with minor adjustments) quite closely to the experimental 220 MHz spectra of the denatured protein. We shall illustrate some of these results in later discussion.

The least shielded protons (Table XIV-1) are those of tryptophan indole NH (about 0.0τ), peptide NH (1.5–2.0τ), and C-2 histidine and arginine NH protons (2–3τ). The peptide NH protons will in general not be seen in D_2O solution, and in H_2O will broaden or disappear at higher *p*H because of base-catalyzed exchange (Chapter XIII, Section 3G). Aromatic and C-4 histidine protons appear next, near 2.5–3.2τ, followed by a pronounced "window" which is observed at about 3.5 to 5τ in all protein spectra. Between 4 and 6τ appear the broad and usually undifferentiated resonances of the α-CH protons. These tend to be obscured to varying degrees—often nearly completely—by H_2O or HDO peaks. Then come a wide variety of side-chain methylene and methyl resonances. At highest field, near 9τ, are the methyl groups of the aliphatic side chains of valine, leucine, and isoleucine; these form a single prominent and rather broad resonance in random-coil spectra, but may be split into a number of peaks in the native state because of a variety of environmental influences. These peaks and those at the other extreme of the spectrum (and contact-shifted resonances) have been those most intensively studied.

In the following subsections, we shall detail the effects of *p*H (or *p*D), temperature, solvent, oxidation state, and small-molecule binding upon the spectra of individual proteins, including those with heme units and other prosthetic groups, the resonances and influences of which we have not yet discussed. Ring-current shifts (Chapter I, Section 11), contact interactions (Chapter I, Section 11 and Chapter XIII, Section 2E), hydrogen bonding, and local charge variations are principal contributors to the more significant spectral phenomena. A number of other contributing factors, as well as other techniques of observation which are best considered in specific applications, will also be discussed. Most of the spectra shown are the result of the accumulation of a number (sometimes 100 or more) of CAT (Chapter I, Section 18C) scans.

B. Lysozyme

Lysozyme is found very widely distributed in many bodily secretions, including tears and nasal mucus. It is particularly plentiful in hen egg whites. Hen egg white (HEW) lysozyme has a molecular weight of 14,600. There are 129 amino acid residues in a single chain containing four disulfide bridges. Human lysozyme differs in sequence from HEW lysozyme at some fifty residues but is very similar in form and function.† The structure of HEW lysozyme is known from x-ray diffraction.[13–18] It has a roughly spherical form with a very deep cleft in which the mucopolysaccharide substrate (see Section 3) is bound. The cleft divides the molecule into two parts: the sequences 1–40 and 101–129, which are predominantly helical; and an irregular part, residues 41–87. Paralleling the cleft is a further segment of helix, 88–100. About 31 residues are involved in α-helices, some portions of which are considerably distorted. A considerable fraction of the residues 41–61 are involved in β-structures of varied regularity. The great majority of the hydrophobic groups (Ala, Leu, Ilu, Val, Phe, Pro) are buried in the interior of the structure and nearly all the charged polar groups are on the surface—a common structural pattern in globular proteins. The protein tends to dimerize at *p*H above about 7 and so is best studied at *p*H 6 or lower.

Both HEW and human lysozyme have been studied by NMR,[7–11, 19–28] the major share of the effort being devoted to the former. Earlier work[7, 8] established the general characteristics of the spectrum in D_2O and indicated a relatively high stability against thermal unfolding up to 65°, probably due in part to the four disulfide linkages. Measurements at 220 MHz have shown small changes in the 35–60° range, but in general support this conclusion.[24–27] On heating from 65 to 75°, however, denaturation occurs and marked spectral changes are observed. (The origin of these changes will be discussed in more detail a little later; for the present we merely note that they occur.) In Fig. XIV-2, the complete spectra are shown for 10% solutions in D_2O (*p*D 5.5) at 30°, 54°, and 80°. In Fig. XIV-3 the low-field and high-field regions are shown in more detail at temperatures just below (65°) and just above (80°) denaturation, and are compared with a reconstructed spectrum based primarily on the data of Table XIII-1, and therefore appropriate for the random-coil state, under the assumption of about 10 Hz line width.[11, 24] The spectrum at 80° closely approximates that of the random coil, except for the restraints imposed by the disulfide linkages. When these are ruptured, the lines narrow

† For a convenient summary of sequence and structural data on this and other proteins discussed in this chapter, see Dickerson and Geis,[12] and Dayhoff.[12a] The reader is referred to these sources for such information, particularly with regard to the environments of specific amino acid residues, in connection with the subsequent discussion.

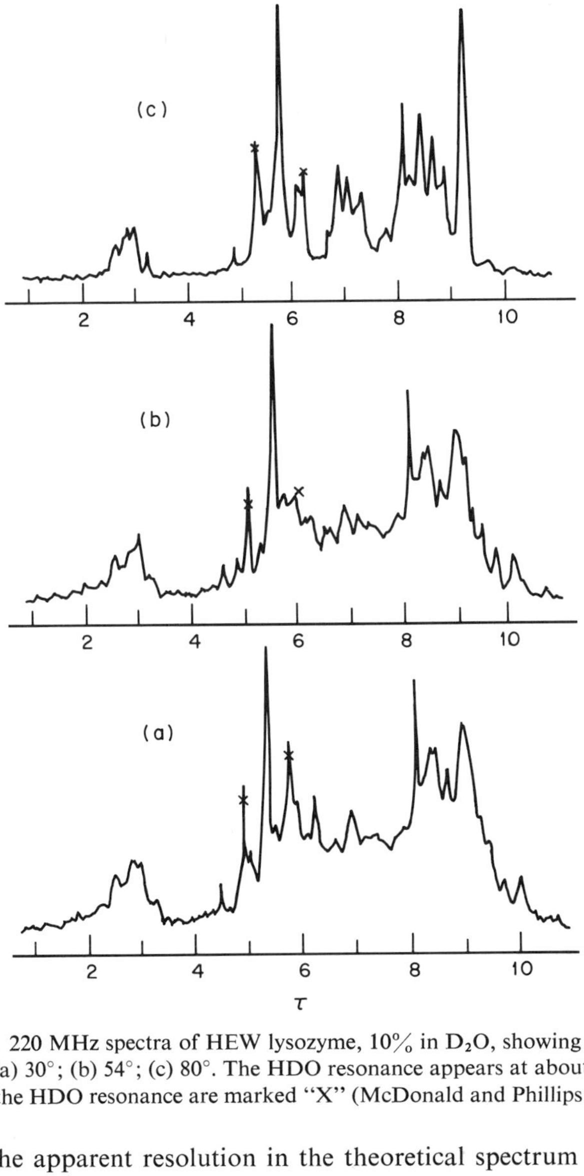

FIG. XIV-2. 220 MHz spectra of HEW lysozyme, 10% in D_2O, showing dependence on temperature: (a) 30°; (b) 54°; (c) 80°. The HDO resonance appears at about 5.5τ. Spinning side bands to the HDO resonance are marked "X" (McDonald and Phillips[10]).

further.[22] The apparent resolution in the theoretical spectrum exceeds even that of the latter, because of the triangle approximation used in constructing it.[11]

It will be noted that in one detail, spectra (a) and (b) in Fig. XIV-3 fail to agree: the theoretical spectrum shows a histidine C-2 proton peak at 1740 Hz which does not appear in the experimental spectrum. This proton is exceptional among carbon-bonded protons in being slowly exchanged with the solvent; at 80° it was completely exchanged in the time required to obtain the spectrum in Fig. XIV-3(b).

The unfolding represented by Fig. XIV-3(b) is entirely reversible. Upon cooling through the denaturation temperature range, spectrum (c) is restored.

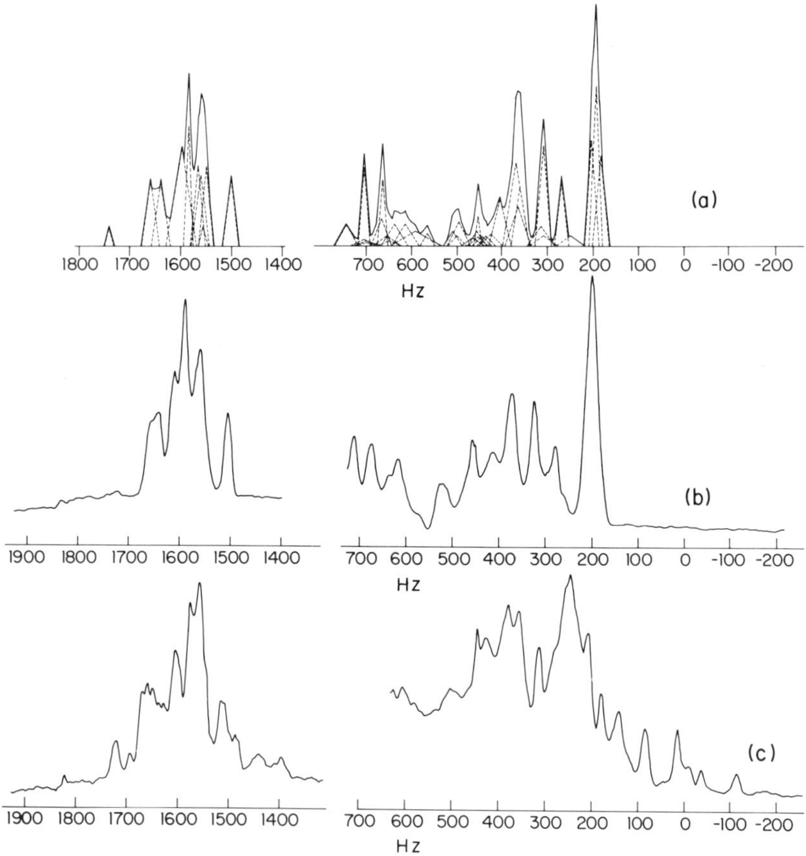

FIG. XIV-3. Low-field and high-field portions of the HEW lysozyme spectrum (220 MHz) in D_2O: (a) calculated spectrum of denatured protein (see text); (b) observed spectrum at 80°; (c) observed spectrum at 65°. Chemical shifts expressed in hertz, referred to DSS (McDonald and Phillips[11,24]).

An apparent deviation from reversibility is observed in the NH region. The native protein which has never been heated above 65° in D_2O shows a number of small discrete NH resonances which disappear in the "renatured" protein.[25] This is not due to a failure of refolding, however, because these peaks also disappear at 50°, i.e., well below denaturation. They evidently arise from interior NH groups which exchange only when the protein molecule "breathes" and exposes them to the solvent.

A careful examination of lysozyme spectra at temperatures between the extremes of the native and completely unfolded structures represented by Fig. XIV-3 shows that the alterations in peak positions and intensities do not occur by gradual steps, but that upon raising the temperature, new peaks corresponding to the random-coil state "grow in" while the native protein peaks shrink in intensity.[25, 27] Upon lowering the temperature, these changes occur in reverse. The spectra of intermediate states may be reproduced by an appropriate summation of the native and random-coil spectra. A clear example of this behavior is shown in Fig. XIV-4. The C-2 proton peaks of the His-15 residue (see below) of HEW lysozyme appears at 1.04τ in the native spectrum (a) and at 1.41τ in the denatured state (c); in the intermediate state represented by (b), these peaks appear with equal intensity and *without broadening*. This behavior is strongly reminiscent of the helix–coil transition in polypeptides discussed in Chapter XIII, Section 4. It indicates that the folded–unfolded transition can be represented by a two-state scheme representing the entire protein molecule (a conclusion also indicated by other physical measurements[29–35]) and that the transition is slow on the NMR time scale. One may postulate any number of intermediate partially unfolded states which exchange rapidly with each other and with the native molecule on the one hand or with the random-coil molecule on the other

$$\text{native} \rightleftarrows \text{A} \rightleftarrows \text{B} \rightleftarrows \cdots \text{Y} \rightleftarrows \text{Z} \rightleftarrows \text{random coil}$$

but at some point in this progression at least one slow step must occur. The NMR data would also appear to indicate that the concentration of the A, B, ..., X, Y states must be small compared to the initial and final states.[28] At present there is no experimental evidence, at least in the case of lysozyme, for any recognizable intermediate states. The position is not necessarily so simple for other proteins. From the measurement of the intensities of suitably chosen low-field and high-field peaks (see below) as a function of temperature, McDonald *et al.*[28] deduced values of $\Delta H = 73.5$ kcal-mole^{-1} and $\Delta S = 215$ cal-deg^{-1}-mole^{-1} for the thermal denaturation.

Let us now consider in more detail some of the significant high-field and low-field resonances of lysozyme, and their dependence on chain conformation and small-molecule binding.

1. *The NH and Aromatic Region*

The titration behavior of the lone His residue, which occurs at position 15 in HEW lysozyme and at position 77 in human lysozyme (well removed spatially), has been studied.[21, 23] The C-2 and C-4 protons both become more shielded as the ring is deprotonated (Figs. XIII-4 and XIII-5). Rather surprisingly, the *pK* of this residue in human lysozyme is higher (7.1) than in HEW lysozyme (5.8) and even higher than for free histidine (6.8). The higher *pK* in human lysozyme may indicate proximity to a negative charge or strong hydrogen bonding to an acceptor group, while the relatively low *pK* in HEW

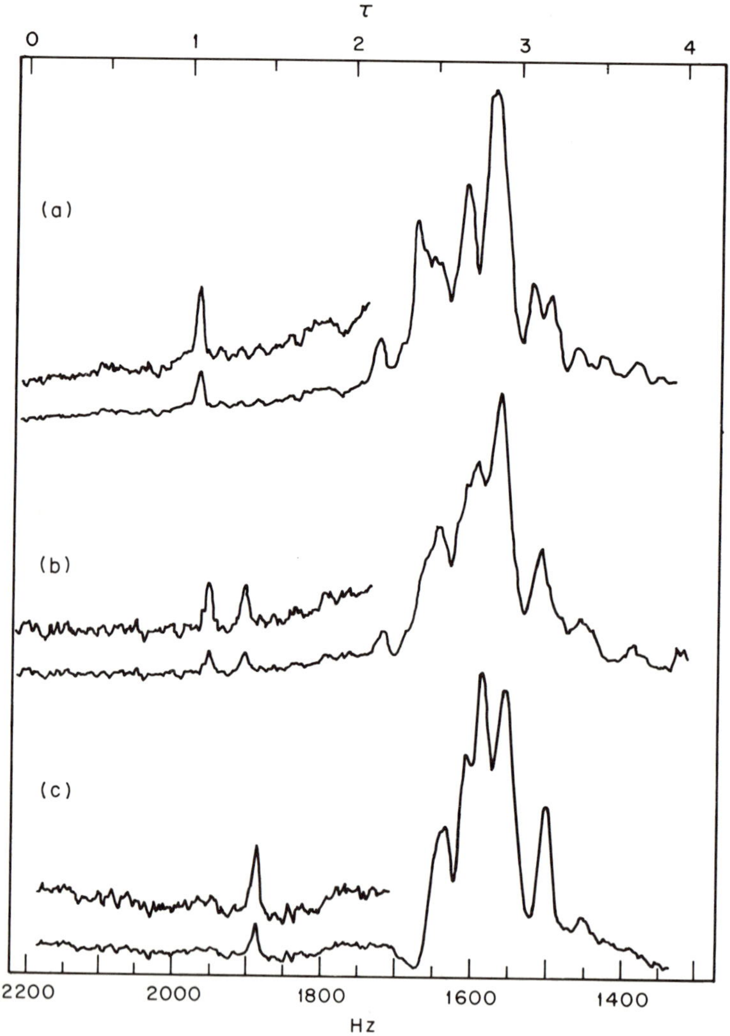

FIG. XIV-4. 220 MHz spectra of the aromatic resonances of HEW lysozyme in D_2O: (a) 60°; (b) 69°; (c) 80°. The C-2 resonance of His-15 appears at 1.0–1.4τ (1880–1950 Hz from DSS, lower scale) (McDonald and Phillips,[24] and Glickson *et al.*[26]).

lysozyme may be due to a neighboring positive charge or hydrogen bonding to a donor group.

In both cases, the His units are external rather than buried and their relatively narrow (3–5 Hz) resonance lines suggest that they may be free to rotate. Binding studies (Section 3) show that the His-15 resonance in HEW lysozyme is unaffected by the presence of substrate and is therefore not involved in the active site although it is not very far away from it.

Figure XIV-5 shows the region from −1.2 to 3.0τ for HEW lysozyme (observed at 220 MHz) at (a) *p*H 3.3 in H_2O at 71°; (b) at 35°; and (c) in D_2O at 35°.[24, 25] The spectra differ from those in Fig. XIV-3 chiefly in that the peptide NH protons are now present, or at most only partially exchanged for deuterons (c); their resonances appear in the region from 1.4 to 2.2τ in the denatured protein (a). When the protein is refolded (b), these peaks become spread out over a more extended range down to about 0.3τ and present a more complex appearance. At this *p*H, the peaks are narrow because exchange with the solvent is slow on the NMR time scale, even for the denatured protein. The process is base-catalyzed, however (Chapter XIII, Section 3G), so that at a *p*H equal to or greater than 7, these resonances broaden and disappear. Figure XIV-(5c) shows that CONH protons exchange with D_2O at different rates (as known from other studies), since under conditions of partial exchange their peak pattern changes. Individual assignments in this region have not yet been made.

The resonances in the 2.2–3.0τ range are those of the aromatic CH protons of one His (just discussed), three Phe, three Tyr, and six Trp residues. The Phe and Tyr residues have not been individually assigned, but Cohen and Jardetzky[22] were able to detect a downfield shift of peaks attributed to them upon thermal or urea denaturation. This is consistent with a mutual shielding (Chapter I, Section 11) among the Phe-3, Phe-34, and Phe-38 residues, which are clustered together in the folded state, and a similar mutual shielding of Tyr-20 and Tyr-23. Tyr-53 is relatively isolated from other aromatic residues, and could be expected to show smaller effects.

At very low field, appearing as a single peak near 0.0τ in the denatured HEW lysozyme spectrum [Fig. XIV-5(a)] and as five separate peaks in the native spectrum [Fig. XIV-5(b)], are the indole NH protons of five of the six Trp residues.[24, 26] The sixth has not yet been identified; it probably occurs in the complex peptide NH region. These resonances correspond to exchangeable protons and are thus best observed at slightly acid *p*H or *p*D (about 3.3).[28] (In the spectrum of human lysozyme, which has only five Trp residues, five separate resonances can be resolved.[24, 26]) The two at lowest field, designated I and II in Fig. XIV-5(b) and (c), exchange more slowly with D_2O than the others, and are therefore assigned to Trp-28 and Trp-108, which are known[15, 16] to be hydrogen bonded to the peptide carbonyl groups of Tyr-23 and Leu-56, respectively. The assignment of peak II to Trp-108 is further strengthened by the observation[26] that binding of the inhibitors *N*-acetylglucosamine (NAG; p. 377) and *N*,*N*′-diacetylglucosamine further reduces its rate of exchange; from the x-ray structure it appears that only Trp-108 should be further obstructed by inhibitor binding. It is also observed that Trp-111 is

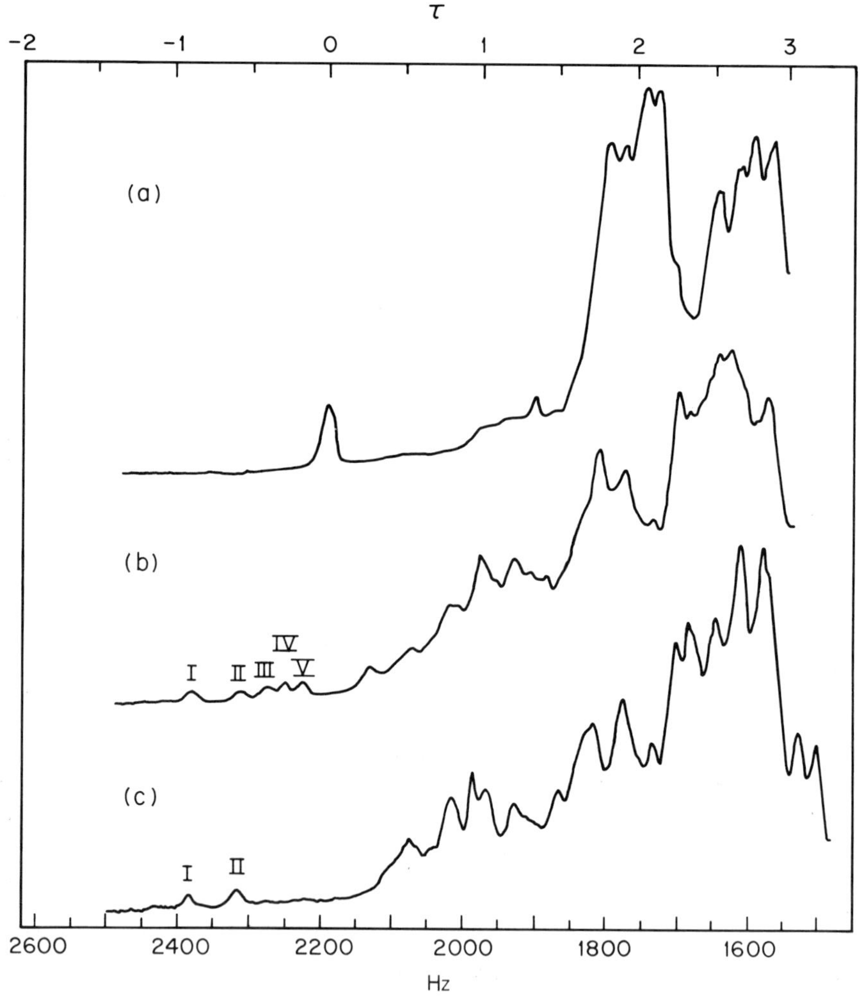

FIG. XIV-5. 220 MHz spectra of the -1.2 to 3.0τ resonances of HEW lysozyme spectrum: (a) *p*H 3.3 in H_2O at 71°; (b) same at 35°; (c) in D_2O (*p*D 3.3) at 35°. The peaks marked with roman numerals are those of the indole NH of tryptophan residues. Bottom scale in hertz from DSS (McDonald and Phillips[24,25]).

located in a relatively inaccessible interior position, and so assignment of peak I to this residue cannot be excluded at present.

Like Trp-108, Trp-62 and Trp-63 are also situated at the substrate binding cleft containing the active site; the latter two are believed to participate in binding substrates. Treatment with *N*-bromosuccinimide removes peak III[26]; this provides its assignment to Trp-26, which is known to be selectively oxidized to an oxindole under these conditions.[36] Inhibitor binding perturbs the positions only of peaks III and V. This probably arises from hydrogen bonding of the indole NH protons to the inhibitor, and permits the assignment of peak V to Trp-63.

The sixth Trp NH resonance cannot yet be specifically identified, but by logical elimination must be assigned to Trp-111 (or possibly, considering the ambiguity of peak I, to Trp-28). The assignments are summarized in Table XIV-2.

TABLE XIV-2

Peak	I	II	III	IV	V	VI
Residue	28 or 111	108	62	123	63	111 or 28

It is evident that such information can be very valuable in precise studies of the binding of substrates and inhibitors to the active site of the enzyme. Unfortunately, this approach is not applicable to the next-most-studied enzyme, ribonuclease (see Section 2C), because this protein contains no tryptophan.

2. *High-Field Resonances*

The high-field portion ($\geqslant 8.50\tau$) of the lysozyme spectrum has been interpreted principally in terms of ring-current shifts (Chapter I, Section 11) arising from proximity of CH_3 and CH_2 groups to aromatic rings.[9, 10, 20, 24, 28] Such shifts no doubt occur in the less shielded parts of the spectrum as well, but are difficult to detect because of the large number of overlapping resonances. Ring-current shifts may of course be of either sign. This is evidenced in the lysozyme spectrum by the observation that the methyl resonances of the denatured protein form a single large peak at 9.1τ [Fig. XIV-3(b)] whereas in the folded state [Fig. XIV-3(c) and Fig. XIV-6] this peak is largely replaced by smaller peaks both upfield and downfield.

On the basis of the x-ray structure[13–18] and the Johnson–Bovey ring-current shift tables, Sternlicht and Wilson[20] have calculated the increased shielding which would be experienced by the methyl and methylene protons of aliphatic side chains which are appropriately positioned with respect to aromatic rings. On the same basis, McDonald and Phillips[9, 10, 24, 28] have assigned many of the more prominent aliphatic proton peaks (mainly CH_3) in the high-field region of the spectrum to specific residues. These two studies agree quite well for the most part, but diverge in that in each some peaks are assigned that are

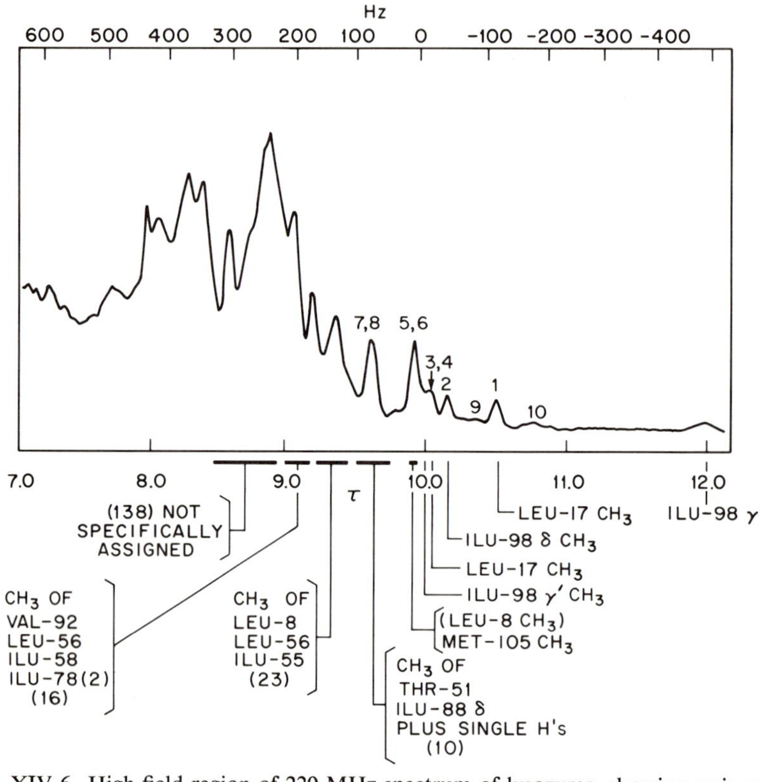

FIG. XIV-6. High-field region of 220 MHz spectrum of lysozyme, showing assignments made by McDonald and Phillips.[24] Numbers in parentheses indicate the number of protons assigned in each peak or group of peaks. See Table XIV-3. Upper scale in hertz from DSS.

omitted in the other. For example, the Met-105 CH_3 peak at 9.93τ, considered the most securely assigned by McDonald and Phillips—in part on the basis of its narrowness (since there is no appreciable coupling of these protons to any others)—is not assigned by Sternlicht and Wilson, whereas the latter assigned three Thr-51 and eight Lys CH_2 protons, predicted to occur at about $9.2–10.4\tau$ and not specifically considered by the other investigators. Of the total of 40 high-field methyl resonances in lysozyme (8 Leu, 6 Ilu, and 6 Val), about half appear to be shifted upfield from their "normal" or random-coil positions,[11] according to the assignments of McDonald and Phillips (Table XIV-1); the remainder of the 72 abnormally shielded protons are assigned to side chain CH and CH_2 groups. Nine protons are at the unperturbed Ilu methyl position of 9.12τ.

Downfield shifts of Leu, Ilu, Val, and Thr methyl resonances, caused by proximity to the edges rather than the faces of aromatic rings, are not expected to exceed about 0.23–0.24 ppm for close approach; most will show lesser perturbations. The 86 protons in the 8.64–8.99τ region (Table XIV-3) are considered to arise from such methyl groups, as well as from protons that are at lower fields in the random-coil spectrum and are shifted upfield into this region by ring currents.

TABLE XIV-3

HIGH-FIELD RESONANCES IN HEW LYSOZYME[a]

τ	Estimated number of protons	Assignment	Major ring-current contributor
11.98	3	(Ilu-98 γ)	Trp-63
10.84	1	—	—
10.56	3	Leu-17 CH_3	Tyr-28 (Tyr-20 minor)
10.27	1	—	—
10.17	3	Ilu-98 δ-CH_3	Trp-63
10.04	3	Leu-17 CH_3	Trp-28 (Tyr-20 minor)
10.00	3	Ilu-98 γ-CH_3	Trp-63 Trp-108
9.93	3	(Leu-8 CH_3)	Phe-3
9.93	3	Met-105 CH_3	Trp-28 Trp-108 Trp-111 Trp-23
9.81	2	—	—
9.74–9.26	10	2 or 3 CH_3 (Thr-51 CH_3, Ilu-88 δ-CH_3) plus single protons	Tyr-53 Phe-3 His-15
9.74–9.26	23	(CH_3 of Leu-8, Ilu-56, Ilu-55)	—
9.26–9.12	16	[CH_3 of Val-92, Ilu-56, Ilu-58, Ilu-78(2)]	—
9.12–8.99	26	—	—
8.99–8.64	86	—	—
8.64–8.50	26	—	—

[a] See McDonald and Phillips[24] and McDonald *et al.*[28]

The addition of Co^{2+} ion to solutions of lysozyme has been found[25] to perturb many of the resonances of lysozyme, some being moved upfield and others downfield, due to pseudocontact effects on the chemical shifts. These effects arise, like contact shifts (Chapter XIII, Section 2E), from the presence of unpaired electron(s) but are due not to delocalization of spin, but to direct interaction with the magnetic dipole of the unpaired electron. They are invoked to explain shifts of protons in aliphatic chains, in which no delocalization of spin would be expected, as in the present case. Methyl resonances in the region from 9.9 to 10.5τ which are normally grouped together (Fig. XIV-6) are found to be split apart. On the assumption (for which there is evidence) that the Co^{2+} is specifically coordinated at Glu-35, peaks 1 and 3 are assigned to methyl groups of Leu-17; peaks 2 and 4, to the δ and γ' methyls of Ilu-98 (peaks 3 and 4 are split apart); peaks 5 and 6 (also split, 5 being the more shielded), to methyls of Leu-8 and Met-105, respectively. These assignments are in agreement with those shown in Fig. XIV-6 for the unperturbed spectrum.

McDonald *et al.*[28] have observed the acid denaturation of lysozyme (*p*D 3.3–1.6 at 55°). The spectral changes in general parallel those occurring during thermal denaturation, except that in the presence of HDO new peaks corresponding to the NH groups of the 11 Arg residues and to the peptide NH protons appear as the *p*D (*p*H) is lowered, because of the retardation of exchange with the solvent.

Ion, substrate, and inhibitor binding to lysozyme are discussed in Section 3.

C. Ribonuclease

The function of ribonuclease is of course the hydrolytic scission of ribonucleic acids and oligonucleotides. As we have seen, it was the subject of the earliest NMR investigation of a protein, although the spectra then attainable revealed no distinctive details. It is similar in size (13,700 molecular weight, 124 amino acid residues) and shape to lysozyme and is therefore favorable for NMR study. It has four disulfide crosslinks, contains eighteen basic amino acids (10 Lys, 4 Arg, and 4 His) and only ten acidic amino acids (5 Glu and 5 Asp), and is therefore positively charged in solution at neutral *p*H. It has somewhat less α-helical and more β structure (residues 42–49, 71–92, and 94–110) than lysozyme. In addition to 4 His, it has also 6 Tyr and 3 Phe but no Trp residues. The complete three-dimensional structure is known from the x-ray studies of two groups.[37, 38, 38a] It is roughly spheroidal in form with a large groove in one side to accommodate the substrate. His-12, His-119, and Lys-7 are clustered together on one side of this groove while Lys-41 is positioned on the other side. All four of these residues are known from extensive chemical studies to be involved in the active site.

Ribonuclease can be split between residues 20 and 21 by subtilisin. The resulting "S-peptide" remains attached to the rest of the protein, which is termed the "S-protein." The S-protein is fully active and has an x-ray structure differing only in minor details from that of the unmodified protein. It can be dissociated in urea solution, with loss of activity. Activity is restored on reassociation.

Kowalsky[4, 5] and Mandel[7, 8] reported marked changes in the 60 and 100 MHz spectra of ribonuclease, chiefly peak narrowing, upon scission of the disulfide crosslinks or upon raising its aqueous solution above 65°, the classical denaturation temperature. At 220 MHz it becomes evident, as McDonald and Phillips have shown,[9, 10, 11, 24] that when the denatured protein (with disulfide bonds intact) is cooled through the renaturing temperature, refolding of the chains introduces many new resonances and a gross alteration in the intensity distribution of the resonances throughout the spectrum. This is illustrated in Fig. XIV-7. Here, (a) is the spectrum of bovine pancreatic ribonuclease constructed from those of the constitutent amino acids, in the same manner as was shown for lysozyme in Fig. XIV-3(a). Spectrum (b) is that of the thermally denatured protein in D_2O. Spectra (a) and (b) match fairly closely, again supporting the general conclusion that the chemical shifts of amino acid residues are perturbed only to a minor extent by their neighbors in the linear sequence when secondary and tertiary structure is lost. Spectrum (c), however, differs markedly from (a) and (b). Spectrum (c) differs from the corresponding spectrum of lysozyme [Fig. XIV-3(c)] in that there is no appreciable intensity above 10.0τ. Evidently, there are no methyl or methylene protons in close proximity to the faces of aromatic rings.

The nature of the native–denatured equilibrium, whether two-state or multistate, has not been explored by NMR as it has been for lysozyme (Section 2B). Other physical evidence suggests that below pH 6, it is probably a two-state equilibrium but that a higher pH it may be more complex.

1. *Titration of Histidine Residues*

The four His residues of ribonuclease have been individually titrated and assigned. We have seen that two of these, His-12 and His-119, are involved in the active site. Bradbury and Scheraga[6] first showed (at 60 MHz) that the C-2 protons of the His residues can be discriminated in this protein and that the titration of the imidazole rings can be followed in the manner already described in Chapter XIII (Section 2) for free histidine and in Section 2C of this chapter for the single His of lysozyme. Jardetzky and his co-workers have extended these findings.[21, 39–42] Figure XIV-8 shows the 100 MHz spectra[21] of the aromatic region of ribonuclease in D_2O with deuteroacetate buffer at four different values of pH encompassing the pK's of the histidine imidazole rings. The C-2 protons appear as four spearate peaks at the left in (a) (pH

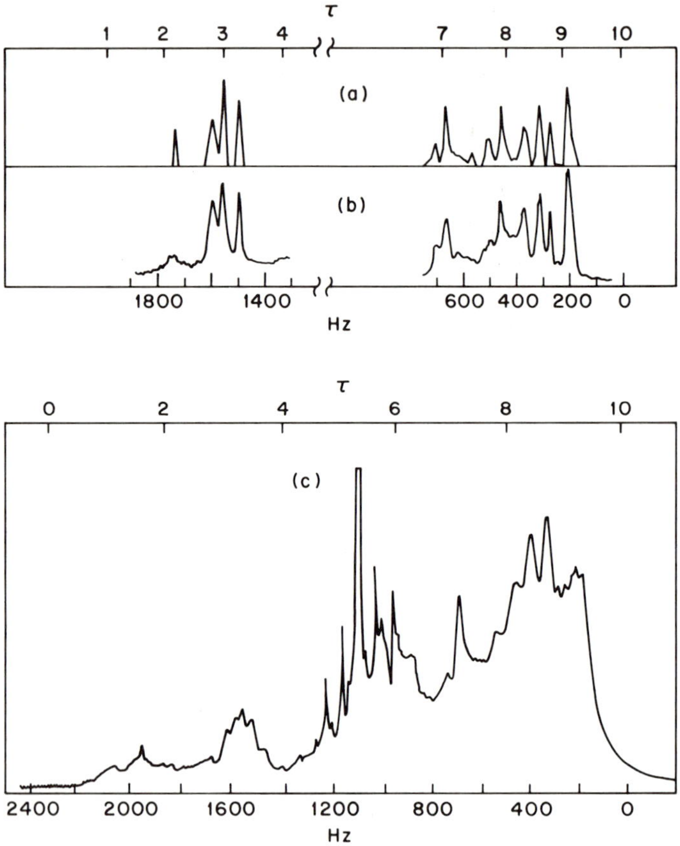

FIG. XIV-7. 220 MHz spectra of bovine pancreatic ribonuclease: (a) calculated spectrum of denatured protein (see text); (b) observed spectrum in D_2O at *p*D 7.0, 67°; (c) observed spectrum of native protein, *p*D 6.8, 22°. Scales expressed in hertz are referred to DSS (McDonald and Phillips[10, 11]).

4.95) which move upfield as the *p*H is increased, becoming clustered together near 810 Hz at *p*H 8.12 (c). A single histidine C-4 proton resonance (marked 5) appears at higher field. The other C-4 protons are masked by the large Phe and Tyr ring resonances centered at about 720 Hz. The titration curves of these five protons are shown in Fig. XIV-9 (note that the chemical shift scale runs in the opposite direction to that of Fig. XIII-5). The *pK*'s to be associated with each proton are inset; it is evident that the C-4 proton of peak 5 is to be associated with the C-2 proton of peak 1, and it is so indicated. Peak 4 behaves anomalously, and is somewhat broader than the others.

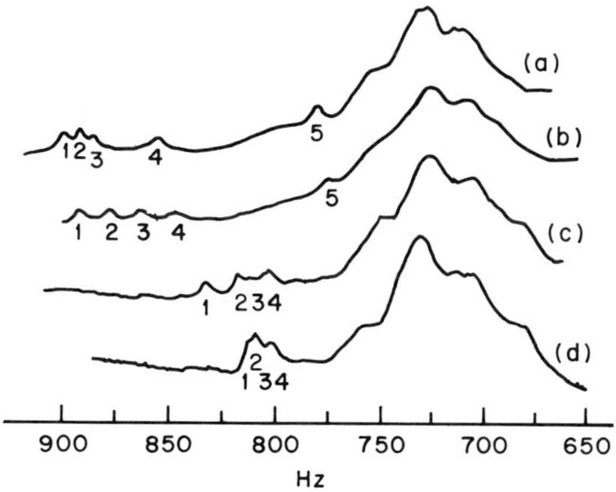

FIG. XIV-8. 100 MHz spectra of 0.012 *M* ribonuclease in D_2O with deuteroacetate buffer: (a) *p*H 4.95, 100 CAT scans; (b) *p*H 5.37, 14 CAT scans; (c) *p*H 6.93, 51 CAT scans; (d) *p*H 8.12, 57 CAT scans. Peaks marked 1–4 are C-2 imidazole peaks of the four histidine residues. Peak 5 is a C-4 imidazole resonance. The envelope of larger peaks at the right includes three other C-4 imidazole protons as well as peaks from six tyrosine and three phenylalanine residues. The chemical shift scale is referred to external HMDS (Meadows *et al.*[21]).

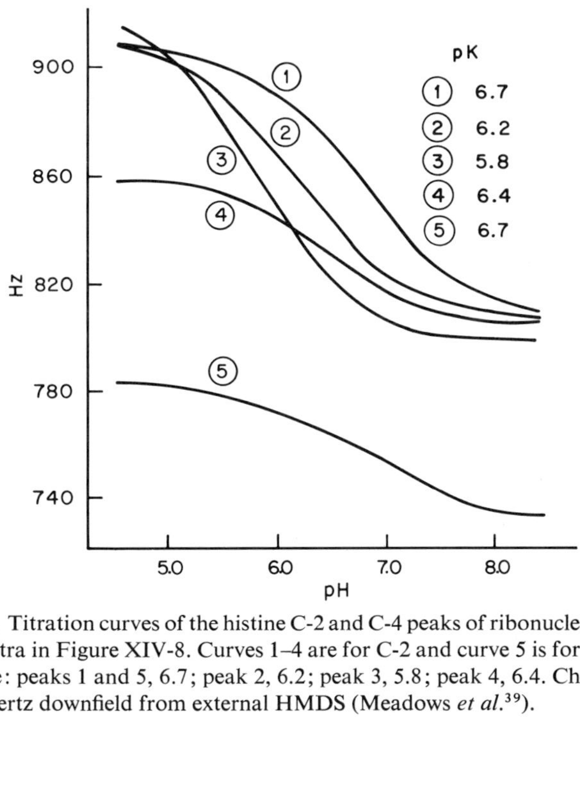

FIG. XIV-9. Titration curves of the histine C-2 and C-4 peaks of ribonuclease, corresponding to the spectra in Figure XIV-8. Curves 1–4 are for C-2 and curve 5 is for C-4. Calculated *pK*'s at 32° are: peaks 1 and 5, 6.7; peak 2, 6.2; peak 3, 5.8; peak 4, 6.4. Chemical shifts are expressed in hertz downfield from external HMDS (Meadows *et al.*[39]).

These resonances were assigned in the following way.[21, 39–42] It was observed that a competitive inhibitor, 3′-cytidine monophosphate (3′-CMP), shifts peaks 2 and 3 downfield, suggesting their assignment to His-12 and His-119. To make specific assignments, the C-2 proton of His-12 was replaced by deuterium by exposing the S-peptide to D_2O at 40° for five days. The active enzyme was then reconstituted and it was observed that peak 2 had disappeared. This then arises from His-12 and peak 3 can be assigned to His-119. It was further observed that carboxymethylation of *either* His-12 or His-119 shifts *both* peaks, but does not affect peaks 1 and 4—strong evidence for the close spatial proximity of the residues corresponding to peaks 2 and 3. The anomalous peak 4 is assigned to His-48, which the x-ray structure indicates to be buried within the protein molecule. Peak 1 is assigned by default to the remaining residue, His-105.

Determination of the His *pK*'s as a function of temperature has shown[41] that while the entropy and enthalpy of ionization of His-105, which is on the exterior of the molecule and exposed to solvent, are "normal," i.e., similar to those of histidine itself, these quantities are very abnormal for His-12 and His-105. Their ionization enthalpies appear to change from abnormally small to abnormally large as the temperature is increased, while the entropies go from negative to positive. The temperature of this "transition" is about 32°, and it is suggested that a minor conformational adjustment of the protein structure takes place at this temperature. Further study will be needed to confirm and more fully interpret these findings.

The anomalous breadth of the resonance of the buried His-48 residue has been the subject of some speculation[39–42]; the most probable hypothesis at present appears to be that it reflects a moderately rapid local conformational isomerization which places this proton in two sites characterized by different chemical shifts. In NaCl solution, this process appears to be slower than in acetate buffer, for the His-48 C-2 proton peak becomes so broad that it cannot be seen.

2. *Slow-Exchanging Protons in Ribonuclease*

Ribonuclease is known to have about 20 protons which exchange very slowly with the aqueous environment (see Wishnia and Saunders[43] and references therein). These are usually attributed to buried peptide NH groups, but Wishnia and Saunders[43] have employed NMR to show that guanidinium protons appear to exchange quite slowly in the *p*D 2.8–4.5 range, and have suggested that these 24 ± 5 slowly exchanging protons may correspond to the four arginine residues. This conclusion has not been generally accepted and remains to be further supported.

D. Heme Proteins

Heme proteins are present in all living organisms and play an essential part in oxygen transport, as oxidation–reduction carriers, and as enzymes. The heme group is present at the active site of all these proteins, but the nature of its function depends upon the nature of the attached ligands and on the structure and conformation of the surrounding polypeptide chains with which it interacts, as well as the oxidation state of the iron atom at the center of the porphyrin ring. The heme group in myoglobin and hemoglobin is a protoporphyrin iron(III) or "protoheme IX" unit, in which the iron is bound to four nitrogen atoms. One of the axial ligands is a histidyl residue of the

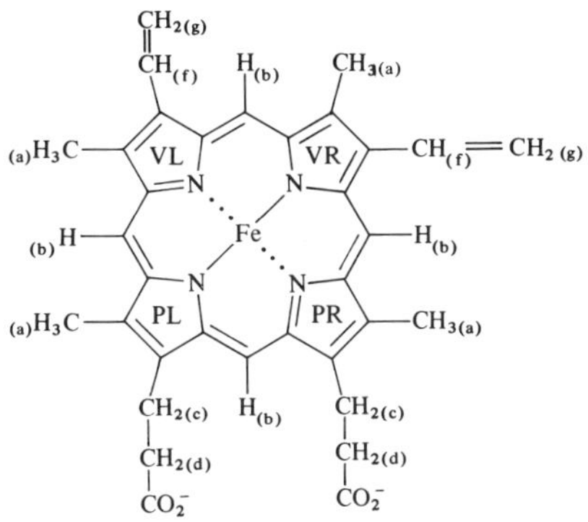

polypeptide chain. The sixth site varies with the function and state of the

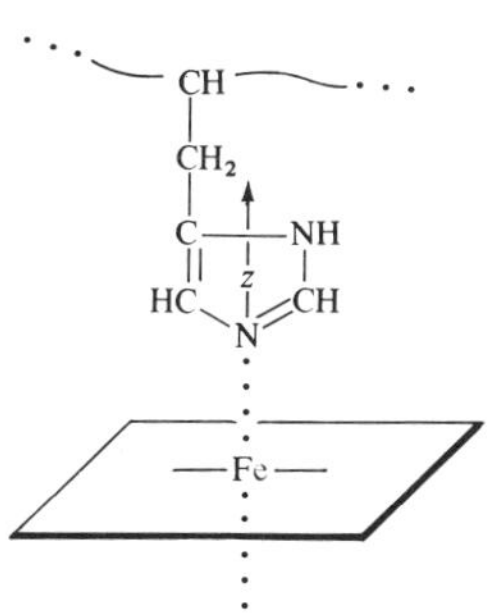

protein and provides the active site. The heme substituents are referred to as ring methyl groups (a), meso protons (b), propionate groups [protons (c) and (d)], and vinyl groups [protons (f) and (g)]. [In cytochromes, the vinyl groups are modified by addition of cysteinyl thiol groups (Section 2D3).]

The heme iron will vary in spin S depending upon its state of oxidation and the nature of the sixth ligand, if any. States of interest are summarized in Table XIV-4. In the ferrous state, the heme units of myoglobin and hemoglobin are capable of binding molecular oxygen, becoming then diamagnetic. The ferric heme units of ferrimyoglobin and ferrihemoglobin are in a high spin state. The electronic spin–lattice relaxation time T_{1e} (p. 258) is about 2×10^{-10} sec, and the ligand field about Fe^{3+} is "weak-field"; these factors make the ligand protons difficult to observe because of excessive line broadening. Upon complexing with cyanide ion (last column of Table XIV-4), the spin is reduced to 1/2; the effect in relaxing ligand protons, since it depends on $S(S+1)$, is reduced by a factor of 35/3; T_{1e} is reduced by two orders of magnitude; and the ligand field becomes "strong-field." (In ferricytochrome C two axial ligands are normally present; see Section 2D3.) In this low spin state, the unpaired electron occupies molecular orbitals of π symmetry. The contact interaction constant A of an aromatic proton [Eq. (XIII-4)] is related to ρ_c^π, the integrated spin density centered on the π orbital of the neighboring carbon atom, by

$$A = Q\rho_c^\pi \tag{XIV-1}$$

For the meso protons of the heme ring, $Q = \sim -6.3 \times 10^7$ Hz; for the protons of the methyl groups attached to the ring, $Q = \sim 7 \times 10^7$ Hz.[45, 46] The observed hyperfine shifts to be discussed in the following sections correspond to a total delocalization of twenty percent of an unpaired electron among the carbon and nitrogen atoms of the protoporphyrin IX ring of MbCN and HbCN.

1. *Myoglobin*

Myoglobin binds oxygen and stores it in muscle until required for metabolic oxidation. For evident reasons it is especially plentiful in the flesh of diving mammals such as the whale and porpoise. Each molecule is composed of a single chain of 153 amino acid residues (molecular weight of 17,800) and contains one heme unit. There are no disulfide crosslinks. The sequence[47] and conformation[48–51] of sperm whale myoglobin are known. This was the first protein the structure of which was determined by x-ray diffraction. It is distinguished from lysozyme and ribonuclease in that 121 of the residues are in α-helical regions.[12] It may be regarded as a basket composed of eight helical supporting members which hold the heme unit buried in a hydrophobic pocket; only the edge to which the hydrophilic propionic acid chains are attached is exposed. It must be supposed that this arrangement helps to keep the iron in the Fe^{2+} state necessary for oxygen binding.

Myoglobins from different species differ in sequence at as many as 35 residues, particularly near the N-terminal.[12, 48, 52–55] Differences can be inferred for other myoglobins for which complete sequences are not yet known (see later discussion).

TABLE XIV-4

STATE OF IRON IN HEME PROTEINS[a]

Protein		Myoglobin (Mb) Hemoglobin (Hb)	Oxymyoglobin (MbO_2) Oxyhemoglobin (HbO_2) Ferrocytochrome C	Ferrimyoglobin ($Hb \cdot H_2O$) (metmyoglobin) Ferrihemoglobin ($Fe \cdot H_2O$) (methemoglobin)	Cyanoferrimyoglobin (MbCN) Cyanoferrihemoglobin (HbCN) Ferricytochrome C Cyanoferricytochrome C
Oxidation state		Fe^{2+}	Fe^{2+}	Fe^{3+}	Fe^{3+}
Spin state		$S = 2$	$S = 0$	$S = \frac{5}{2}$	$S = \frac{1}{2}$
Electronic structure	$d_{x^2-y^2}$	↑	—	↑	—
	d_{z^2}	↑	—	↑	—
	d_{xy}, d_{xz}, d_{yz}	↑	↑↓	↑	↑
		↑	↑↓	↑	↑↓
		↑↓	↑↓	↑	↑↓

[a] After Wüthrich and Shulman.[44]

As we have seen, one of the axial ligands of the heme unit in heme proteins is a His residue; in myoglobin, this is His-93 (or His-8 in the F-helix[12]). There is another his residue, His-64 (or His-7 in the E-helix[12]), in a similar axial position on the other side of the heme, but its imidazole ring is too far away to coordinate with the iron. There is a gap which is filled by O_2 in oxymyoglobin or by other ligands such as CO and NO under appropriate conditions (CO binds about 50 times more strongly than O_2). In the absence of such small molecules, this sixth coordination site remains empty in ferromyoglobin. In ferrimyoglobin this space is occupied by a water molecule. The heme group is not covalently bound to the protein and by careful manipulation can be removed and replaced by other types of heme units, as we shall see.

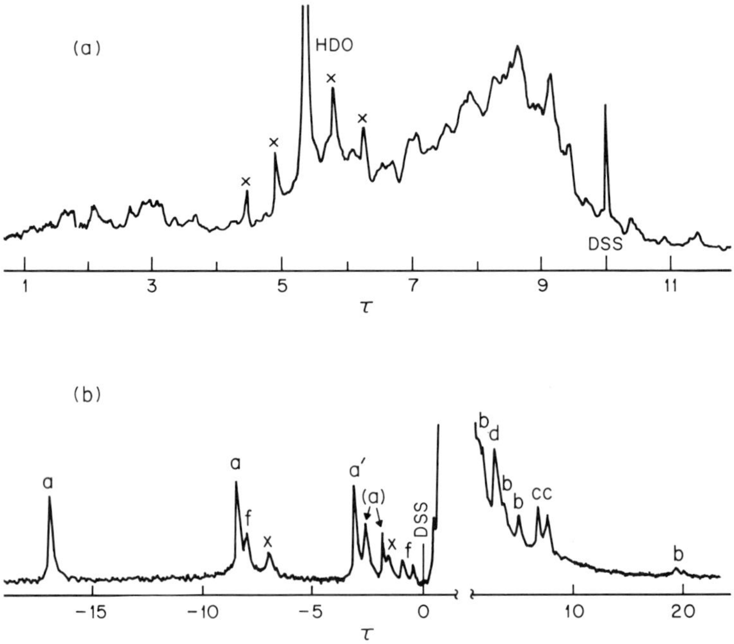

FIG. XIV-10. 220 MHz spectra of sperm whale cyanoferrimyoglobin in D_2O, pD 7.0, 25°; (a) "normal" spectral range; (b) expanded low-field and high-field regions, showing contact-shifted resonances, identified in text. Peaks marked "X" are spinning side bands to the HDO peak (Wüthrich *et al.*[56,58] and Shulman *et al.*[57]).

The high resolution NMR spectra of cyanoferrimyoglobins has been reported and interpreted by Shulman and his co-workers.[44,56–61] Figure XIV-10 shows the 220 MHz spectrum of sperm whale cyanoferrimyoglobin, observed in D_2O at neutral pD; (a) is the "normal" 1.0–11.0τ range, where all the resonances of myoglobin in the diamagnetic state are found; (b) shows the low-field (about −20 to 0τ) and high-field (+10 to +20τ) ranges, with most of the normal part omitted. The DSS reference (Chapter I, Section 17B) appears at 10τ and the HDO peak at 5.35τ.

The low-field lines are hyperfine-shifted resonances of the heme and His-93 (His-F8) imidazole protons. They have been assigned on the basis of their intensities, by comparison to the spectra of reconstituted myoglobins having heme units containing protons and ethyl groups in place of the vinyl groups of protoporphyrin IX,[59] and by the observation of the effect of xenon and cyclopropane upon the MbCN spectrum.[61] These molecules, although inert, are capable of associating with myoglobin, becoming bound in a hydrophobic pocket on the proximal side of the heme unit, i.e., on the His-F8 side. Steric and cyclopropane ring-current arguments, which we shall not discuss here, have been invoked to aid the assignments. The three-proton peaks marked *a* are those of the heme methyl groups. One methyl resonance is split into two peaks. The methyl resonance marked *a′* is believed[56] to be that of the VR pyrrole of the heme (see p. 357). The one-proton peaks marked *f* are assigned to the CH proton of the ring vinyl group. The peaks marked *x* are very probably those of the C-2 and C-4 protons of the imidazole ring of the axial histidyl group. That all of these peaks are contact-shifted is confirmed by the observation that, in accordance with Eq. (XIII-4), the magnitude of the shift is inversely proportional to the reciprocal of the absolute temperature. This is shown in Fig. XIV-11, in which the plots for each resonance are identified by the same letters.

In the upfield portion of the spectrum, there are six distinguishable lines close to 10τ, the positions of which suggest that they may arise either from peptide methyl groups and other protons shifted by aromatic ring currents, as in lysozyme (Section 2B), or from hyperfine-shifted heme protons. The temperature dependence shown in Fig. XIV-12 demonstrates that this second alternative is the correct one, for the ring-current shifts would not be expected to be appreciably temperature dependent. The other lines have marked slopes and are assigned to heme meso protons *b*, and propionic acid methylenes *c* and *d*; (see structural formula on p. 357).

According to Eqs. (XIII-4) and (XIV-1), the $1/T$ dependence of hyperfine-shifted lines should be approximately proportional to the extent of shift. It can be seen from the slopes of the lines in Figs. XIV-11 and XIV-12 that this is approximately but not precisely true for methyl and meso protons. The vinyl CH protons *f* behave anomalously in MbCN, and are conspicuous in prophyrin iron(III) cyanide spectra for actually shifting away from their diamagnetic positions with increasing temperature.[58]

The contribution of pseudocontact interactions, i.e., effects arising from the anisotropy of electron magnetic dipoles apart from spin delocalization (Chapter I, Section 11D), to these effects is not negligible but not very

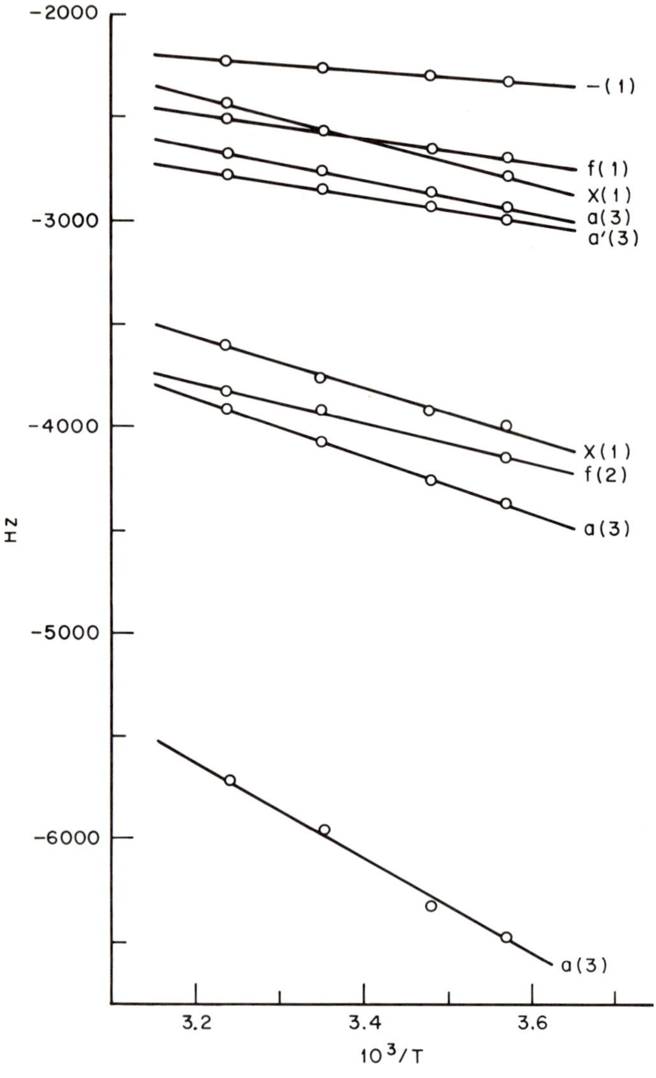

FIG. XIV-11. Dependence of the downfield contact-shifted peaks in Fig. XIV-10(b) on reciprocal temperature (Wüthrich *et al.*[56]).

large.[62, 63] It cannot be distinguished from that of the contact interactions by their temperature dependences, which are, to a first approximation, the same. It has been possible to distinguish the pseudocontact contributions theoretically[63] and from a study of model systems.

Wüthrich *et al.*[60] have compared the 220 MHz spectra of the cyano derivatives of myoglobins from harbor seal, porpoise, and horse. These are shown in Fig. XIV-13 together with the spectrum of sperm whale myoglobin. It can be seen that the hyperfine-shifted portions of the spectra are all very similar, reflecting the fact that of the 35 known positions at which variations in the amino acid residues occur in the myoglobins of these species,

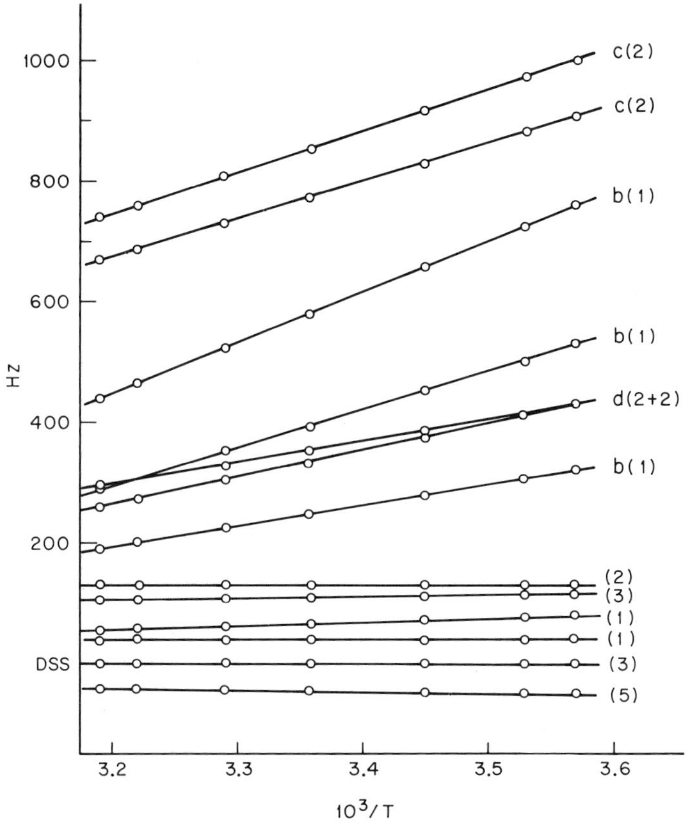

FIG. XIV-12. Dependence of the upfield contact-shifted peak in Fig. XIV-10(b) on reciprocal temperature. The lower group of six lines shows the dependence of the most shielded peaks (and DSS) in Fig. XIV-10(a) (Wüthrich *et al.*[56]).

only two are within 10 Å of the heme unit and do not greatly perturb the conformation of the polypeptide chain in its vicinity, or indeed at any point. It was further demonstrated that alkylation of the accessible histidine residues (7 out of a total of 12 in sperm whale myoglobin) with bromoacetate had little effect on the spectrum, consistent with all of these residues being more than 10 Å removed from the heme group. Those in van der Waals contact with the heme (His-64, His-93, and His-97) are not alkylated. Both lines of evidence in this study indicate that protein modifications outside of the van der Waals distance from the porphyrin ring do not affect the spin density distribution of the heme group. Evolutionary structural changes have tended to leave

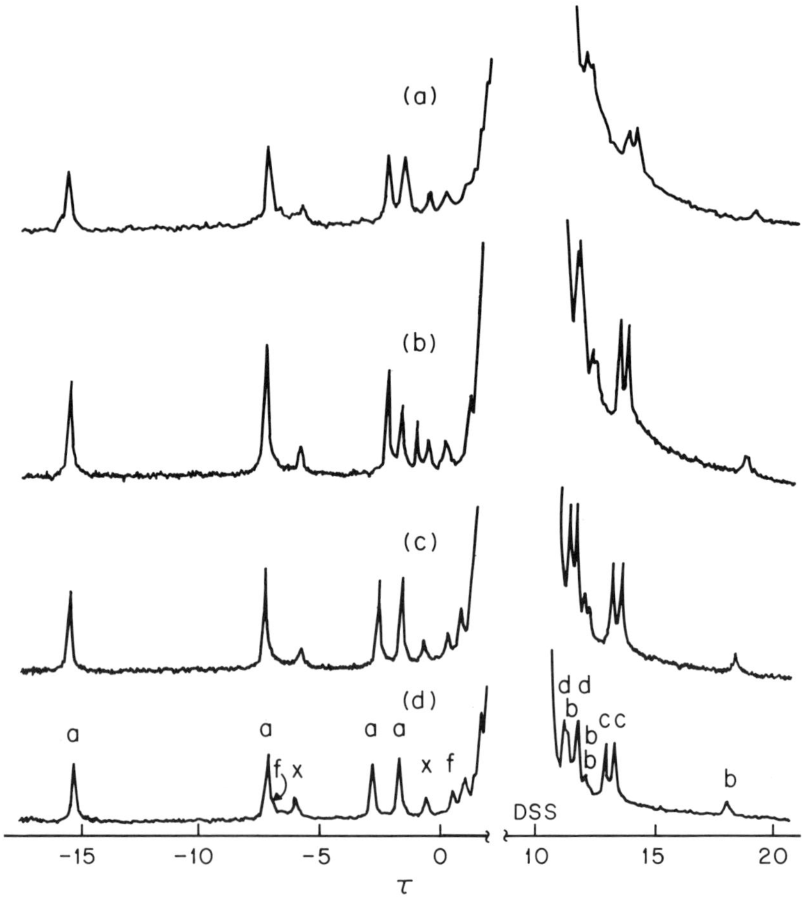

FIG. XIV-13. 220 MHz spectra of cyanoferrimyoglobin from four different species, observed in D_2O at *p*D 7.0, 40°: (a) harbor seal; (b) sperm whale; (c) porpoise; (d) horse (Wüthrich *et al.*[60]).

the portion of the molecule within this distance essentially undisturbed, thus preserving its function undisturbed.

Shulman *et al.*[64] have reported the 220 MHz spectra of deoxymyoglobin and oxymyoglobin. Deoxymyoglobin shows three lines grouped near 10.3τ which are not present in the oxymyoglobin spectrum. Two of these move with temperature and are therefore presumably contact-shifted heme resonances. The third, which has an intensity of six protons, does not show temperature dependence and probably corresponds to ring-current-shifted methyl resonances. In the oxymyoglobin spectrum a peak appears at 12.84τ which does not appear in the deoxymyoglobin spectrum. This can only be a ring-current-shifted peak. These observations imply that the oxygenation of myoglobin causes conformational changes which affect the relative positions of some of the aromatic rings and their neighboring protons.

A significant aspect of the NMR study of myoglobin is the observation of exchangeable low-field protons by comparison of the spectra in D_2O and H_2O, as already described for lysozyme (Section 2B). In myoglobin and oxymyoblogin, Patel *et al.*[64a] have reported resonances in the range of 0 to -5τ, which were identified by considerations of x-ray structure, chemical modification, and species variation as corresponding to the indole NH protons of the two tryptophan residues as well as one resonance from an arginine and one from a histidine residue. Sheard *et al.*[62] have observed four resonances in the 0 to -4τ range in H_2O solutions of cyanoferrimyoglobin; these had relatively small temperature dependence and are of uncertain assignment. They observed three additional peaks in the -3 to -14τ region which were strongly temperature dependent. Of these latter, the least deshielded was assigned to the hyperfine-shifted peptide NH resonance of His-F8, and the other two to hydrogen-bonded histidine imidazole NH resonances (His-F8 and His-E7), although one of them did not titrate.

2. *Hemoglobin*

Hemoglobin is made up of two α-chains and two β-chains, which differ in small details from each other and from myoglobin, to which they have a family resemblance. Each of the four units contains a heme group in a hydrophobic pocket very similar to that of myoglobin. The α-chain has 141 amino acid residues and the β-chain 146, giving a molecular weight of about 64,500. The individual chains are not covalently bonded together but appear to be held largely by hydrophobic bonding. As in myoglobin, there are species variations in the chain sequences and in addition many variant and pathological human hemoglobins are known. Dickerson and Geis[12] should be consulted for detailed information concerning sequence and structure.

The packing of the α- and β-chains brings side chains of unlike chains into close contact, but permits very little contact between like chains. There is a true twofold axis in the molecule. Because of the close similarity of the α- and β-chains there are also two others which are very nearly twofold axes; these are mutually perpendicular and perpendicular to the true twofold axis.

An important property of hemoglobin, which arises from the multiple heme groups, is the form of the oxygen binding curve. Instead of the simple hyperbolic saturation curve observed for myoglobin, hemoglobin shows a sigmoidal curve. This means that at moderate oxygen pressures, the affinity for oxygen increases with pressure. This characteristic enables hemoglobin to take up oxygen effectively at the moderate pressure of the lungs but to yield it up to myoglobin at the low oxygen pressure of the tissues. Upon loss of oxygen to yield deoxyhemoglobin, the subunits of the molecule slide past each other and rotate slightly so that the two β-hemes move farther apart by about 6.5 Å. This conformational adjustment is no doubt closely associated with the cooperative interaction of the hemes which enables them to bind oxygen more strongly when the molecule is already partially oxygenated, but the details of this interaction are not yet clear. In high-salt solutions (e.g., 4 *M* NaCl), the hemoglobin molecule can be cleaved into the two separate halves (Section 3H).

Another important characteristic of hemoglobin is that upon binding four oxygen molecules it becomes a stronger acid and releases protons to the environment. Conversely, as the medium becomes more basic, the equilibrium shifts in favor of the oxy form. This phenomenon, known as the Bohr effect, enables deoxyhemoglobin to take up CO_2 in the veins and release it in the lungs. The detailed mechanism is again obscure. It is possible that NMR may make an important contribution to our understanding of the mechanism of both O_2 and CO_2 binding.

The NMR spectra of hemoglobins in paramagnetic states show hyperfine-shifted peaks similar to those shown by myoglobin, but differing in detail.[65–68] Since hemoglobin has a molecular weight about four times that of myoglobin, it would be expected and is indeed observed that the individual resonances are broadened by approximately the same factor. This makes measurement and interpretation correspondingly more difficult.

Figure XIV-14 shows the 220 MHz spectrum of human cyanoferrihemoglobin A, abbreviated $\alpha_2(Fe^{3+}CN^-)\beta_2(Fe^{3+}CN^-)$. The spectrum is presented in three parts: (a) is the normal spectrum, embracing the 3500 protons appearing between about 0.0 and 11.0τ; (b) is the low-field portion of the spectrum to -13.0τ (the signal-to-noise ratio has been improved by 12 hours of signal averaging); (c) is the high-field portion of the spectrum to $+15.0\tau$. The peaks marked *a* (three protons each) are attributed to heme methyl groups; the others are not assigned with certainty. That the heme methyl peaks are hyperfine-shifted is demonstrated by their temperature dependence (not shown).

The high-field peak at 10.0τ corresponds to six protons per subunit and is not temperature dependent. It therefore probably corresponds to ring-current-shifted methyl groups of valine or leucine residues which are in contact with porphyrin rings or aromatic amino acid side chains. The remaining resonances in Fig. XIV-14(c) could correspond to meso, vinyl, and propionic acid side-chain protons. Not all of the protons expected to be found here can actually be observed, probably because some are not sufficiently shifted. All the observable ones show the characteristic temperature dependence of heme protons.

There remains the question whether the α- and β-chains give identical heme proton resonances. This was answered[69] by partial oxidation of hemoglobin to give $\alpha_2(Fe^{2+}O_2)\beta_2(Fe^{3+}CN^-)$ (i.e., with the α-chains now diamagnetic) as well as by making the complementary compound $\alpha_2(Fe^{3+}CN)\beta_2(Fe^{2+}O_2)$. A comparison of the $\alpha(Fe^{3+}CN)$ and $\beta(Fe^{3+}CN)$

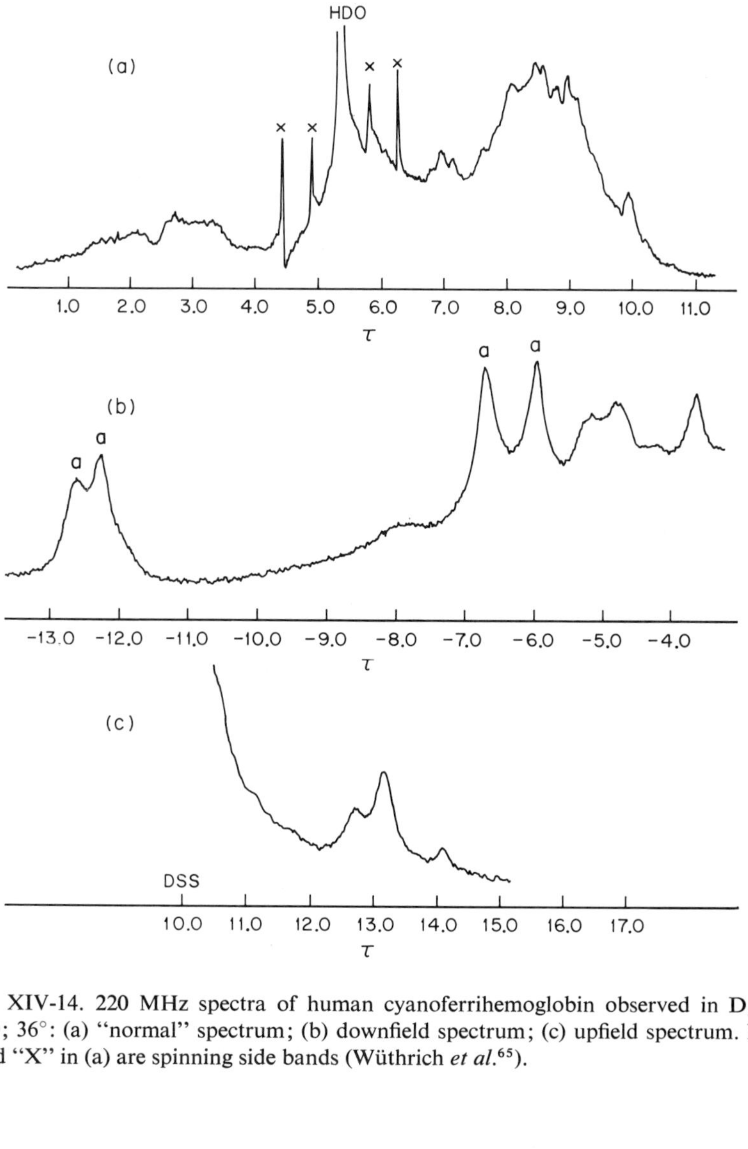

FIG. XIV-14. 220 MHz spectra of human cyanoferrihemoglobin observed in D_2O at pD 7.0; 36°: (a) "normal" spectrum; (b) downfield spectrum; (c) upfield spectrum. Peaks marked "X" in (a) are spinning side bands (Wüthrich *et al.*[65]).

spectra showed small but definite differences between them in the two mixed oxidation state samples while normal $\alpha_2(Fe^{3+}CN)\beta_2(Fe^{3+}CN)$ had a spectrum which was the exact superposition of the two chains. Hence it has been possible to distinguish the α- and β-chain resonances.

The spectra of hemoglobins from other species (bovine, sheep, rabbit, horse) differ observably from each other,[66, 68] in contrast to myoglobins, (Section 2D1), in which the heme spin density distribution is insensitive to species variations in amino acid sequence. This may indicate that in the more imperfectly packed and flexible hemoglobin structure sequence, variations remote from the heme units may nonetheless influence its spin density. It was also observed[67] that the variant human hemoglobin known as hemoglobin J Capetown, in which one amino acid in the α-chain, arginine-α92 (FG4), is replaced by glutamine, gives the same spectrum as normal human hemoglobin A. This hemoglobin is known to have a higher oxygen affinity than hemoglobin A. It is evident that this behavior is not caused by a change in the electronic structure of the hemes.

In a study paralleling that of myoglobin (Section 2D1) Patel *et al.*[69a] have observed peaks in the 0 to -5τ range in H_2O solutions of oxy- and deoxyhemoglobin. In the free β-chains, exchangeable resonances of the A12 and C3 tryptophan indole NH protons were detected at -0.2 and -0.5τ; these were virtually unaffected by the state of oxidation of the heme units, or upon association in the noncooperative β_4 tetramer. No resonance was detected below -0.6τ in any state of the β-chains, in contrast to the behavior of myoglobin (previous section). (The A12 tryptophan of the α-chain showed no comparable exchangeable resonance in this region.) It is of particular significance that the tryptophan residues are in fact too far from the heme units to be expected to show appreciable hyperfine or ring-current shifts, in harmony with these observations. Nevertheless, the $(\alpha\beta)_2$ cooperative tetramer, i.e., intact hemoglobin, shows marked spectral differences between the oxy- and deoxy-forms. It was suggested that a one-proton peak at -2.18τ in HbO_2 which moves to -4.14τ upon deoxygenation is indicative of the alteration in quaternary structure accompanying this change (p. 366) or of a tertiary structural change related to this alteration.

3. *Cytochrome C*

Cytochrome C is essential to the function of all aerobic cells and is therefore very widely distributed in both plants and animals. Its role is to serve as an electron carrier, i.e., to be alternately oxidized and reduced, in the terminal oxidation chain which represents the last steps in the oxidative breakdown of nutrients in the mitochondria of the cell. It has a single polypeptide chain of 104 residues, a molecular weight of 12,400, and a single heme group. This heme group is similar to that of myoglobin and hemoglobin but differs, as we have already seen (p. 357), in that it is covalently bonded through two cysteine units by addition of their thiol groups to the heme vinyl double bonds. The amino acid sequence has been very thoroughly studied and is known for about 35 species (see Dickerson and Geis[12]). All variant versions seem to be extremely similar in function although, as we shall see, species variations in heme electron spin density can be detected. As might be expected, although species variations occur in the Lys-Cys-Ala-Gln-Cys-His-Thr sequence shown (this sequence is found in horse, pig, bovine, sheep, and wheat cytochrome C), the Cys-14, Cys-17, and His-18 residues are never varied. The His-18 residue

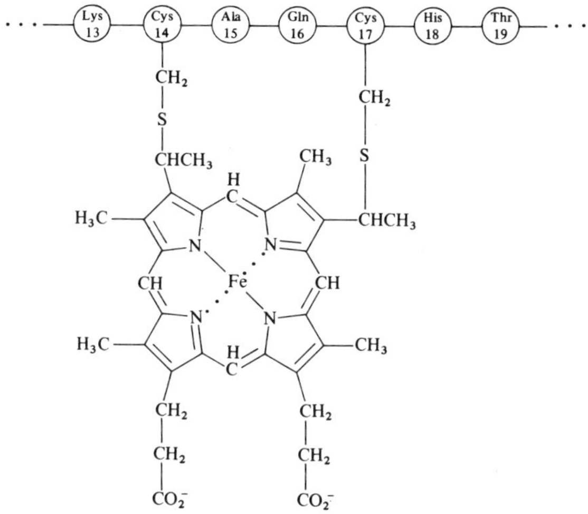

supplies one of the axial ligands. The sixth ligand is not established with certainty, but is believed to be a methionine residue, Met-80; this supposition is supported by the NMR results, to be discussed later in this section. In the

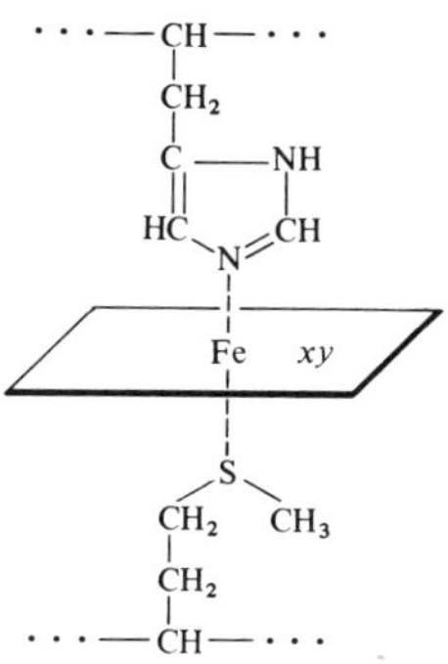

(Met-80)

presence of cyanide ion, the Met-80 ligand is displaced by CN^-. In both cases, the heme has a spin S of $\frac{1}{2}$. Ferrocytochrome C is diamagnetic (see Table XIV-4).

The x-ray structure of cytochrome C has so far been established only at low resolution.[70] This is sufficient to reveal an ovoid molecule with very little α-helical structure. The heme unit is positioned in a crevice with the xy plane normal to the surface of the protein. The 13–19 sequence shown previously is at an outer edge of the crevice. The Met-80 unit is presumably brought into position by a loop of polypeptide chain at the opposite edge of the crevice.

The presence of contact-shifted heme resonances in the NMR spectrum of cytochrome C was first observed by Kowalsky.[5] Subsequent work by Wüthrich and co-workers[71, 73] and by McDonald and co-workers[10, 72] have provided further information concerning these resonances and significant ring-current-shifted peaks as well. The spectrum of ferricytochrome C of guanaco [Fig. XIV-15(a)] shows six contact-shifted heme methyl peaks, marked a, and a

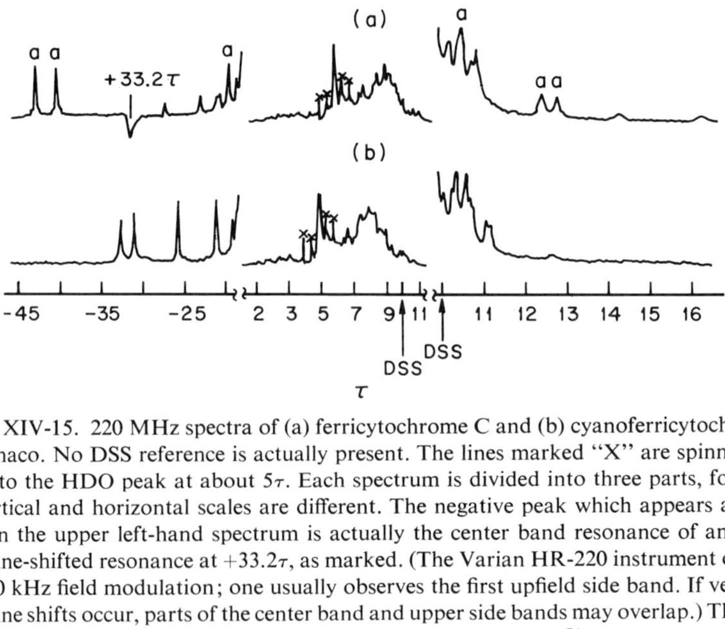

FIG. XIV-15. 220 MHz spectra of (a) ferricytochrome C and (b) cyanoferricytochrome C of guanaco. No DSS reference is actually present. The lines marked "X" are spinning side bands to the HDO peak at about 5τ. Each spectrum is divided into three parts, for which the vertical and horizontal scales are different. The negative peak which appears at about -32τ in the upper left-hand spectrum is actually the center band resonance of an upfield hyperfine-shifted resonance at $+33.2\tau$, as marked. (The Varian HR-220 instrument operates with 10 kHz field modulation; one usually observes the first upfield side band. If very large hyperfine shifts occur, parts of the center band and upper side bands may overlap.) The peaks marked a are hyperfine-shifted heme methyl resonances (Wüthrich[71]).

number of one-proton lines corresponding to heme meso protons and vinyl protons of His-18. Of particular interest is the five-proton peak at $+33.2\tau$, which actually appears as a negative peak in Fig. XIV-15(a) for reasons explained in the caption. When the ferricytochrome C is complexed with CN^- [Fig. XIV-15(b)], this peak disappears, presumably moving into the 0–10τ

range. This exchange of ligands still leaves the heme group in the spin-$\frac{1}{2}$ state, as we have seen (Table XIV-4); the contact-shifted resonances may be expected to shift position, as indeed they are observed to do, but a reversion to the diamagnetic position is difficult to understand. A reasonable conclusion[71] is that this high-field resonance corresponds to the $—CH_2—S—CH_3$ protons of the Met-80 residue. Strong confirmation is provided by the spectra of cyano-ferricytochrome C upon reduction with dithionite.[71, 73] As reduction becomes complete (this occurs in a few minutes), all contact-shifted resonances disappear, because the heme group is now diamagnetic. However, new resonances appear in the 11.9–13.7τ region, in particular a three-proton peak at about 13.2τ, which appears in spectra of cytochrome from a variety of species[73] (Fig. XIV-16). This latter is believed[71, 73] to be the Met-80 methyl resonance, shifted upfield by the porphyrin ring current. It appears slowly as the CN^- ligand is displaced by the methionine ligand.

It will be noted in Fig. XIV-16 that the ferrocytochrome C spectra show little species variation except for that of the mold *candida krusei,* in which the three resonances in the 10.5–10.7τ region, believed due to methyl groups, split farther apart and move to higher field. It appears significant that in the other cytochromes, the Thr-19 and Ilu-81 residues provide methyl groups appropriately positioned with respect to the porphyrin ring to experience shifts of the observed magnitude, while in the *Candida krusei* cytochrome the Ilu-18 is replaced by an alanine residue.[70]

4. *Ferredoxin*

Ferredoxin is an iron-containing electron-transport protein similar in function to cytochrome C but differing from it in that the iron is not bound in a heme group but is coordinated with the sulfur atoms of cysteine residues. It contains six to eight atoms of iron per protein molecule and an equal number of cysteines. The primary sequence is known.[74] It has a molecular weight of about 6000. The 220 MHz spectrum[75, 76] of thc ferric form of ferredoxin from *Clostridium pasteurianum* shows eight one-proton resonances in the −0.4 to −0.9τ region; these were established as hyperfine-shifted by their temperature dependence. (There are also a number of hyperfine-shifted peaks in the 0–4τ region.) It has been suggested[75, 76] that these correspond to one methylene proton of each of eight cysteinyl residues coordinated to iron:

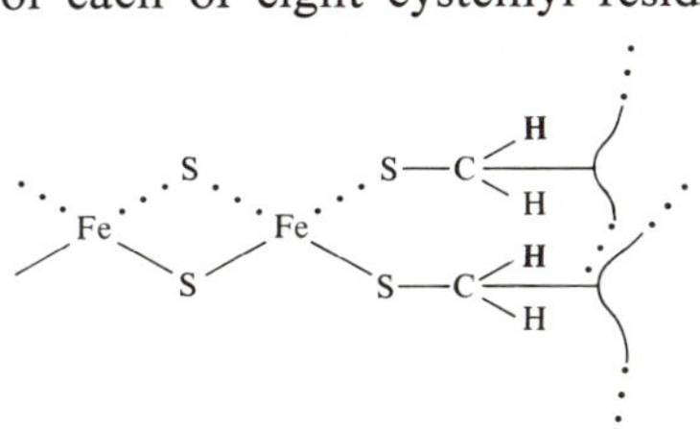

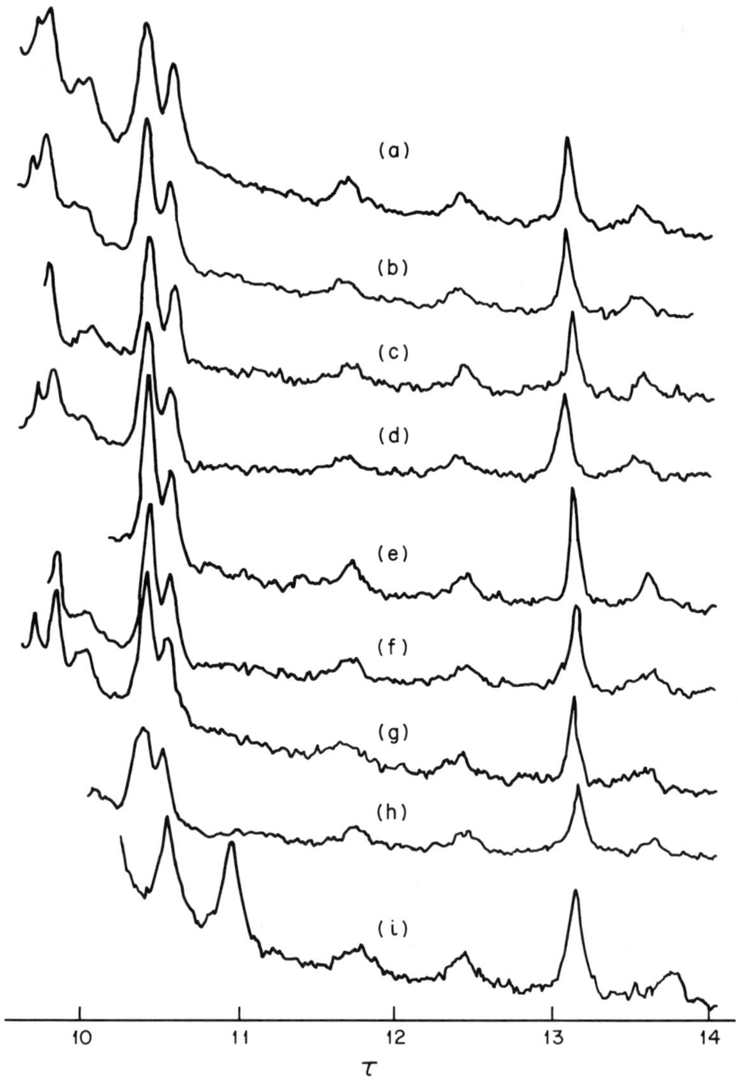

FIG. XIV-16. Upfield portion of 220 MHz spectra of ferrocytochromes C from different species, observed in D_2O at pD 6.0–7.5, 40°; internal DSS reference: (a) horse, (b) bovine, (c) hog, (d) chicken, (e) turkey, (f) duck, (g) pigeon, (h) tuna, (i) *Candida krusei* (mold) (McDonald *et al.*[72]).

It should be recalled (Chapters III, XIII) that the methylene protons are nonequivalent because of the presence of the center of asymmetry at the α-carbon atom. The conformation about the C—S bond may be such that one proton is much nearer the nodal plane of the nonbonding sulfur p orbital than the other, and hence senses the electron spin density in this orbital less than the other.[76] Such a situation is well known to occur in hyperfine interactions in ESR spectra.

E. Other Proteins

1. *α-Chymotrypsin*

The high-field portion of the 100 MHz spectrum of α-chymotrypsin has been observed by Hollis *et al.*[77] The molecule is roughly spherical[12] and has a molecular weight of about 25,000. Thermal unfolding was followed by measurement of the line width of the aliphatic side-chain protons as a function of temperature. The results were interpreted as consistent with the two-state denaturation process (p. 345) indicated by other physical measurements. As yet, however, only minimal information concerning this protein has been supplied by NMR.

2. *Insulin*

Insulin contains 51 amino acids in two chains: an A chain of 21 residues and a B chain of 30 residues. These are joined by two disulfide bridges. The A chain contains a third disulfide bond in an intrachain loop embracing six residues. The sequence[78] and x-ray structure[79] have been determined. The A chain has a much-folded structure with short quasi-helical stretches. There is a stretch of α-helix in the B chain between the disulfide bridges. The low molecular weight of 5780 would appear to make insulin an attractive subject for NMR study. Nevertheless, no study of the native molecule has been published. This may be in part because it has no recognized "active site"; its well-known hormonal function of decreasing blood sugar levels is not understood in the chemical sense. It also has a strong tendency to form polymers, the dimer and hexamer being well characterized.[79] The dimer exhibits an interesting boxlike arrangement of B-26 Tyr and B-24 Phe and their twofold-axis-related partners in the second member of the dimer; this clearly may be of interest at 220 MHz.

Bak *et al.* have reported the 60 MHz[80] and 220 MHz[81] spectra of porcine and bovine insulin in their unfolded states in CF_3CO_2H and CF_3CO_2D, and have partially sorted out the individual resonances, using the amino acid spectra (Chapter XIII, Section 2) as a guide.

3. *Glucagon*

Like insulin, glucagon is also secreted by the pancreas. Its function is to raise blood sugar levels: glucagon and insulin act in dynamic opposition.

Glucagon is a borderline or "miniature" protein. It has 29 amino acid residues in a known[82] sequence, shown for the porcine hormone.

1 5 10 15
His-Ser-Gln-Gly-Thr-Phe-Thr-Ser-Asp-Tyr-Ser-Lys-Tyr-Leu-Asp-
16 20 25 29
Ser-Arg-Arg-Ala-Gln-Asp-Phe-Val-Gln-Trp-Leu-Met-Asn-Thr

The question of its secondary structure is at present in some doubt. It has been suggested,[83] on the basis of NMR and ORD spectra, that it exists in a helix–coil equilibrium with a helix content of about 10% at room temperature. It aggregates in aqueous solution; a monomer–dimer–hexamer equilibrium has been suggested[84] based on ultraviolet difference spectra, gel filtration, and sedimentation. Patel[85] has analyzed the 220 MHz spectra of glucagon in solutions of 10% CD_3CO_2D–D_2O. Sharp side-chain resonances were observed for the residues 1 through 13, while broadened resonances were observed for residues 14 through 29. The broadening was concentration dependent, suggesting that its origin is not internal organization but aggregation, very possibly as "tail-to-tail" dimer:

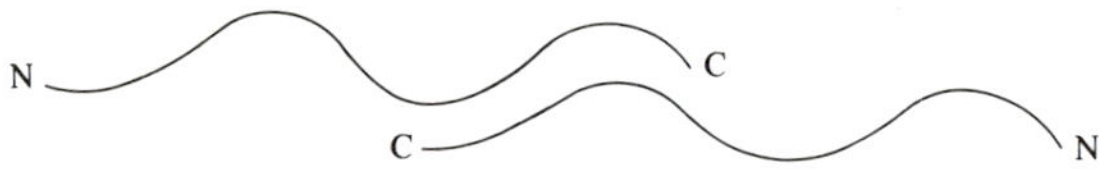

4. *Bacterial Nuclease*

Staphylococcus aureus produces an extracellular enzyme which is similar in molecular weight to ribonuclease (about 16,500) but has a broader specificity, attacking both RNA and DNA. The sequence of the V8 strain is known. Meadows *et al.*[21] have reported the 100 MHz spectra of the product of another strain which, like ribonuclease, has four His residues instead of the three of the V8 enzyme. The His C-2 proton peaks can be resolved. Their *p*H titrations are similar to those of ribonuclease, except that one is specifically shifted downfield by Ca^{2+} ion when the imidazole ring is in the uncharged state. It is known that Ca^{2+} is necessary for enzyme activity.

5. *Secretin*

The gastrointestinal polypeptide hormone secretin has been investigated by ORD and CD.[86] It has the structure:

1 6 10
His-Ser-Asp-Gly-Thr-Phe-Thr-Ser-Glu-Leu-
11 16 20
Ser-Arg-Leu-Arg-Asp-Ser-Ala-Arg-Leu-Gln-
21 26
Arg-Leu-Leu-Gln-Gly-Leu-Val-NH_2

Porcine Secretin

A proton NMR study of secretin and its fragments in aqueous acid media at 220 MHz has shown[87] that the methyl resonance of Ala-17 is a sensitive indicator of conformation. The application of addition and difference NMR spectroscopy in the NH and α-CH regions to the intact molecule and to the 1–10 and 11–27 fragments suggests the existence of a preferred conformation, probably arising from interaction of the 1–14 residues with those of the 15–27 sequence. It was not possible to specify this conformation in detail, in particular, to tell whether an α-helix is involved.

6. ^{31}P *in* α_s-*Casein and Phosvitin*

Ho and co-workers[88,89] have observed the ^{31}P resonance in α_s-casein (molecular weight about 27,300) and in phosvitin. The latter protein occurs in hen egg yolk. Its molecular weight is uncertain, but a minimum value is about 45,000. It contains approximately 10% phosphorus, or about 119 phosphate groups, and a nearly equal number of serine residues. From the deshielding of the ^{31}P resonance with increasing *p*H and the magnitude of ^{1}H–^{31}P *J* coupling, it was concluded that in both proteins the ^{31}P occurs as the serine monoester.

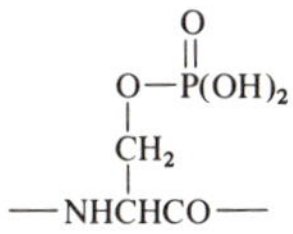

F. Observation of Partially Deuterated Proteins

We have already seen an example of the deuteration of a specific His residue in ribonuclease (Section 2C) made possible by the removal, D_2O exchange, and replacement of a specific portion of the protein structure. This procedure is not generally possible, but nevertheless deuterium substitution can be achieved by other means and has proved useful for the observation of specific amino acid residues.[90 92] Crespi *et al.* have incorporated normal L-leucine,[90,92] L-methionine,[92] L-phenylalanine,[92] L-alanine,[92] and L-valine[92] into fully deuterated C-phycocyanin, C-phycoerythrin, and cytochrome C by the growth of blue-green algae in 99.8% D_2O culture media to which the specific amino acids were added. Only the α-CH is replaced by deuterium during the growth of the algae, the side chains remaining protonated. The broad resonances in the characteristic regions of each type of side-chain proton indicated a variety of environments within the protein structure. Markley *et al.*[91] prepared staphylococcal nuclease with all positions fully deuterated except tryptophan, methionine, tyrosine (in ring positions 2 and 6), histidine (C-2 position), aspartic acid and asparagine (β-methylenes), and glutamic acid and glutamine (γ-methylenes). The spectrum was, as might be expected, strikingly simplified. Resonances of known position such as His C-2 were unchanged, showing that massive substitution of deuterium does not appeciably alter the folding of the polypeptide chain. In the methionine S-methyl region near 8.3τ, sharp separate resonances of the four Met residues, normally engulfed in the large number and variety of peaks in this region, could be clearly seen. The effect of Ca^{2+} (see Section 2E4) and of the inhibitor 3′,5′-thymidine diphosphate could be readily observed.

3. Direct Observations of the Binding of Small Molecules to Proteins

We have seen (Sections 2B and 2C) that the binding of substrates and inhibitors to lysozyme and ribonuclease perturbs the resonances of the peptide aromatic groups which are located near the binding site. The information provided in this way by the histidine and tryptophan resonances of lysozyme is particularly significant. We now consider the direct observation of the resonance of the small molecule itself. Chemical shift and line-width alterations can be detected for a number of systems, and have proved at least as useful as measurements of the protein resonances in delineating the active site. However, because our focus in this treatise is principally on polymers themselves, our discussion of this aspect of protein NMR will be relatively compact.

The effect of binding to a large molecule on the relaxation and resonance line width of the nuclei of a small molecule can be a complex phenomenon. In Chapter I we have discussed the contributions to T_1 and T_2 on the assumption that the motions of the nuclei can be described by a single correlation time. This clearly cannot be an adequate description for any except the simplest molecules, and, as we have seen, is a particularly poor description for the protons of the side chains of random-coil polymer chains in solution. Consider a molecule such as a drug or enzyme substrate, which is relatively small, yet large enough to be articulated—to have, for example, phenyl or methyl groups attached to a chain of several atoms. When the molecule is tumbling more or less isotropically in solution, the approximation of a single correlation time for each type of nucleus is a crude but possibly adequate one. When this molecule is bound to a large protein, certain of its nuclei—say those near its binding group—may become attached firmly enough to share the motion of the whole protein molecule. Since the latter is usually roughly spherical, a single and now much longer correlation time will perhaps describe its motion sufficiently well, and will be reflected in a shortened T_2 and a greater line width, which can be reasonably interpreted. But other protons of the bound molecule, such as those of methyl groups, may still be free to undergo considerable motion of their own; this motion is highly anisotropic and clearly cannot be described by one correlation time. Yet at the present state of development of this field, it is customary to make this assumption. The conclusions drawn in such circumstances can at best be only qualitative.

One suggested way to assure that at least the peak width does provide a measure of T_2, since other factors may contribute to its broadening, is to measure T_1 as well, on the grounds that they should be equal under these conditions. This is possible in principle by some appropriate method such as progressive saturation (Chapter I, Section 9), but can be awkward if the spectrum is complex. Furthermore, it does not necessarily provide adequate assurance on this point. There is a notable disparity between T_1 and T_2 for vinyl polymer protons even in solution [93] (for solid polymers the disparity can be large, as is well known); the higher the observing frequency, the larger the discrepancy may be. Thus both experiment and interpretation are considerably beclouded in this area. This subject has been reviewed.[94–96]

A. Lysozyme

The substrate of lysozyme is the mucopolysaccharide of bacterial cell walls. This is a 1:1 alternating copolymer of *N*-acetylglucosamine (NAG) and *N*-acetylmuramic acid (NAM) joined by β-1,4 linkages. Lysozyme acts at the

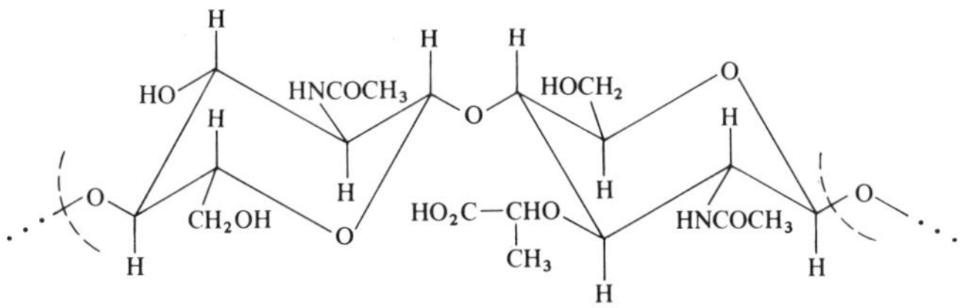

oxygen linkage indicated by the dashed lines. Chitin (poly-β-NAG) is cut at any such bond; NAG itself, the oligosaccharides $(NAG)_2$, $(NAG)_3$, and $(NAG)_4$ (the latter also known as chitobiose, chitotriose, and chitotetraose, respectively) all act as inhibitors. The trisaccharide is known from x-ray measurements[17] to bind in the right half of the crevice. It is concluded that the active site is located in this region. Oligosaccharides of five or more units are cut beyond the fourth unit; $(NAG)_6$ appears to occupy the entire crevice.

Thomas[97, 98] and Raftery and co-workers[99–102] have observed broadening and chemical shift perturbations of the acetamido methyl resonances of these inhibitors and substrates in the presence of the enzyme. Raftery and co-workers studied NAG, $(NAG)_2$, $(NAG)_3$, and $(NAG)_4$, together with their α- and β-methyl glycosides. The equilibrium concerned is

$$E + S \rightleftarrows ES,$$

where E and S are the enzyme and substrate (or inhibitor), respectively, and ES is the complex formed between them. The dissociation constants of the complexes, K_s, are known. It was assumed that exchange of free and bound substrate molecules was sufficiently rapid that the resonance observed represented an average of the chemical shifts and line widths of both environments, weighted according to the substrate:enzyme mole ratio, which was always at least 4. In some cases, exchange appeared too slow for this assumption to be valid and the rate had to be increased by *p*H or temperature adjustments. It is particularly noteworthy that under slow exchange conditions the acetamido peak was *greatly broadened*, a contribution to the apparent value of $1/T_2$ which has nothing to do with molecular motion.

Figure XIV-17 shows the acetamido methyl resonances of NAG methyl-β-glycoside in the free state and at six different substrate–enzyme ratios. As this

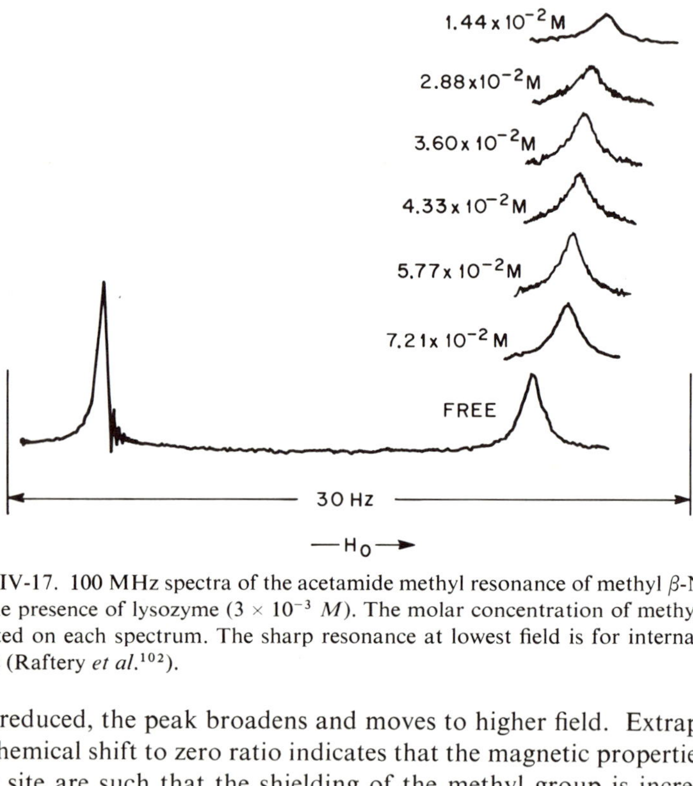

FIG. XIV-17. 100 MHz spectra of the acetamide methyl resonance of methyl β-NAG free and in the presence of lysozyme (3×10^{-3} *M*). The molar concentration of methyl β-NAG is indicated on each spectrum. The sharp resonance at lowest field is for internal acetone reference (Raftery *et al.*[102]).

ratio is reduced, the peak broadens and moves to higher field. Extrapolation of the chemical shift to zero ratio indicates that the magnetic properties of the binding site are such that the shielding of the methyl group is increased by about 0.5 ppm over that of the free molecule. The α-glycoside and the α- and β-anomers of NAG all show a similar shift and nearly identical binding properties. The acetamido methyls of the reducing end (i.e., corresponding to NAG itself) of $(NAG)_2$, $(NAG)_3$, and $(NAG)_4$ showed similar behavior. The acetamido methyls of the other rings in these oligomers appeared at lower field than that of the reducing end and exhibited little or no alteration in chemical shift. This might be interpreted as indicating weaker binding, but could also be attributed to a differing magnetic environment at their binding subsite. From the NMR spectral behavior of these and other related molecules it was possible to delineate three contiguous binding subsites on the enzyme binding surface, each with its particular magnetic properties. The results in general correlated well with the conclusions from x-ray studies.[17]

B. Chymotrypsin

α-Chymotrypsin is reversibly inhibited by tryptophan. The effect of this interaction on the NMR spectrum of tryptophan was investigated by Gerig.[103] There is a marked broadening of the lines of all of the protons, probably indicating a binding of all parts of the molecule. By this criterion, the D isomer was more strongly bound than the L isomer. Treatment of the enzyme with diisopropyl fluorophosphate, which is known to inactivate it, or with *N*-bromoacetyl-3-trifluoromethylaniline, which reacts to place a bulky carbamylmethylsulfonium bromide group at the Met-192 residue, causes a marked diminution of the broadening effect. This suggests that binding occurs at the active site.

Spotswood *et al.*[104] reported the spectra of *N*-acetyl-DL-*p*-fluorophenylalanine in the presence of α-chymotrypsin. The use of ^{19}F as a probe has the advantage that its chemical shift range and sensitivity to environmental effects are considerably greater than those of protons (Chapter I, Section 16B). The spectrum of the free DL compound in an achiral solvent is a pentuplet with finer structure; in the presence of the enzyme this spectrum splits into two pentuplets. Study of the individual enantiomers showed that the D-isomer's ^{19}F nucleus has become deshielded, the L-isomer being unaffected. Similar results were obtained with the *m*-fluoro derivative; even *N*-acetyl-DL-phenylalanine showed a splitting of the *N*-acetyl peak, although it was much smaller. In the presence of lysozyme, no effect was observable. It was concluded that the aromatic ring and acetyl group are bound in a hydrophobic pocket at the active site.

Zeffren and Reavill[105] reported a similar study of the binding of *N*-trifluoroacetyl-DL-phenylalanine. Both enantiomers showed some downfield shift of the ^{19}F resonance in the presence of α-chymotrypsin.

Marshall[106] found that the ^{35}Cl resonance of a *p*-mercuribenzenesulfonyl chloride group, attached to Ser-195 at the active site through the mercury atom, is broadened when an adjacent Met-180 or Met-192 residue is alkylated with benzyl bromide. This broadening is eliminated when the protein is unfolded in 8 *M* urea, suggesting that the restriction of the motion of the Cl-containing group depends upon maintaining intact secondary and tertiary structure.

Although not within the scope of this discussion, it should be mentioned that Sykes[107] has employed measurements of ^{19}F spin–lattice relaxation and spin–spin relaxation to determine the rate of exchange of trifluoroacetyl-D-phenylalanine with α-chymotrypsin, phosphorylated chymotrypsin, and chymotrypsinogen.

C. Binding of Penicillin to Serum Albumin

Fischer and Jardetzky,[108] in a pioneering study, showed that in the presence of serum albumin, the resonance peaks of penicillin were markedly broadened. The phenyl peak appeared to be more greatly broadened than the others, but since this is really a complex, unresolved multiplet and since binding could affect the relative chemical shifts of the *o*, *m*, and *p* aromatic protons, this conclusion must be treated with reserve. An enhancement of broadening at higher salt concentrations suggested hydrophobic binding rather than attachment through ionic groups.

D. Binding of Cofactors and Substrates to Alcohol Dehydrogenase

Jardetzky *et al.*[109] reported that in the presence of a molar excess of yeast alcohol dehydrogenase, the pyridine resonance of DPN broadened to the vanishing point while the resonances of the purine moiety (Chapter XV, Section 2) were still detectable. However, Hollis *et al.*[110] and Hollis[111] reported observations at variance with this. They also studied the binding of ethanol and acetaldehyde to the enzyme alone and in the presence of the cofactors DPN and DPNH, with partially contradictory conclusions. The subject is at present somewhat confused and calls for more definitive treatment.

E. Aspartate Transcarbamylase

Aspartate transcarbamylase catalyzes the transfer of a carbamyl group from carbamyl phosphate to L-aspartic acid (the intial step in the synthesis of cytidine triphosphate). The molecular weight of the enzyme is approximately

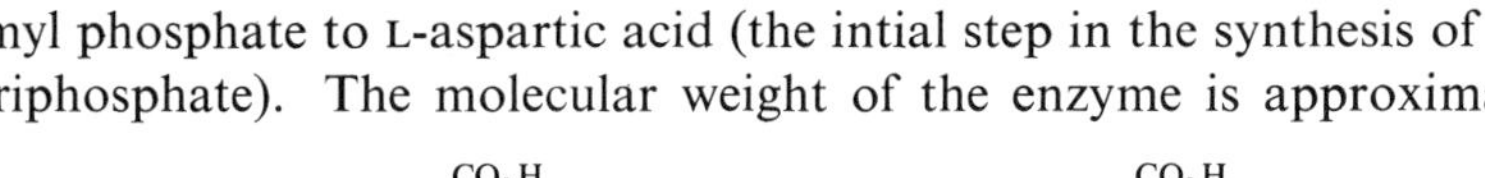

300,000. It contains a number of subunits, one type of which contains the entire catalytic activity. There are apparently four such catalytic subunits per molecule. Schmidt *et al.*[112] have reported a careful study of the effect of binding to the enzyme upon the methyl and methylene proton resonance of a number of analogs of carbamyl phosphate (the H_2N protons of carbamyl phosphate itself are not suitable for measurement because of broadening due to ^{14}N coupling; see Chapter XIII, Section 3B), including phosphonacetamide $H_2NCOCH_2PO_3^{2-}$, *N*-methylphosphonacetamide

$$\begin{matrix} CH_3 \diagdown \\ \quad NCOCH_2PO_3^{2-} \\ H \diagup \end{matrix}$$

and methyl phosphonate $CH_3PO_3^{2-}$. The CH_2 protons of phosphonacetamide are greatly broadened in the presence of the catalytic subunit, while those of the other two compounds are not. Since all are bound, this indicates that binding of the H_2NCO group inhibits rotation about the P—C bond, but that the binding of the other compounds is such as to permit rotation at this bond.

The binding of succinate ion, a competitive inhibitor for asparate, was also investigated. Broadening of its methylene resonance was induced by the simultaneous bindings of analogs of carbamyl phosphate, provided these analogs were no larger than carbamyl phosphate and contained both a carbonyl and a phosphate group.

F. Carboxypeptidase A

Shulman and co-workers[113–115] have investigated the active site of carboxypeptidase A by observation of the relaxation of associated small molecules. Carboxypeptidase A is a proteolytic metalloenzyme which catalyzes the hydrolysis of C-terminal peptide bonds in peptides and proteins. It has a molecular weight of 34,600 and contains one zinc atom per molecule which is essential for activity but which can be replaced by Mn^{2+} or Co^{2+} with retention of activity.[116] The crystal structure of the enzyme is known;[117, 118] there are three protein ligands to the metal atom, and sites for at least two more ligands. The binding of solvent (H_2O),[113] inhibitors,[114] or fluoride ion[115] to the active site of the Mn^{2+}-enzyme causes relaxation of the bound nuclei not only because of the long correlation time of the protein but because of the paramagnetic metal ion. The broadening of the resonance lines of the solvent was interpreted as indicating that one water molecule is bound to the Mn^{2+} ion. Measurement of the broadening of methylene (or methyl) proton peaks of the competitive inhibitors indoleacetic acid, *tert*-butylacetic acid, bromoacetic acid, and methoxyacetic acid, taken together with an estimate of the correlation time of the inhibitor proton–metal interaction, led to the conclusion that the relaxation is primarily determined by the exchange time of the inhibitor–protein complex. On this basis, and using the known Michaelis–Menten constants, one can determine all the ordinary rate constants of the enzyme–substrate reaction.

G. Hapten–Antibody Complex Formation

An antigen prepared from a hapten can often bind small molecules if they are of similar size and nature to the binding group on the hapten. Thus, an antigen prepared by injecting into rabbits bovine serum albumin to which diazotized 4-aminophenoxycholine

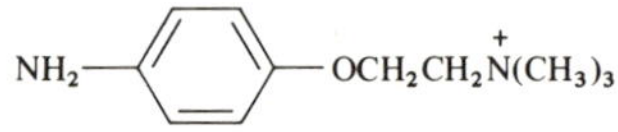

has been attached, will bind tetramethylammonium ions. Burgen *et al.*[119] observed a marked broadening of the methyl resonance of the latter in the presence of the antibody. That this was not due to nonspecific binding was shown by the absence of broadening in the presence of nonantibody gamma globulin. There were indications that the hindrance to methyl rotation in the bound state increased with the strength of the hapten–antibody binding.

H. Binding of Ions

In Section 3F, we have indicated the use of fluoride ion as a probe of the active site of carboxypeptidase A. Of the other ions that have been employed in this way, chloride has received the major share of study.[120–122] The ^{35}Cl nucleus has a spin of $\frac{3}{2}$ and therefore has a quadrupole moment (Chapter I, Section 10). As with ^{14}N (Chapter XIII, Section 3B), its spin–lattice relaxation time will depend on the symmetry of the electrostatic environment. The Cl^- ion in water is symmetrically hydrated and its resonance line has a width of the order of 10–15 Hz, while in CCl_4 the resonance width is of the order of 14.5 kHz. When bound to a protein (usually through an Hg atom, itself bonded to a suitable site such as a sulfhydryl group), its resonance may be expected to be broad, but the line width will be reduced in proportion to the time spent in the aqueous phase, assuming a fast exchange between the bound and hydrated states. (These questions are carefully discussed by Haugland *et al.*,[122] to whom the reader is referred for details.)

Stengle and Baldeschwieler[121] investigated the nature of the unreactive SH groups of hemoglobin (the ones which will not react with, for example, mercuric compounds when the protein is in the intact state) by observing the ^{35}Cl resonance of HgCl groups introduced by reaction with mercuric chloride. From the broadening of the lines as compared with that of Cl^-, they obtained values of the correlation time at the Hg—Cl bond in the protein. When the hemoglobin molecule was dissociated into halves, as occurs in 4 *M* NaCl solution (p. 366), the unreactive SH groups remain unreactive, showing that they are not exposed upon dissociation and therefore probably are not involved in bonding the halves together. Furthermore, the ^{35}Cl correlation time at the two reactive SH sites actually becomes *longer* upon dissociation. This apparently points to conformational changes that restrict the motion of the ^{35}Cl more greatly at these sites.

Ward and Happe employed ^{35}Cl resonance to study zinc adenosine diphosphate complexes[123] and to explore the active site of bovine carbonic anhydrase,[124] which contains a zinc atom.

Kayne and Reuben[125] used the resonance of ^{205}Tl as a biological probe. This isotope has a high natural abundance of 70.5% and a spin of $\frac{1}{2}$. Broadening of Tl^+ resonance upon binding to protein is thus, like that of ^{19}F, not dependent upon quadrupole relaxation. They used it to explore the metal binding properties of rabbit muscle pyruvate kinase, which is known to require both a monovalent and a divalent metal ion for activity. There was an easily measurable but relatively small broadening of the Tl^+ resonance in the presence of protein, and no measurable perturbation of the Tl^+ chemical shift, known to be very sensitive to chemical bonding. It was concluded that exchange was so fast that the broadening must be due to the neighboring Mn^{2+} ion, bound at the active site, and that the Tl^+ and Mn^{2+} binding sites must be very close together.

I. Relaxation of Solvent by Bound Paramagnetic Ions

There is a large literature on the measurement of solvent (H_2O) relaxation by paramagnetic ions bound to the active sites of enzymes or in heme proteins. However, this work is outside the scope of this book. The reader is referred to Sheard and Bradbury,[95] Cohen,[96] and Cohn and Reuben[126] for recent reviews and an entry into this literature.

REFERENCES

1. M. Saunders, A. Wishnia, and J. G. Kirkwood, *J. Amer. Chem. Soc.* **79**, 3289 (1957)
2. O. Jardetzky and C. D. Jardetzky, *J. Amer. Chem. Soc.* **79**, 5322 (1957); *J. Biol. Chem.* **233**, 383 (1958).
3. F. A. Bovey, G. V. D. Tiers, and G. Filipovich, *J. Polym. Sci.* **38**, 73 (1959).
4. A. Kowalsky, *J. Biol. Chem.* **237**, 1807 (1962).
5. A. Kowalsky, *Biochemistry* **4**, 2382 (1965).
6. J. H. Bradbury and H. A. Scheraga, *J. Amer. Chem. Soc.* **88**, 4240 (1966).
7. M. Mandel, *Proc. Nat. Acad. Sci. U.S.* **52**, 736 (1964).
8. M. Mandel, *J. Biol. Chem.* **240**, 1586 (1965).
9. C. C. McDonald and W. D. Phillips, *Proc. Int. Conf. Magnetic Resonance Biol., 2nd.* Pergamon Press, Oxford, 1967.
10. C. C. McDonald and W. D. Phillips, *J. Amer. Chem. Soc.* **89**, 6332 (1967).
11. C. C. McDonald and W. D. Phillips, *J. Amer. Chem. Soc.* **91**, 1513 (1969).
12. R. E. Dickerson and I. Geis, "The Structure and Action of Proteins." Harper, New York, 1969.

12a. M. O. Dayhoff, ed., "Atlas of Protein Sequence and Structure," Vol. 4 and succeeding volumes. National Biomedical Research Foundation, Silver Spring, Maryland, 1969.

13. C. C. F. Blake, D. F. Koenig, G. A. Mairs, A. C. T. North, D. C. Phillips, and V. R. Sarma, *Nature* **206**, 757 (1965).
14. L. N. Johnson and D. C. Phillips, *Nature* **206**, 759 (1965).
15. D. C. Phillips, *Proc. Nat. Acad. Sci. U.S.* **57**, 484 (1967).
16. C. C. F. Blake, G. A. Mair, A. C. T. North, D. C. Phillips, and V. R. Sarma, *Proc. Roy. Soc.* (*London*) **B167**, 365 (1967).
17. C. C. F. Blake, L. N. Johnson, G. A. Mair, A. C. T. North, D. C. Phillips, and V. R. Sarma, *Proc. Roy. Soc.* (*London*) **B167**, 378 (1967).
18. L. N. Johnson, *Proc. Roy. Soc.* (*London*) **B167**, 439 (1967).
19. J. H. Bradbury and P. Wilairat, *Biochem. Biophys. Res. Commun.* **29**, 84 (1967).
20. H. Sternlicht and D. Wilson, *Biochemistry* **6**, 2881 (1967).
21. D. H. Meadows, J. L. Markley, J. S. Cohen, and O. Jardetzky, *Proc. Nat. Acad. Sci. U.S.* **58**, 1307 (1967).
22. J. S. Cohen and O. Jardetzky, *Proc. Nat. Acad. Sci. U.S.* **60**, 92 (1968).
23. J. S. Cohen, *Nature* **223**, 43 (1969).
24. C. C. McDonald and W. D. Phillips, *in* "Biological Macromolecules" (G. Fasman, ed.). Dekker, New York, 1970.
25. C. C. McDonald and W. D. Phillips, *Biochem. Biophys. Res. Commun.* **35**, 43 (1969).
26. J. D. Glickson, C. C. McDonald, and W. D. Phillips, *Biochem. Biophys. Res. Commun.* **35**, 492 (1969).
27. J. H. Bradbury and N. L. R. King, *Nature* **223**, 1154 (1969).
28. C. C. McDonald, W. D. Phillips, and J. D. Glickson, *J. Amer. Chem. Soc.* **93**, 235 (1971).
29. A. J. Sophianopoulos and B. J. Weiss, *Biochemistry* **3**, 1920 (1964).
30. J. F. Brandts, *J. Amer. Chem. Soc.* **86**, 4291 (1964).
31. J. Hermans, Jr. and G. Acampora, *J. Amer. Chem. Soc.* **89**, 1547 (1967).
32. C. Tanford, R. H. Pain, and N. S. Otchin, *J. Mol. Biol.* **15**, 489 (1966).
33. K. C. Aune, A. Salahaddin, M. H. Zarlengo, and C. Tanford, *J. Biol. Chem.* **242**, 4486 (1967).
34. K. C. Aune and C. Tanford, *Biochemistry* **8**, 4579, 4586 (1969).
35. C. Tanford and K. C. Aune, *Biochemistry* **9**, 206 (1970).
36. K. Hayashi, T. Imoto, G. Funatsu, and M. Funatsu, *J. Biochem.* (*Tokyo*) **58**(3), 227 (1965).
37. G. Kartha, J. Bello, and D. Harker, *Nature* **213**, 862 (1967).
38. H. W. Wyckoff, K. D. Hardman, N. M. Allewell, T. Inagami, L. N. Johnson, and F. M. Richards, *J. Biol. Chem.* **242**, 3984 (1967).

38a. H. W. Wyckoff, D. Tsernoglou, A. W. Hanson, J. R. Knox, B. Lee, and F. M. Richards, *J. Biol. Chem.* **245**, 305 (1970).

39. D. H. Meadows, O. Jardetzky, R. M. Epand, H. H. Ruterjans, and H. A. Scheraga, *Proc. Nat. Acad. Sci. U.S.* **60**, 766 (1968).
40. D. H. Meadows and O. Jardetzky, *Proc. Nat. Acad. Sci. U.S.* **61**, 406 (1968).
41. G. C. K. Roberts, D. H. Meadows, and O. Jardetzky, *Biochemistry* **8**, 2053 (1969).
42. D. H. Meadows, G. C. K. Roberts, and O. Jardetzky, *J. Mol. Biol.* **45**, 491 (1969).
43. A. Wishnia and M. Saunders, *J. Amer. Chem. Soc.* **84**, 4235 (1962).
44. K. Wüthrich and R. G. Shulman, *Physics Today* **23**, 43 (April, 1970).
45. H. M. McConnell and D. B. Chesnut, *J. Chem. Phys.* **28**, 107 (1958).
46. A. D. McLachlan, *Mol. Phys.* **3**, 233 (1960).
47. A. B. Edmundson, *Nature* **205**, 883 (1965).
48. J. C. Kendrew, R. E. Dickerson, B. E. Strandberg, R. G. Hart, D. R. Davies, D. C. Phillips, and V. C. Shore, *Nature* **185**, 422 (1960).

49. J. C. Kendrew, H. C. Watson, B. E. Strandberg, R. E. Dickerson, D. C. Phillips, and V. C. Shore, *Nature* **190**, 665 (1961).
50. J. C. Kendrew, *Science* **139**, 1259 (1963) (Nobel Prize address).
51. M. F. Perutz, *Science* **140**, 863 (1963) (Nobel Prize address).
52. R. A. Bradshaw and F. R. N. Gurd, *J. Biol. Chem.* **244**, 2167 (1969).
53. M. Dautrevaux, K. Han, Y. Boulanger, and G. Biserte, *Bull. Soc. Chim. Biol.* **49**, 1973 (1967); **50**, 505 (1968).
54. P. Edman and G. Begg, *Eur. J. Biochem.* **1**, 80 (1967).
55. J. Hermans, Jr. and L. W. Lu, *Arch. Biochem. Biophys.* **122**, 331 (1967).
56. K. Wüthrich, R. G. Shulman, and J. Peisach, *Proc. Nat. Acad. Sci. U.S.* **60**, 373 (1968).
57. R. G. Shulman, K. Wüthrich, and J. Peisach, *Proc. Nat. Acad. Sci. U.S.* **61**, 1190 (1968).
58. K. Wüthrich, R. G. Shulman, B. J. Wyluda, and W. S. Caughey, *Proc. Nat. Acad. Sci. U.S.* **62**, 636 (1969).
59. R. G. Shulman, K. Wüthrich, T. Yamane, E. Antonini, and M. Brunori, *Proc. Nat. Acad. Sci. U.S.* **63**, 623 (1969).
60. K. Wüthrich, R. G. Shulman, T. Yamane, B. J. Wyluda, T. E. Hugli, and F. R. N. Gurd, *J. Biol. Chem.* **245**, 1947 (1970).
61. R. G. Shulman, J. Peisach, and B. J. Wyluda, *J. Mol. Biol.* **48**, 517 (1970).
62. B. Sheard, T. Yamane, and R. G. Shulman, *J. Mol. Biol.* **53**, 35 (1970).
63. R. G. Shulman, S. H. Glarum, and M. Karplus, *J. Mol. Biol.* **57**, 93 (1971).
64. R. G. Shulman, K. Wüthrich, T. Yamane, D. J. Patel, and W. E. Blumberg, *J. Mol. Biol.* **53**, 143 (1970).
64a. D. J. Patel, L. Kampa, R. G. Shulman, T. Yamane, and B. J. Wyluda, *Proc. Nat. Acad. Sci. U.S.* **67**, 1109 (1970).
65. K. Wüthrich, R. G. Shulman, and T. Yamane, *Proc. Nat. Acad. Sci. U.S.* **61**, 1199 (1968).
66. T. Yamane, K. Wüthrich, R. G. Shulman, and S. Ogawa, *J. Mol. Biol.* **49**, 197 (1970).
67. S. Ogawa, R. G. Shulman, P. A. M. Kynoch, and H. Lehmann, *Nature* **225**, 1042 (1970).
68. D. G. Davis, N. L. Mock, V. R. Laman, and C. Ho, *J. Mol. Biol.* **40**, 311 (1969).
69. S. Ogawa and R. G. Shulman, *Biochem. Biophys. Res. Commun.* **42**, 9 (1971).
69a. D. J. Patel, L. Kampa, R. G. Shulman, T. Yamane, and M. Fujiwara, *Biochem. Biophys. Res. Commun.* **40**, 1224 (1970).
70. R. E. Dickerson, M. L. Kapka, J. E. Weinzierl, J. C. Varnum, D. Eisenberg, and E. Margoliash, *J. Biol. Chem.* **242**, 3015 (1967).
71. K. Wüthrich, *Proc. Nat. Acad. Sci. U.S.* **63**, 1071 (1969).
72. C. C. McDonald, W. D. Phillips, and S. N. Vingradov, *Biochem. Biophys. Res. Commun.* **36**, 442 (1969).
73. K. Wüthrich and R. G. Shulman, paper presented at 4th Middle Atlantic Regional Meeting of Amer. Chem. Soc., February, 1969.
74. M. Tanaka, T. Nakashima, A. Benson, M. Mower, and K. T. Yasanobu, *Biochemistry* **5**, 1666 (1966).
75. M. Poe, W. D. Phillips, C. C. McDonald, and W. Lovenberg, *Proc. Nat. Acad. Sci. U.S.* **65**, 797 (1970).
76. W. D. Phillips, M. Poe, C. C. McDonald, and R. G. Bartsch, *Proc. Nat. Acad. Sci. U.S.* **67**, 682 (1970).
77. D. P. Hollis, G. McDonald, and R. I. Biltonen, *Proc. Nat. Acad. Sci. U.S.* **58**, 758 (1967).
78. A. P. Ryle, F. Sanger, L. F. Smith, and R. Kitai, *Biochem. J.* **60**, 541 (1955).

79. M. J. Adams, T. L. Blundell, E. J. Dodson, G. G. Dodson, M. Vijayan, E. N. Baker, M. M. Harding, D. C. Hodgkin, B. Rimmer, and S. Sheat, *Nature* **224**, 491 (1969).
80. B. Bak, E. J. Pedersen, and F. Sundby, *J. Biol. Chem.* **242**, 2637 (1967).
81. B. Bak, C. Dambmann, F. Nicolaisen, and E. J. Pedersen, *J. Mol. Spectrosc.* **26**, 78 (1968).
82. W. W. Bromer, L. G. Sinn, A. Staub, and O. K. Behrens, *J. Amer. Chem. Soc.* **78**, 3858 (1956).
83. W. B. Gratzer, G. H. Beaver, H. W. E. Rattle, and E. M. Bradbury, *Eur. J. Biochem.* **3**, 276 (1968).
84. J. C. Swann and G. G. Hammes, *Biochemistry* **8**, 1 (1969).
85. D. J. Patel, *Macromolecules* **3**, 448 (1970).
86. A. Bodanszky, M. A. Ondetti, V. Mutt, and M. Bodanszky, *J. Amer. Chem. Soc.* **91**, 944 (1969).
87. D. J. Patel, M. Bodanszky, and M. A. Ondetti, *Macromolecules* **3**, 694 (1970).
88. C. Ho and R. J. Kurland, *J. Biol. Chem.* **241**, 3002 (1966).
89. C. Ho, J. A. Magnuson, J. B. Wilson, N. S. Magnuson, and R. J. Kurland, *Biochemistry* **8**, 2074 (1969).
90. H. L. Crespi, R. M. Rosenberg, and J. J. Katz, *Science* **161**, 795 (1968).
91. J. L. Markley, I. Putter, and O. Jardetzky, *Science* **161**, 1249 (1968).
92. H. L. Crespi and J. J. Katz, *Nature* **224**, 560 (1969).
93. W. P. Slichter and D. D. Davis, *Macromolecules* **1**, 47 (1969).
94. O. Jardetzky, *Advan. Chem. Phys.* **7**, 499 (1964).
95. B. Sheard and E. M. Bradbury, "Progress In Biophysics and Molecular Biology." Pergamon Press, Oxford, 1970.
96. J. S. Cohen, *J. Clin. Pharmacol. J. New Drugs* **9**, 72 (1969).
97. E. W. Thomas, *Biochem. Biophys. Res. Commun.* **24**, 611 (1966).
98. E. W. Thomas, *Biochem. Biophys. Res. Commun.* **29**, 628 (1968).
99. M. A. Raftery, F. W. Dahlquist, S. I. Chan, and S. M. Parsons, *J. Biol. Chem.* **243**, 4175 (1968).
100. F. W. Dahlquist and M. A. Raftery, *Biochemistry* **7**, 3269, 3277 (1968).
101. F. W. Dahlquist and M. A. Raftery, *Biochemistry* **8**, 713 (1969).
102. M. A. Raftery, F. W. Dahlquist, S. M. Parsons, and R. G. Wolcott, *Proc. Nat. Acad. Sci. U.S.* **62**, 44 (1969).
103. J. T. Gerig, *J. Amer. Chem. Soc.* **90**, 2681 (1968).
104. T. McL. Spotswood, J. M. Evans, and J. H. Richards, *J. Amer. Chem. Soc.* **89**, 5052 (1967).
105. E. Zeffren and R. E. Reavill, *Biochem. Biophys. Res. Commun.* **32**, 73 (1968).
106. A. G. Marshall, *Biochemistry* **7**, 2450 (1968).
107. B. D. Sykes, *J. Amer. Chem. Soc.* **91**, 949 (1969).
108. J. J. Fischer and O. Jardetzky, *J. Amer. Chem. Soc.* **87**, 3237 (1965).
109. O. Jardetzky, N. G. Wade, and J. J. Fischer, *Nature* **197**, 183 (1963).
110. D. P. Hollis, J. L. Bolen, and J. M. Kellum, *Biochem. Biophys. Res. Commun.* **22**, 135 (1966).
111. D. P. Hollis, *Biochemistry* **6**, 2080 (1967).
112. P. G. Schmidt, G. R. Stark, and J. D. Baldeschwieler, *J. Biol. Chem.* **244**, 1860 (1969).
113. R. G. Shulman, G. Navon, B. J. Wyluda, D. C. Douglass, and T. Yamane, *Proc. Nat. Acad. Sci. U.S.* **56**, 39 (1966).
114. G. Navon, R. G. Shulman, B. J. Wyluda, and T. Yamane, *Proc. Nat. Acad. Sci. U.S.* **60**, 86 (1968).
115. G. Navon, R. G. Shulman, B. J. Wyluda, and T. Yamane, in press.

116. B. L. Vallee, *Fed. Proc. Fed. Amer. Soc. Exp. Biol.* **32**, 8 (1964).
117. M. L. Ludwig, J. A. Hartsuck, T. A. Steitz, H. Muirhead, J. C. Coppola, G. N. Reek, and W. N. Lipscomb, *Proc. Nat. Acad. Sci. U.S.* **57**, 511 (1967).
118. T. A. Steitz, M. L. Ludwig, F. A. Quiocho, and W. N. Lipscomb, *J. Biol. Chem.* **242**, 4662 (1967).
119. A. S. V. Burgen, O. Jardetzky, J. C. Metcalfe, and N. Wade-Jardetzky, *Proc. Nat. Acad. Sci. U.S.* **58**, 447 (1967).
120. T. R. Stengle and J. D. Baldeschwieler, *Proc. Nat. Acad. Sci.* **55**, 1020 (1966).
121. T. R. Stengle and J. D. Baldeschwieler, *J. Amer. Chem. Soc.* **89**, 3045 (1967).
122. R. P. Haugland, L. Stryer, T. R. Stengle, and J. D. Baldeschwieler, *Biochemistry* **6**, 498 (1967).
123. R. L. Ward and J. A. Happe, *Biochem. Biophys. Res. Commun.* **28**, 785 (1967).
124. R. L. Ward, *Biochemistry* **8**, 1879 (1969).
125. F. J. Kayne and J. Reuben, *J. Amer. Chem. Soc.* **92**, 220 (1970).
126. M. Cohn and J. Reuben, *Accounts Chem. Res.* **4**, 214 (1971).

Chapter XV

NUCLEIC ACIDS

1. Introduction

The application of high resolution NMR to the study of nucleic acids has not been so well developed as that of polypeptides and proteins. Although the NMR literature on the nucleic acid bases, i.e., substituted purines and pyrimidines, and on the nucleosides and nucleotides is fairly extensive, the study of the nucleic acids themselves is rather limited and the significant conclusions drawn from this work still more so. Our discussion of these polymers and their constituent units will therfore be considerably briefer than that contained in Chapters XIII and XIV.

2. The Structure of Nucleic Acids

The nucleic acid chain, as is now well known, is composed of pentose units connected by phosphate links, with purine and pyrimidine bases attached to the sugar units; the base-sugar-phosphate unit is known as a *nucleotide* unit:

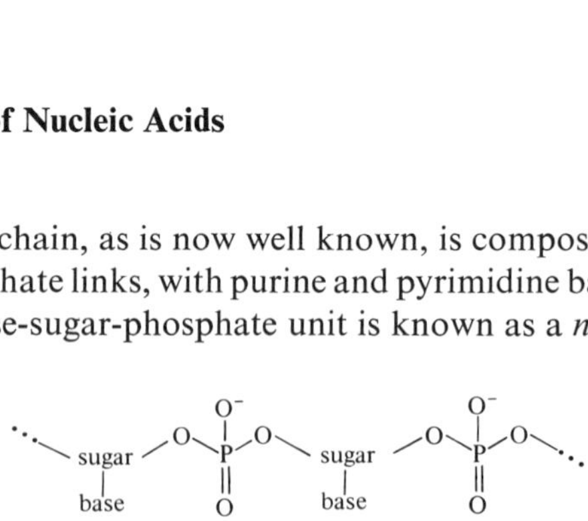

Two classes of nucleic acids have long been recognized: those in which the sugar is D-ribose, called ribonucleic acids or RNA, and those in which the sugar is 2-deoxy-D-ribose, called deoxyribonucleic acids or DNA. DNA occurs in cell nuclei and is the genetic material which transmits information for protein synthesis from one generation to the next. (In certain viruses, RNA plays this role.) The DNA replicates upon cell division and, in addition,

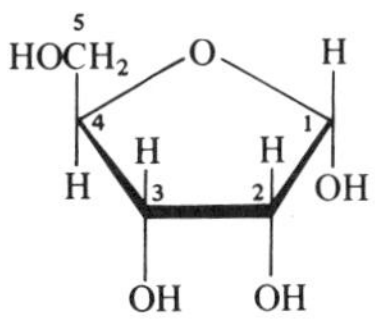

D-ribose

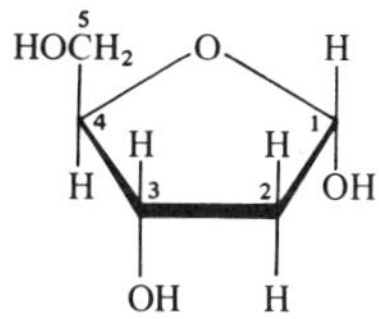

2-deoxy-D-ribose

performs the essential function of serving as a template for the transcription of this information onto *messenger* RNA (mRNA). The mRNA molecule then moves to the *ribosome*, a large complex nucleoprotein composed of two subunits containing *ribosomal* RNA (rRNA). The messenger RNA serves as an assembly line where amino acid residues are joined in proper sequence to form finished protein molecules. This process requires a third type of RNA, *transfer* RNA (tRNA, formerly designated as *soluble* or sRNA), which recognizes specific amino acids and adds them at the proper point as directed by the messenger RNA.†

Messenger RNA composes only a small fraction of the RNA present in the cell cytoplasm, most of which is tRNA and rRNA. It is continually degraded and renewed. Its molecular weight is polydisperse and is in the range 200,000–500,000 and higher. Ribosomal RNA is the predominant type; in bacteria it may constitute as much as 80% of the total. Its molecular weight varies from about 560,000 to 1,100,000, depending upon the ribosomal subunit in which it occurs. Transfer RNA molecules contain only 77–85 nucleotide units (molecular weight about 28,000–30,000), and should thus be most suitable for study by NMR because of minimal dipolar broadening. We shall consider the detailed structure of tRNA later in the discussion (Section 6B).

† It is not within the scope of this discussion to consider these matters in detail. A convenient source of information is "The Molecular Basis of Life," a compilation of articles from the *Scientific American*.[1] For a discussion of the sequence and structure of tRNA's see Zachau.[1a]

The pyrimidine bases occurring in DNA are thymine and cytosine. (The initials in parentheses are the commonly employed abbreviations for the bases and also for the nucleotide units containing them in polynucleotide chains.) In RNA, uracil almost entirely replaces thymine:

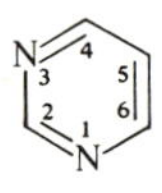

pyrimidine

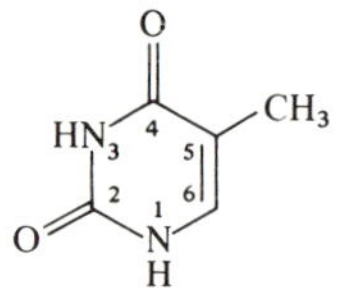

thymine (T)
(5-methyl-2,4-dioxy-pyrimidine)

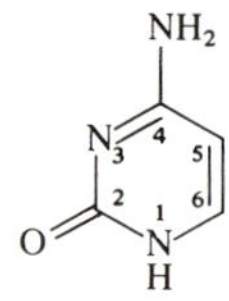

cytosine (C)
(2-oxy-4-amino-pyrimidine)

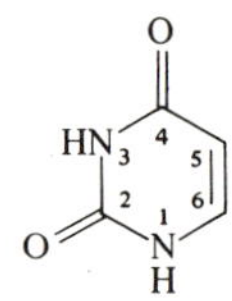

uracil (U)
(2,4-dioxypyrimidine)

Other pyrimidine bases occurring in minor proportions in RNA are:

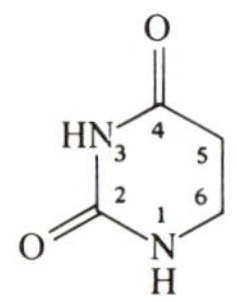

dihydrouracil ($U^{\underline{h}}$, D, or diHU)
(5,6-dihydro-2,4-dioxypyrimidine)

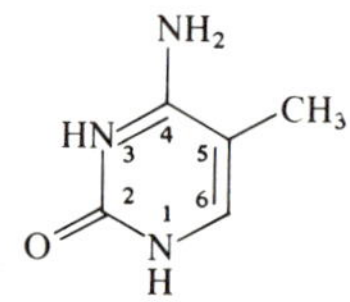

5-methylcytosine (5 MeC or C^{5-Me})
(5-methyl-2-oxy-4-aminopyrimidine)

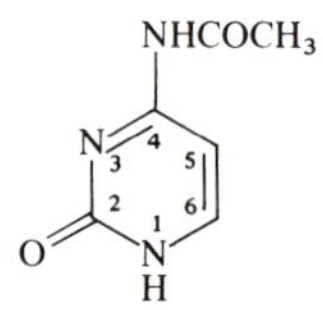

N^4-acetylcytosine (AcC)

For model studies, alkylated pyrimidines such as 1-methylcytosine and 1,3-dimethylcytosine have been employed (see Section 2):

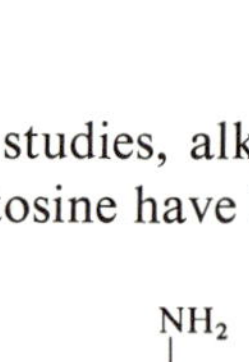

1-methylcytosine

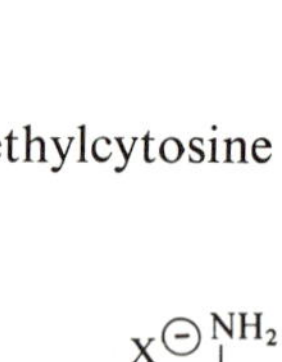

1,3-dimethylcytosine

The pyrimidine bases may exist in both lactam and lactim tautomers:

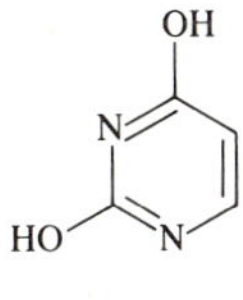

lactim

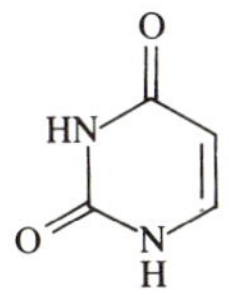

lactam

The lactam form, which is nonaromatic, strongly predominates at neutral and acid *p*H and is the only form normally considered present. Ring-current effects in pyrimidines are therefore expected to be insignificant. It is also possible that the amino group in cytosines may actually exist as the imino tautomer, but this has been shown not to be the case (Section 3).

The purine bases occurring in DNA are adenine and guanine. The actual

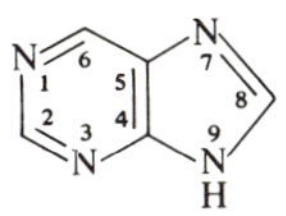

purine

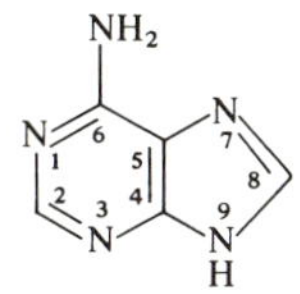

adenine (A)
(6-aminopurine)

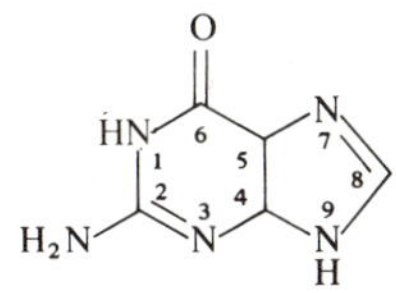

guanine (G)
(2-amino-6-oxypurine)

position of the hydrogen atom on the five-membered ring is not known with certainty. Again, lactim forms (weak acids) can exist and in some cases do, as in uric acid (2,6,8-trioxypurine); these will not be considered further here. In RNA, a number of other purine bases occur in small proportions; some of these are shown:

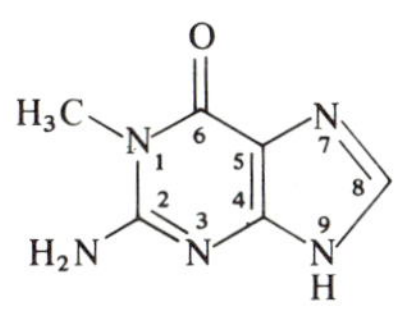

1-methylguanine
(G^m, G^{1Me}, MeG, 1MeG)

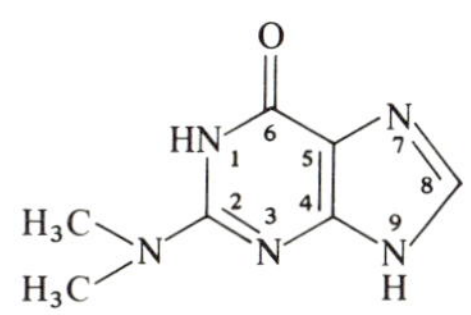

N,*N*′-dimethylguanine
($G^{\underline{m}}$, G^{2Me}, diMeG, 2MeG)

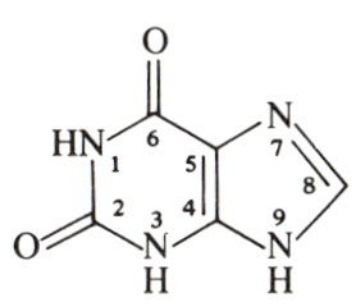

xanthine (X)
(2,6-dioxypurine)

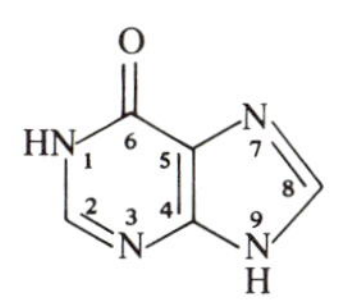

hypoxanthine (I)
[6-oxypurine; the base of inosine (p. 400)]

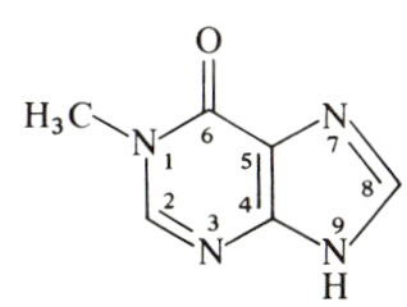

1-methylhypoxanthine
(I^m, I^{Me}, 1MeI; the base of 1-methylinosine)

In DNA and RNA, the purine bases adenine and guanine are attached to the chain by a bond between the 9-nitrogen of the purine and the 1-carbon (designated 1′) of the pentose sugar. The pyrimidine bases cytosine and thymine are attached by their 1-nitrogen to the 1-carbon of the pentose. The phosphate links between the pentose units connect the 5′-carbon of one to the 3′-carbon of its neighbour. A segment of a DNA chain may thus be represented as shown, bearing in mind that the proportions and sequence of the bases are not necessarily as indicated. The chain is conventionally written with the terminal unit having a free 5′-OH at the left and that with a free 3′-OH at the right.

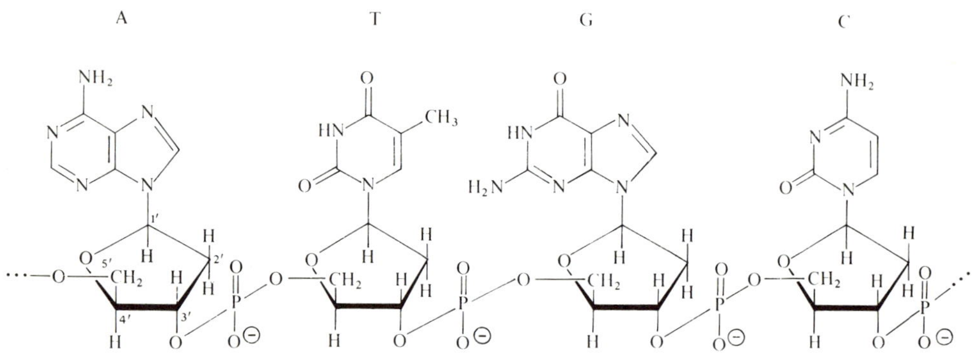

Segment of DNA chain

In the DNA molecule, two such chains are wound in a helical pattern about a common axis. As shown in Fig. XV-1, the two chains are linked together by hydrogen bonds between the bases, the planes of which are perpendicular to the helical axis. The steric requirements are such that an adenine must be paired with a thymine at the same level in the other chain and a guanine with a cytosine:

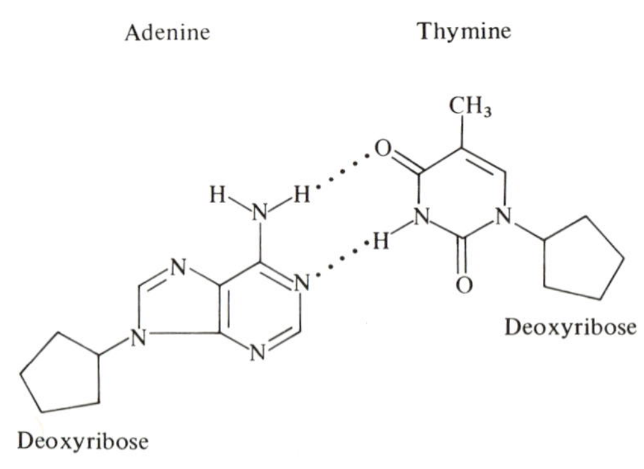

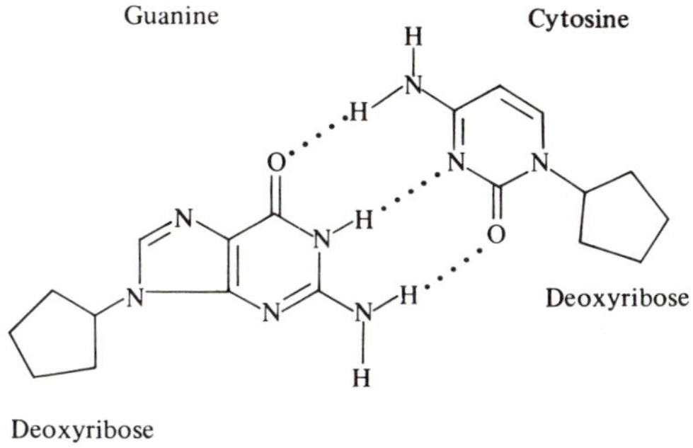

All the bases are paired off in this manner throughout the entire helix. The chains are thus necessarily complementary in sequence. They also run in opposite directions, the sequence of main-chain atoms going one way in one and the opposite way in the other. Upon heating, the twin helix is denatured; the hydrogen bonds between base pairs are disrupted and the two chains form separate single strands. Upon cooling, the helix is partially reconstituted.

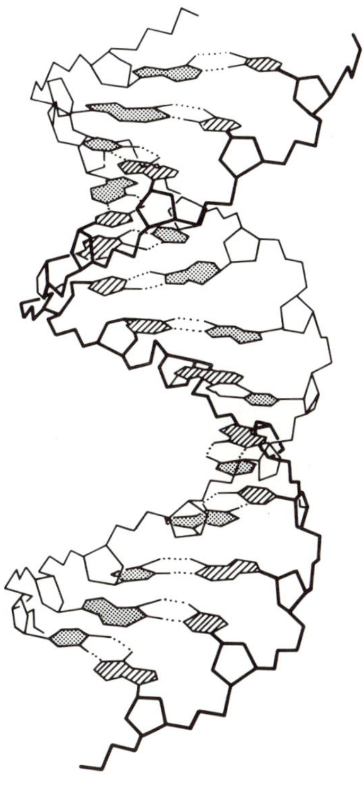

FIG. XV-1. The Watson–Crick double helix structure of DNA.[1b] The main-chain bonds of one of the chains are drawn more heavily so that the two may be more readily distinguished.

The conformation of RNA is much less certain. It is known that viral RNA may exist either as a double helix or in single-stranded form. Messenger RNA is single-stranded. Native ribosomal RNA is combined with protein, as we have seen (p. 389). Both probably have coiled structures in which the base units are "stacked" (see Sections 3 and 6) to some degree. The conformation of transfer RNA has been the subject of many investigations by a variety of physical methods, but the results are still inconclusive. The principal information concerning the three-dimensional structures of tRNA's has been deduced from a knowledge of nucleotide sequence and the observation that if certain folded shapes are assumed, the most popular of which is the "clover-leaf" model (see Fig. XV-13, p. 418), the functional portions (e.g., the amino acid binding site and the anticodon sequence) of the chains of all known tRNA molecules consistently appear in corresponding parts of the model. For further discussion, see "The Molecular Basis of Life"[1] (p. 72 *et seq.*), Zachau,[1a] and Section 6B.

Synthetic RNA-like polynucleotides may be prepared by treating ribonucleoside diphosphates (Section 4) with the enzyme *polynucleotide phosphorylase* in the presence of magnesium ion. The polymers of a single repeating unit are known as polyadenylic acid (poly-A), polyuridylic acid (poly-U), polycytidylic acid (poly-C), and polyinosinic acid (poly-I). The conformations of these polymers are discussed in Section 6A. If two or more ribonucleoside diphosphates are employed, copolymers are produced.

Corresponding DNA-like polymers can be prepared only if all four DNA bases (i.e., A, T, G, and C) are present as their nucleoside triphosphates in the presence of denatured DNA as primer. If the primer DNA is A–T copolymer, the synthesized DNA will contain only A and T.

3. Bases

There have been extensive studies of the proton resonances of purines and pyrimidines.† Not all of this work is immediately relevant to the understanding of polynucleotide spectra, and we shall for the most part limit our discussion to that which is.

A. Pyrimidines

Assignments in the spectrum of pyrimidine itself [Fig. XV-2(a)] are straightforward[3–5]: the C-2 proton appears at lowest field, the C-4 and C-6 protons

† For a review with much detailed data, see Rowe *et al.*[2] An authoritative review of the NMR of nucleosides, nucleotides, and nucleic acids is that of Yamane.[2a]

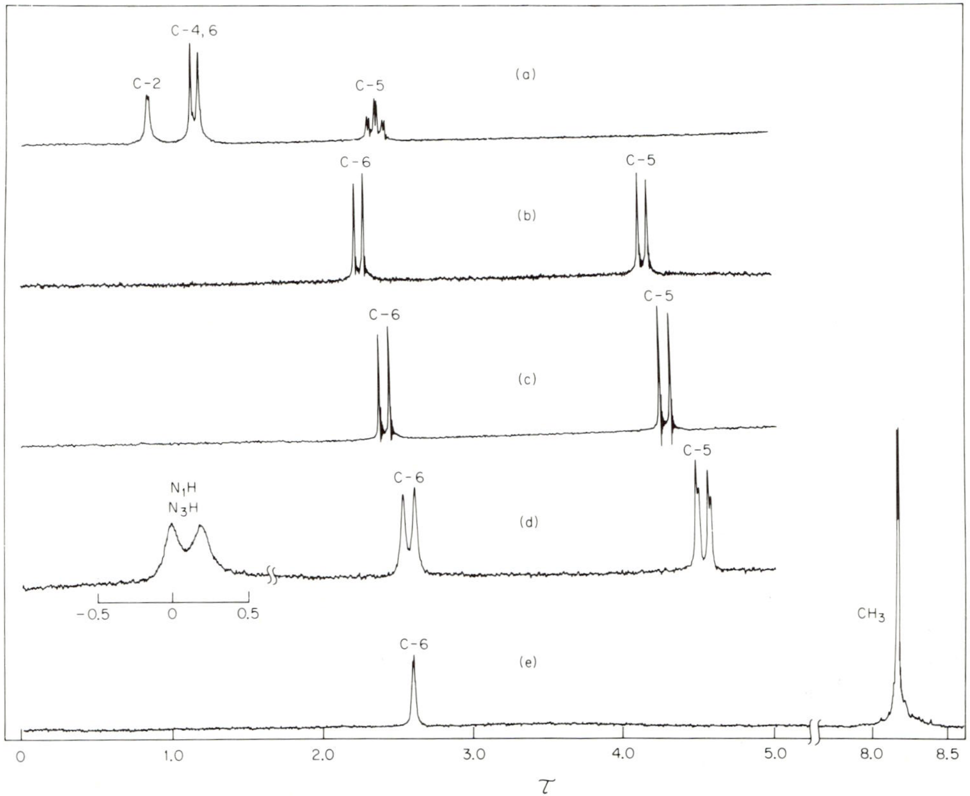

FIG. XV-2. 100 MHz spectra of (a) pyrimidine in D_2O, (b) cytosine in D_2O, (c) uracil in D_2O, (d) uracil in d_6-DMSO, (e) thymine in D_2O. All measurements at about 25° using 5% w/v solutions; aqueous solutions near neutral pH or at somewhat higher pH for more difficultly soluble bases (Bovey and Brewster[8]).

(equivalent by symmetry) appear as a doublet at an intermediate position and the C-5 proton appears at highest field ($J_{45} = J_{56} \simeq 5$ Hz; $J_{25} \simeq 1.5$ Hz).

The spectra of cytosine[2,4,6–13] [Fig. XV-2(b)] and of 1-methylcytosine and 1,3-dimethylcytosine[12,13] show doublets for the C-5 and C-6 protons (J_{56} about 8 Hz), the latter being at lower field. In water the resonance of the amino protons coalesces with that of the solvent except at very low *p*H; in DMSO at room temperature, it appears as a broad peak at about 3τ. When ^{14}N is replaced by ^{15}N in the amino group of 1-methylcytosine, this resonance splits into a 90 Hz doublet due to direct ^{15}N–H coupling. This observation demonstrates that the compound exists as the amino tautomer.[12,13] The amino resonance of 1-methylcytosine hydrochloride in SO_2 at $-60°$ shows separate resonances for each of the amino protons and a new resonance at about -1.4τ. The latter must arise from a proton on N-3 being coupled to the C-5 proton (by 2.5 Hz), as clearly demonstrated by double resonance. Upon replacing ^{14}N by ^{15}N, each of the amino proton resonances splits to an approximately 90 Hz doublet. These observations are explained by the following structure, in which protonation takes place at N-3 and the C—N bond of the amino group has sufficient double bond character to inhibit rotation of the group, as in amides (Chapter XIII, Section 3). Substitution

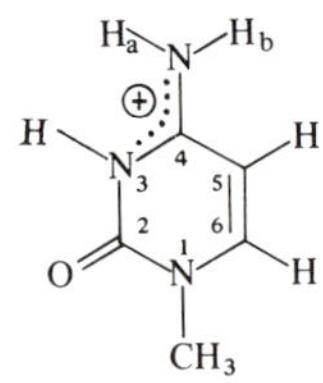

of either H_a or H_b by a methyl group leads to two different geometrical isomers which are distinguishable at low temperature. Roberts *et al.*[14] have confirmed protonation at N-3 in 1-methylcytosine by observations on a molecule having all three ^{14}N nuclei replaced by ^{15}N.

The spectrum of uracil in D_2O or H_2O shows doublets at 4.2τ and 2.3τ for the C-5 and C-6 protons, respectively [Fig. XV-2(c)], and is similar to that of cytosine. In d_6-DMSO [Fig. XV-2(d)] separate resonances for the N-1 and N-3 resonances are observed near -1.0τ[8] and a small further splitting (about 1.5 Hz) of the C-5 proton resonance is observable. Gronowitz *et al.*[15] report a double doublet for the C-6 protons and a doublet of triplets for the C-5 protons. Evidently as a result of trace impurities, the present author[8] was not able to retard the NH exchange rate sufficiently to reproduce this observation. It appears that the N-1 proton must exchange considerably more rapidly than the N-3 proton, which couples measurably to the C-5 proton. From these measurements[8, 15] we may conclude that $J_{H_5-H_6} \simeq 7.5$ Hz, $J_{N_1H-H_6} \simeq 5.6$ Hz, $J_{N_3H-H_5}$ and $J_{N_1H-H_5}$ are both about 1.5 Hz, and $J_{N_3H-H_6} \simeq 0$ Hz.

The spectrum of thymine in D_2O [Fig. XV-2(e)] differs from that of uracil principally in having a methyl resonance at 8.2τ. We shall see in later discussion that this resonance is of particular significance in polynucleotide spectra.

The state of protonation of the pyrimidine ring has a significant effect on the chemical shifts. Thus, in the spectrum of 1-methylcytosine hydrochloride in d_6-DMSO, the amino proton resonances, in addition to splitting to a doublet, move an average of nearly 3 ppm downfield.[12, 13] The C-5 and C-6 proton resonances become less shielded by about 0.5 ppm. It is important to note, however, that in aqueous solution the pyrimidine bases (with the exception of the amino group of cytosine) are not significantly protonated even at low *p*H, differing in this respect from the purines (next subsection).

B. Purines

In aqueous solution, the proton spectrum of purine (about 6.0%) shows three singlet resonances at about 1.0, 1.2, and 1.4τ [Fig. XV-3(a)].[8] The

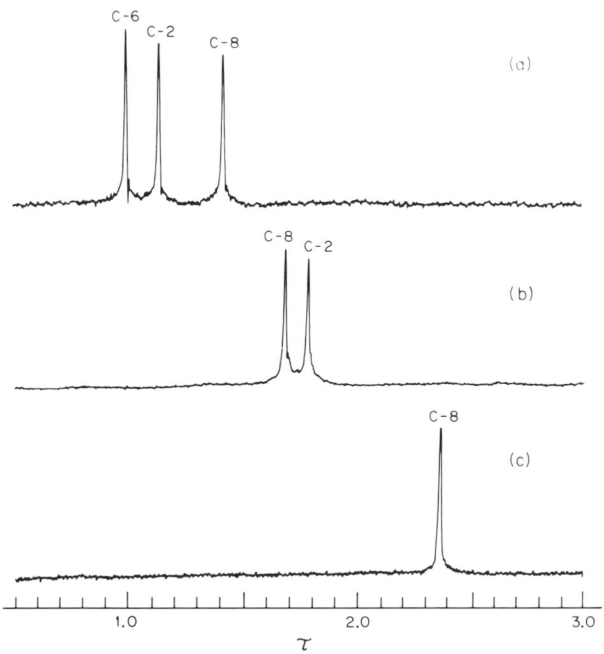

FIG. XV-3. 100 MHz spectra in D_2O (5% w/v; about 25°) of (a) purine, (b) adenine, (c) guanine (Bovey and Brewster[8]).

N-9 proton exchanges too rapidly with the solvent to be separately observed. These peaks are assigned to the C-6, C-2, and C-8 protons in order of increasing shielding.[16–18] These assignments are based principally on observations of specifically deuterated purines and are secure. Earlier erroneous assignments[3, 19] were based on values obtained from substituted purines and related heterocyclic compounds, and illustrate the inadequacy of chemical shift arguments so based. Pugmire *et al.*[20] have assigned the ^{13}C resonances of purine in aqueous solution, and have observed that they increase in shielding in the order C-4, C-2, C-8, C-6, C-5. Thus C-8 is less shielded than C-6 while the attached protons show the reverse order. There is a general correlation of ^{13}C shielding with π-electron density, as proposed by Spiesecke and Schneider.[21] This correlation is discussed further in Section 4.

The spectra of adenine [Fig. XV-3(b)] and guanine [Fig. XV-3(c)] in aqueous solution are very simple, exhibiting peaks at low field for the C-8 protons; in adenine the C-2 proton is somewhat more shielded than the C-8. The C-8 proton of guanine is more shielded than that of adenine.

Upon protonation, the C-2, C-6, and C-8 protons of purine move downfield by nearly equal amounts,[18, 20, 22–25] suggesting protonation in some degree at all the nitrogens. The direct ^{13}C–H couplings at all three positions increase somewhat with decreasing pH, indicating an increase in sp^2 bond hybridization. Interpretation of these latter results in terms of empirical correlations leads to the conclusion that in protonated purine 54% of the charge is on N-1, 12% on N-3, and 29% on the imidazole ring.[2] The titration behavior of all three protons yields a pK_a value of 2.75,[22] somewhat higher than values of about 2.4 obtained by other means. In adenine, more strongly predominant protonation on N-1 is to be expected by analogy to the behavior of 1-methylcytosine.

The weak coupling of the C-2 and C-6 protons, unobservable in Fig. XV-3(a), increases with the extent of protonation, reaching a maximum of about 1 Hz; these resonances exhibit a splitting in aqueous solution below a pH of about 2.5[20, 22, 25] and in trifluoroacetic acid.

The hydrogen-bonded pairing of bases characteristic of the double helix DNA structure is also observable for the free bases in appropriate solvents. Shoup *et al.*[26] and Katz and Penman[27] observed downfield shifts for the NH and NH_2 resonances when the appropriate pairs of bases are mixed in solvents such as dimethyl sulfoxide and chloroform which do not compete too strongly with solute–solute hydrogen bonds. For equimolar mixtures of 9-ethylguanine and 1-methylcytosine, the shifts are twice as great for the guanine NH protons as for the cytosine and guanine NH_2 protons.[26] This is attributed to the fact that each of the protons can participate in a hydrogen bond only half the time, the pairing being assumed to be that shown on p. 393. This phenomenon is discussed further in Section 4.

Chan *et al.*[23] have concluded that another feature of the DNA helical structure—the stacking of bases—may be observed in aqueous solution. They reported an upfield shift of the C-2, C-6, and C-8 protons of purine and of the C-2, C-8, and methyl protons of 6-methyl purine with increasing concentration in water, suggesting a face-to-face association with ring-current shielding. However, in these experiments external chloroform was used as a reference (Chapter I, Section 17B) without adequate correction for the change of magnetic susceptibility of the solution with concentration. When internal DSS is employed as reference,[8] the upfield shift of the C-2 and C-6 protons with concentration is very small and the C-8 proton resonance actually shifts slightly downfield. In Fig. XV-4, these results are compared to

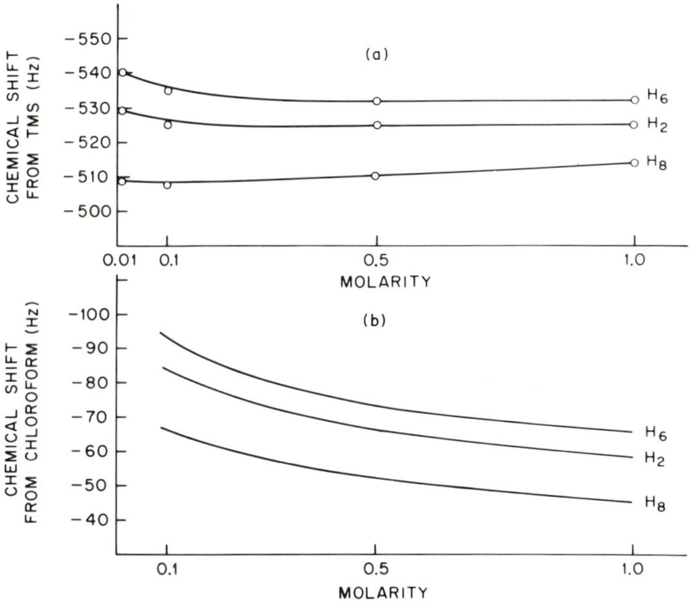

FIG. XV-4. Chemical shifts of the C-2, C-6, and C-8 protons of purine as a function of concentration in D_2O with (a) an internal reference, DSS (Bovey and Brewster[8]), and with (b) external chloroform as reference (Chan *et al.*[23]).

those reported by Chan *et al.*[3] There appears to be no real evidence for base stacking in this system. This question is also discussed further in Section 4.

4. Nucleosides and Mononucleotides

Closer models of the nucleic acid chains than the bases are the nucleosides, which are fragments containing both base and sugar. As indicated on p. 389, the structures occurring in RNA are the β-ribofuranosides.

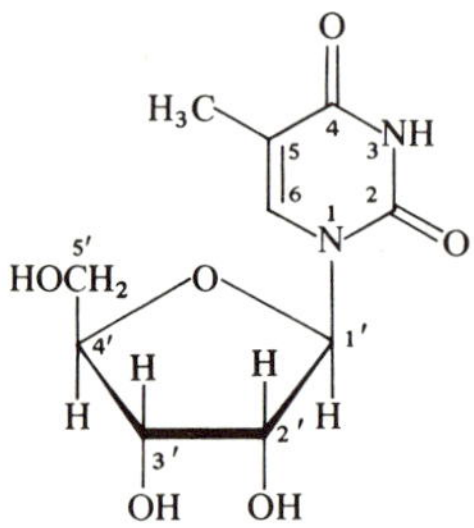

ribothymidine (rT)

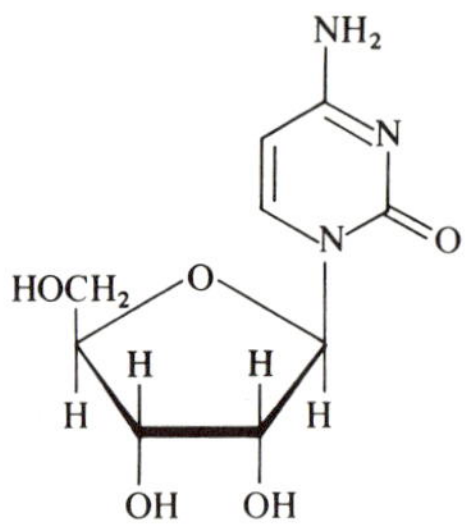

cytidine (C)

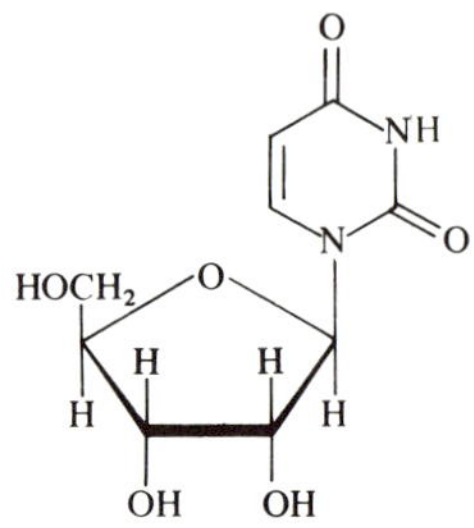

uridine (U)

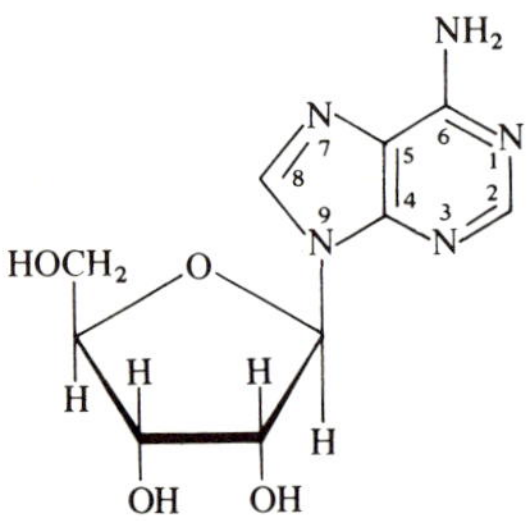

adenosine (A)

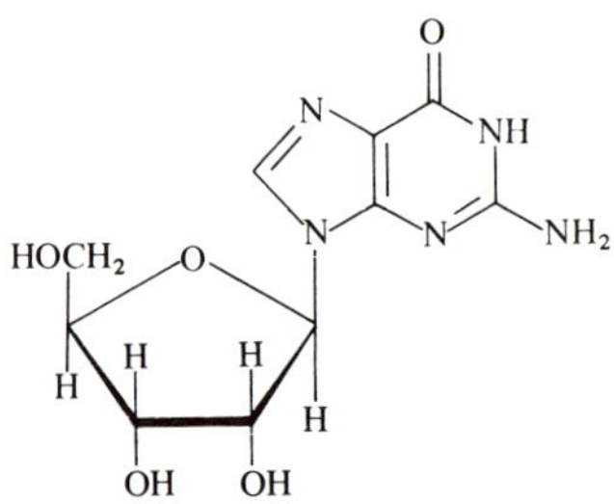

guanosine (G)

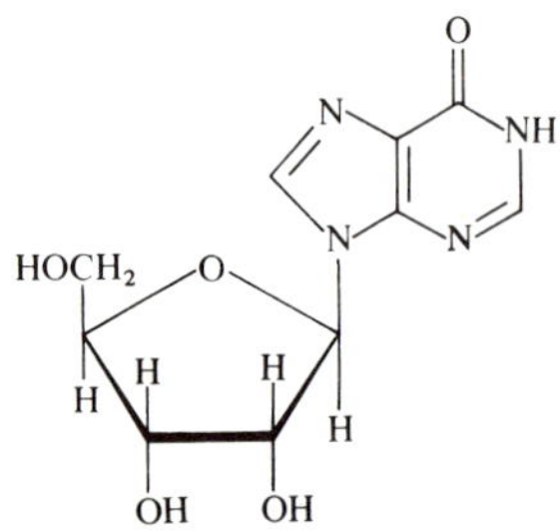

inosine (I)
(base is hypoxanthine)

In addition to the nucleosides of the several variant bases shown on pp. 390–391, we may mention pseudouridine (ψ) which differs from uridine in having the uracil attached to the ribose at C-5; it occurs in all transfer RNA species sequenced to date. The corresponding DNA nucleosides are designated

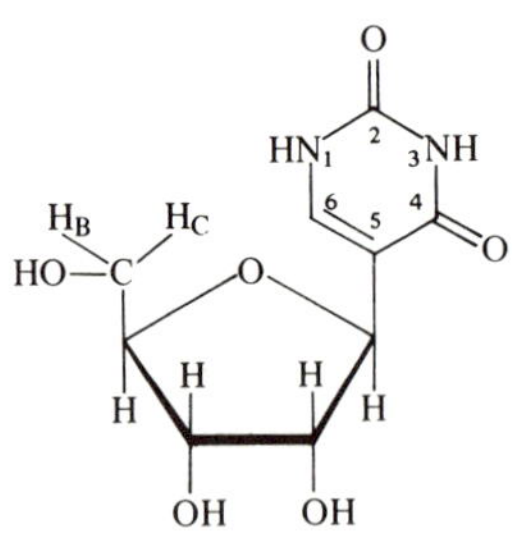

pseudouridine (ψ)

thymidine (T), deoxycytidine (dC), deoxyadenosine (dA), and deoxyguanosine (dG).

The mononucleotides are phosphoric acid monoesters of the nucleosides. The phosphate linkage in ribonucleotides may occur at the 2′, 3′, or 5′ position; the last two correspond to the naturally occurring structural units. In deoxyribonucleotides, only the 3′ and 5′ positions are available. The structures of adenosine-3′-monophosphate and adenosine-5′-monophosphate are:

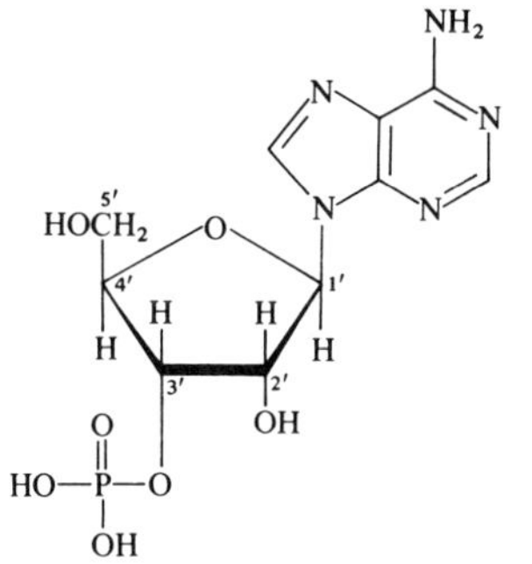

adenosine-3′-monophosphate
(3′-AMP)

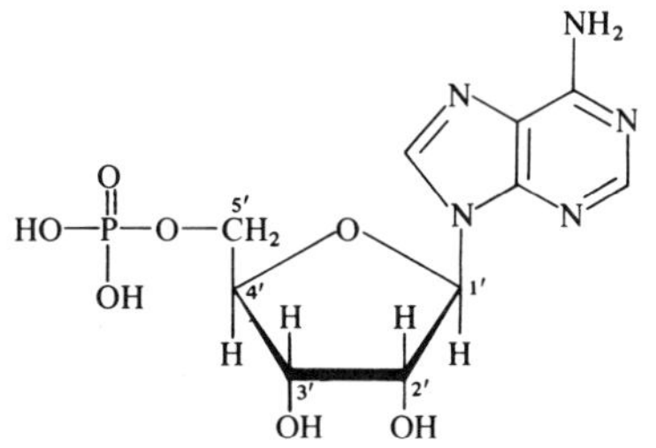

adenosine-5′-monophosphate
(AMP), adenylic acid

The other 5′-monophosphates are:

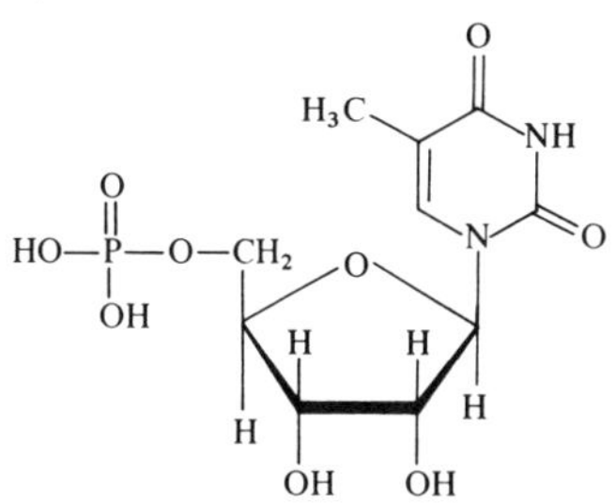

ribothymidine-5′-monophosphate
(rTMP) thymidylic acid

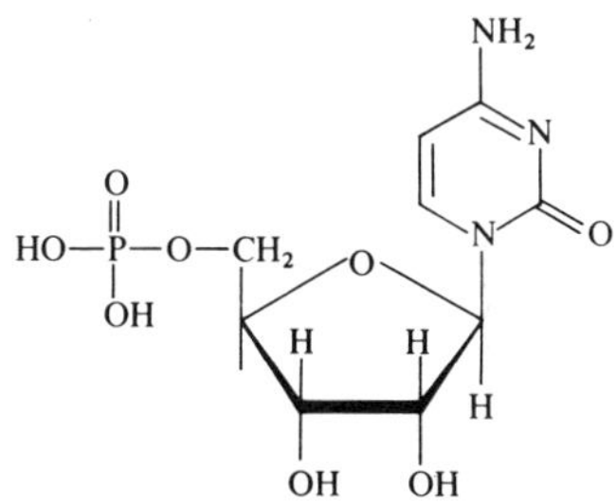

cytidine-5′-monophosphate
(CMP), cytidylic acid

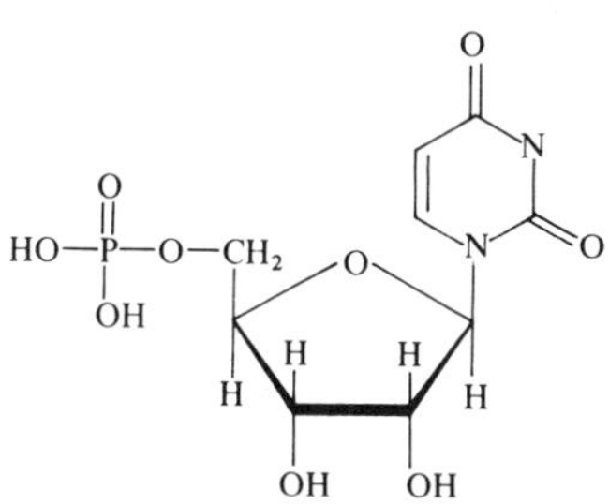

uridine-5′-monophosphate
(UMP), uridylic acid

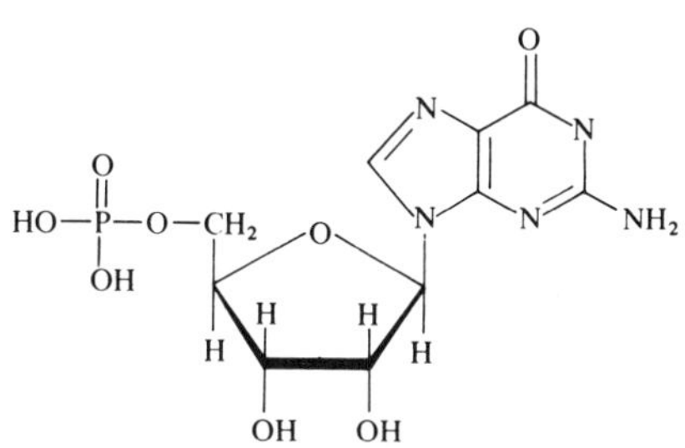

guanosine-5′-monophosphate
(GMP), guanylic acid

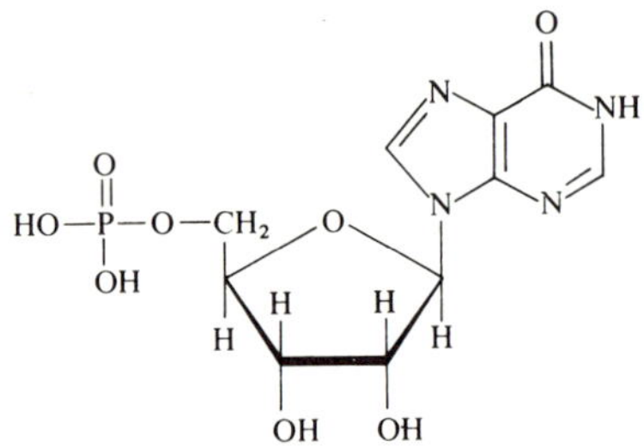

inosine-5′-monophosphate
(IMP), inosinic acid

The diphosphate and triphosphate 5′-esters, in which the phosphate group structures are as shown, are also well known. They are of great importance

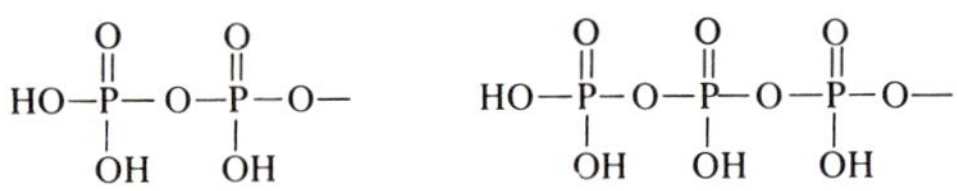

as metabolic intermediates (particularly adenosine-5′-triphosphate, ATP) and in the synthesis of nucleic acid chains (p. 394), but will not be considered further in this discussion.

The pK_a of (a) is about 2 and that of (b) about 6. Therefore in neutral and

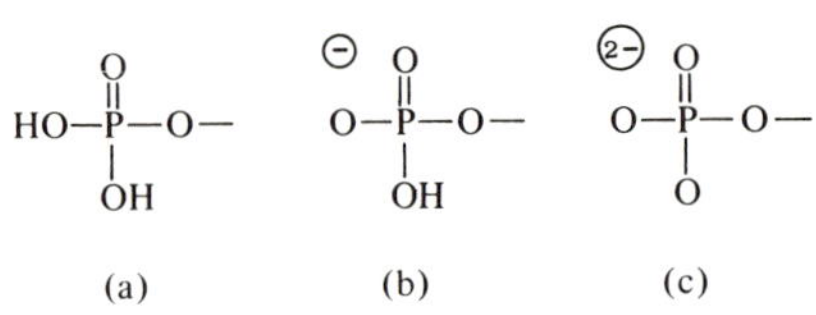

basic aqueous solution, the phosphate group of a mononucleotide is in form (b) or (c). Upon increasing the pH from a low value, one observes a decrease in shielding of the base proton resonances in purine nucleotides due to the first ionization of the phosphate group, followed by a shielding increase for purine and nucleotides in the pH (or pD) range 3–5 due to deprotonation of the base. There may, however, be special deshielding effects at pH of about 5 and above, arising from the $^{2-}O_3PO$ group; this we shall discuss shortly.

A. Pairwise Base Interaction

The pairwise hydrogen-bonded interaction discussed in the previous section for the free bases can be observed also for nucleotides in solvents which are not too strongly hydrogen bonded, such as dimethylsulfoxide and chloroform.

Katz and Penman[27] observed a dramatic downfield shift of the NH resonance of guanosine and a smaller deshielding of the NH_2 resonances of guanosine and cytidine when the two are mixed in d_6-DMSO. Figure XV-5[8, 27] shows

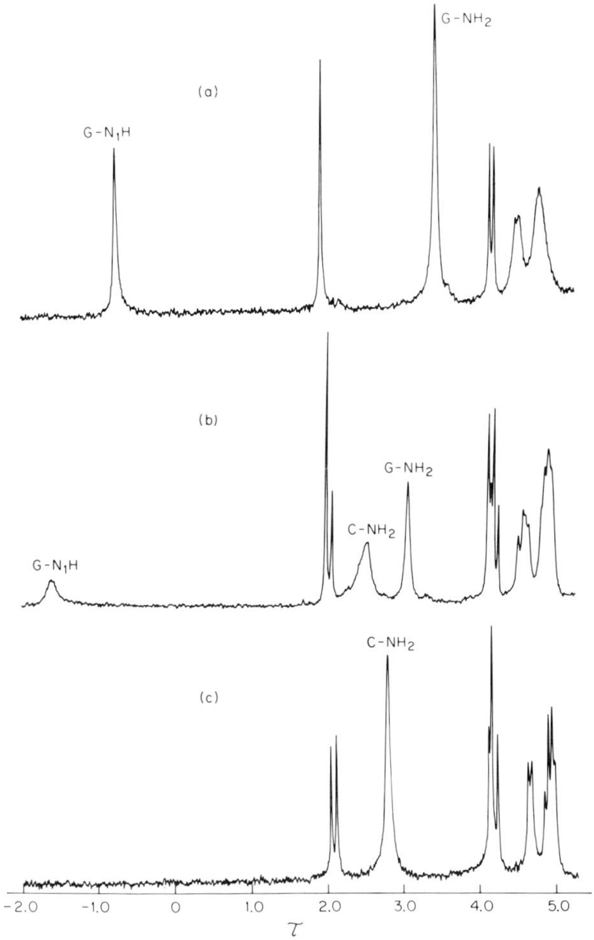

FIG. XV-5. 100 MHz spectra in D_2O (10% w/v of each nucleoside, about 25°) of (a) guanosine (G), (c) cytidine (C), (b) an equimolar mixture of G and C (DSS reference). The singlet near 2τ in (a) is that of the guanine C-8 proton. The doublets near 2 and 4τ in (c) are those of the cytidine C-6 and C-5 protons, respectively (Bovey and Brewster[8] and Katz and Penman[27]).

the spectra of the separate nucleosides and of a 1:1 molar mixture. For the complementary base pair adenosine and uridine, however, no interaction was detectable, reflecting the known fact that this pair are more weakly bonded.[28]

A marked effect was observed for the pair 9-ethyladenine and 1-cyclohexyluracil in $CDCl_3$, which is much more weakly competitive than DMSO in hydrogen bond formation.

B. Base Stacking

Upfield shifts with increasing concentration in aqueous solution, analogous to those discussed in Section 3 for free purine bases, have been reported for purine nucleosides and nucleotides as well,[23, 29–33] and attributed to the same cause, i.e., stacking of the bases. These conclusions are, however, subject to the criticism, discussed in Section 3, that an external reference was used without adequate susceptibility correction. For example, T'so *et al.*[33] reported downfield shifts with increasing temperature for the C-5, C-6, and C-1′ protons of UMP, using an external tetramethylsilane reference [Fig. XV-6(b)], without,

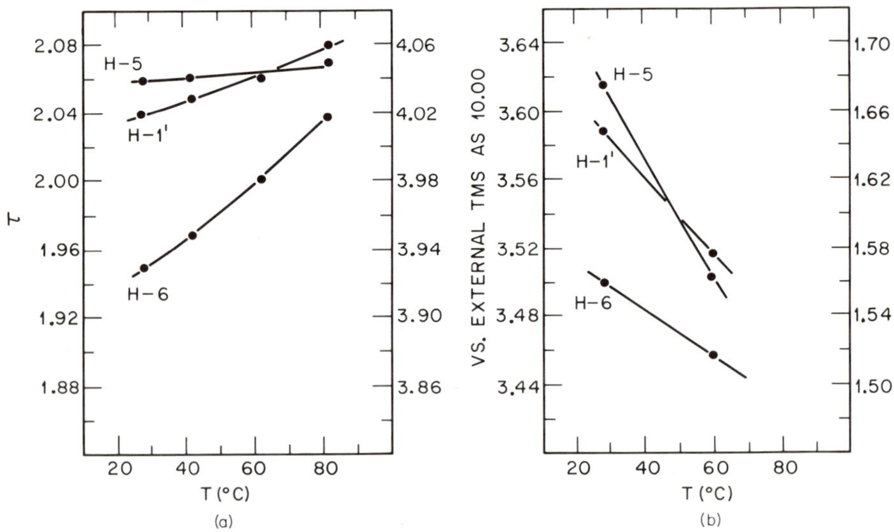

FIG. XV-6. 100 MHz spectra in D_2O of uridine-5′-monophosphate (UMP) showing the chemical shifts of the C-5, C-6, and C-1′ protons as a function of temperature with (a) internal DSS reference, and (b) external TMS reference. In (a) the UMP was 0.1 *M*, *p*D 7.0; in (b) the UMP was 0.045 *M*, *p*D 6.3 (Blackburn *et al.*[34]).

however, having allowed for the changing difference in susceptibility of solvent and reference with temperature. Blackburn *et al.*[34] demonstrated that when internal DSS is employed the trend is actually in the opposite direction [Fig. XV-6(a)]. Stacking may yet occur, but is not proved by such findings.

Stacking of bases is also believed to occur in the pyrimidine nucleosides cytidine and uridine on the basis of osmotic pressure observations[35] but cannot be observed even in principle by NMR for such nucleosides because of the absence of a ring current. However, their protons are reported to move upfield in the presence of purine, suggesting purine–pyrimidine stacking.[30] Similar effects are observed for pseudouridine.[36]

C. CONFORMATION

1. *Rotational State about the Base–Sugar Bond*

Examination of molecular models strongly suggests that in nucleosides and nucleotides the base will be either *syn* or *anti* with respect to the sugar, as shown for AMP and TMP, intermediate torsional states (allowing for some libration) being discouraged by steric interactions. (The representation here is somewhat more realistic with respect to the ribose bond angles than in the formulas on previous pages, all of which are shown as *anti*.) Steric considerations further

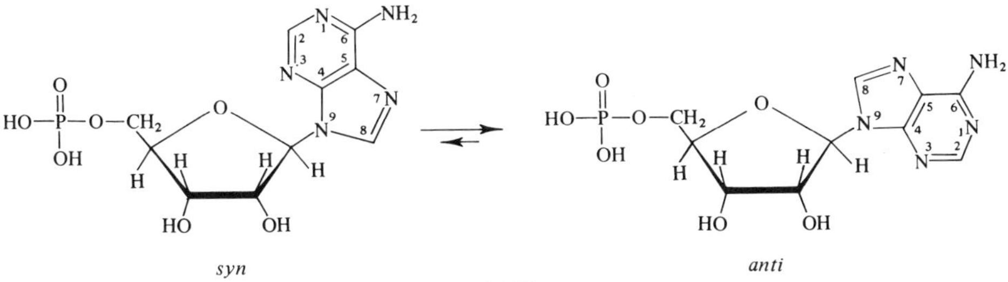

AMP

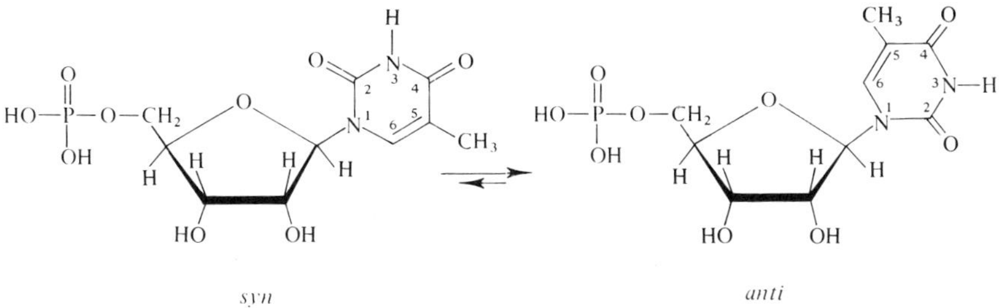

TMP

suggest that the *anti* conformation will be energetically preferred over the *syn*, and that this preference will be stronger for purine bases. In the *anti* conformation of the pyrimidine bases, the C-6 proton is much closer to the phosphate group than in the *syn* conformation. In *anti*-AMP, the C-8 proton is very much closer to the phosphate group than the C-2 proton; the reverse holds for *syn*-AMP. A deshielding effect of the charged phosphate group on the C-6 proton of UMP and CMP was demonstrated by Schweizer *et al.*[32]; there was no comparable effect in 3′-UMP and 3′-CMP. It has been further shown[31] that upon titration of TMP from very low *p*D, the C-6 proton moves to low field in two steps with transitions at *p*D of about 1.5 and 6.0, corresponding to the first and second ionizations of the phosphate group. The C-8 proton of AMP shows an even more striking effect.[31, 32] On increasing the *p*D from about 0, there is at first a decrease in shielding due to the first ionization of the phosphate group; this is followed by an increase in shielding arising from deprotonation of the purine ring (as we have seen, this does not occur for any of the ring protons of pyrimidine nucleotides because the ring is not protonated); above *p*D 5, this increase is reversed and the peak moves downfield again. All the other protons show increasing shielding with smooth sigmoidal curves as the *p*D is increased. These results are undoubtedly due to a field effect[37] arising from the development first of a single charge and then a double charge on the phosphate group. This effect operates directly through space. The presence of the charge distorts the electron cloud about the C-6 proton, resulting in less effective shielding (Chapter I, Section 11). Thus, the NMR findings confirm the indications from models that at biological temperatures the *anti* conformation of the base is strongly preferred in nucleotides. The phosphate deshielding effect decreases with increasing temperature because of increasing participation of the *syn* form.

The evidence of this sort is less convincing for nucleosides,[38] but from steric considerations the preference would appear to be nearly equally strong.

A method for the determination of the base conformation that promises to be of considerable power is that employing the nuclear Overhauser effect (Chapter I, Section 18B). A number of nucleosides and nucleotides have been studied by this means by Hart and collaborators.[39, 39a] Here, as we have seen, the magnitude of the enhancement depends upon the inverse sixth power of the spatial separation of the observed and irradiated nuclei and can be highly sensitive to structure in favorable cases. As an example of their observations, it was found that in isopropylidene adenosine (a) there is a substantial en-

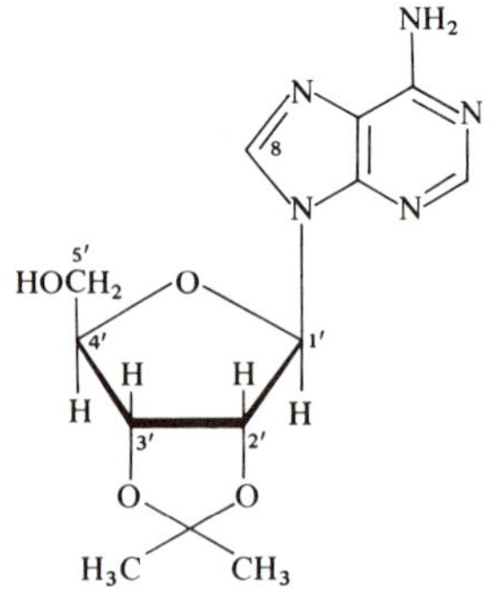

(a)

hancement of the C-8 proton resonance when the C-1′ proton is irradiated (a result to be expected regardless of the orientation of the base) and that there are definite but smaller enhancements when the C-2′, C-3′, C-4′, and C-5′ protons are irradiated. These latter enhancements are considerably larger when the same experiments are performed on isopropylidene guanosine, indicating a stronger *trans* preference here. On the other hand, there was negligible enhancement of the C-8 proton resonance upon irradiation of the C-2 and C-3 protons in isopropylidene cycloguanosine (b) which can only assume

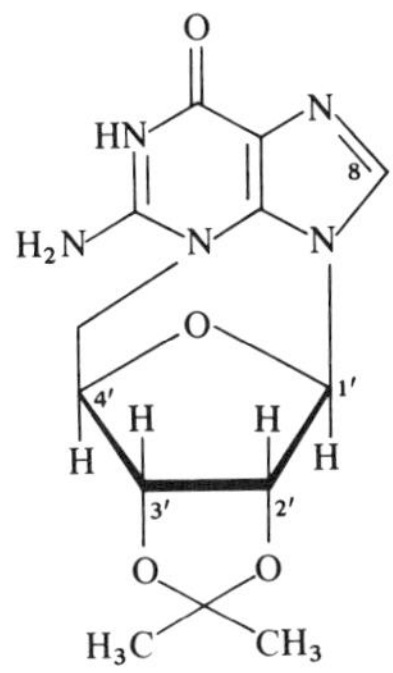

(b)

a *syn* conformation.

There is support for these conclusions from x-ray crystallography[40, 41] and from optical studies.[42–46]

2. *Conformation of the Furanose Ring*

The couplings between adjacent carbon-bonded protons of the furanose ring of nucleosides (and nucleotides) can be interpreted in terms of Karplus functions of the H–C–C–H dihedral angle (see Chapter I, Section 14) to yield the conformation of the ring. Early work[47, 48] yielded approximate values of the coupling of the C-1′ and C-2′ protons, $J_{1'2'}$, but the other couplings are needed as well in order to draw reliable conclusions. Because of spectral overlap these are less easy to determine. Well-resolved spectra of uridine,[49, 50] pseudouridine,[36] and nucleoside 3′,5′-cyclic monophosphates, e.g., 3′,5′-uridine cyclomonophosphate,[51] have been reported. The observed and

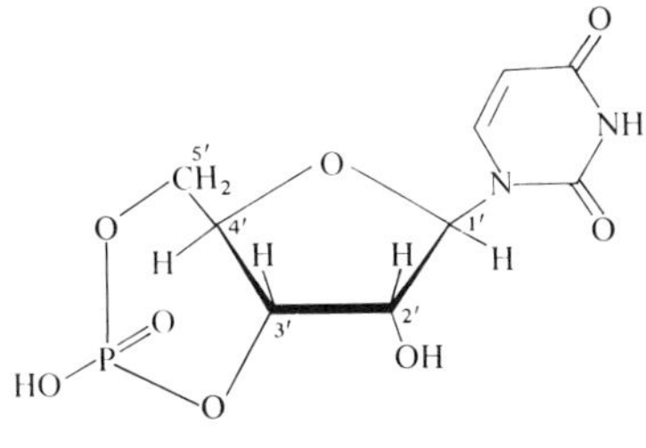

simulated 220 MHz spectra obtained by Hruska *et al.*[36] for pseudouridine are shown in Fig. XV-7. Such simulations are, as we have seen in earlier

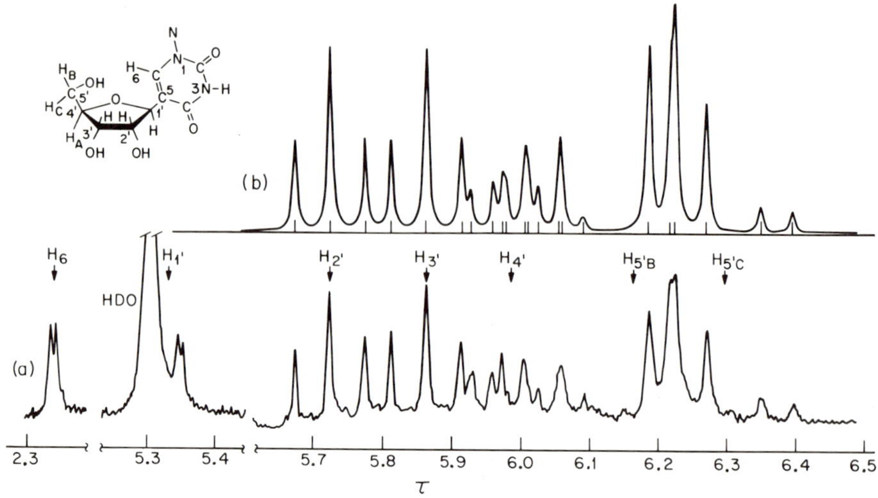

FIG. XV-7. (a) 100 MHz spectrum of β-pseudouridine in D_2O, pD 6.7, 0.12 M, 30°, DSS reference. A second doublet of the quartet due to the C–$H_{1'}$ proton ($J_{H_{1'}-H_{2'}} = 5.0$ Hz; $J_{H_{1'}-H_{6'}} = 0.8$ Hz) is obscured by the HDO resonance near 5.3τ. (b) Computer-simulated spectrum of the region due to the ribose protons C-2′ through C-5′ (Hruska *et al.*[36]).

chapters, the most reliable way to extract parameters from strong-coupled, complex spectra. All the ribose vicinal couplings were in the range 5.0–5.2 Hz, with the exception of $J_{4'5'B}$ and $J_{4'5'C}$ (see the pseudouridine structural formula on p. 400 and in Fig. XV-7; it should be noted that the C-5′ protons H_B and H_C were not specifically assigned), which were 3.2 and 4.6 Hz, or vice versa. They were almost independent of temperature. These values indicate that no one conformation is strongly preferred; if this were the case, inspection of models shows that the values of $J_{1'2'}$ and $J_{3'4'}$ would both have to be attributed to dihedral angles of about 136°, which is in conflict with steric requirements. Furthermore, the C-2′ and C-3′ protons cannot be eclipsed because in this case $J_{2'3'}$ would be about 9–10 Hz (see Fig. I-17). The necessary conclusion is that all of the conformations shown in Fig. XV-8 contribute in some degree.

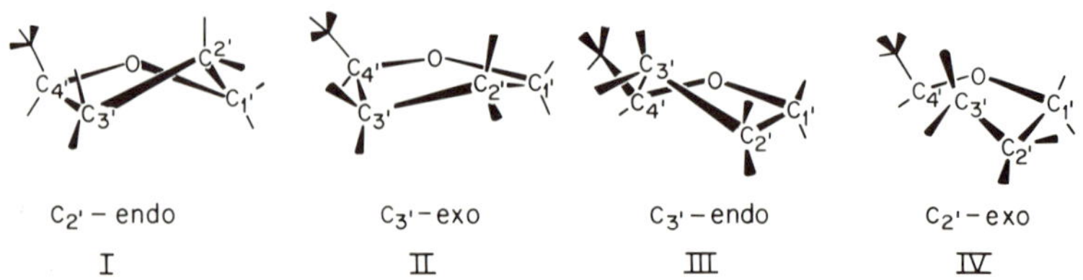

FIG. XV-8. Possible puckered conformations of the ribose ring of β-pseudouridine (Hruska *et al.*[36]).

It was felt that the I $\rightleftarrows$ III equilibrium, i.e., between C-2′-*endo* and C-3′-*endo* is probably preferred.† It is also possible that the equilibrium

$$\text{C-2}'\textit{-endo}, \text{C-3}'\textit{-exo} \rightleftarrows \text{C-2}'\textit{-exo}, \text{C-3}'\textit{-endo}$$

may contribute. In any event, whatever the principal conformations, the absence of temperature dependence shows that they must differ very little in energy.

Virtually identical parameters and conclusions were reported for the conformations of uridine[50] as for pseudouridine.

Prestegard and Chan[52] have reported that "solvent-structure breaking" electrolytes such as $Mg(ClO_4)_2$ and $NaClO_4$ induce spectral changes in uracil nucleotides and nucleosides in aqueous solution suggestive of alterations in both the rotational state about the base–sugar bond and the puckering of the furanose ring.

3. *Rotational State about the* C-4′—C-5′ *Bond*

The conformation of the exocyclic —CH_2O group can be determined from the values of $J_{4'5'B}$ and $J_{4'5'C}$. These were obtained for pseudouridine[36] and uridine[50] by analyzing the $H_{4'}H_{5'B}H_{5'C}$ system as an ABC subspectral set. We have just seen that these values are small and similar in magnitude for both compounds. This points to a major contribution of the *gauche–gauche* conformer (I in Fig. XV-9) in which the dihedral angles are about 60° for

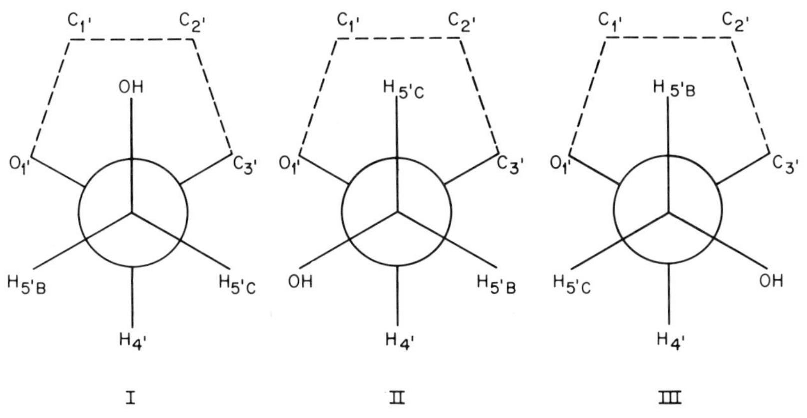

FIG. XV-9. Rotational isomers about the C-4′—C-5′ bond of β-pseudouridine (Hruska *et al.*[36]).

† The *endo* conformation is defined as that in which the designated carbon is puckered out of the plane of the other rings and in a direction toward the C-4′—C-5′ bond; in the *exo* conformation, this carbon puckers away from the C-4′—C-5′ bond.

both H_B and H_C (see Fig. I-17). In order to determine the relative contributions of II and III, it would be necessary to assign the chemical shifts of H_B and H_C, and this, as we have seen, has not yet been done. The population of either II (or III) is about twice that of III (or II). If divergence from exact staggerng by as much as 15° is permitted, the calculated rotamer populations become more nearly equal. As the temperature is increased, the couplings would be expected to tend to converge to a common value, corresponding to equal conformer populations. However, for pseudouridine this is not actually observed. (The situation is in some respects similar to that of the α- and β-protons in phenylalanine anion; see Chapter XIII, Section 2D).

4. *Rotational State about the* H—C-5′—O—P *Bonds*

Model compound studies[53] indicate a strong, "Karplus-like" (Chapter I, Section 14) dependence of the ^{31}P–H coupling on the H—C—O—P angle in the diester fragment O_3P—O—C—H. The coupling is about 20 Hz when the angle is 180° and about 2–3 Hz when it is 60°. (The geometrical relationship here, however, may be more complicated than in an H—C—C—H fragment, because the state of rotation about the O—P bond may affect the coupling even when the dihedral angle is held constant.) Tsuboi and co-workers observed that in inosinic acid, the coupling of the phosphorus is small (about 4–5 Hz), probably indicating that the phosphate group has the folded conformation shown. This indeed is the only sterically reasonable conformation

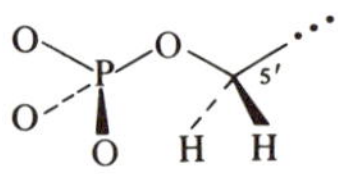

if the state of rotation about the C-4′—C-5′ bond is predominantly *gauche*, as reported for pseudouridine (Section 4C2).

D. ^{13}C SPECTRA

The natural abundance ^{13}C spectra of eleven pyrimidine and purine nucleosides have been reported by Jones *et al.*[54] Representative spectra in d_6-DMSO are shown in Fig. XV-10. The resonances are divided into two groups: those at high field corresponding to the pentose carbons and those at low field corresponding to the base carbons. The sugar resonances all show a similar pattern but are somewhat dependent on the nature of the base, e.g., the C-1′ and C-4′ resonances, always the least shielded of this group, may exchange relative shielding positions. In deoxyribose moieties, the C-2′ proton is the most shielded, being about 30 ppm upfield from its position in ribose units. The base carbon resonances (assigned by intercomparisons and reference to model compounds) are independent of whether the sugar is ribose or deoxyribose (compare adenosine and deoxyadenosine in Fig. XV-10). It was shown that

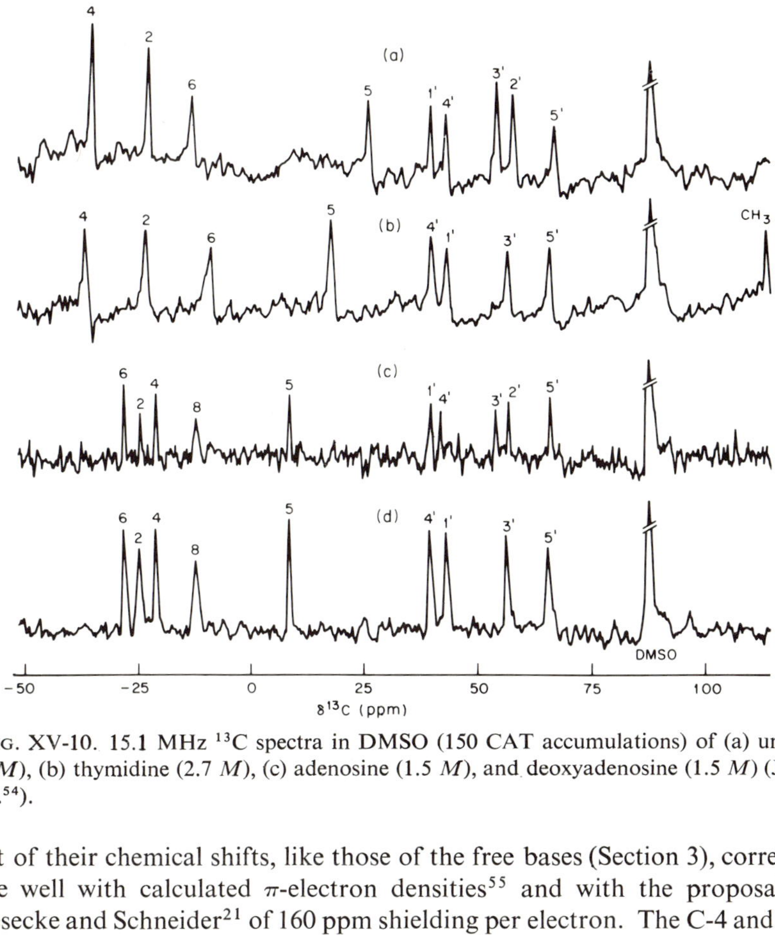

FIG. XV-10. 15.1 MHz ^{13}C spectra in DMSO (150 CAT accumulations) of (a) uridine (2.7 *M*), (b) thymidine (2.7 *M*), (c) adenosine (1.5 *M*), and deoxyadenosine (1.5 *M*) (Jones *et al.*[54]).

most of their chemical shifts, like those of the free bases (Section 3), correlate quite well with calculated π-electron densities[55] and with the proposal of Spiesecke and Schneider[21] of 160 ppm shielding per electron. The C-4 and C-5 quaternary carbon atoms of the purines and the C-5 carbons of uridine and thymidine are less shielded than expected on this basis.

Dorman and Roberts[56] have reported the spectra of nine nucleotides in aqueous solution, including ATP. In general the patterns of peaks for both furanose ring and base are very similar to those of the nucleosides. The 5′-carbon, always the most shielded, is slightly more shielded in ATP than in AMP, and shows a weak splitting (about 4 Hz) by the phosphorus nucleus. It is unexpected (but not unprecedented) that the 4′-carbon shows a substantially larger phosphorus coupling of about 8 Hz, corresponding to a three-bond ^{31}P–^{13}C coupling. This splitting permits the C-4′ resonance to be assigned with confidence. The geometrical dependence of this coupling should be explored, because it may furnish information concerning the rotational state of the P—O—C-5′—C-4′ bond system.

An attempt to observe base pairing in mixtures of, for example, AMP and TMP, and of CMP and GMP, yielded negative results, there being no appreciable changes in any of the chemical shifts. This was probably due largely to the choice of water as solvent.

Some effects of *p*H on base carbon chemical shifts were observed. In general, those next to a more strongly basic nitrogen, e.g., N-1 in CMP, are shifted downfield on acidification whereas those next to a weakly basic nitrogen, N-3 in CMP, are shifted upfield.

E. Binding of Small Molecules

Studies of the binding of dye molecules[57] and of metal ions, both diamagnetic and paramagnetic,[58–66] to nucleosides and nucleotides have been reported. The binding of acridine orange was found[57] to lead to upfield shifts of nucleoside and nucleotide resonances, suggesting the formation of mixed stacked aggregates. The ions Mn^{2+} and Co^{2+} are observed to bind principally to the phosphate groups in AMP and ATP.[58–66]

5. Dinucleotides and Oligonucleotides

Nucleotides containing two or more bases are commonly specified using the single-letter base code (preceded by d for deoxyriboside units) and the lower case letter p for either a linking phospho-diester bond or, if present, a terminal phosphate. The sequence implied is that shown for DNA on p. 392, i.e., the furanose units are represented with the 5′-linkage on the left and the 3′-linkage on the right. If the phosphate position is other than 3′→5′, it is specifically indicated. Thus ApA designates adenylyl(3′,5′)adenosine:

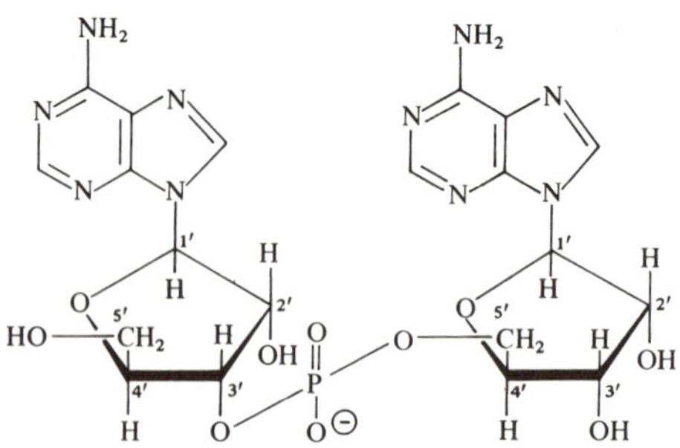

ApA

while dUpTp represents:

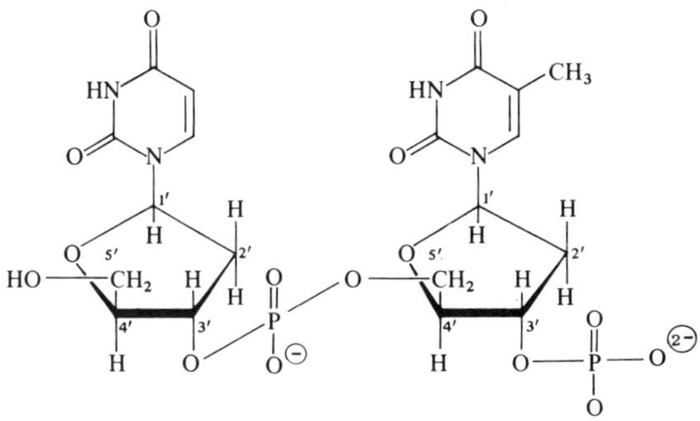

dUpTp

NMR studies have been reported for TpdU,[67] dUpT,[67] ApGp,[68] CpGp,[68] UpGp,[68] dApT, dTpA,[70] dGpT,[69] ApAp,[70] ApA,[33, 71–73] ApC,[72] (2′-5′)-ApC,[33] CpA,[33] ApU,[33] UpA,[33] ApG,[33] GpA,[33, 72] GpC,[33] CpG,[33] GpU,[33] UpG,[33] CpU,[33] UpC,[33] CpC,[33] UpU,[33, 73] TpT,[69, 33] GpG,[33] and ApAcp[71] [adenylyl (3′→5′) adenosine cyclic 2′,3′-monophosphate]. A principal object of such studies has been the determination of the degree of intramolecular stacking of the base units. Here one may compare the spectra of the dinucleotide and corresponding mononucleotide, and therefore the results are not vitiated by possible referencing difficulties (pp. 399, 404). In most dinucleotides, a substantial degree of stacking is believed to occur (a possible exception may be UpGp[68]). For example, in ApA, the C-8 and C-1′ protons of the 5′-linked adenylyl unit are shifted about 0.20 ppm upfield from their positions in 5′-methyl AMP.[71, 72] It is believed that the conformations of both base units are *trans*, as in DNA. In order to achieve maximal stacking, the bases are actually turned somewhat away from the normal mononucleotide position in such a manner that the C-8 proton is closer to the C-2′ proton. This conformational change is signalled by a marked reduction of the magnitude of $J_{1'2'}$,[70] compared to the mononucleotide (2.3 Hz instead of 5.3 Hz), consistent with a change in the ribose conformation from 2′-*endo* toward 3′-*endo*, i.e., the C-2′ carbon turns away from the base, form III in Fig. XV-8 evidently becoming favored. As the temperature is increased, coupling and chemical shift anomalies both disappear, indicating progressive "destacking." Measurement of the chemical shift temperature dependence in ApAp suggests a base–base interaction energy of about -8.2 ± 1.0 kcal-mole^{-1}, comparable with results obtained from optical measurements.[74–76] Charging of the bases of purine dinucleotides at low *p*H tends to disrupt the stacking.

In TpT, the thymine C-6 proton and methyl resonances appear at the same position for both rings[67]; the methyl resonances are not distinguishable even at 220 MHz.[69] However upon addition of purine to a TpT solution, the methyl and C-6 resonances are moved upfield and split apart, the extent of splitting increasing with the purine:TpT ratio.[67] This is attributed to intercalation of the purine ring between the T units, resulting in ring-current shifts with a differential shielding of the methyls.

A particularly significant shielding effect on thymine methyl groups is observed on comparing this resonance in dTpA and in dApT.[69] In dTpA, and indeed in all measured dinucleotides dTpX in which thymine is in the 5′-neighbor position (i.e., has a free 5′-hydroxyl group), the methyl resonance is at about 8.18τ. Its position is similar in dXpT when X is a pyrimidine nucleoside; but when X is a purine nucleoside, as in dApT (or dGpT) its position moves upfield to about 8.39τ. As we shall see in later discussion, this provides important sequence information in DNA spectra. If one assumes that the dinucleotides have a conformation similar to that of the DNA double helix, models show that in the dApT or dGpT sequence the T methyl is in the shielding zone of the purine ring current, whereas in the reverse sequence it is not.

These findings were confirmed for the deoxyribotrinucleotides RpTpTpT, RpTpdGpT, and RpTpTpdA[77] (where R designates an ethyl group). In the first and last of these, the thymine methyl resonance is a single peak at 8.05τ, whereas in RpTpdGpT, peaks appear at 8.06 and 8.16τ, the latter undoubtedly due to the 3′-OH (i.e., right-hand) terminal unit.

Smith *et al.*[38] showed from the temperature dependence of the C-6 protons (see Section 4C1) that in the trinucleotides UpUpC and UpUpU the bases are all in the *anti* conformation. These two sequences constitute the "wobble" codon pair for phenylalanine [see "The Molecular Basis of Life,"[1] p. 222, and Fig. XV-13 (p. 418)] but no special conformational feature of the base in the 3′-OH terminal position was observed.

The coupling of the ^{31}P nucleus to the C-3′ and C-5′ protons (and possibly to the C-2′ and C-4′ protons as well) may be expected to provide conformational information for dinucleotides analogous to that described in Section 4C4 for mononucleotides. An attempt to observe such couplings in ApA and UpU has been reported by Tsuboi *et al.*,[73] but the ^{31}P spectrum reported for ApA had an inexplicable lack of symmetry and also a signal-to-noise ratio so low as to preclude satisfactory interpretation. The possibility of sizable four-bond P–H couplings, well documented in the literature,† should also be considered.

6. Polynucleotides

A. Polyribonucleotides

Polyribonucleotides having a single type of base unit (Section 2) would be expected to give the simplest and most readily interpretable spectra, and this is indeed the case. A number of studies of poly-A, poly-U, poly-I, and poly-C

† See, for example, Chapter 4 of Crutchfield *et al.*[78]

have been reported.[79–87] At the outset, it must be said that, although several phenomena of interest have been uncovered, the full potential of NMR to determine the characteristics, particularly the conformations, of these intriguing and important polymers has by no means been realized. Some of the work is also considerably lessened in value because of the use of external references. In Fig. XV-11 are shown 100 MHz spectra of poly-A and poly-U at 5°, 28°

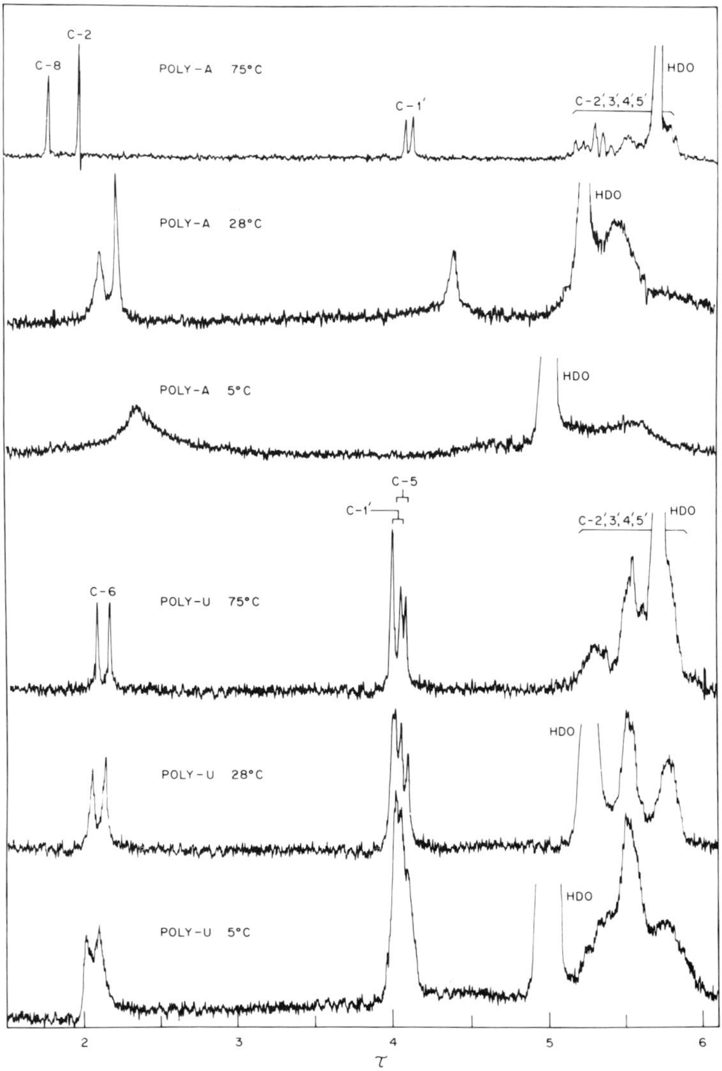

FIG. XV-11. 100 MHz spectra of poly-A and poly-U at 5°, 28° and 75° in D_2O (Bovey and Brewster[8]).

and 75°.[8] It is observed[79, 80, 85] that the peaks in the poly-A spectrum are considerably broadened and diminished in intensity at 25°, due to the development of internal structure and possibly due to aggregation. There are also upfield shifts of the C-1′, C-2, and C-8 protons, presumably arising from base stacking. These findings support other indications,[88–93] principally ultraviolet hypochromicity, of an organized structure having stacked bases in poly-A in neutral aqueous solution. The "melting" of this structure as judged from the intensity of the NMR resonances[81, 85] (Fig. XV-12) takes place in

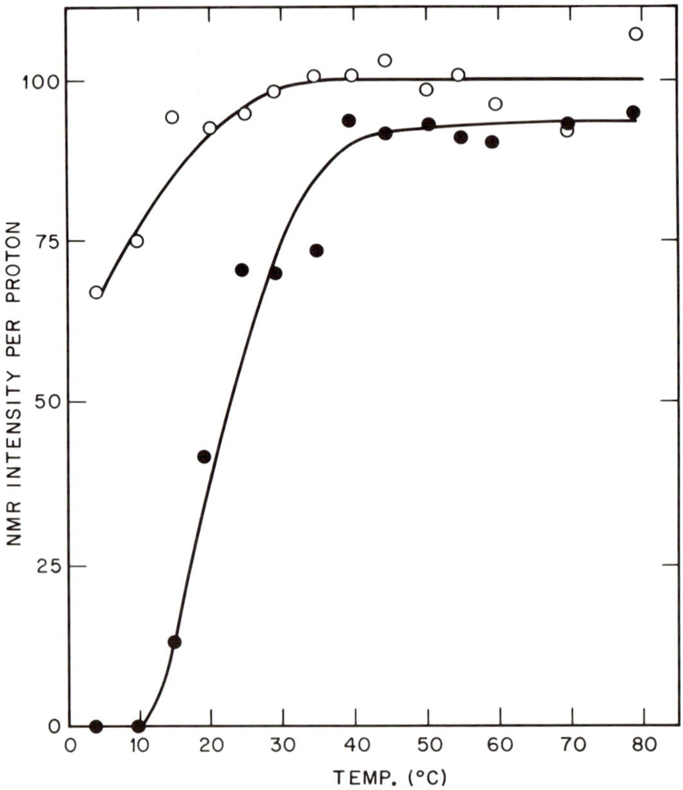

FIG. XV-12. "Melting" of poly-A in D_2O solution, expressed as the apparent intensities of the resonances of C-2, C-8, and C-1′ protons. The open circles are combined intensities of the C-2 and C-8 resonances. The plotted intensities of the C-1′ ribose protons (filled circles) represent the experimental intensities multiplied by 2. Intensities are expressed as percentages of the combined intensity exhibited by the C-2 and C-8 resonances above 35° (McDonald *et al.*[81] and McDonald and Phillips[85]).

the range 0–30°C. (It should be noted that since the structure is apparently not cooperatively formed,[76, 92] "melting" may not be an apt description.)

The ribose C-1′ proton peak decreases more rapidly in intensity than the base proton peaks as the temperature is decreased, reaching essentially zero intensity at 10°. There is a differential broadening of the C-8 and C-2 peaks for which there is no obvious explanation. This suggests that the ribose rings may be involved in a structure which is more stable than that involving the bases. At lower *p*H, the peaks decrease in intensity still further,[80] possibly because of the onset of double helix formation.[93, 94]

Poly-U evidently has no (or at least much less) such internal structure and may approximate a random coil under these conditions; there are indications of structure at 0°.[85] Poly-I[80, 81, 85] behaves like poly-A, but with much less broadening and no peak shifting, whereas poly-C[80, 85] parallels poly-U. The presence of purine bases thus appears necessary to stabilize this internal structure.

It is well known[95, 96] that when poly-A and poly-U are mixed in 1:1 and 1:2 molar proportions they form 1:1 and 1:2 double helical complexes. The formation of these complexes can be quantitatively followed by NMR[79, 80] because dipolar broadening in these complexed forms is so great that no high resolution spectrum can be detected. In aqueous solution, the 1:1 complex melts over the range 35–45°; the presence of 0.1 *M* NaCl raises the transition temperature by about 20°, and 1.0 *M* NaCl raises it another 20°. By analogy with the findings for tRNA (see next subsection) it is possible that these NMR observations reflect at least in part an increase in salt-induced aggregation rather than an increase in intramolecular structure.

There is NMR evidence of a strong association not only of poly-U with purine polynucleotides such as poly-A but also of poly-U with free purine bases,[82, 86] and of poly-U and poly-C with nucleosides, 5′-nucleotides, and nucleoside triphosphates.[84] Broadening and upfield shifts are, as usual, the principal evidence. Of particular interest and significance is the observation by Chan and Kreishman[86] that purine is so intimately intercalated in the poly-U structure that upon irradiation of the ribose C-2′ protons, positive nuclear Overhauser effects (Chapter I, Section 18B) are observed for the C-6 and C-8 (but not for the C-2) protons of the purine.

Kearns *et al.*[87] have reported that a 2:1 mixture (by weight) of poly-U and ApA exhibits a single peak for hydrogen-bonded base-paired protons at -1.0τ whereas a 2:1 mixture of poly-U and $(Ap)_6$ exhibits peaks at -1.0 and -3.3τ. Resonances in this region very probably are those of hydrogen bonded NH protons rather than NH_2 protons, since the latter are not shifted this far downfield by hydrogen bonding. It has been shown[97] that the system $(Ap)_6$ plus poly-U forms both two- and three-stranded helices whereas ApA plus poly-U forms only a two-stranded helix. It is possible that the -1.0τ peak is due to the two-stranded structure and that at -3.3τ to the three-stranded structure. Whether this proves to be the case or not, such results demonstrate the great sensitivity of such resonances to structural details even when only A–U pairs can be present.

B. Transfer RNA

We have seen (Section 2) that transfer RNA molecules are of low molecular weight (28,000–30,000) and of known primary structure, 14 having been sequenced to the present date.[1a, 98] The structures of the two species which have been principally studied by NMR, $tRNA^{ala}_{yeast}$ and $tRNA^{phe}_{yeast}$, are shown in Fig. XV-13, in which the clover-leaf structure[1, 1a, 99] is shown in

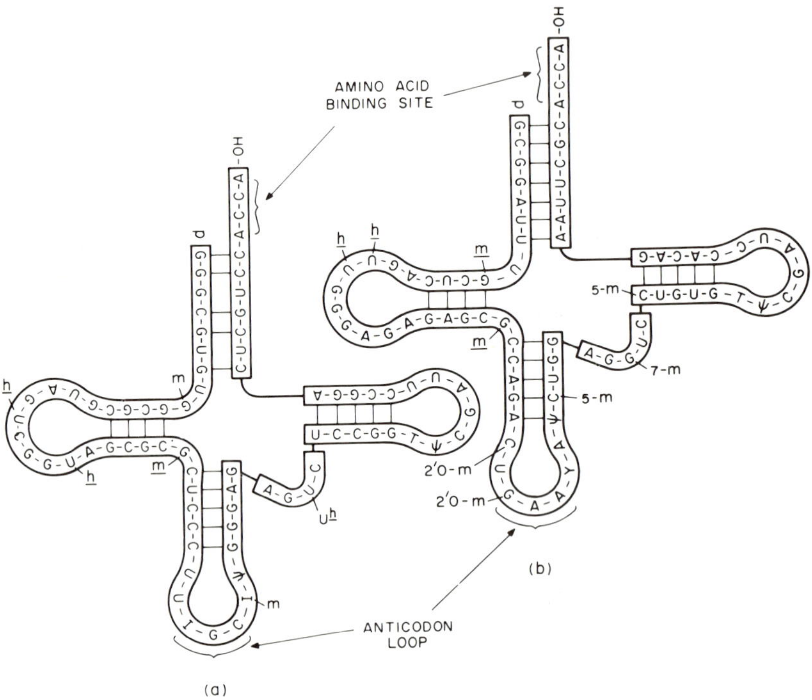

Fig. XV-13. Sequences and "clover-leaf" structures of (a) alanine tRNA of yeast, and (b) phenylalanine tRNA of yeast.

planar form, although it is very probably folded.[100] The sequences have much in common; the 3′-OH terminal sequence ACCA, which binds the amino acid, is invariant, and the $\binom{G}{A}C\Psi T$ sequence in the right-hand or "Ψ" loop occurs in all. The three-base-pair portion of the anticodon loop (at the bottom), which recognizes the proper position on the mRNA template, of course varies in each amino acid tRNA and shows species variations in tRNA's for a given amino acid.[1a] There are 20 to 24 base pairs in the duplex regions of the structure, the exact number depending mainly on the total length of the chain (which varies from 77 to 85 nucleotide units), but also on the occasional occurrence of pairs such as U–U which cannot form hydrogen bonds.

The nonexchangeable protons of both pure and unfractionated tRNA give broad resonances, due both to overlapping chemical shifts and to dipolar broadening.[79–81, 85, 101, 102] The low-field region of the spectrum shows two regions, designated I and II in Fig. XV-14. Region II, at lowest field, is due to C-8 protons of guanosine, C-2 and C-8 protons of adenosine, and C-6 protons of cytidine and uridine. Region I is due to the C-1′ (ribose) protons of all the common nucleotides and the C-5 protons of cytidine and uridine. It will be observed in Fig. XV-14 that at 30° there is no "improvement" in apparent resolution on going from 60 MHz to 100 MHz for unfractionated yeast tRNA, or even at 220 MHz for purified yeast alanine tRNA. Upon raising the temperature above the melting transition (about 60–70° for $\mathrm{tRNA}^{\mathrm{ala}}_{\mathrm{yeast}}$), region II becomes partially resolved into three peaks, shown in Fig. XV-15(a). The assignments of these peaks are indicated in the caption to Fig. XV-15. At high field [Fig. XV-15(b)], very broad resonances can be distinguished corresponding to methyl and dihydro protons of the rare nucleotides rT, diHU, *Ψ*, 1MeI, diMeG, and MeG (pp. 390, 391). These have been tentatively assigned.[102]

McDonald and Phillips and co-workers[79, 81, 85] have reported "melting" behavior for unfractionated yeast tRNA which appears to parallel that for poly-A (Fig. XV-12), in particular the appearance of the ribose C-1′ proton at temperatures higher than that at which the base protons appear. Smith *et al.*,[101] however, showed that the *areas* of these peaks do not monitor the melting of secondary structure, but reflect rather aggregation induced by the high concentration (1.0 *M*) of NaCl. Peak *widths* do monitor the secondary structure, but these were not found to vary differently in Region I and Region II. It is therefore probable that at least in tRNA the base and sugar units are equally involved in maintaining the internal structure.

The information thus available from the resonances of the carbon-bonded protons is rather limited. Considerably more rewarding are the NH proton resonances of the Watson–Crick base pairs in the duplex regions. These can be observed between 0 and -5τ in H_2O solution,[87] well below even the Region II protons, due to the large deshielding which we have already seen to be characteristic of such protons. Figure XV-16 exhibits these peaks at 24°, 35°, and 59°. Loss of resolution and signal are apparent at 59°; at 61.5° these peaks have entirely disappeared because of the disruption of the duplex region.[103] Figure XV-16(a) shows the spectrum of a sample that was dialyzed against EDTA to remove metal ions, which are known to affect the conformation of tRNA.

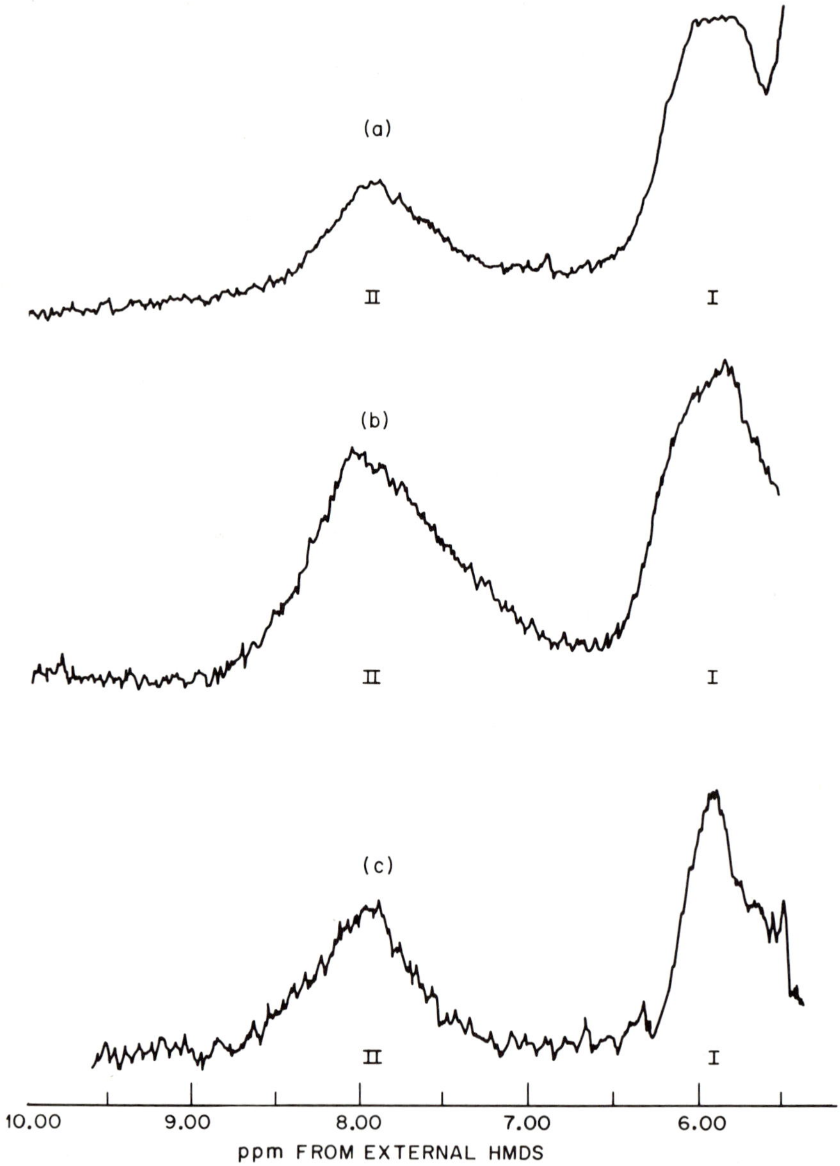

FIG. XV-14. The 60 MHz NMR spectra at 30° of (a) unfractionated yeast tRNA, 100 mg/ml, pD 7.0, 0.1 M NaCl in D_2O; (b) same as (a) but at 100 MHz; (c) alanine tRNA, 40 mg/ml, pD 7.1, 0.5 M NaCl in D_2O at 220 MHz (all extensively time averaged). Chemical shifts are expressed versus external HMDS as 0.00 (Smith *et al.*[102]).

The total area of the −2 to−5τ region of the $tRNA^{phe}_{yeast}$ spectrum is 21 ± 3 protons, which corresponds satisfactorily to the NH protons of 21 Watson–Crick pairs in this structure [Fig. IV-13(b)]. At least eight peaks can be discriminated in this region (those between 0 and −2τ may be due to NH_2 protons and so are not included in this count). As yet, it is not possible to assign them

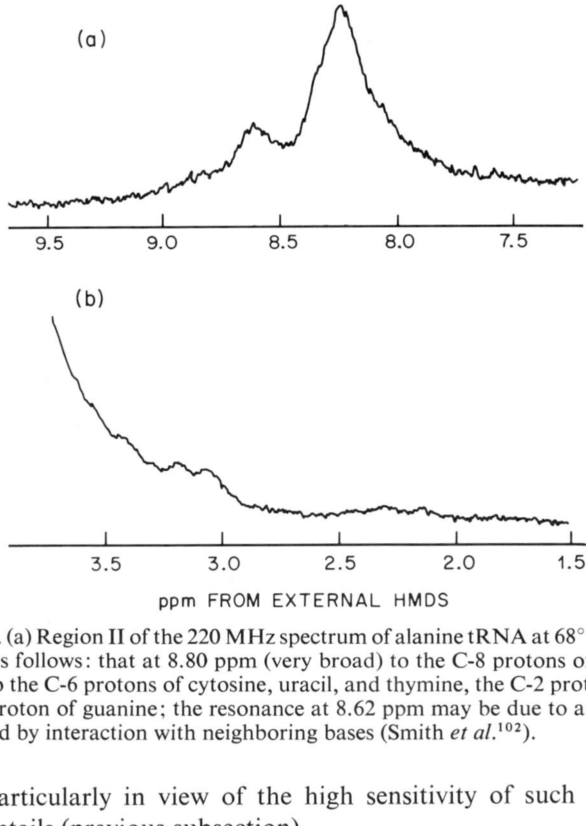

FIG. XV-15. (a) Region II of the 220 MHz spectrum of alanine tRNA at 68°; the resonances are assigned as follows: that at 8.80 ppm (very broad) to the C-8 protons of adenines; that at 8.25 ppm to the C-6 protons of cytosine, uracil, and thymine, the C-2 protons of adenine, and the C-8 proton of guanine; the resonance at 8.62 ppm may be due to any of the above protons shifted by interaction with neighboring bases (Smith *et al.*[102]).

in detail, particularly in view of the high sensitivity of such resonance to structural details (previous subsection).

The exchange of these protons with solvent is probably slow enough at 24–35° so that the line width is determined by dipolar broadening. As the temperature is increased, exchange broadening dominates, particularly with the onset of "melting."

C. DNA

The rigid double helical structure of DNA gives rise to dipolar broadening so marked that there is no observable resonance below the thermal transition point.[69, 79] This temperature, above which the DNA is a single-strand random

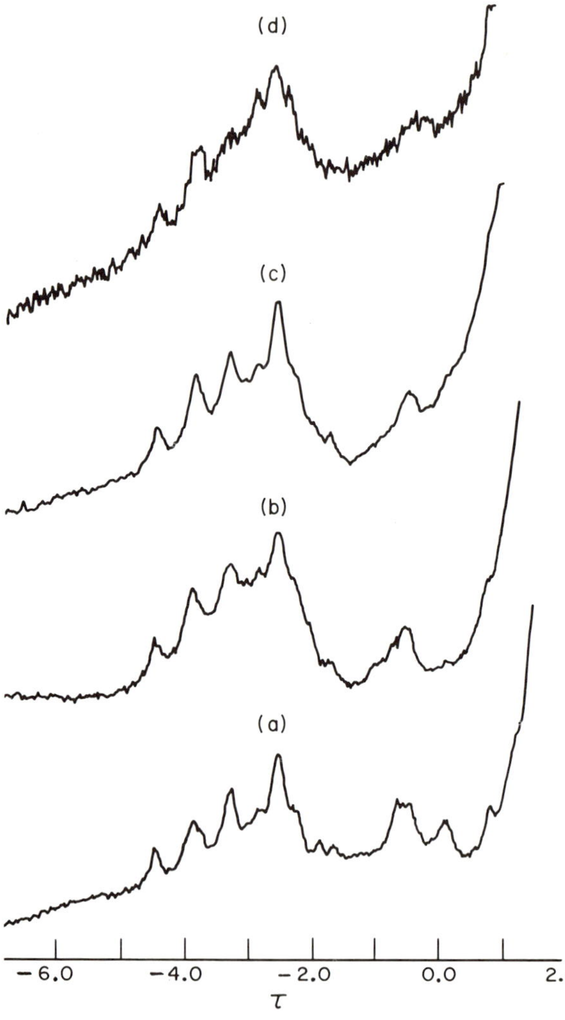

FIG. XV-16. Low-field regions of 220 MHz spectra of yeast phenylalanine tRNA in H_2O (80 mg/ml): (a) dialyzed sample, 24°; (b)–(d) sample in aqueous solution before dialysis, containing about 0.4 *M* salt, (b) 24°, (c) 35°, (d) 59° (Kearns *et al.*[87]).

coil, varies from about 80 to 100° in DNA from various species. It increases with the mole percentage of G–T pairs,[104] presumably because these contain three hydrogen bonds whereas A–T pairs contain only two (pp. 392, 393). At temperatures above the transition, at least seven spectral regions can be distinguished and assigned by reasonable deductions from the spectra of mono- and oligonucleotides.[69] These are shown in Fig. XV-17. By observation of

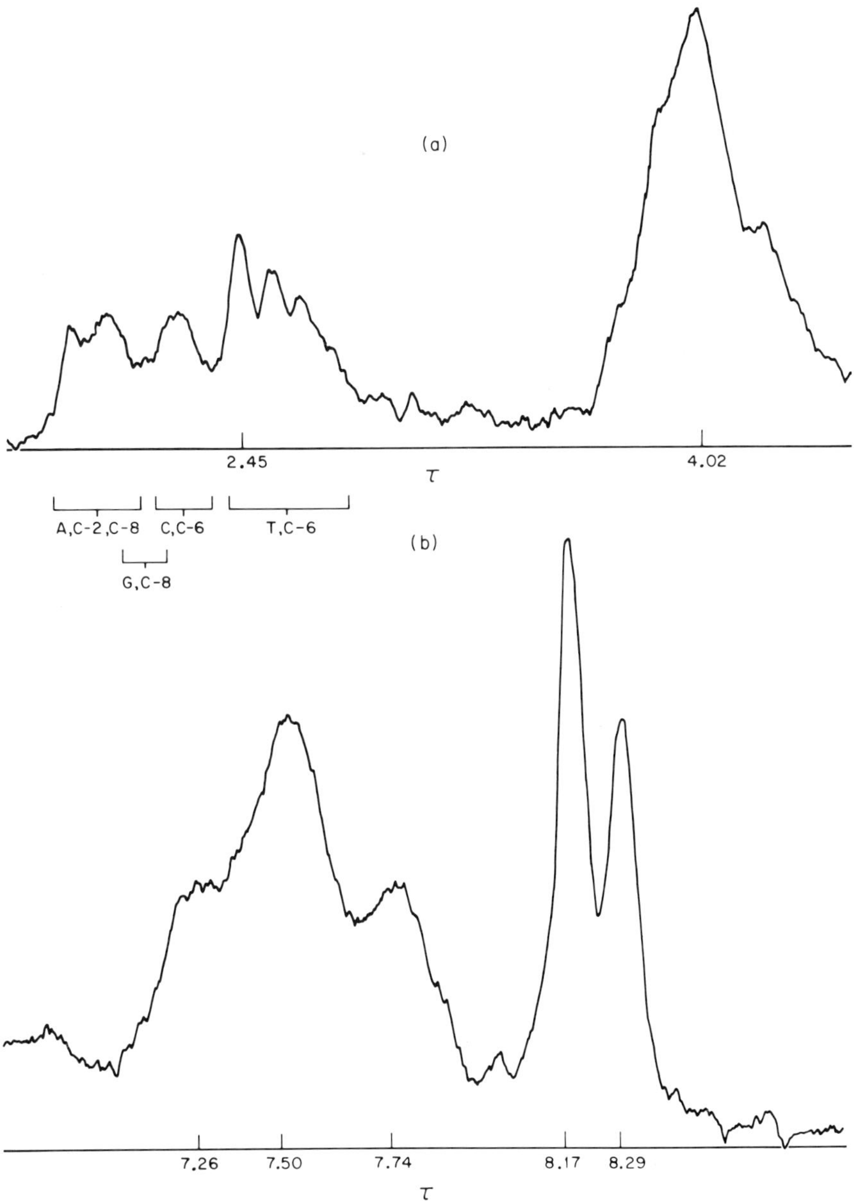

FIG. XV-17. 220 MHz spectra of calf thymus DNA in D_2O, pD 7.0, 93°. (a) Low-field region (30 mg/ml, 50 CAT scans), assignments as shown. The large peak near 4τ is probably due to C-5 and C-1′ cytidine protons. (b) High-field region (20 mg/ml, 20 CAT scans). The peaks at 8.17 and 8.29τ are due to thymidine methyl protons, as discussed in the text (McDonald *et al.*[69]).

these peaks as a function of temperature one can, at least in principle, follow the course of disruption and reformation of the helical structure in some detail. The most interesting feature of these spectra is the occurrence of two thymine methyl resonances at 8.17 and 8.29τ. The origin of these peaks has already been described in Section 5. They appear in the same positions but with varying relative intensity in DNA from different species. It is readily seen from the previous discussion (Section 5) that

$$\frac{I_{8.29\tau}}{I_{8.17\tau}} = \frac{\text{ApT} + \text{GpT}}{\text{TpT} + \text{CpT}}$$

where I denotes the area of the T methyl resonance. Comparison with the determinations of this ratio by Kornberg and co-workers[105, 106] using chemical means showed qualitative but not quantitative agreement (Table XV-1). The NMR method is certainly more direct and it is possible that it is also more accurate.

TABLE XV-1

RELATIVE THYMINE METHYL INTENSITIES IN CALF, SALMON, AND PHAGE DNA

DNA source	$I_{8.29\tau}/I_{8.17\tau}$	$\frac{\text{ApT} + \text{GpT}^a}{\text{TpT} + \text{CpT}}$
Calf thymus (Sigma)	0.67	0.77
Calf thymus (CalBiochem)	0.66	
Calf marrow	0.69	
Calf spleen	0.69	
φX-174	0.61	0.82
C_{16}	0.75	
Salmon sperm	0.79	0.85
T_6	0.85	0.92
T_4	0.89	0.95

[a] Calculated from Kornberg and co-workers.[105,106]

REFERENCES

1. "The Molecular Basis of Life." Freeman, San Francisco, 1968.

1a. H. G. Zachau, *Angew. Chem. Int. Ed. Engl.* **8**, 711 (1969).

1b. J. D. Watson and F. H. C. Crick, *Nature* (*London*) **171**, 964 (1953).

2. J. J. M Rowe, J. Hinton, and K. L. Rowe, *Chem. Rev.* **70**, 1 (1970).

2a. T. Yamane, *in* "Procedures in Nucleic Acid Research," Vol. 2 (D. Davies, ed.). Harper, New York, 1971.

3. C. D. Jardetzky and O. Jardetzky, *J. Amer. Chem. Soc.* **82**, 222 (1960).

4. G. S. Reddy, R. T. Hobgood, and J. H. Goldstein, *J. Amer. Chem. Soc.* **84**, 336 (1962).
5. J. P. Kokko, L. Mandell, and J. H. Goldstein, *J. Amer. Chem. Soc.* **84**, 1042 (1962).
6. A. R. Katritzky and A. J. Waring, *J. Chem. Soc. London* **1963**, 3046 (1963).
7. O. Jardetzky, *Biopolymers* **1**, 501 (1964).
8. F. A. Bovey and A. I. Brewster, unpublished observations.
9. L. Gatlin and J. C. Davis, Jr., *J. Amer. Chem. Soc.* **84**, 4464 (1962).
10. O. Jardetzky, P. Pappas, and N. G. Wade, *J. Amer. Chem. Soc.* **85**, 1657 (1963).
11. H. T. Miles, *J. Amer. Chem. Soc.* **85**, 1007 (1963).
12. H. T. Miles, R. B. Bradley, and E. D. Becker, *Science* **142**, 1569 (1963).
13. E. D. Becker, H. T. Miles, and R. B. Bradley, *J. Amer. Chem. Soc.* **87**, 5575 (1965).
14. B. W. Roberts, J. B. Lambert, and J. D. Roberts, *J. Amer. Chem. Soc.* **87**, 5439 (1965).
15. S. Gronowitz, B. Norrman, B. Gestblom, B. Mathiasson, and R. A. Hoffman, *Ark. Kemi* **22**, 65 (1964).
16. N. C. Li, E. Doody, and J. M. White, *J. Amer. Chem. Soc.* **79**, 5859 (1957).
17. M. P. Schweizer, S. I. Chan, G. K. Helmkamp, and P. O. P. T'so, *J. Amer. Chem. Soc.* **86**, 696 (1964).
18. J. F. Bullock and O. Jardetzky, *J. Org. Chem.* **29**, 1988 (1964).
19. G. S. Reddy, L. Mandell, and J. H. Goldstein, *J. Chem. Soc. London* **1962**, 1414 (1962).
20. R. J. Pugmire, D. M. Grant, R. K. Robins, and G. W. Rhodes, *J. Amer. Chem. Soc.* **87**, 2225 (1965).
21. H. Spiesecke and W. G. Schneider, *Tetrahedron Lett.* **1961**, 468 (1961).
22. J. M. Read and J. H. Goldstein, *J. Amer. Chem. Soc.* **87**, 3440 (1965).
23. S. I. Chan, M. P. Schweitzer, P. O. P. T'so, and G. K. Helmkamp, *J. Amer. Chem. Soc.* **86**, 4182 (1964).
24. W. C. Coburn, Jr., M. C. Thorpe, J. A. Montgomery, and K. Hewson, *J. Org. Chem.* **30**, 1110 (1965).
25. W. C. Coburn, Jr., M. C. Thorpe, J. A. Montgomery, and K. Hewson, *J. Org. Chem.* **30**, 1114 (1965).
26. R. R. Shoup, H. T. Miles, and E. D. Becker, *Biochem. Biophys. Res. Commun.* **23**, 194 (1966).
27. L. Katz and S. Penman, *J. Mol. Biol.* **15**, 220 (1966).
28. J. Marmur and P. Doty, *J. Mol. Biol.* **5**, 109 (1962).
29. P. O. P. T'so, G. K. Helmkamp, and C. Sander, *Proc. Nat. Acad. Sci. U.S.* **48**, 686 (1962).
30. M. P. Schweizer, S. I. Chan, and P. O. P. T'so, *J. Amer. Chem. Soc.* **87**, 5241 (1965).
31. S. S. Danyluk and F. E. Hruska, *Biochemistry* **7**, 1938 (1968).
32. M. P. Schweizer, A. D. Broom, P. O. P. T'so, and D. P. Hollis, *J. Amer. Chem. Soc.* **90**, 1042 (1968).
33. P. O. P. T'so, N. S. Kondo, M. P. Schweizer, and D. P. Hollis, *Biochemistry* **8**, 997 (1969).
34. B. J. Blackburn, F. E. Hruska, and I. C. P. Smith, *Can. J. Chem.* **47**, 4491 (1969).
35. P. O. P. T'so and S. I. Chan, *J. Amer. Chem. Soc.* **86**, 4176 (1964).
36. F. E. Hruska, A. A. Grey, and I. C. P. Smith, *J. Amer. Chem. Soc.* **92**, 214, 4088 (1970).
37. A. D. Buckingham, *Can. J. Chem.* **38**, 300 (1960).
38. I. C. P. Smith, B. J. Blackburn, and T. Yamane, *Can. J. Chem.* **47**, 513 (1969).
39. P. A. Hart and J. P. Davis, *J. Amer. Chem. Soc.* **91**, 512 (1969).
39a. P. E. Shirmer, J. H. Noggle, J. P. Davis, and P. A. Hart, *J. Amer. Chem. Soc.* **92**, 3266 (1970).
40. M. Sundaralingam and L. H. Jensen, *J. Mol. Biol.* **13**, 930 (1965).
41. A. V. E. Haschmeyer and A. Rich, *J. Mol. Biol.* **27**, 69 (1967).

42. T. L. V. Ulbricht, J. P. Jennings, P. M. Scopes, and W. Klyne, *Tetrahedron Lett.* **1964**, 695 (1964).
43. T. R. Emerson and T. L. V. Ulbricht, *Chem. Ind.* (*London*) **1964**, 2129 (1964).
44. T. R. Emerson, R. J. Swan, and T. L. V. Ulbricht, *Biochemistry* **6**, 843 (1967).
45. W. H. Inskeep, D. W. Miles, and H. Eyring, *J. Amer. Chem. Soc.* **92**, 3866 (1970).
46. D. W. Miles, W. H. Inskeep, M. J. Robins, M. W. Winkley, R. K. Robins, and H. Eyring, *J. Amer. Chem. Soc.* **92**, 3872 (1970).
47. C. D. Jardetzky, *J. Amer. Chem. Soc.* **82**, 299 (1960).
48. C. D. Jardetzky, *J. Amer. Chem. Soc.* **83**, 62 (1961).
49. S. Fujiwara and M. Uetsuki, *in* "Recent Developments in Magnetic Resonance in Biological Systems" (S. Fujiwara and L. H. Piette, eds.). Hirokawa Publ., Tokyo, Japan, 1968.
50. B. J. Blackburn, A. A. Grey, I. C. P. Smith, and F. E. Hruska, *Can. J. Chem.* **48**, 2866 (1970).
51. M. Smith and C. D. Jardetzky, *J. Mol. Spectrosc.* **28**, 70 (1968).
52. J. H. Prestegard and S. I. Chan, *J. Amer. Chem. Soc.* **91**, 2843 (1969).
53. M. Tsuboi, M. Kainosho, and A. Nakamura, *in* "Recent Developments in Magnetic Resonance in Biological Systems" (S. Fujiwara and L. H. Piette, eds.), p. 43. Hirokawa Publ., Tokyo, Japan, 1968.
54. A. J. Jones, M. W. Winkley, D. M. Grant, and R. K. Robins, *Proc. Nat. Acad. Sci. U.S.* **65**, 27 (1970).
55. J. I. Fernandez-Alonso, *Advan. Chem. Phys.* **7**, 3 (1964).
56. D. E. Dorman and J. D. Roberts, *Proc. Nat. Acad. Sci. U.S.* **65**, 19 (1970).
57. F. E. Hruska and S. S. Danyluk, *Biochim. Biophys. Acta* **161**, 250 (1968).
58. P. W. Schneider and H. Brintzinger, *Helv. Chim. Acta* **47**, 1717 (1964).
59. H. Moll, P. W. Schneider, and H. Brintzinger, *Helv. Chim. Acta* **47**, 1837 (1964).
60. S. M. Wang and N. C. Li, *J. Amer. Chem. Soc.* **88**, 4592 (1966).
61. G. L. Eichhorn, P. Clark, and E. D. Becker, *Biochemistry* **5**, 245 (1966).
62. R. G. Shulman, H. Sternlicht, and B. J. Wyluda, *J. Chem. Phys.* **43**, 3116 (1965).
63. H. Sternlicht, R. G. Shulman, and E. W. Anderson, *J. Chem. Phys.* **43**, 3123 (1965).
64. H. Sternlicht, R. G. Shulman, and E. W. Anderson, *J. Chem. Phys.* **43**, 3133 (1965).
65. R. G. Shulman and H. Sternlicht, *J. Mol. Biol.* **13**, 952 (1965).
66. H. Sternlicht, D. E. Jones, and K. Kustin, *J. Amer. Chem. Soc.* **90**, 7110 (1968).
67. S. I. Chan, B. W. Bangerter, and H. H. Peter, *Proc. Nat. Acad. Sci. U.S.* **55**, 720 (1966).
68. Y. Inoue and S. Aoyagi, *Biochem. Biophys. Res. Commun.* **28**, 973 (1967).
69. C. C. McDonald, W. D. Phillips, and J. Lazar, *J. Amer. Chem. Soc.* **89**, 4166 (1967).
70. F. E. Hruska and S. S. Danyluk, *Biochim. Biophys. Acta* **157**, 238 (1968).
71. F. E. Hruska and S. S. Danyluk, *J. Amer. Chem. Soc.* **90**, 3266 (1968).
72. S. I. Chan and J. H. Nelson, *J. Amer. Chem. Soc.* **91**, 168 (1969).
73. M. Tsuboi, S. Takahashi, Y. Kyosoku, H. Hayatsu, T. Ukita, and M. Kainosho, *Science* **166**, 1504 (1969).
74. D. Poland, J. N. Vournakis, and H. A. Scheraga, *Biopolymers* **4**, 223 (1966).
75. J. Brahms, A. M. Michelson, and K. E. Van Holde, *J. Mol. Biol.* **12**, 726 (1965).
76. M. Leng and G. Felsenfeld, *J. Mol. Biol.* **15**, 455 (1966).
77. K.-H. Scheit, F. Cramer, and A. Franke, *Biochim. Biophys. Acta* **145**, 21 (1967).
78. M. M. Crutchfield, C. H. Dungan, J. H. Letcher, V. Mark, and J. R. Van Wazer, "P^{31} Nuclear Magnetic Resonance." Wiley (Interscience), New York, 1967.
79. C. C. McDonald, W. D. Phillips, and S. Penman, *Science* **144**, 1234 (1964).
80. J. P. McTague, V. Ross, and J. H. Gibbs, *Biopolymers* **2**, 163 (1964).
81. C. C. McDonald, W. D. Phillips, and J. Penswick, *Biopolymers* **3**, 609 (1965).

82. B. W. Bangerter and S. I. Chan, *Biopolymers* **6**, 983 (1968).
83. B. W. Bangerter and S. I. Chan, *Proc. Nat. Acad. Sci. U.S.* **60**, 1144 (1968).
84. P. O. P. T'so and M. P. Schweizer, *Biochemistry* **7**, 2963, (1968).
85. C. C. McDonald and W. D. Phillips, *in* "Magnetic Resonance in Biological Systems" (B. G. Malmstrom and T. Vanngard, eds.). Pergamon Press, Oxford, 1967.
86. S. I. Chan and G. P. Kreishman, *J. Amer. Chem. Soc.* **92**, 1102 (1970).
87. D. R. Kearns, D. J. Patel, and R. G. Shulman, *Nature* **229**, 338 (1971).
88. V. Luzzati, A. Mathis, F. Masson, and J. Witz, *J. Mol. Biol.* **10**, 28 (1964).
89. D. N. Holcomb and I. Tinoco, Jr., *Biopolymers* **3**, 121 (1965).
90. K. E. Van Holde, J. Brahms, and A. M. Michelson, *J. Mol. Biol.* **12**, 726 (1965).
91. A. M. Michelson, T. L. V. Ulbricht, T. R. Emerson, and R. J. Swan, *Nature* (*London*) **209**, 873 (1966).
92. J. Brahms, A. M. Michelson, and K. E. Van Holde, *J. Mol. Biol.* **15**, 467 (1966).
93. J. R. Fresco and P. Doty, *J. Amer. Chem. Soc.* **79**, 3928 (1957).
94. A. Rich, D. R. Davies, F. H. C. Crick, and J. D. Watson, *J. Mol. Biol.* **3**, 71 (1961).
95. R. C. Warner, *Fed. Proc. Fed. Amer. Soc. Exp. Biol.* **15**, 379 (1956).
96. A. Rich and D. R. Davies, *J. Amer. Chem. Soc.* **78**, 3548 (1956).
97. M. N. Lipsett, L. A. Heppel, and D. F. Bradley, *J. Biol. Chem.* **236**, 587 (1961).
98. S. Kowalski, T. Yamane, and J. R. Fresco, *Science* **172**, 384 (1971).
99. R. W. Holley, J. Apgar, G. A. Everett, J. T. Madison, M. Marquisee, S. H. Merrill, J. R. Penswick, and A. Zamir, *Science* **147**, 1462 (1965).
100. M. Levitt, *Nature* (*London*) **224**, 759 (1969).
101. I. C. P. Smith, T. Yamane, and R. G. Shulman, *Science* **159**, 1360 (1968).
102. I. C. P. Smith, T. Yamane, and R. G. Shulman, *Can. J. Biochem.* **47**, 480 (1969).
103. F. Cramer, H. Doepner, F. van der Haar, E. Schlimme, and H. Seidel, *Proc. Nat. Acad. Sci. U.S.* **61**, 1384 (1968).
104. P. Doty, *in* "The Structure and Biosynthesis of Macromolecules" (D. J. Bell and J. K. Grant, eds.). Cambridge Univ. Press, London and New York, 1962.
105. J. Josse, A. D. Kaiser, and A. Kornberg, *J. Biol. Chem.* **236**, 864 (1961).
106. M. N. Swartz, T. A. Trautner, and A. Kornberg, *J. Biol. Chem.* **237**, 1961 (1962).

AUTHOR INDEX

Numbers in parentheses are reference numbers and indicate that an author's work is referred to, although his name is not cited in the text. Numbers in italics show the page on which the complete reference is listed.

B

C

D

L

M

T

Y

Z

SUBJECT INDEX

E

F

G

H